STUDENT SOLUTIONS MANUAL

BEVERLY FUSFIELD

TO ACCOMPANY
FINITE MATHEMATICS
AN APPLIED APPROACH
11TH EDITION

MICHAEL SULLIVAN
Chicago State University

WILEY

JOHN WILEY & SONS, INC.

COVER PHOTO: Terry Husebye/Stone/Getty Images

ISBN   978-0-470-45828-0

V10006488_112818

Printed and bound by Quad/Graphics.

CONTENTS

CONTENTS

Chapter 1 Linear Equations

1.1 Lines

1. True

3. True

5. If the slope of a line is undefined, the line is <u>vertical</u>.

7. The point-slope form of the equation of a line with slope m containing the point (x_1, y_1) is

$$\underline{y - y_1 = m(x - x_1).}$$

9. $A = (4, 2)$; $B = (6, 2)$; $C = (5, 3)$; $D = (-2, 1)$
$E = (-2, -3)$; $F = (3, -2)$; $G = (6, -2)$
$H = (5, 0)$

11. The set of points of the form, $(2, y)$, where y is a real number, is a vertical line passing through 2 on the x-axis. The equation of the line is $x = 2$.

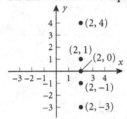

13. $y = 2x + 4$

x	0	-2	2	-2	4	-4
y	4	0	8	0	12	-4

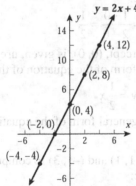

15. $2x - y = 6$

x	0	3	2	-2	4	-4
y	-6	0	-2	-10	2	-14

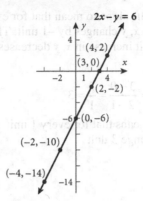

17.a. The vertical line containing the point $(2, -3)$ is $x = 2$.

b. The horizontal line containing the point $(2, -3)$ is $y = -3$.

c. $y + 3 = 5(x - 2)$
$y + 3 = 5x - 10$
$\quad 13 = 5x - y$
The line with a slope of 5 containing the point $(2, -3)$ is $5x - y = 13$.

19.a. The vertical line containing the point $(-4, 1)$ is $x = -4$.

b. The horizontal line containing the point $(-4, 1)$ is $y = 1$.

c. $y - 1 = 5(x + 4)$
$y - 1 = 5x + 20$
$\quad -21 = 5x - y$
The line with a slope of 5 containing the point $(-4, 1)$ is $5x - y = -21$.

21.a. The vertical line containing the point $(0, 3)$ is $x = 0$.

b. The horizontal line containing the point $(0, 3)$ is $y = 3$.

c. $y - 3 = 5x$
$\quad -3 = 5x - y$
The line with a slope of 5 containing the point $(0, 3)$ is $5x - y = -3$.

23. $m = \dfrac{y_2 - y_1}{x_2 - x_1} = \dfrac{1 - 0}{2 - 0} = \dfrac{1}{2}$

We interpret the slope to mean that for every 2 unit change in x, y changes 1 unit. That is, for every 2 units x increases, y increases by 1 unit.

25. $m = \dfrac{y_2 - y_1}{x_2 - x_1} = \dfrac{3-1}{-1-1} = -1$

We interpret the slope to mean that for every 1 unit change in x, y changes by -1 unit. That is, for every 1 unit increase in x, y decreases by 1 unit.

27. $m = \dfrac{y_2 - y_1}{x_2 - x_1} = \dfrac{3-0}{2-1} = \dfrac{3}{1} = 3$

A slope of 3 means that for every 1 unit change in x, y will change 3 units.

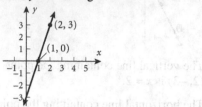

29. $m = \dfrac{y_2 - y_1}{x_2 - x_1} = \dfrac{1-3}{2-(-2)} = \dfrac{-2}{4} = -\dfrac{1}{2}$

A slope of $-\dfrac{1}{2}$ means that for every 2 unit increase in x, y will decrease -1 unit.

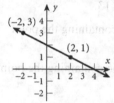

31. $m = \dfrac{y_2 - y_1}{x_2 - x_1} = \dfrac{(-1)-(-1)}{2-(-3)} = \dfrac{0}{5} = 0$

A slope of zero indicates that regardless of how x changes, y remains constant.

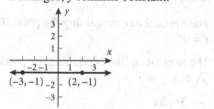

33. $m = \dfrac{y_2 - y_1}{x_2 - x_1} = \dfrac{(-2)-2}{(-1)-(-1)} = \dfrac{-4}{0}$

The slope is not defined.

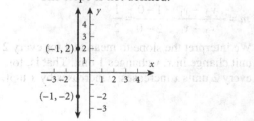

35.

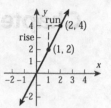

37.

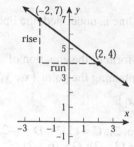

39.

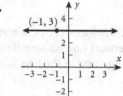

41.

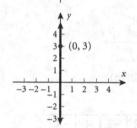

43. Use the points $(0, 0)$ and $(2, 1)$ to compute the slope of the line:

$m = \dfrac{y_2 - y_1}{x_2 - x_1} = \dfrac{1-0}{2-0} = \dfrac{1}{2}$

Since the y-intercept, $(0, 0)$, is given, use the slope-intercept form of the equation of the line:

$y = \dfrac{1}{2}x + 0 \Rightarrow y = \dfrac{1}{2}x$

Then write the general form of the equation:
$x - 2y = 0$.

45. Use the points $(1, 1)$ and $(-1, 3)$ to compute the slope of the line:

$m = \dfrac{y_2 - y_1}{x_2 - x_1} = \dfrac{3-1}{(-1)-1} = \dfrac{2}{-2} = -1$

Now use the point $(1, 1)$ and the slope $m = -1$ to write the point-slope form of the equation of the line:

$y - y_1 = m(x - x_1)$
$y - 1 = (-1)(x - 1)$
$y - 1 = -x + 1$
$x + y = 2$

This is the general form of the equation.

47. Since the slope and a point are given, use the point-slope form of the line:

$$y - y_1 = m(x - x_1)$$
$$y - 1 = 2(x - (-4))$$
$$y - 1 = 2x + 8$$
$$2x - y = -9$$

This is the general form of the equation.

49. Since the slope and a point are given, use the point-slope form of the line:

$$y - y_1 = m(x - x_1)$$
$$y - (-1) = -\frac{2}{3}(x - 1)$$
$$3y + 3 = -2(x - 1)$$
$$3y + 3 = -2x + 2$$
$$2x + 3y = -1$$

This is the general form of the equation.

51. Since we are given two points, (1, 3) and (−1, 2), first find the slope.

$$m = \frac{3 - 2}{1 - (-1)} = \frac{1}{2}$$

Then use the slope, one of the points, (1, 3), and the point-slope form of the line:

$$y - y_1 = m(x - x_1)$$
$$y - 3 = \frac{1}{2}(x - 1)$$
$$2y - 6 = x - 1$$
$$x - 2y = -5$$

This is the general form of the equation.

53. Since we are given the slope $m = -2$ and the y-intercept (0, 3), we use the slope-intercept form of the line:

$$y = mx + b$$
$$y = -2x + 3$$
$$2x + y = 3$$

This is the general form of the equation.

55. We are given the slope $m = 3$ and the x-intercept (−4, 0), so we use the point-slope form of the line:

$$y - y_1 = m(x - x_1)$$
$$y - 0 = 3(x - (-4))$$
$$y = 3x + 12$$
$$3x - y = -12$$

This is the general form of the equation.

57. We are given the slope $m = \frac{4}{5}$ and the point (0, 0), which is the y-intercept. So, we use the slope-intercept form of the line.

$$y = mx + b$$
$$y = \frac{4}{5}x + 0$$
$$5y = 4x$$
$$4x - 5y = 0$$

This is the general form of the equation.

59. We are given two points, the x-intercept (2, 0) and the y-intercept (0, −1), so we need to find the slope and then use the slope-intercept form of the line to get the equation.

$$\text{slope} = \frac{0 - (-1)}{2 - 0} = \frac{1}{2}$$
$$y = mx + b$$
$$y = \frac{1}{2}x - 1$$
$$2y = x - 2$$
$$x - 2y = 2$$

This is the general form of the equation.

61. Since the slope is undefined, the line is vertical. The equation of the vertical line containing the point (1, 4) is $x = 1$.

63. Since the slope = 0, the line is horizontal. The equation of the horizontal line containing the point (1, 4) is $y = 4$.

65. $y = 2x + 3$
slope: $m = 2$; y-intercept: (0, 3)

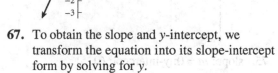

67. To obtain the slope and y-intercept, we transform the equation into its slope-intercept form by solving for y.

$$\frac{1}{2}y = x - 1$$
$$y = 2x - 2$$

slope: $m = 2$; y-intercept: (0, −2)

69. To obtain the slope and y-intercept, we transform the equation into its slope-intercept form by solving for y.

$$2x - 3y = 6$$
$$y = \frac{2}{3}x - 2$$

slope: $m = \frac{2}{3}$; y-intercept: $(0, -2)$

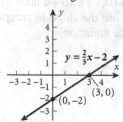

71. To obtain the slope and y-intercept, we transform the equation into its slope-intercept form by solving for y.

$$x + y = 1$$
$$y = -x + 1$$

slope: $m = -1$; y-intercept: $(0, 1)$

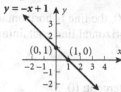

73. The slope is not defined; there is no y-intercept. So the graph is a vertical line.

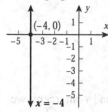

75. slope: $m = 0$; y-intercept: $(0, 5)$

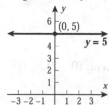

77. To obtain the slope and y-intercept, we transform the equation into its slope-intercept form by solving for y.

$$y - x = 0$$
$$y = x$$

slope: $m = 1$; y-intercept: $(0, 0)$

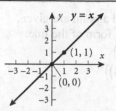

79. To obtain the slope and y-intercept, we transform the equation into its slope-intercept form by solving for y.

$$2y - 3x = 0$$
$$2y = 3x$$
$$y = \frac{3}{2}x$$

slope: $m = \frac{3}{2}$; y-intercept: $(0, 0)$

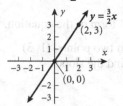

81. To graph an equation on a graphing utility, first solve the equation for y.

$$1.2x + 0.8y = 2$$
$$0.8y = -1.2x + 2$$
$$y = -1.5x + 2.5$$

Window: Xmin = -10; Xmax = 10
Ymin = -10; Ymax = 10

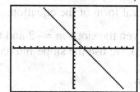

The x-intercept is $(1.67, 0)$.
The y-intercept is $(0, 2.50)$.

83. To graph an equation on a graphing utility, first solve the equation for y.

$$21x - 15y = 53$$
$$15y = 21x - 53$$
$$y = \frac{21}{15}x - \frac{53}{15} = \frac{7}{5}x - \frac{53}{15}$$

Window: Xmin = -10; Xmax = 10
Ymin = -10; Ymax = 10

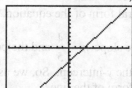

The x-intercept is $(2.52, 0)$.
The y-intercept is $(0, -3.53)$.

85. To graph an equation on a graphing utility, first solve the equation for *y*.

$$\frac{4}{17}x + \frac{6}{23}y = \frac{2}{3}$$

$$\frac{6}{23}y = -\frac{4}{17}x + \frac{2}{3}$$

$$y = \frac{23}{6}\left(-\frac{4}{17}x + \frac{2}{3}\right) = -\frac{46}{51}x + \frac{23}{9}$$

Window: Xmin = –10; Xmax = 10
Ymin = –10; Ymax = 10

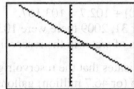

The *x*-intercept is (2.83, 0).
The *y*-intercept is (0, 2.56).

87. To graph an equation on a graphing utility, first solve the equation for *y*.

$$\pi x - \sqrt{3}y = \sqrt{6}$$

$$\sqrt{3}y = \pi x - \sqrt{6}$$

$$y = \frac{\pi}{\sqrt{3}}x - \sqrt{2}$$

Window: Xmin = –10; Xmax = 10
Ymin = –10; Ymax = 10

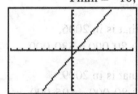

The *x*-intercept is (0.78, 0).
The *y*-intercept is (0, –1.41).
The *y*-intercept is (0, 1.23).

89. The graph passes through the points (0, 0) and (4, 8). We use the points to find the slope of the line:

$$m = \frac{y_2 - y_1}{x_2 - x_1} = \frac{8 - 0}{4 - 0} = \frac{8}{4} = 2$$

The *y*-intercept (0, 0) is given, so we use the *y*-intercept and the slope *m* = 2, to obtain the slope-intercept form of the line.

$$y = mx + b$$
$$y = 2x + 0 \Rightarrow y = 2x$$

This is choice (b).

91. The graph passes through the points (0, 0) and (2, 8). We use the points to find the slope of the line:

$$m = \frac{y_2 - y_1}{x_2 - x_1} = \frac{8 - 0}{2 - 0} = \frac{8}{2} = 4$$

The *y*-intercept (0, 0) is given, so we use the *y*-intercept and the slope *m* = 4, to obtain the slope-intercept form of the line.

$$y = mx + b$$
$$y = 4x + 0 \Rightarrow y = 4x$$

This is choice (d).

93. Using the intercepts (–2, 0) and (0, 2),

$$m = \frac{y_2 - y_1}{x_2 - x_1} = \frac{2 - 0}{0 - (-2)} = \frac{2}{2} = 1$$

Slope-intercept form: *y* = *x* + 2
General form: *x* – *y* = –2

95. Using the intercepts (3, 0) and (0, 1),

$$m = \frac{y_2 - y_1}{x_2 - x_1} = \frac{1 - 0}{0 - 3} = \frac{1}{-3} = -\frac{1}{3}$$

Slope-intercept form: $y = -\frac{1}{3}x + 1$

General form:

$$3y = 3\left(-\frac{1}{3}x + 1\right)$$
$$3y = -x + 3$$
$$x + 3y = 3$$

97. a. The equation is *C* = 0.54*x*, where *x* is the number of miles the car is driven.

    b. *x* = 15,000
      C = 0.54(15,000) = 8100
      It costs $8100 to drive the car 15,000 miles.

    c.

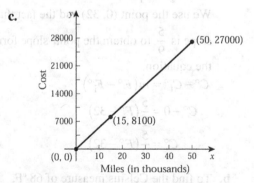

Miles (in thousands)

    d. For every unit increase in *x*, the number of miles driven, *C*, the cost, increases by 0.54*x*. This represents the additional cost of driving a standard-sized car another mile.

99.a. The fixed cost of electricity for the month is $8.23. In addition, the electricity costs $0.10438 (10.438 cents) for every kilowatt-hour (KWH) used. If x represents the number of KWH of electricity used in a month, the total monthly charge is represented by the equation
$$C = 0.10438x + 8.23, \quad 0 \le x \le 400.$$

b.
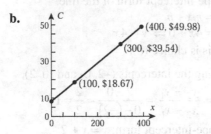

c. The charge for using 100 KWH of electricity is found by substituting 100 for x in part (a):
$$C = 0.10438(100) + 8.23$$
$$= 10.438 + 8.23 = 18.668 \approx \$18.67$$

d. The charge for using 300 KWH of electricity is found by substituting 300 for x in part (a):
$$C = 0.10438(300) + 8.23$$
$$= 31.314 + 8.23 = 39.544 \approx \$39.54$$

e. The slope of the line, $m = 0.10438$, indicates that for every extra KWH used (up to 400 KWH), the electric bill increases by 10.438 cents.

101.a. Since we are told the relationship is linear, we will use the two points to get the slope of the line:
$$m = \frac{C_2^\circ - C_1^\circ}{F_2^\circ - F_1^\circ} = \frac{100 - 0}{212 - 32} = \frac{100}{180} = \frac{5}{9}$$

We use the point (0, 32) and the fact that the slope is $\frac{5}{9}$ to obtain the point-slope form of the equation.
$$C^\circ - C_1^\circ = m\left(F^\circ - F_1^\circ\right)$$
$$C^\circ - 0 = \frac{5}{9}\left(F^\circ - 32\right)$$
$$C^\circ = \frac{5}{9}\left(F^\circ - 32\right)$$

b. To find the Celsius measure of 68 °F, substitute 68 for F in the equation and simplify:
$$C^\circ = \frac{5}{9}(68 - 32) = 20^\circ$$

103.a. If $t = 0$ represents December 21, then January 20 is represented by $t = 30$. Use the points (0, 102.7) and (30, 104.1) to find the slope of the equation.
$$m = \frac{104.1 - 102.7}{30 - 0} = \frac{1.4}{30} = 0.0467$$
We will use the slope and the point (0, 102.7) to write the point-slope form of the equation.
$$A - 102.7 = 0.0467(t - 0)$$
$$A = 0.0467t + 102.7$$

b. $A = 0.0467(10) + 102.7 = 103.167$. On December 31, 2009 there were 103.167 billion gallons of water in the reservoir.

c. The slope indicates that the reservoir gains 0.0467 billion (or 46.7 million) gallons of water a day.

d. $A = 0.0467(41) + 102.7 = 104.6147$. The model predicts that there will be 104.6147 billion gallons of water in the reservoir on January 31, 2010.

e. The reservoir will be full when $A = 265.5$.
$$0.0467t + 102.7 = 265.5$$
$$0.0467t = 162.8$$
$$t \approx 3486.1$$
The reservoir will be full after day 3486, or about 9.55 years.

105.a. When $x = 0$, that is in 2006,
$$S = 5000(0) + 80{,}000 = \$80{,}000$$

b. When $x = 3$, that is in 2009,
$$S = 5000(3) + 80{,}000 = \$95{,}000$$

c. If the trend continues, sales in 2012 should be equal to S when $x = 6$,
$$S = 5000(6) + 80{,}000$$
$$S = \$110{,}000$$

d. If the trend continues, sales in 2015 should be equal to S when $x = 9$.
$$S = 5000(9) + 80{,}000$$
$$S = \$125{,}000$$

107. a. Writing 5% as a decimal, 5% = 0.05, we form the equation $S = 0.05x + 400$.

b. If $x = \$4000.00$, then Dan's earnings will be $S = 0.05(4000) + 400 = 200 + 400 = \$600$.

c. Let S equal the median earnings, and solve for x.

$857.31 = 0.05x + 400$

$457.31 = 0.05x$

$9146.2 = x$

Dan would need to have sales that generate $9146.20 in profit to earn the median amount.

109. a. Since the rate of increase is constant, we use the points to find the slope of the line.

$$m = \frac{S_2 - S_1}{t_2 - t_1} = \frac{496 - 475}{2009 - 1999} = \frac{21}{10} = 2.1$$

Then we choose a point, say $(1999, 475)$, and write the point-slope form of the line.

$$S - S_1 = m(t - t_1)$$
$$S - 475 = 2.1(t - 1999)$$
$$S = 2.1(t - 1999) + 475$$
$$S = 2.1t - 3722.9$$

b. If the trend continues, in 2011,

$S = 2.1(2011) - 3722.9 = 500.2$

The projected SAT mathematics score is about 500.

111. a. Since we assume the rate of increase is constant, we use the points to find the slope of a line.

$$m = \frac{P_2 - P_1}{t_2 - t_1} = \frac{29.4 - 24.4}{2008 - 1998} = \frac{5}{10} = 0.5$$

Now we choose a point, say $(24.4, 1998)$, and write the point-slope form of the line.

$$P - P_1 = m(t - t_1)$$
$$P - 24.4 = 0.5(t - 1998)$$
$$P = 0.5(t - 1998) + 24.4$$
$$P = 0.5t - 974.6$$

b. To find the predicted percentage of people who will have a bachelor's degree, let $t = 2011$.

$P = 0.5(2011) - 974.6 = 30.9$

By 2011 it is predicted that 30.9% of people over 25 years of age will have a bachelor's degree or higher.

c. The slope is the annual average increase in the percentage of people over 25 years of age who have a bachelor's degree or higher. For every unit change in the year t, P increases by 0.5.

113. a. Since the cost of the houses is linear, we first use the points to find the slope of the line.

$$m = \frac{C_2 - C_1}{t_2 - t_1} = \frac{88,280 - 94,823}{2009 - 2008} = -6543$$

Next, choose a point, say $(2008, 94823)$, and write the point slope form of the line.

$$C - C_1 = m(t - t_1)$$
$$C - 94,823 = -6543(t - 2008)$$
$$C = -6543(t - 2008) + 94,823$$
$$C = -6543t + 13,233,167$$

b. The projected average cost of a house in 2011 is found by letting $t = 2011$ in the equation.

$C = -6543(2011) + 13,233,167 = \$75,194$

115. a. First we find the slope of the line.

$$m = \frac{S_2 - S_1}{t_2 - t_1} = \frac{264.908 - 214.091}{2008 - 2007} = 50.817$$

Then we use the point $(2007, 214.091)$ to write the point-slope form of the equation of the line.

$$S - S_1 = m(t - t_1)$$
$$S - 214.091 = 50.817(t - 2007)$$
$$S = 50.817(t - 2007) + 214.091$$
$$S = 50.817t - 101,775.628$$

b. $S = 50.817(2011) - 101,775.628 = 417.359$

The model predicts that the total sales and other operating income for Chevron Corporation will be $417.359 billion in 2011.

117. a. If the Smiths drive x miles in a year, and their car averages 17 miles per gallon of gasoline, then they will use about $\dfrac{x}{17}$ gallons of gasoline per year. In 2010, the Smith's annual fuel cost is given by

$$C = 2.739\left(\frac{x}{17}\right) = \frac{2.739}{17}x \approx 0.1611x.$$

b. In 2009, the Smiths annual fuel cost is given by $C = 1.847\left(\dfrac{x}{17}\right) = \dfrac{1.847}{17}x \approx 0.1086x.$

c. Assuming that the Smiths drove 15,000 miles, their fuel cost in 2010 was

$$C = 2.739\left(\frac{15,000}{17}\right) \approx \$2417.$$

d. Assuming that the Smiths drove 15,000 miles, their fuel cost in 2009 was

$$C = 1.847\left(\frac{15,000}{17}\right) \approx \$1630.$$

e. The difference in the annual costs is about $2417 - $1630 = $787. The Smiths spend $787 more at the 2010 price than at the 2009 price.

119.a. First we compute the slope of the line.

$$m = \frac{N_2 - N_1}{t_2 - t_1} = \frac{1147.5 - 954.7}{3 - 0}$$

$$= \frac{192.8}{3} \approx 64.3$$

Then we use the slope and the point $(0, 954.7)$ to write the point-slope form of the line.

$$N - N_1 = m(t - t_1)$$
$$N - 954.7 = 64.3(t - 0)$$
$$N = 64.3t + 954.7$$

b. The slope indicates that credit and debit cards in force are increasing at an average rate of 64.3 million cards per year.

c. In 2011, $t = 5$, and
$N = 64.3(5) + 954.7 = 1276.2$
There will be an estimated 1.276 billion credit cards in force at the end of the first quarter of 2011.

d. To estimate the year that the number of credit cards will first exceed 1.5 billion, let $N = 1500$ (since the equation is written in millions), and solve for t.
$$1500 = 64.3t + 954.7$$
$$545.3 = 64.3t$$
$$t = \frac{545.3}{64.3} \approx 8.5$$
Since $t = 0$ represented the year 2006, $t = 8.5$ is in the year 2014.
The number of credit and debit cards will surpass 1.5 billion in 2014.

121. From the graph we can see that the line has a negative slope and a y-intercept of the form $(0, b)$ where b is a positive number. Put each of the equations into slope-intercept form and choose those with negative slope and positive y-intercept.

(a) $y = -\frac{2}{3}x + 2$ (c) $y = -\frac{3}{4}x + 3$

(f) $y = -2x + 1$ (g) $y = -\frac{1}{2}x + 10$

123. A vertical line cannot be written in slope-intercept form since its slope is not defined.

125. Two lines that have equal slopes and equal y-intercepts have equivalent equations and identical graphs.

127. If two lines have the same slope, but different x-intercepts, they cannot have the same y-intercept. If Line 1 has x-intercept $(a, 0)$ and Line 2 has x-intercept $(c, 0)$,

but both have the same slope m, write the equation of each line using the point-slope form then change them to slope-intercept form and compare the y-intercepts:

Line 1: $y - 0 = m(x - a)$
           $y = mx - ma$
           y-intercept is $(0, -ma)$

Line 2: $y - 0 = m(x - c)$
           $y = mx - mc$
           y-intercept is $(0, -mc)$

129. The line $y = 0$ has infinitely many x-intercepts, so yes, a line can have two distinct x-intercepts or a line can have infinitely many x-intercepts.

131. monter *(t.v.)* to go up; to ascend, to mount; to climb; to embark; to rise, to slope up, to be uphill; to grow up; to shoot; to increase. (source: *Cassell's French Dictionary*).
We use m to represent slope. The French verb *monter* means to rise, to climb or to slope up.

1.2 Pairs of Lines

1. parallel

3. To determine whether the pair of lines is parallel, coincident, or intersecting, rewrite each equation in slope-intercept form, compare their slopes, and, if necessary, compare their y-intercepts.

L: $x + y = 10$
          $y = -x + 10$
slope: $m = -1$; y-intercept: $(0, 10)$

M: $3x + 3y = 6$
          $3y = -3x + 6$
          $y = -x + 2$
slope: $m = -1$; y-intercept: $(0, 2)$
The slopes of the two lines are the same, but the y-intercepts are different, so the lines are parallel.

5. To determine whether the pair of lines is parallel, coincident, or intersecting, rewrite each equation in slope-intercept form, compare their slopes, and, if necessary, compare their y-intercepts.

L: $2x + y = 4 \Rightarrow y = -2x + 4$
slope: $m = -2$; y-intercept: $(0, 4)$

M: $2x - 2y = 8$
          $-2y = -2x + 8 \Rightarrow y = x - 4$
slope: $m = 1$; y-intercept: $(0, -4)$
The slopes of the two lines are the different, so the lines intersect.

7. To determine whether the pair of lines is parallel, coincident, or intersecting, rewrite each equation in slope-intercept form, compare their slopes, and, if necessary, compare their y-intercepts.

$L: \quad -x + y = 2 \Rightarrow y = x + 2$

slope: $m = 1$; y-intercept: $(0, 2)$

$M: \quad 2x - 2y = -4$
$\qquad\quad -2y = -2x - 4$
$\qquad\qquad\quad y = x + 2$

slope: $m = 1$; y-intercept: $(0, 2)$

Since both the slopes and the y-intercepts of the two lines are the same, the lines are coincident.

9. To determine whether the pair of lines is parallel, coincident, or intersecting, rewrite each equation in slope-intercept form, compare their slopes, and, if necessary, compare their y-intercepts.

$L: \quad 2x - 3y = -8$
$\qquad\quad -3y = -2x - 8$
$\qquad\qquad\quad y = \dfrac{2}{3}x + \dfrac{8}{3}$

slope: $m = \dfrac{2}{3}$; y-intercept: $\left(0, \dfrac{8}{3}\right)$

$M: \quad 6x - 9y = -2$
$\qquad\quad -9y = -6x - 2$
$\qquad\qquad\quad y = \dfrac{2}{3}x + \dfrac{2}{9}$

slope: $m = \dfrac{2}{3}$; y-intercept: $\left(0, \dfrac{2}{9}\right)$

The slopes of the two lines are the same, but the y-intercepts are different, so the lines are parallel.

11. To determine whether the pair of lines is parallel, coincident, or intersecting, rewrite each equation in slope-intercept form, compare their slopes, and, if necessary, compare their y-intercepts.

$L: \quad 3x - 4y = 1$
$\qquad\quad -4y = -3x + 1$
$\qquad\qquad\quad y = \dfrac{3}{4}x - \dfrac{1}{4}$

slope: $m = \dfrac{3}{4}$; y-intercept: $\left(0, -\dfrac{1}{4}\right)$

$M: \quad x - 2y = -4$
$\qquad\quad -2y = -x - 4$
$\qquad\qquad\quad y = \dfrac{1}{2}x + 2$

slope: $m = \dfrac{1}{2}$; y-intercept: $(0, 2)$

The slopes of the two lines are the different, so the lines intersect.

13. $L: \quad x = 3$; slope: not defined; no y-intercept
$M: \quad y = -2$; slope: $m = 0$, y-intercept: $(0, -2)$
Since the slopes of the two lines are different, the lines intersect.

15. To find the point of intersection of two lines, first put the lines in slope-intercept form.

$L: \; x + y = 5 \qquad M: \; 3x - y = 7$
$\quad\;\; y = -x + 5 \qquad\qquad\;\; y = 3x - 7$

Since the point of intersection, (x_0, y_0), must be on both L and M, we set the two equations equal to each other and solve for x_0. Then we substitute the value of x_0 into the equation of one of the lines to find y_0.

$-x_0 + 5 = 3x_0 - 7 \Rightarrow 12 = 4x_0 \Rightarrow x_0 = 3$

$y_0 = -(3) + 5 = 2$

The point of intersection is $(3, 2)$.

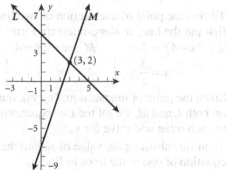

17. To find the point of intersection of two lines, first put the lines in slope-intercept form.

$L: \; x - y = 2 \qquad M: \; 2x + y = 7$
$\quad\;\; y = x - 2 \qquad\qquad\;\; y = -2x + 7$

Since the point of intersection, (x_0, y_0), must be on both L and M, we set the two equations equal to each other and solve for x_0. Then we substitute the value of x_0 into the equation of one of the lines to find y_0.

$x_0 - 2 = -2x_0 + 7$
$\quad 3x_0 = 9 \Rightarrow x_0 = 3$

$y_0 = 3 - 2 = 1$

The point of intersection is $(3, 1)$.

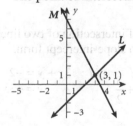

19. To find the point of intersection of two lines, first put the lines in slope-intercept form.

$L: \quad 4x + 2y = 4 \qquad M: \quad 4x - 2y = 4$

$\qquad \quad y = -2x + 2 \qquad\qquad\quad y = 2x - 2$

Since the point of intersection, (x_0, y_0), must be on both L and M, we set the two equations equal to each other and solve for x_0. Then we substitute the value of x_0 into the equation of one of the lines to find y_0.

$-2x_0 + 2 = 2x_0 - 2$

$\qquad\quad 4 = 4x_0$

$\qquad\quad x_0 = 1$

$y_0 = -2(1) + 2 = 0$

The point of intersection is $(1, 0)$.

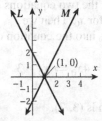

21. To find the point of intersection of two lines, first put the lines in slope-intercept form.

$L: \quad 3x - 4y = 2 \qquad M: \quad x + 2y = 4$

$\qquad \quad y = \dfrac{3}{4}x - \dfrac{1}{2} \qquad\qquad y = -\dfrac{1}{2}x + 2$

Since the point of intersection, (x_0, y_0), must be on both L and M, we set the two equations equal to each other and solve for x_0.

Then we substitute the value of x_0 into the equation of one of the lines to find y_0.

$\dfrac{3}{4}x_0 - \dfrac{1}{2} = -\dfrac{1}{2}x_0 + 2$

$3x_0 - 2 = -2x_0 + 8$

$\qquad 5x_0 = 10$

$\qquad\;\; x_0 = 2$

$y_0 = -\dfrac{1}{2}(2) + 2 = 1$

The point of intersection is $(2, 1)$.

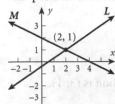

23. To find the point of intersection of two lines, first put the lines in slope-intercept form.

$L: \quad 3x - 2y = -5 \qquad M: \quad 3x + y = -2$

$\qquad\qquad y = \dfrac{3}{2}x + \dfrac{5}{2} \qquad\qquad\quad y = -3x - 2$

Since the point of intersection, (x_0, y_0), must be on both L and M, we set the two equations equal to each other and solve for x_0. Then we substitute the value of x_0 into the equation of one of the lines to find y_0.

$\dfrac{3}{2}x_0 + \dfrac{5}{2} = -3x_0 - 2$

$3x_0 + 5 = -6x_0 - 4$

$\quad 9x_0 = -9 \Rightarrow x_0 = -1$

$y_0 = -3(-1) - 2 = 1$

The point of intersection is $(-1, 1)$.

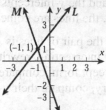

25. L is the vertical line on which the x-value is always 4. M is the horizontal line on which y-value is always -2. The point of intersection is $(4, -2)$.

27. L is parallel to $y = 2x$, so the slope of L is $m = 2$. We are given the point $(3, 3)$ on line L. Use the point-slope form of the line.

$y - y_1 = m(x - x_1)$

$y - 3 = 2(x - 3)$

$y - 3 = 2x - 6$

slope-intercept form: $y = 2x - 3$

general form: $2x - y = 3$

29. We want a line parallel to $y = 4x$. So our line will have slope $m = 4$. It must also contain the point $(-1, 2)$. Use the point slope form of the equation of a line:

$y - y_1 = m(x - x_1)$

$y - 2 = 4(x + 1)$

$y - 2 = 4x + 4$

slope-intercept form: $y = 4x + 6$

general form: $4x - y = -6$

31. We want a line parallel to $2x - y = -2$. Find the slope of the line and use the given point, $(0, 0)$ to obtain the equation. Since the y-intercept $(0, 0)$ is given, use the slope-intercept form of the equation of a line.
Original line:
$$2x - y = -2$$
$$y = 2x + 2$$
$$m = 2$$
Parallel line:
$$y = mx + b$$
$$y = 2x + 0$$
$$y = 2x$$
general form: $2x - y = 0$

33. We want a line parallel to the line $x = 3$. This is a vertical line so the slope is not defined. A parallel line will also be vertical, and it must contain the point $(4, 2)$. The parallel line will have the equation $x = 4$.

35. To find the equation of the line, we must first find the slope of the line containing the points $(-2, 9)$ and $(3, -10)$:
$$m = \frac{y_2 - y_1}{x_2 - x_1} = \frac{9 - (-10)}{(-2) - 3} = \frac{19}{-5} = -\frac{19}{5}$$
The slope of a line parallel to the line containing these points is also $-\dfrac{19}{5}$. Use the slope and the point $(-2, -5)$ to write the point-slope form of the parallel line.
$$y - y_1 = m(x - x_1)$$
$$y + 5 = -\frac{19}{5}(x + 2)$$
Solve for y to get the slope-intercept form:
$$y = -\frac{19}{5}x - \frac{38}{5} - 5 \Rightarrow y = -\frac{19}{5}x - \frac{63}{5}$$
Multiply both sides by 5 and rearrange terms to obtain the general form of the equation:
$$19x + 5y = -63$$

37. We will let x = the number of caramels the box of candy, and y = the number of creams in the box of candy. Since there are a total of 50 pieces of candy in a box, we have $x + y = 50$, or $y = 50 - x$. Each caramel costs \$0.10 to make, and each cream costs \$0.20 to make. So, the cost of making a box of candy is given by the equation:
$$C = 0.1x + 0.2y$$
$$= 0.1x + 0.2(50 - x)$$

The box of candy sells for \$8.00. So to break even, we need
$$R = C$$
$$8 = 0.1x + 0.2(50 - x)$$
$$8 = 0.1x + 10 - 0.2x$$
$$8 = 10 - 0.1x$$
$$0.1x = 2$$
$$x = 20$$
$$y = 50 - x = 50 - 20 = 30$$
To break even, put 20 caramels and 30 creams into each box. If the candy shop owner increases the number of caramels to more than 20 (and decreases the number of creams) the owner will obtain a profit since the caramels cost less to produce than the creams.

39. Investment problems are simply mixture problems involving money. We will use a table to organize the information. Let x denote the amount Mr. Nicholson invests in AA bonds, and y denote the amount he invests in S & L Certificates.

Investment	Amount Invested	Interest Rate	Interest Earned
AA Bonds	x	0.10	$0.10x$
S&L Certificates	$y = 150,000 - x$	0.05	$0.05(150,000 - x)$
Total	$x + y = 150,000$		10,000

The last column gives the information we need to set up the equation solve since the sum of the interest earned on the two investments must equal the total interest earned.
$$0.1x + 0.05(150,000 - x) = 10,000$$
$$0.1x + 7500 - 0.05x = 10,000$$
$$0.05x + 7500 = 10,000$$
$$0.05x = 2500$$
$$x = 50,000$$
$$y = 150,000 - x$$
$$= 100,000$$
Mr. Nicholson should invest \$50,000 in AA Bonds and \$100,000 in Savings and Loan Certificates in order to earn \$10,000 per year.

41. Let x denote the amount of Kona coffee and y denote the amount of Columbian coffee in the mix. We will use the hint and assume that the total weight of the blend is 100 pounds.

Coffee	Amount Mixed	Price per Pound	Total Value
Kona	x	22.95	$22.95x$
Columbian	$y = 100 - x$	6.75	$6.75(100 - x)$
Mixture	$x + y = 100$	10.80	$10.80(100)$

(continued on next page)

(*continued*)

The last column gives the information necessary to write the equation, since the sum of the values of each of the two individual coffees must equal the total value of the mixture.

$$22.95x + 6.75(100 - x) = 10.80(100)$$
$$22.95x + 675 - 6.75x = 1080$$
$$16.2x = 405$$
$$x = 25$$
$$y = 100 - 25 = 75$$

Mix 25 pounds of Kona coffee with 75 pounds of Columbian coffee to obtain a blend worth $10.80 per pound.

43. Let x represent the amount of Acid A used and let y represent the amount of Acid B used in the solution. We will use a table to organize the information. We will use a table to organize the information.

	Amount Mixed	Protein Content	Amount of Protein
Acid A	x	0.15	$0.15x$
Acid B	$y = 100 - x$	0.05	$0.05(100 - x)$
Mixture	$x + y = 100$	0.08	$0.08(100)$

The last column gives the information necessary to write the equation, since the sum of the two individual solutions must equal the total solution.

$$0.15x + 0.05(100 - x) = 8$$
$$0.15x + 5 - 0.05x = 8$$
$$0.1x = 3$$
$$x = 30$$
$$y = 100 - 30 = 70$$

We should mix 30 cubic centimeters of 15% solution with 70 cubic centimeters of 5% solution to obtain a 100 cubic centimeters of 8% solution.

45. a. The realized gain, y, is the difference between the 2006 value and the 2010 value of the gold.
$y = 1112.30x - 549.86x = 562.44x$.

b. The point of intersection is the point (x_0, y_0) that satisfies both equation $y = 562.44x$ and equation $y = 10,000$.
$$10,000 = 562.44x$$
$$x = \frac{10,000}{562.44} \approx 17.8$$
The individual would have had to bought and sold about 17.8 ounces of gold to realize a gain of $10,000.00.

47. a. Assuming that the rate of growth is constant, we find the slope of the line passing through the points (0, 1615) and (365, 1877).
$$m = \frac{y_2 - y_1}{x_2 - x_1} = \frac{1877 - 1615}{365 - 0} = \frac{262}{365}$$
Next use the slope and the point (0, 1615) to write the point-slope form of the line.
$$y - y_1 = m(x - x_1)$$
$$y - 1615 = \frac{262}{365}(x - 0)$$
$$y = \frac{262}{365}x + 1615$$

b.
$$2000 = \frac{262}{365}x + 1615$$
$$385 = \frac{262}{365}x$$
$$x \approx 536.4$$
There were 2000 HD radio stations 537 days after February 1, 2008 or on July 22, 2009.

1.3 Applications to Business and Economics

1. False. The break even point is the intersection of the revenue graph and the cost graph.

3. The break-even point is the point where the revenue and the cost are equal. Setting $R = C$, we find
$$30x = 10x + 600$$
$$20x = 600$$
$$x = 30$$
That is, 30 units must be sold to break even. The break-even point is $x = 30$, $R = 30(30) = 900$ or (30, 900).

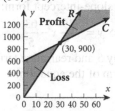

5. The break-even point is the point where the revenue and the cost are equal. Setting $R = C$, we find
$$0.30x = 0.20x + 50$$
$$0.10x = 50$$
$$x = 500$$
That is, 500 units must be sold to break even. Break-even point: (500, 150)

(*continued on next page*)

(*continued*)

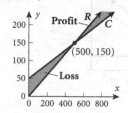

7. The market price is the price at which the supply and the demand are equal.

$$S = D$$
$$p + 1 = 3 - p$$
$$2p = 2$$
$$p = 1$$

At a price of $1.00 supply and demand are equal, so $1.00 is the market price.

9. The market price is the price at which the supply and the demand are equal.

$$S = D$$
$$20p + 500 = 1000 - 30p$$
$$50p = 500$$
$$p = 10$$

At a price of $10.00 supply and demand are equal, so $10.00 is the market price.

11. The break-even point is the point where the revenue and the cost are equal. Cost is given by the variable cost of producing x pennants at $0.75 per pennant, plus the fixed operational overhead of $300 per day.

$$C = \$0.75x + \$300$$

Revenue is the product of price of each pennant ($1) and the number of pennants sold.

$$R = \$1x$$

Setting $R = C$, we find

$$1x = 0.75x + 300$$
$$0.25x = 300$$
$$x = 1200$$

1200 pennants must be sold each day to break even.

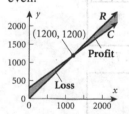

13. a. The market price is the price at which the supply and the demand are equal.

$$S = D$$
$$0.7p + 0.4 = -0.5p + 1.6$$
$$1.2p = 1.2 \Rightarrow p = 1$$

The market price is $1.00 per pound.

b. To find the quantity supplied at market price, let $p = 1$ and solve for S:

$$S = 0.7(1) + 0.4 = 1.1$$

So 1.1 million pounds are demanded at $1.00.

c.

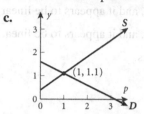

d. The point of intersection called the market equilibrium. It is the price where the quantity supplied equals the quantity demanded.

15. At the market price of $3.00, $S = 2(3) + 5 = 11$, so 11 units of the commodity are supplied. Since at the market price the supply and demand are equal, the point (3, 11) satisfies the demand equation. In addition, we are told that at a price of $1.00, 19 units are demanded. To find the demand equation, use the points (3, 11) and (1, 19) to find the slope.

$$m = \frac{D_2 - D_1}{p_2 - p_1} = \frac{19 - 11}{1 - 3} = \frac{8}{-2} = -4$$

We then use the point (1, 19), the slope $m = -4$, and the point-slope form of the equation.

$$D - D_1 = m(p - p_1)$$
$$D - 19 = -4(p - 1)$$
$$D = -4p + 23$$

17. Let x denote the number of DVDs purchased. The cost of purchasing x DVDs from the club is

$$C = 0.49(4) + 17.95(x - 4) + 2.31x$$
$$= 1.96 + 17.95x - 71.8 + 2.31x$$
$$= 20.26x - 69.84$$

The cost of purchasing x DVDs from the discount retailer is $D = 14.95(1.07)x = 16.00x$. We want to buy as many DVDs as possible from the club, while keeping the cost below the retailers. So we set $C = D$ and solve for x.

$$20.26x - 69.84 = 16.00x$$
$$4.26x = 69.84$$
$$x = 16.39$$

At most sixteen DVDs can be ordered from the club to keep the price lower than that of the discount retailer.

1.4 Scatter Diagrams; Linear Curve Fitting

1. True

3. A relation exists, and it appears to be linear.

5. A relation exists, and it appears to be linear.

7. No relation exists.

9. a.

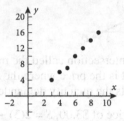

b. We select points (3, 4) and (9, 16). The slope of the line containing these points is:

$$m = \frac{y_2 - y_1}{x_2 - x_1} = \frac{16 - 4}{9 - 3} = \frac{12}{6} = 2$$

The equation of the line is:

$$y - y_1 = m(x - x_1)$$
$$y - 4 = 2(x - 3)$$
$$y - 4 = 2x - 6$$
$$y = 2x - 2$$

c. The line on the scatter diagram will vary depending on the choice of points in part (b).

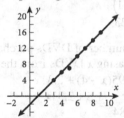

d.

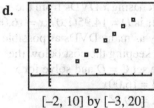

[–2, 10] by [–3, 20]

e.

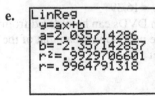

$$y = 2.0357x - 2.357$$

f.

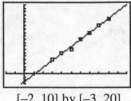

[–2, 10] by [–3, 20]

11. a.

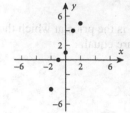

b. We select points (–2, –4) and (2, 5). The slope of the line containing these points is:

$$m = \frac{y_2 - y_1}{x_2 - x_1} = \frac{5 - (-4)}{2 - (-2)} = \frac{9}{4}$$

The equation of the line is:

$$y - y_1 = m(x - x_1)$$
$$y - 5 = \frac{9}{4}(x - 2)$$
$$y - 5 = \frac{9}{4}x - \frac{9}{2}$$
$$y = \frac{9}{4}x + \frac{1}{2}$$

c. The line on the scatter diagram will vary depending on the choice of points in part (b).

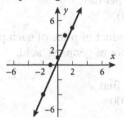

d.

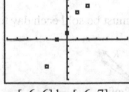

[–6, 6] by [–6, 7]

e.

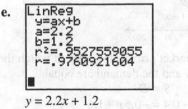

$$y = 2.2x + 1.2$$

f.

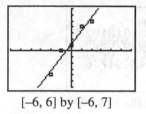

$[-6, 6]$ by $[-6, 7]$

f.

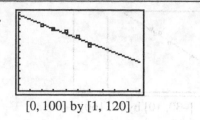

$[0, 100]$ by $[1, 120]$

13. a.

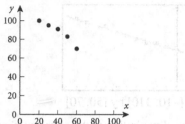

15. a.

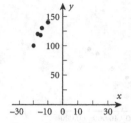

b. We select points $(20, 100)$ and $(60, 70)$. The slope of the line containing these points is:

$$m = \frac{y_2 - y_1}{x_2 - x_1} = \frac{70 - 100}{60 - 20} = \frac{-30}{40} = -\frac{3}{4}$$

The equation of the line is:

$$y - y_1 = m(x - x_1)$$
$$y - 100 = -\frac{3}{4}(x - 20)$$
$$y - 100 = -\frac{3}{4}x + 15$$
$$y = -\frac{3}{4}x + 115$$

c. The line on the scatter diagram will vary depending on the choice of points in part (b).

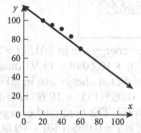

d.

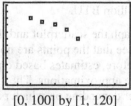

$[0, 100]$ by $[1, 120]$

e.

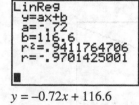

LinReg
y=ax+b
a=-.72
b=116.6
r²=.9411764706
r=-.9701425001

$y = -0.72x + 116.6$

b. We select points $(-20, 100)$ and $(-10, 140)$. The slope of the line containing these points is:

$$m = \frac{y_2 - y_1}{x_2 - x_1} = \frac{140 - 100}{(-10) - (-20)} = \frac{40}{10} = 4$$

The equation of the line is:

$$y - y_1 = m(x - x_1)$$
$$y - 100 = 4(x - (-20))$$
$$y - 100 = 4x + 80$$
$$y = 4x + 180$$

c. The line on the scatter diagram will vary depending on the choice of points in part (b).

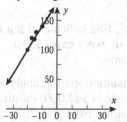

d.

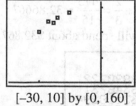

$[-30, 10]$ by $[0, 160]$

e.

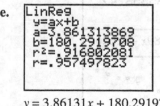

LinReg
y=ax+b
a=3.861313869
b=180.2919708
r²=.916802081
r=.957497823

$y = 3.86131x + 180.29197$

f.

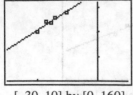

[−30, 10] by [0, 160]

17. a.

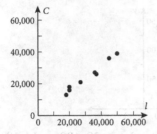

b. Answers will vary. We select points (20, 16) and (50, 39) with numbers in thousands. The slope of the line containing these points is:

$$m = \frac{C_2 - C_1}{I_2 - I_1} = \frac{39 - 16}{50 - 20} = \frac{23}{30}$$

The equation of the line is:

$$C - C_1 = m(I - I_1)$$

$$C - 16 = \frac{23}{30}(I - 20)$$

$$C - 16 = \frac{23}{30}I - \frac{46}{3}$$

$$C = \frac{23}{30}I + \frac{2}{3}$$

c. The slope of this line indicates that a family will spend $23 of every extra $30 of disposable income.

d. To find the consumption of a family whose disposable income is $42,000, substitute 42 for x in the equation from part (b).

$$C = \frac{23}{30}(42) + \frac{2}{3} = \frac{473}{15} \approx 32.86667$$

The family will spend about $32,867.

e.

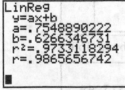

$y = 0.75489x + 0.62663$

19. a.

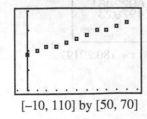

[−10, 110] by [50, 70]

b.

$y = 0.07818x + 59.0909$

c.

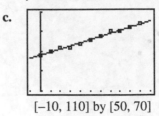

[−10, 110] by [50, 70]

d. The slope indicates the apparent change in temperature in a 65°F room for every percent increase in relative humidity.

e. To determine the apparent temperature when the relative humidity is 75%, evaluate the equation of the line of best fit when $x = 75$.

$y = 0.07818(75) + 59.0909 = 64.95$

When the relative humidity is 75%, the temperature of the room will appear to be 65°F.

21. a. Using the LinReg function, the line of best fit is $y = 0.0651t + 10.6049$.

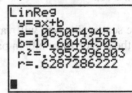

The predicted energy use in 2015, $t = 20$, is $y = 0.0651(20) + 10.6049 \approx 11.91$ quadrillion BTU. The predicted energy use in 2030, $t = 35$, is $y = 0.0651(35) + 10.6049 \approx 12.88$ quadrillion BTU. The predicted energy use in 2035, $t = 40$, is $y = 0.0651(40) + 10.6049 \approx 13.21$ quadrillion BTU.

b. When we graph the scatterplot and the line of best fit, we see that the points are not strictly linear. Therefore, estimates based on the line of best fit are approximations. The line of best fit may differ depending on which points are chosen or a linear model may not have been used.

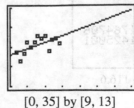

[0, 35] by [9, 13]

Chapter 1 Review Exercises

1.

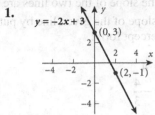

$y = -2x + 3$

3.

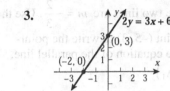

$2y = 3x + 6$

5. a. $m = \dfrac{y_2 - y_1}{x_2 - x_1} = \dfrac{4 - 2}{(-3) - 1} = \dfrac{2}{-4} = -\dfrac{1}{2}$

A slope of $-\dfrac{1}{2}$ means that for every 2 unit change in x, y will change (-1) unit. That is, for every 2 units x moves to the right, y will move down 1 unit.

b. Use the point $(1, 2)$ and the slope to get the point-slope form of the equation of the line:

$y - y_1 = m(x - x_1)$

$y - 2 = -\dfrac{1}{2}(x - 1)$

Simplifying and solving for y gives the slope-intercept form:

$y - 2 = -\dfrac{1}{2}x + \dfrac{1}{2}$

$y = -\dfrac{1}{2}x + \dfrac{5}{2}$

Rearranging terms gives the general form of the equation: $x + 2y = 5$

c.

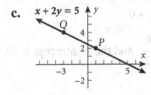

$x + 2y = 5$

7. a. $m = \dfrac{y_2 - y_1}{x_2 - x_1} = \dfrac{5 - 3}{(-1) - (-2)} = \dfrac{2}{1} = 2$

A slope of 2 means that for every one unit change in x, y changes by 2 units. That is, for every 1 unit x moves to the right, y moves up 2 units.

b. Use the point $(-1, 5)$ and the slope to get the point-slope form of the equation of the line:

$y - y_1 = m(x - x_1)$

$y - 5 = 2(x - (-1))$

Simplifying and solving for y gives the slope-intercept form:

$y - 5 = 2x + 2$

$y = 2x + 7$

Rearranging terms gives the general form of the equation: $2x - y = -7$

c.

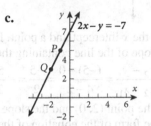

$2x - y = -7$

9. Since we are given the slope $m = -3$ and a point, we get the point-slope equation of the line:

$y - y_1 = m(x - x_1)$

$y - (-1) = -3(x - 2)$

Solving for y puts the equation into the slope-intercept form:

$y + 1 = -3x + 6$

$y = -3x + 5$

Rearranging terms gives the general form of the equation:

$3x + y = 5$

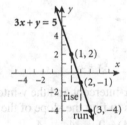

$3x + y = 5$

11. Since we are given the slope $m = 0$ and a point on the line. We either use the point-slope formula or recognize that this is a horizontal line and the equation of a horizontal line $y = b$. The general form of the equation is $y = 4$.

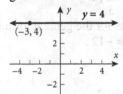

$y = 4$

13. We are told that the line is vertical, so the slope is not defined. We also know the line contains the point (8, 5). The general equation of a vertical line is $x = a$, so the general form of this equation is $x = 8$.

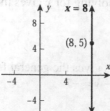

15. We are given the x-intercept and a point. First we find the slope of the line containing the two points. $m = \dfrac{y_2 - y_1}{x_2 - x_1} = \dfrac{(-5) - 0}{4 - 2} = \dfrac{-5}{2} = -\dfrac{5}{2}$

    We then use the point (2, 0) and the slope to get the point-slope form of the equation of the line.

    $$y - y_1 = m(x - x_1)$$
    $$y - 0 = -\frac{5}{2}(x - 2)$$

    To get the slope-intercept form, solve for y:

    $$y = -\frac{5}{2}x + 5$$

    Rearrange the terms for the general form of the equation: $5x + 2y = 10$.

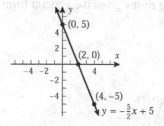

17. We are given the x-intercept and the y-intercept. Use the two points to find the slope of the line.

    $m = \dfrac{y_2 - y_1}{x_2 - x_1} = \dfrac{(-4) - 0}{0 - (-3)} = \dfrac{-4}{3} = -\dfrac{4}{3}$

    Since one of the points is the y-intercept, use it and the slope to write the slope-intercept form of the equation: $y = -\dfrac{4}{3}x - 4$

    Rearrange the terms for the general form of the equation: $4x + 3y = -12$.

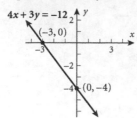

19. Since the line we are seeking is parallel to $2x + 3y = -4$, the slope of the two lines are the same. Find the slope of the given line by putting it into slope-intercept form:

    $$2x + 3y = -4$$
    $$3y = -2x - 4$$
    $$y = -\frac{2}{3}x - \frac{4}{3}$$

    The slopes of the two lines are $m = -\dfrac{2}{3}$. Use the slope and the point (–5, 3) to write the point-slope form of the equation of the parallel line.

    $$y - y_1 = m(x - x_1)$$
    $$y - 3 = -\frac{2}{3}(x - (-5))$$
    $$y - 3 = -\frac{2}{3}(x + 5)$$

    To put the equation into slope-intercept form, solve for y.

    $$y - 3 = -\frac{2}{3}x - \frac{10}{3}$$
    $$y = -\frac{2}{3}x - \frac{1}{3}$$

    Rearrange the terms to obtain the general form of the equation: $2x + 3y = -1$

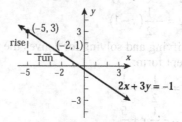

21. To find the slope and y-intercept of the line, put the equation into the slope-intercept form.

    $$9x + 2y = 18$$
    $$2y = -9x + 18$$
    $$y = -\frac{9}{2}x + 9$$

    The slope is $-\dfrac{9}{2}$, and the y-intercept is (0, 9).

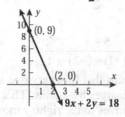

23. To find the slope and y-intercept of the line, put the equation into the slope-intercept form.

$$4x + 2y = 9$$
$$2y = -4x + 9$$
$$y = -2x + \frac{9}{2}$$

The slope is -2, and the y-intercept is $\left(0, \dfrac{9}{2}\right)$.

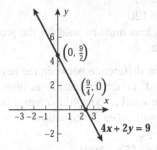

25. To find the slope and y-intercept of the line, put the equation into the slope-intercept form.

$$\frac{1}{2}x + \frac{1}{3}y = \frac{1}{6}$$
$$\frac{1}{3}y = -\frac{1}{2}x + \frac{1}{6}$$
$$y = -\frac{3}{2}x + \frac{1}{2}$$

The slope is $-\dfrac{3}{2}$, and the y-intercept is $\left(0, \dfrac{1}{2}\right)$.

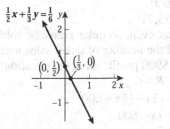

27. Put each equation into slope-intercept form:

$$3x - 4y = -12 \qquad\qquad 6x - 8y = -9$$
$$-4y = -3x - 12 \qquad\qquad -8y = -6x - 9$$
$$y = \frac{3}{4}x + 3 \qquad\qquad y = \frac{6}{8}x + \frac{9}{8}$$
$$\qquad\qquad\qquad\qquad\qquad y = \frac{3}{4}x + \frac{9}{8}$$

Since both lines have the same slope but different y-intercepts, the lines are parallel.

29. Put each equation into slope-intercept form:

$$x - y = -2 \qquad\qquad 3x - 4y = -12$$
$$-y = -x - 2 \qquad\qquad -4y = -3x - 12$$
$$y = x + 2 \qquad\qquad y = \frac{3}{4}x + 3$$

Since the lines have different slopes, they intersect.

31. Put each equation into slope-intercept form:

$$4x + 6y = -12 \qquad\qquad 2x + 3y = -6$$
$$6y = -4x - 12 \qquad\qquad 3y = -2x - 6$$
$$y = -\frac{2}{3}x - 2 \qquad\qquad y = -\frac{2}{3}x - 2$$

Since both lines have the same slope and the same y-intercept, the lines are coincident.

33. To find the point of intersection of two lines, first put the lines in slope-intercept form:

$$L: \quad x - y = 4 \qquad\qquad M: \ x + 2y = 7$$
$$\qquad y = x - 4 \qquad\qquad\qquad y = -\frac{1}{2}x + \frac{7}{2}$$

Since the point of intersection, (x_0, y_0), must be on both L and M, we set the two equations equal to each other and solve for x_0. Then we substitute the value of x_0 into the equation of one of the lines to find y_0.

$$x_0 - 4 = -\frac{1}{2}x_0 + \frac{7}{2} \qquad y_0 = x_0 - 4$$
$$2x_0 - 8 = -x_0 + 7 \qquad\quad y_0 = 5 - 4$$
$$3x_0 = 15 \qquad\qquad\qquad\ y_0 = 1$$
$$x_0 = 5$$

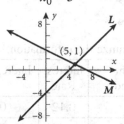

35. To find the point of intersection of two lines, first put the lines in slope-intercept form:

$$L: \quad x - y = -2 \qquad\qquad M: \ x + 2y = 7$$
$$\qquad y = x + 2 \qquad\qquad\qquad y = -\frac{1}{2}x + \frac{7}{2}$$

Since the point of intersection, (x_0, y_0), must be on both L and M, we set the two equations equal to each other and solve for x_0. Then we substitute the value of x_0 into the equation of one of the lines to find y_0.

$$x_0 + 2 = -\frac{1}{2}x_0 + \frac{7}{2} \qquad y_0 = x_0 + 2$$
$$2x_0 + 4 = -x_0 + 7 \qquad\quad y_0 = 1 + 2$$
$$3x_0 = 3 \qquad\qquad\qquad\ y_0 = 3$$
$$x_0 = 1$$

The point of intersection is $(1, 3)$.

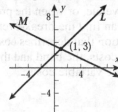

37. To find the point of intersection of two lines, first put the lines in slope-intercept form:

$L: 2x - 4y = -8$ 　　　 $M: 3x + 6y = 0$

$$-4y = -2x - 8$$ 　　　 $$6y = -3x$$

$$y = \frac{1}{2}x + 2$$ 　　　 $$y = -\frac{1}{2}x$$

Since the point of intersection, (x_0, y_0), must be on both L and M, we set the two equations equal to each other and solve for x_0. Then we substitute the value of x_0 into the equation of one of the lines to find y_0.

$$\frac{1}{2}x_0 + 2 = -\frac{1}{2}x_0 \qquad y_0 = -\frac{1}{2}x_0$$

$$x_0 + 4 = -x_0 \qquad y_0 = -\frac{1}{2} \cdot (-2)$$

$$2x_0 = -4 \qquad y_0 = 1$$

$$x_0 = -2$$

The point of intersection is (–2, 1).

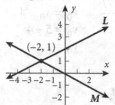

39. Use a table to organize the information.

	Amount Invested	Interest Rate	Interest Earned
B-Bonds	x	0.12	$0.12x$
Bank	$y = 90,000 - x$	0.05	$0.05(90,000 - x)$
Total	$x + y = 90,000$		10,000

The last column gives the information needed for the equation since the sum of the interest earned on the individual investments must equal the total interest earned.

$$0.12x + 0.05(90,000 - x) = 10,000$$

$$0.12x + 4500 - 0.05x = 10,000$$

$$0.07x = 5,500$$

$$x \approx 78,571.43$$

$$y = 90,000 - 78,571.43$$

$$= 11,428.57$$

Karen should invest $78,571.43 in B-rated bonds and $11,428.57 in the well-known bank in order to achieve their investment goals.

41.a. The break-even point is the point where the cost equals the revenue, or when the profit is zero. Before we can find the break-even point we need the equation that describes cost. We are told the fixed costs, the band and the advertising, and the variable costs.

If we let x denote the number of tickets sold, the cost of the dance is described by the equation: $C = 500 + 100 + 5x = 5x + 600$. The revenue is given by the equation, $R = 10x$, since each ticket costs $10. Setting $C = R$, and solving for x, will tell how many tickets must be sold to break even.

$$5x + 600 = 10x$$

$$600 = 5x$$

$$x = 120$$

So, 120 tickets must be sold for the group to break even.

b. Profit is the difference between the revenue and the cost. To determine the number of tickets that need to be sold to clear a profit of $900, we will solve the equation:

$$P = R - C$$

$$900 = 10x - (5x + 600)$$

$$900 = 5x - 600$$

$$1500 = 5x$$

$$x = 300$$

The church group must sell 300 tickets to realize a profit of $900.

c. If tickets cost $12, the break-even point will come from the equation

$$R = C$$

$$12x = 5x + 600$$

$$7x = 600$$

$$x = 85.71$$

To break even, 86 tickets must be sold.
To find the number of ticket sales needed to have a $900 profit, solve the equation:

$$P = R - C$$

$$900 = 12x - (5x + 600)$$

$$900 = 7x - 600$$

$$1500 = 7x$$

$$x = 214.29$$

So the church group needs to sell 215 tickets at $12 each to realize a profit of $900.

43.

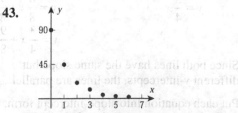

This relation does not appear to be linear.

45. a. The market price is the price for which supply equals demand. To find market price set $S = D$ and solve for p.

$$0.8p + 0.2 = -0.4p + 1.8$$
$$1.2p = 1.6$$
$$p = 1.33$$

The market price is $1.33 per bushel.

b. When $p = 1.33$, $S = 0.8(1.33) + 0.2 = 1.264$. There will be 1.264 million bushels of corn supplied at the market price of $1.33.

c.

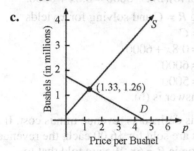

d. At a price of $1.33 per bushel the supply and the demand for corn are equal.

47. a.

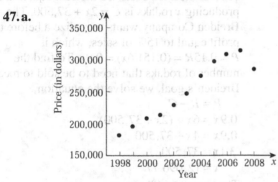

b. $m = \dfrac{p_2 - p_1}{x_2 - x_1} = \dfrac{228,700 - 181,900}{2002 - 1998}$

$= \dfrac{46,800}{4} = 11,700$

c. The mean price of houses sold in the United States increased by an average of $11,700 per year from 1998 through 2002.

d. $m = \dfrac{p_2 - p_1}{x_2 - x_1} = \dfrac{292,600 - 228,700}{2008 - 2002}$

$= \dfrac{63,900}{6} = 10,650$

e. The mean price of houses sold in the United States increased by an average of $10,650 per year from 2002 through 2008.

f.

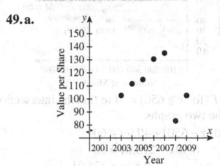

The slope of the line of best fit is $a = 13,960$.

g. The mean price of houses sold in the United States increased by an average of $13,960 per year from 1998 through 2008.

h. The average annual increase in mean price of houses sold in the United States is not constant. The slope depends on the points used to calculate it.

i. The average price of houses sold in the United States increased from 1998 though 2007 and then started to decline.

49. a.

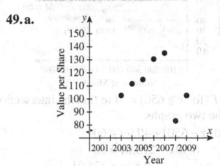

b. The data do not appear to be linearly related.

c. Since the data are not linearly related, the prediction in Problem 48 is not valid.

d. Answers will vary.

51. a. This is the equation of a vertical line that includes the point (0, 0). Its graph is the y-axis.

b. The graph of $y = 0$ is the x-axis. It is a horizontal line and has a slope of zero.

c. This is an equation of a line with a slope of -1 and a y-intercept of (0, 0). Its graph is a negatively sloped diagonal line through the origin.

Chapter 1 Project

1. Since Metro PCS charges a flat rate that does not depend on the number of minutes, x, the equation is that of a horizontal line.
$M = 60$

3.

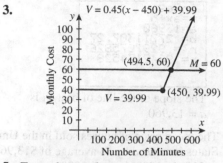

5. From the graph, we can see that if you talk more than 494.5 minutes per month, the Metro PCS plan is more economical.

7.

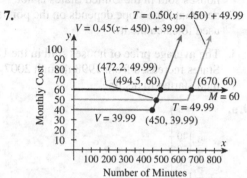

Set T (for $x \geq 650$) $= M$ to find the intersection of the two graphs.
$$0.50(x - 650) + 49.99 = 60$$
$$0.50x - 275.01 = 60$$
$$0.50x = 335.01$$
$$x \approx 670$$
Set V (for $x \geq 450$) $= T = 49.99$ to find the intersection of the two graphs.
$$0.45(x - 450) + 39.99 = 49.99$$
$$0.45x - 162.51 = 49.99$$
$$0.45x = 212.5$$
$$x \approx 472.2$$

9. T-Mobile is cheapest if you talk between 472 and 670 minutes per month. Metro PCS is cheapest if you talk more than 670 minutes per month.

11. The monthly charge for Metro PCS is $70 including the phone.
$$0.45(x - 450) + 39.99 = 70$$
$$0.45x - 162.51 = 70$$
$$0.45x = 232.51$$
$$x \approx 516.7$$
Verizon is the better deal if you talk less than 517 minutes per month.

Mathematical Questions from Professional Exams

1. The break-even point is the value of x for which the revenue equals cost. If x units are sold at price of $2.00 each, the revenue is $R = 2x$. Cost is the total of the fixed and variable costs. We are told that the fixed costs are $6000, and that the variable cost per item is 40% of the price. So the cost is given by the equation:
$$C = (0.40)(2)x + 6000 = 0.8x + 6000.$$
Setting $R = C$ and solving for x yields
$$R = C$$
$$2x = 0.8x + 6000$$
$$1.2x = 6000$$
$$x = 5000$$
The answer is (b).

2. Profit is defined as revenue minus cost. If x rodaks are sold for $6.00 each, the revenue equation is $R = 6x$. We are told that to manufacture rodaks costs $2.00 per unit and $37,500 in fixed costs. So the cost equation for producing x rodaks is $C = 2x + 37,500$. The Breiden Company wants to realize a before tax profit equal to 15% of sales, which is $P = 0.15R = (0.15)(6x) = 0.9x$. To find the number of rodaks that need to be sold to meet Breiden's goal, we solve the equation
$$P = R - C$$
$$0.9x = 6x - (2x + 37,500)$$
$$0.9x = 4x - 37,500$$
$$3.1x = 37,500$$
$$x = 12,096.77$$
The answer is (d).

3. The break-even point is the number of units that must be sold for revenue to equal cost. Using the notation given, we have $R = SPx$ and $C = VCx + FC$. To find the sales level necessary to break even, we set $R = C$ and solve for x.
$$R = C$$
$$SPx = VCx + FC$$
$$SPx - VCx = FC$$
$$(SP - VC)x = FC$$
$$x = \frac{FC}{SP - VC}$$
The answer is (d)

4. At the break-even point of $x = 400$ units sold, cost equals revenue. We are told cost, $C = \$400 + 200 = 600$, so revenue $R = \$600$. Since $R = \text{price} \cdot \text{quantity}$, the price of each item is $\dfrac{\$600}{400} = \$1.50$. The variable cost per unit is $\$1.00$, so the 401st unit contributes $\$1.50 - \$1.00 = \$0.50$ to profits.
   The answer is (b).

5. Since straight-line depreciation remains the same over the life of the property, its expense over time will be a horizontal line. Sum-of-year's-digits depreciation expense decreases as time increases.
   The answer is (c).

6. The answer is (c). $Y = \$1000 + \$2X$ is a linear relationship.

7. The answer is (b). Y is an estimate of total factory overhead.

8. The answer is (b). In the equation $\$2$ is the estimate of variable cost per direct labor hour.

Chapter 2 Systems of Linear Equations

2.1 Systems of Linear Equations: Substitution; Elimination

1. True

Solve each equation for y and compare their slopes and y-intercepts.

$3x - 4y = 24$
$$-4y = -3x + 24$$
$$y = \frac{3}{4}x - 6$$
$3x - 4y = 12$
$$-4y = -3x + 12$$
$$y = \frac{3}{4}x - 3$$

Since the slopes are the same but the y-intercepts are different, the lines are parallel.

3. Since both equations are equal to y, they are equal to each other. First solve for x, then use the result to solve for y.
$$60x - 900 = -15x + 2850$$
$$75x = 3750$$
$$x = 50$$
$$y = 60x - 900$$
$$y = 60(50) - 900 = 2100$$

The point of intersection is $(50, 2100)$.

5. False. It is possible that a system has no solutions.

7. To be a solution to the system of equations, the given values of the variables must solve each equation. So we evaluate each equation at $x = 2$ and $y = -1$.
$$\begin{cases} 2x - y = 5 & (1) \\ 5x + 2y = 8 & (2) \end{cases}$$
(1) $\quad 2(2) - (-1) = 4 + 1 = 5$ ✓
(2) $\quad 5(2) + 2(-1) = 10 + (-2) = 8$ ✓

Since both equations are satisfied, $x = 2$, $y = -1$ is a solution to the system.

9. To be a solution to the system of equations, the given values of the variables must solve each equation. So we evaluate each equation at $x = 2$ and $y = \frac{1}{2}$.
$$\begin{cases} 3x + 4y = 4 & (1) \\ \frac{1}{2}x - 3y = -\frac{1}{2} & (2) \end{cases}$$

(1) $\quad 3(2) + 4\left(\frac{1}{2}\right) = 6 + 2 = 8 \neq 4$

Since the first equation was not satisfied, $x = 2$, $y = \frac{1}{2}$ is not a solution to the system.

11. To be a solution to the system of equations, the given values of the variables must solve each equation. So we evaluate each equation at $x = 1$, $y = -1$, and $z = 2$.
$$\begin{cases} 3x + 3y + 2z = 4 & (1) \\ x - y - z = 0 & (2) \\ 2y - 3z = -8 & (3) \end{cases}$$
(1) $\quad 3(1) + 3(-1) + 2(2) = 3 - 3 + 4 = 4$ ✓
(2) $\quad 1 - (-1) - 2 = 0$ ✓
(3) $\quad 2(-1) - 3(2) = -2 - 6 = -8$ ✓

Since all three equations are satisfied, $x = 1$, $y = -1$, $z = 2$ is a solution to the system.

13. To be a solution to the system of equations, the given values of the variables must solve each equation. So we evaluate each equation at $x = 2$, $y = -2$, and $z = 2$.
$$\begin{cases} 3x + 3y + 2z = 4 & (1) \\ x - 3y + z = 10 & (2) \\ 5x - 2y - 3z = 8 & (3) \end{cases}$$
(1) $\quad 3(2) + 3(-2) + 2(2) = 6 - 6 + 4 = 4$ ✓
(2) $\quad 2 - 3(-2) + 2 = 2 + 6 + 2 = 10$ ✓
(3) $\quad 5(2) - 2(-2) - 3(2) = 10 + 4 - 6 = 8$ ✓

Since all three equations are satisfied, $x = 2$, $y = -2$, $z = 2$ is a solution to the system.

15. Using substitution:
$$\begin{cases} 5x - y = 13 & (1) \\ 2x + 3y = 12 & (2) \end{cases}$$

Solve equation (1) for y, then substitute the expression for y into equation (2) and solve for x.
$$5x - y = 13 \Rightarrow y = 5x - 13$$
$$2x + 3(5x - 13) = 12$$
$$2x + 15x - 39 = 12$$
$$17x = 51$$
$$x = 3$$
Now substitute $x = 3$ into equation (1) and solve for y.
$$5(3) - y = 13$$
$$y = 2$$
Solution: $x = 3$, $y = 2$, or $(3, 2)$

17. Using elimination:

$$\begin{cases} 3x - 6y = 2 & (1) \\ 5x + 4y = 1 & (2) \end{cases}$$

Multiply equation (1) by 2 and equation (2) by 3.

$$\begin{cases} 6x - 12y = 4 & (3) \\ 15x + 12y = 3 & (4) \end{cases}$$

Now add equations (3) and (4) to eliminate y. Then solve for x.

$$21x = 7$$
$$x = \frac{1}{3}$$

Substitute $x = \frac{1}{3}$ into equation (1) and solve for y.

$$3\left(\frac{1}{3}\right) - 6y = 2$$
$$1 - 6y = 2$$
$$-6y = 1$$
$$y = -\frac{1}{6}$$

Solution: $x = \frac{1}{3}$, $y = -\frac{1}{6}$ or $\left(\frac{1}{3}, -\frac{1}{6}\right)$

19. Using elimination:

$$\begin{cases} 2x + y = 1 & (1) \\ 4x + 2y = 3 & (2) \end{cases}$$

Multiply equation (1) by -2, then add the resulting equation to equation (2).

$$\begin{cases} -4x - 2y = -2 & (3) \\ \underline{4x + 2y = 3} & (2) \\ 0 = 1 \end{cases}$$

This is a contradiction, so there is no solution to this system of equations. The system is inconsistent.

21. Using elimination:

$$\begin{cases} x + 2y = 4 & (1) \\ 2x + 4y = 8 & (2) \end{cases}$$

Multiply equation (1) by -2, then add the resulting equation to equation (2).

$$\begin{cases} -2x - 4y = -8 & (3) \\ \underline{2x + 4y = 8} & (2) \\ 0 = 0 \end{cases}$$

The equations are dependent, so there are infinitely many solutions to the system of equations.

They can be written as $x = -2y + 4$ where y is any real number, or as $y = -\frac{1}{2}x + 2$ where x is any real number, or using ordered pairs, we write $\{(x, y) \mid x = -2y + 4,\ y \text{ is any real number}\}$ or $\left\{(x, y) \mid y = -\frac{1}{2}x + 2,\ x \text{ any real number}\right\}$.

23. Using substitution:

$$\begin{cases} 2x - 3y = -1 & (1) \\ 10x + y = 11 & (2) \end{cases}$$

Solve equation (2) for y, then substitute the expression for y into equation (1) and solve for x.

$$10x + y = 11 \Rightarrow y = -10x + 11$$
$$2x - 3(-10x + 11) = -1$$
$$2x + 30x - 33 = -1$$
$$32x = 32$$
$$x = 1$$

Substitute $x = 1$ into equation (2) and solve for y.

$$10(1) + y = 11$$
$$y = 1$$

Solution: $x = 1$, $y = 1$, or $(1, 1)$

25. Using elimination:

$$\begin{cases} 2x + 3y = 6 & (1) \\ x - y = \frac{1}{2} & (2) \end{cases}$$

Multiply equation (2) by -2, then add the resulting equation to equation (1) and solve for the remaining variable.

$$\begin{cases} 2x + 3y = 6 & (1) \\ \underline{-2x + 2y = -1} & (3) \\ 5y = 5 \\ y = 1 \end{cases}$$

Substitute 1 for y in equation (1) and solve for x.

$$2x + 3(1) = 6$$
$$2x = 3$$
$$x = \frac{3}{2}$$

Solution: $x = \frac{3}{2}$, $y = 1$, or $\left(\frac{3}{2}, 1\right)$

(continued on next page)

27. Using elimination:

$$\begin{cases} \dfrac{1}{2}x + \dfrac{1}{3}y = 3 & (1) \\ \dfrac{1}{4}x - \dfrac{2}{3}y = -1 & (2) \end{cases}$$

Multiply equation (1) by 24 and equation (2) by 12, then add the resulting equations and solve for the remaining variable.

$$\begin{aligned} 12x + 8y &= 72 \quad (3) \\ 3x - 8y &= -12 \quad (4) \\ \hline 15x &= 60 \Rightarrow x = 4 \end{aligned}$$

Substitute 4 for x in equation (2) and solve for y.

$$\frac{1}{4}(4) - \frac{2}{3}y = -1$$
$$-\frac{2}{3}y = -2 \Rightarrow y = 3$$

Solution: $x = 4$, $y = 3$, or $(4, 3)$

29. $\begin{cases} x - y = 6 & (1) \\ 2x - 3z = 16 & (2) \\ 2y + z = 4 & (3) \end{cases}$

Multiply equation (1) by -2, then add the resulting equation to equation (2).

$$\begin{aligned} -2x + 2y &= -12 \quad (4) \\ 2x - 3z &= 16 \quad (2) \\ \hline 2y - 3z &= 4 \quad (5) \end{aligned}$$

Multiply equation (5) by -1, then add the resulting equation to equation (3) and solve for z.

$$\begin{aligned} 2y + z &= 4 \quad (3) \\ -2y + 3z &= -4 \quad (6) \\ \hline 4z &= 0 \Rightarrow z = 0 \end{aligned}$$

Substitute 0 for z in equation (2) and solve for x.

$$2x - 3(0) = 16 \Rightarrow x = 8$$

Substitute 8 for x in equation (1) and solve for y.

$$8 - y = 6 \Rightarrow y = 2$$

Solution: $x = 8$, $y = 2$, $z = 0$, or $(8, 2, 0)$

31. $\begin{cases} x - 2y + 3z = 7 & (1) \\ 2x + y + z = 4 & (2) \\ -3x + 2y - 2z = -10 & (3) \end{cases}$

Multiply equation (2) by 2, then add the resulting equation to equation (1).

$$\begin{aligned} 4x + 2y + 2z &= 8 \quad 2 \times (2) \\ x - 2y + 3z &= 7 \quad (1) \\ \hline 5x + 5z &= 15 \quad (4) \end{aligned}$$

Add equations (1) and (3).

$$\begin{aligned} x - 2y + 3z &= 7 \quad (1) \\ -3x + 2y - 2z &= -10 \quad (3) \\ \hline -2x + z &= -3 \quad (5) \end{aligned}$$

Multiply equation (5) by -5, then add the resulting equation to equation (4) and solve for the remaining variable.

$$\begin{aligned} 10x - 5z &= 15 \quad -5 \times (5) \\ 5x + 5z &= 15 \quad (4) \\ \hline 15x &= 30 \Rightarrow x = 2 \end{aligned}$$

Substitute 2 for x in equation (5) and solve for z.

$$-2(2) + z = -3 \Rightarrow z = 1$$

Substitute 2 for x and 1 for z in equation (2) and solve for y.

$$2(2) + y + 1 = 4$$
$$5 + y = 4 \Rightarrow y = -1$$

Solution: $x = 2$, $y = -1$, $z = 1$, or $(2, -1, 1)$

33. $\begin{cases} x - y - z = 1 & (1) \\ 2x + 3y + z = 2 & (2) \\ 3x + 2y = 0 & (3) \end{cases}$

Since (3) has no z term we will eliminate z first. Add equations (1) and (2).

$$\begin{aligned} x - y - z &= 1 \quad (1) \\ 2x + 3y + z &= 2 \quad (2) \\ \hline 3x + 2y &= 3 \quad (4) \end{aligned}$$

Multiply equation (4) by -1, then add the resulting equation to equation (3).

$$\begin{aligned} -3x - 2y &= -3 \quad -1 \times (4) \\ 3x + 2y &= 0 \quad (3) \\ \hline 0 &= -3 \end{aligned}$$

This is a contradiction, so the system is inconsistent and there is no solution.

35. $\begin{cases} x - y - z = 1 & (1) \\ -x + 2y - 3z = -4 & (2) \\ 3x - 2y - 7z = 0 & (3) \end{cases}$

Eliminate x by adding equations (1) and (2).

$$\begin{aligned} x - y - z &= 1 \quad (1) \\ -x + 2y - 3z &= -4 \quad (2) \\ \hline y - 4z &= -3 \quad (4) \end{aligned}$$

Multiply equation (1) by -3, then add the resulting equation to equation (3) to eliminate x.

$$\begin{aligned} -3x + 3y + 3z &= -3 \quad -3 \times (1) \\ 3x - 2y - 7z &= 0 \quad (3) \\ \hline y - 4z &= -3 \quad (5) \end{aligned}$$

Multiply equation (5) by -1 and add the resulting equation to equation (4).

$$\begin{aligned} y - 4z &= -3 \quad (4) \\ -y + 4z &= 3 \quad -1 \times (5) \\ \hline 0 &= 0 \end{aligned}$$

(*continued on next page*)

(*continued*)

The original system is equivalent to a system containing 2 equations, so the equations are dependent and the system has infinitely many solutions. If z represents any real number, then substituting $y = 4z - 3$ into (1) gives

$$x - (4z - 3) - z = 1$$
$$x = 5z - 2$$

The solution to the system is $\begin{cases} x = 5z - 2 \\ y = 4z - 3 \end{cases}$ where

z is any real number.

37. $\begin{cases} 2x - 2y + 3z = 6 & (1) \\ 4x - 3y + 2z = 0 & (2) \\ -2x + 3y - 7z = 1 & (3) \end{cases}$

We will eliminate x.
Multiply equation (1) by -2, then add the resulting equation to equation (2).

$$\begin{array}{rl} -4x + 4y - 6z = -12 & -2 \times (1) \\ \underline{4x - 3y + 2z = 0} & (2) \\ y - 4z = -12 & (4) \end{array}$$

Add equations (1) and (3).

$$\begin{array}{rl} 2x - 2y + 3z = 6 & (1) \\ \underline{-2x + 3y - 7z = 1} & (3) \\ y - 4z = 7 & (5) \end{array}$$

Multiply equation (4) by -1, then add the resulting equation to equation (5).

$$\begin{array}{rl} -y + 4z = 12 & -1 \times (4) \\ \underline{y - 4z = 7} & (5) \\ 0 = 19 \end{array}$$

This is a contradiction, so the system is inconsistent and there is no solution.

39. $\begin{cases} x + y - z = 6 & (1) \\ 3x - 2y + z = -5 & (2) \\ x + 3y - 2z = 14 & (3) \end{cases}$

We will eliminate z.
Add equations (1) and (2).

$$\begin{array}{rl} x + y - z = 6 & (1) \\ \underline{3x - 2y + z = -5} & (2) \\ 4x - y = 1 & (4) \end{array}$$

Multiply equation (1) by -2, then add the resulting equation to equation (3).

$$\begin{array}{rl} -2x - 2y + 2z = -12 & -2 \times (1) \\ \underline{x + 3y - 2z = 14} & (3) \\ -x + y = 2 & (5) \end{array}$$

Now add equations (4) and (5), then solve for x.

$$\begin{array}{rl} 4x - y = 1 & (4) \\ \underline{-x + y = 2} & (5) \\ 3x = 3 \\ x = 1 \end{array}$$

Substitute 1 for x in equation (5) and solve for y.
$$-1 + y = 2$$
$$y = 3$$
Now substitute 1 for x and 3 for y in equation (1) and solve for z.
$$1 + 3 - z = 6$$
$$-z = 2$$
$$z = -2$$
The solution to the system is $x = 1$, $y = 3$, $z = -2$, or $(1, 3, -2)$.

41. $\begin{cases} x + 2y - z = -3 & (1) \\ 2x - 4y + z = -7 & (2) \\ -2x + 2y - 3z = 4 & (3) \end{cases}$

We will eliminate x.

Multiply equation (1) by -2, then add the resulting equation to equation (2).

$$\begin{array}{rl} -2x - 4y + 2z = 6 & -2 \times (1) \\ \underline{2x - 4y + z = -7} & (2) \\ -8y + 3z = -1 & (4) \end{array}$$

Multiply equation (1) by 2, then add the resulting equation to equation (3).

$$\begin{array}{rl} 2x + 4y - 2z = -6 & 2 \times (1) \\ \underline{-2x + 2y - 3z = 4} & (3) \\ 6y - 5z = -2 & (5) \end{array}$$

Multiply equation (4) by 3 and equation (5) by 4, then add the resulting equations and solve for z.

$$\begin{array}{rl} -24y + 9z = -3 & 3 \times (4) \\ \underline{24y - 20z = -8} & 4 \times (5) \\ -11z = -11 \\ z = 1 \end{array}$$

Substitute 1 for z in equation (5) and solve for y.
$$6y - 5(1) = -2$$
$$6y = 3$$
$$y = \frac{1}{2}$$

Substitute 1 for z and $\frac{1}{2}$ for y in equation (1) and solve for x.

$$x + 2\left(\frac{1}{2}\right) - 1 = -3$$
$$x = -3$$

The solution to the system is $x = -3$, $y = \frac{1}{2}$, $z = 1$, or $\left(-3, \frac{1}{2}, 1\right)$.

43. We let l represent the length of the floor and w represent its width. We are told the perimeter is 90 feet and that the length is twice the width. We need to solve the system of equations

$$\begin{cases} 2l + 2w = 90 & (1) \\ l = 2w & (2) \end{cases}$$

Using substitution, we have

$$2(2w) + 2w = 90$$
$$6w = 90$$
$$w = 15$$

Substituting 15 for w in (2) gives $l = 2(15) = 30$. The floor has a length of 30 feet and a width of 15 feet.

45. We let x denote the number of acres of corn planted and y denote the number of acres of soybeans planted. We solve the system of equations

$$\begin{cases} x + y = 368 & (1) \\ 512x + 336y = 146,671 & (2) \end{cases}$$

We solve this system by substitution. First, solve equation (1) for y.

$$x + y = 368 \Rightarrow y = -x + 368$$

Now substitute this expression for y in equation (2).

$$512x + 336(-x + 368) = 146,671$$
$$512x - 336x + 123,648 = 146,671$$
$$176x = 23,023$$
$$x = 130.8125$$
$$y = -130.8125 + 368 = 237.1875$$

The farmer should plant 130.8 acres of corn and 237.2 acres of soybeans.

47. Let x represent the number of cashews in the mixture and let y represent the total weight of the mix. To find the amount of cashews needed for the mixture, we solve the system of equations

$$\begin{cases} 30 + x = y & (1) \\ 5.00x + 1.50(30) = 3.00y & (2) \end{cases}$$

We solve this system by substitution, replacing y in (2) with $30 + x$.

$$5x + 1.50(30) = 3.00(30 + x)$$
$$5x + 45 = 90 + 3x$$
$$2x = 45$$
$$x = 22.5$$
$$y = 30 + 22.5 = 52.5$$

The manager should add 22.5 pounds of cashews to the peanuts to make the mixture.

49. Let x represent the cost of liter of milk and y represent the cost of a 330 gram package of tofu. To find the cost of each, we solve the system of equations

$$\begin{cases} 3x + 2y = 1002 & (1) \\ 3y = 8 + 2x & (2) \end{cases}$$

We solve this system by substitution, replacing $\dfrac{8 + 2x}{3}$ for y in equation (1).

$$3x + 2\left(\frac{8 + 2x}{3}\right) = 1002$$
$$9x + 2(8 + 2x) = 3006$$
$$13x = 2990 \Rightarrow x = 230$$
$$y = \frac{8 + 2(230)}{3} = 156$$

A liter of milk costs 230 yen, and a 330 gram package of tofu costs 156 yen.

51. Let x represent the cost of a pound of bacon and y represent the cost of a carton of eggs. To find each item's price, we solve the system of equations

$$\begin{cases} 3x + 2y = 7.45 & (1) \\ 2x + 3y = 6.45 & (2) \end{cases}$$

We solve this system using elimination.

$$\begin{array}{rl} -6x - 4y = -14.90 & -2 \times (1) \\ \underline{6x + 9y = 19.35} & 3 \times (2) \\ 5y = 4.45 \Rightarrow y = 0.89 \end{array}$$

Substitute 0.89 for y in equation (1) to find x.

$$3x + 2(0.89) = 7.45$$
$$3x = 5.67 \Rightarrow x = 1.89$$

Bacon costs \$1.89 per pound and eggs cost \$0.89 per dozen. So the refund when we return 2 pounds of bacon and 2 cartons of eggs will be $2(1.89) + 2(0.89) = \$5.56$.

53. Let x represent the milligrams of liquid 1 and y represent the milligrams of liquid 2 necessary to obtain the desired mixture. To learn the amount of each liquid to use, we solve the system of equations

$$\begin{cases} 0.20x + 0.40y = 40 & (1) \\ 0.30x + 0.20y = 30 & (2) \end{cases} \text{ or }$$

$$\begin{cases} 2x + 4y = 400 & (1) \\ 3x + 2y = 300 & (2) \end{cases}$$

We will solve this system of equations using elimination.

$$\begin{array}{rl} 2x + 4y = 400 & (1) \\ \underline{-6x - 4y = -600} & -2 \times (2) \\ -4x = -200 \Rightarrow x = 50 \end{array}$$

(continued on next page)

(continued)

Substitute 50 for x in equation (1) and solve for y.

$$2(50) + 4y = 400$$
$$4y = 300$$
$$y = 75$$

The pharmacist should mix 50 mg of liquid 1 and 75 mg of liquid 2 to fill the prescription.

55. Let x represent the pounds of rolled oats and y represent the pounds of molasses in the horse's diet. To determine the amount of oats and molasses to feed the horse, we solve the system of equations

$$\begin{cases} 0.41x + 3.35y = 33 & (1) \\ 1.95x + 0.36y = 21 & (2) \end{cases}$$

We will solve the system using substitution. First we solve equation (1) for x:

$$x = \frac{33 - 3.35y}{0.41}$$

Substitute this value for x in equation (2) and solve for y.

$$1.95\left(\frac{33 - 3.35y}{0.41}\right) + 0.36y = 21$$
$$64.35 - 6.5325y + 0.1476y = 8.61$$
$$64.35 - 6.3849y = 8.61$$
$$-6.3849y = -55.74$$
$$y \approx 8.730$$

Substituting into (1) we get

$$x = \frac{33 - 3.35(8.730)}{0.41} \approx 9.157.$$

The farmer should feed the horse about 9.16 pounds of rolled oats and 8.73 pounds of molasses each day.

57. Let x represent the number of orchestra seats, y represent the number of main seats, and z represent the number of balcony seats in the theater. To find the number of each kind of seat solve the system of equations

$$\begin{cases} x + y + z = 500 & (1) \\ 50x + 35y + 25z = 17,100 & (2) \\ 50\left(\frac{1}{2}x\right) + 35y + 25z = 14,600 & (3) \end{cases}$$

or

$$\begin{cases} x + y + z = 500 & (1) \\ 50x + 35y + 25z = 17,100 & (2) \\ 25x + 35y + 25z = 14,600 & (3) \end{cases}$$

We will eliminate x.
Multiply equation (1) by -50 and then add the resulting equation to equation (2).

$$\begin{aligned} -50x - 50y - 50z &= -25,000 \quad -50 \times (1) \\ 50x + 35y + 25z &= \ \ 17,100 \quad (2) \\ \hline -15y - 25z &= \ -7900 \quad (4) \end{aligned}$$

Multiply equation (1) by -25 and then add the resulting equation to equation (3).

$$\begin{aligned} -25x - 25y - 25z &= -12,500 \quad -25 \times (1) \\ 25x + 35y + 25z &= \ \ 14,600 \quad (3) \\ \hline 10y &= \quad 2100 \\ y &= 210 \end{aligned}$$

Substitute 210 for y in equation (4) and solve for z.

$$-15(210) - 25z = -7900$$
$$3150 - 25z = -7900$$
$$-25z = -4750 \Rightarrow z = 190$$

Substitute 210 for y and 190 for z in equation (1) and solve for x.

$$x + 210 + 190 = 500 \Rightarrow x = 100$$

There are 100 orchestra seats, 210 main seats, and 190 balcony seats in the theater.

59. Let x represent the price of a hamburger, y represent the price of fries, and z represent the price of cola. To find the price of each item, we need to solve the system of equations.

$$\begin{cases} 8x + 6y + 6z = 26.10 & (1) \\ 10x + 6y + 8z = 31.60 & (2) \end{cases}$$

By multiplying equation (1) by -1 and adding the resulting equation to equation (2), we eliminate y. Then we will solve for x in terms of z.

$$2x + 2z = 5.50 \Rightarrow x = 2.75 - z$$

Substitute into equation (2) and solve for y.

$$\begin{aligned} 10(2.75 - z) + 6y + 8z &= 31.60 \\ 27.5 - 10z + 6y + 8z &= 31.60 \\ \hline 6y - 2z &= 31.60 - 27.5 \\ 6y &= 4.10 + 2z \\ y &= \frac{41}{60} + \frac{1}{3}z \end{aligned}$$

There is not enough information to determine the price of each food item.

We have $x = 2.75 - z$ and $y = \frac{1}{3}z + \frac{41}{60}$, where z is the parameter. Using these equations and the restrictions $\$1.75 \le x \le \$2.25$, $\$0.75 \le y \le \$1.00$, and $\$0.60 \le z \le \$0.90$, possible prices are

Price of a Hamburger	Price of Fries	Price of Cola
$2.15	$0.88	$0.60
$2.05	$0.92	$0.70
$1.95	$0.95	$0.80
$1.85	$0.98	$0.90

61. a. To find the equilibrium price, solve the system of equations

$$\begin{cases} S = -200 + 50p & (1) \\ D = 1000 - 25p & (2) \end{cases}$$

Setting the equations equal, since at equilibrium supply equals demand, we have

$$-200 + 50p = 1000 - 25p$$
$$50p + 25p = 1000 + 200$$
$$75p = 1200$$
$$p = 16$$

The equilibrium price of the T-shirts is $16.00.

b. Using the equilibrium price in either equation (1) or (2) gives the equilibrium quantity.

$$S = -200 + 50 \cdot 16 = -200 + 800 = 600$$

The equilibrium quantity is 600 T-shirts.

c.

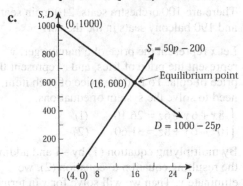

d. If the quantity demanded is greater than the quantity supplies, the price of the T-shirts will increase.

63. $$\begin{cases} 0.06Y - 5000r = 240 & (1) \\ 0.06Y + 6000r = 900 & (2) \end{cases}$$

Using elimination to solve the system, we multiply equation (1) by –1 and add the resulting equation to equation (2).

$$\begin{aligned} -0.06Y + 5000r &= -240 \quad -1 \times (1) \\ 0.06Y + 6000r &= 900 \quad (2) \\ \hline 11000r &= 660 \\ r &= 0.06 \end{aligned}$$

Substitute 0.06 for r in equation (1) and solve for Y.

$$0.06Y - 5000(0.06) = 240$$
$$0.06Y = 540$$
$$Y = 9000$$

At equilibrium, income is $9000 million or $9 billion, and the interest rate is 6%.

65. a. First, set up a system of equations relating the income from the three possible investments. Let x denote the amount invested in Mutual Shares earning 4% per year, y denote the amount invested in Franklin Strategic Income earning 6% per year, and z denote the amount invested in Franklin Small Cap earning 8% per year. There are two equations: one that distributes the $70,000 among the investments and the other that generates the $5000.00 in earnings.

$$\begin{cases} x + y + z = 70,000 & (1) \\ 0.04x + 0.06y + 0.08z = 5,000 & (2) \end{cases}$$

We will use elimination, but first we will subtract the z-term from each side. Doing this allows us to eliminate one of the two variables remaining on the left.

$$\begin{cases} x + y = 70,000 - z & (1) \\ 0.04x + 0.06y = 5,000 - 0.08z & (2) \end{cases}$$

Now multiply equation (1) by 6 and equation (2) by –100 and then add.

$$\begin{aligned} 6x + 6y &= 420,000 - 6z \quad 6 \times (1) \\ -4x - 6y &= -500,000 + 8z \quad -100 \times (2) \\ \hline 2x &= -80,000 + 2z \\ x &= -40,000 + z \quad (3) \end{aligned}$$

Substituting for x in equation (1), we can solve for y in terms of z.

$$(-40,000 + z) + y + z = 70,000 \quad (1)$$
$$-40,000 + 2z + y = 70,000$$
$$y = 110,000 - 2z,$$

where z is a real number between 40,000 and 55,000. (Any other value of z will cause either x or y to be negative.) Using the three restrictions form a table such as the one below.

Mutual Shares $x = z - 40,000$	Strategic Income $y = 110,000 - 2z$	Small Cap z	Income
0	$30,000	$40,000	$5000
$ 5,000	$20,000	$45,000	$5000
$10,000	$10,000	$50,000	$5000
$15,000	0	$55,000	$5000

b. Answers will vary.

67. Answers will vary.

69. Answers will vary.

2.2 Systems of Linear Equations: Gaussian Elimination

1. An m by n rectangular array of numbers is called a <u>matrix</u>.

3. True. An augmented matrix has as many rows as there are equations and as many columns as there are variables plus 1.

5. $\begin{bmatrix} 2 & -3 & | & 5 \\ 1 & -1 & | & 3 \end{bmatrix}$
  7. $\begin{bmatrix} 2 & 1 & | & -6 \\ 3 & 1 & | & -1 \end{bmatrix}$
  9. $\begin{bmatrix} 2 & -1 & -1 & | & 0 \\ 1 & -1 & 1 & | & 1 \\ 3 & -1 & 0 & | & 2 \end{bmatrix}$

11. $\begin{bmatrix} 2 & -3 & 1 & | & 7 \\ 1 & 1 & -1 & | & 1 \\ 2 & 2 & -3 & | & -4 \end{bmatrix}$
  13. $\begin{bmatrix} 4 & -1 & 2 & -1 & | & 4 \\ 1 & 1 & 0 & 0 & | & -6 \\ 0 & 2 & -1 & 1 & | & 5 \end{bmatrix}$
  15. $\begin{bmatrix} 1 & -1 & 1 & -1 & | & 0 \\ 2 & 3 & -1 & 4 & | & 5 \end{bmatrix}$

17. $\begin{bmatrix} 1 & -3 & | & -2 \\ 2 & -5 & | & 5 \end{bmatrix} \xrightarrow{R_2 = -2r_1 + r_2} \begin{bmatrix} 1 & -3 & | & -2 \\ -2(1)+2 & -2(-3)+(-5) & | & -2(-2)+5 \end{bmatrix} = \begin{bmatrix} 1 & -3 & | & -2 \\ 0 & 1 & | & 9 \end{bmatrix}$

19. a. $\begin{bmatrix} 1 & -3 & 4 & | & 3 \\ 2 & -5 & 6 & | & 6 \\ -3 & 3 & 4 & | & 6 \end{bmatrix} \xrightarrow{R_2 = -2r_1 + r_2} \begin{bmatrix} 1 & -3 & 4 & | & 3 \\ -2(1)+2 & -2(-3)-5 & -2(4)+6 & | & -2(3)+6 \\ -3 & 3 & 4 & | & 6 \end{bmatrix} = \begin{bmatrix} 1 & -3 & 4 & | & 3 \\ 0 & 1 & -2 & | & 0 \\ -3 & 3 & 4 & | & 6 \end{bmatrix}$

b. $\begin{bmatrix} 1 & -3 & 4 & | & 3 \\ 2 & -5 & 6 & | & 6 \\ -3 & 3 & 4 & | & 6 \end{bmatrix} \xrightarrow{R_3 = 3r_1 + r_3} \begin{bmatrix} 1 & -3 & 4 & | & 3 \\ 2 & -5 & 6 & | & 6 \\ 3(1)-3 & 3(-3)+3 & 3(4)+4 & | & 3(3)+6 \end{bmatrix} = \begin{bmatrix} 1 & -3 & 4 & | & 3 \\ 2 & -5 & 6 & | & 6 \\ 0 & -6 & 16 & | & 15 \end{bmatrix}$

21. a. $\begin{bmatrix} 1 & -3 & 2 & | & -6 \\ 2 & -5 & 3 & | & -4 \\ -3 & -6 & 2 & | & 6 \end{bmatrix} \xrightarrow{R_2 = -2r_1 + r_2} \begin{bmatrix} 1 & -3 & 2 & | & -6 \\ -2(1)+2 & -2(-3)-5 & -2(2)+3 & | & -2(-6)-4 \\ -3 & -6 & 2 & | & 6 \end{bmatrix} = \begin{bmatrix} 1 & -3 & 2 & | & -6 \\ 0 & 1 & -1 & | & 8 \\ -3 & -6 & 2 & | & 6 \end{bmatrix}$

b. $\begin{bmatrix} 1 & -3 & 2 & | & -6 \\ 2 & -5 & 3 & | & -4 \\ -3 & -6 & 2 & | & 6 \end{bmatrix} \xrightarrow{R_3 = 3r_1 + r_3} \begin{bmatrix} 1 & -3 & 2 & | & -6 \\ 2 & -5 & 3 & | & -4 \\ 3(1)-3 & 3(-3)-6 & 3(2)+2 & | & 3(-6)+6 \end{bmatrix} = \begin{bmatrix} 1 & -3 & 2 & | & -6 \\ 2 & -5 & 3 & | & -4 \\ 0 & -15 & 8 & | & -12 \end{bmatrix}$

23. a. $\begin{bmatrix} 1 & -3 & 1 & | & -2 \\ 2 & -5 & 6 & | & -2 \\ -3 & 1 & 4 & | & 6 \end{bmatrix} \xrightarrow{R_2 = -2r_1 + r_2} \begin{bmatrix} 1 & -3 & 1 & | & -2 \\ -2(1)+2 & -2(-3)-5 & -2(1)+6 & | & -2(-2)-2 \\ -3 & 1 & 4 & | & 6 \end{bmatrix} = \begin{bmatrix} 1 & -3 & 1 & | & -2 \\ 0 & 1 & 4 & | & 2 \\ -3 & 1 & 4 & | & 6 \end{bmatrix}$

b. $\begin{bmatrix} 1 & -3 & 1 & | & -2 \\ 2 & -5 & 6 & | & -2 \\ -3 & 1 & 4 & | & 6 \end{bmatrix} \xrightarrow{R_3 = 3r_1 + r_3} \begin{bmatrix} 1 & -3 & 1 & | & -2 \\ 2 & -5 & 6 & | & -2 \\ 3(1)-3 & 3(-3)+1 & 3(1)+4 & | & 3(-2)+6 \end{bmatrix} = \begin{bmatrix} 1 & -3 & 1 & | & -2 \\ 2 & -5 & 6 & | & -2 \\ 0 & -8 & 7 & | & 0 \end{bmatrix}$

25. $\begin{bmatrix} 1 & 2 & | & 5 \\ 0 & 1 & | & -1 \end{bmatrix}$

a. $\begin{cases} x + 2y = 5 \\ \quad\ y = -1 \end{cases}$

b. The system is consistent. Substituting –1 for y in the first equation gives
$x + 2(-1) = 5$
$x = 7$
The solution is $x = 7$, $y = -1$, or $(7, -1)$.

27. $\begin{bmatrix} 1 & 2 & 3 & | & 1 \\ 0 & 1 & 4 & | & 2 \\ 0 & 0 & 0 & | & 3 \end{bmatrix}$

a. $\begin{cases} x + 2y + 3z = 1 \\ \quad\ y + 4z = 2 \\ \quad\quad\ 0 = 3 \end{cases}$

b. The system is inconsistent since the third equation is a contradiction. There is no solution.

29. $\begin{bmatrix} 1 & 0 & 2 & -1 \\ 0 & 1 & -4 & -2 \\ 0 & 0 & 0 & 0 \end{bmatrix}$

a. $\begin{cases} x + 2z = -1 \\ y - 4z = -2 \\ 0 = 0 \end{cases}$

b. The system is consistent and has an infinite number of solutions. Solving equation (1) for x in terms of z and equation (2) for y in terms of z gives the solutions $x = -2z - 1$, $y = 4z - 2$, where z is any real number.

31. $\begin{bmatrix} 1 & 2 & -1 & 1 & 1 \\ 0 & 1 & 4 & 1 & 2 \\ 0 & 0 & 1 & 2 & 3 \\ 0 & 0 & 0 & 1 & 4 \end{bmatrix}$

a. $\begin{cases} x_1 + 2x_2 - x_3 + x_4 = 1 \\ x_2 + 4x_3 + x_4 = 2 \\ x_3 + 2x_4 = 3 \\ x_4 = 4 \end{cases}$

b. The system is consistent. To find the solution, start with $x_4 = 4$ and back-substitute.

$x_3 = 3 - 2x_4 = 3 - 2(4) = -5$

$x_2 = 2 - 4x_3 - x_4 = 2 - 4(-5) - (4) = 18$

$x_1 = 1 - 2x_2 + x_3 - x_4$

$ = 1 - 2(18) + (-5) - 4 = -44$

The solution is $(-44, 18, -5, 4)$.

33. $\begin{bmatrix} 1 & 2 & 0 & 4 & 2 \\ 0 & 1 & 1 & 3 & 3 \\ 0 & 0 & 1 & 0 & 0 \\ 0 & 0 & 0 & 0 & 0 \end{bmatrix}$

a. $\begin{cases} x_1 + 2x_2 + 4x_4 = 2 \\ x_2 + x_3 + 3x_4 = 3 \\ x_3 = 0 \\ 0 = 0 \end{cases}$

b. The system is consistent. To find the solution, express x_1 and x_2 in terms of x_4.

$x_3 = 0$

$x_2 = -3x_4 - x_3 + 3 = -3x_4 - 0 + 3 = -3x_4 + 3$

$x_1 = -4x_4 - 2x_2 + 2$

$ = -4x_4 - 2(-3x_4 + 3) + 2 = 2x_4 - 4,$

where x_4 is any real number.

35. $\begin{bmatrix} 1 & -2 & 0 & 1 & -2 \\ 0 & 1 & -3 & 2 & 2 \\ 0 & 0 & 1 & -1 & 0 \\ 0 & 0 & 0 & 0 & 0 \end{bmatrix}$

a. $\begin{cases} x_1 - 2x_2 + x_4 = -2 \\ x_2 - 3x_3 + 2x_4 = 2 \\ x_3 - x_4 = 0 \end{cases}$

b. The system is consistent. To find the solution, express x_1, x_2, and x_3 in terms of x_4.

$x_3 = x_4$

$x_2 = -2x_4 + 3x_3 + 2 = -2x_4 + 3x_4 + 2$

$ = x_4 + 2$

$x_1 = -x_4 + 2x_2 - 2 = -x_4 + 2(x_4 + 2) - 2$

$ = x_4 + 2,$

where x_4 is any real number.

37. $\begin{cases} 2x - 3y = 6 \\ 6x - 9y = 10 \end{cases}$

$\begin{bmatrix} 2 & -3 & 6 \\ 6 & -9 & 10 \end{bmatrix} \xrightarrow{R_1 = \frac{1}{2}r_1} \begin{bmatrix} 1 & -\dfrac{3}{2} & 3 \\ 6 & -9 & 10 \end{bmatrix}$

$\xrightarrow{R_2 = -6r_1 + r_2} \begin{bmatrix} 1 & -\dfrac{3}{2} & 3 \\ 0 & 0 & -8 \end{bmatrix}$

The system is inconsistent.

39. $\begin{cases} 2x - 3y = 0 \\ 4x + 9y = 5 \end{cases}$

$\begin{bmatrix} 2 & -3 & 0 \\ 4 & 9 & 5 \end{bmatrix} \xrightarrow{R_1 = \frac{1}{2}r_1} \begin{bmatrix} 1 & -\dfrac{3}{2} & 0 \\ 4 & 9 & 5 \end{bmatrix}$

$\xrightarrow{R_2 = -4r_1 + r_2} \begin{bmatrix} 1 & -\dfrac{3}{2} & 0 \\ 0 & 15 & 5 \end{bmatrix}$

$\xrightarrow{R_2 = \frac{1}{15}r_2} \begin{bmatrix} 1 & -\dfrac{3}{2} & 0 \\ 0 & 1 & \dfrac{1}{3} \end{bmatrix}$

The row-echelon form of the system is

$\begin{cases} x - \dfrac{3}{2}y = 0 \\ \phantom{x - \dfrac{3}{2}}y = \dfrac{1}{3} \end{cases}.$

(*continued on next page*)

(*continued*)

Back-substitute $\frac{1}{3}$ for y in the first equation, giving $x - \frac{3}{2} \cdot \frac{1}{3} = 0$ or $x = \frac{1}{2}$.

The solution of the system of equations is $x = \frac{1}{2}$ and $y = \frac{1}{3}$, or $\left(\frac{1}{2}, \frac{1}{3}\right)$.

41. $\begin{cases} 2x + 6y = 4 \\ 5x + 15y = 10 \end{cases}$

$\begin{bmatrix} 2 & 6 & | & 4 \\ 5 & 15 & | & 10 \end{bmatrix} \xrightarrow{R_1 = \frac{1}{2}r_1} \begin{bmatrix} 1 & 3 & | & 2 \\ 5 & 15 & | & 10 \end{bmatrix} \xrightarrow{R_2 = -5r_1 + r_2} \begin{bmatrix} 1 & 3 & | & 2 \\ 0 & 0 & | & 0 \end{bmatrix}$

This system has an infinite number of solutions. They are $x = 2 - 3y$, where y is any real number, or

$\{(x, y) \,|\, x = -3y + 2, \ y \text{ any real number}\}$. An alternate solution is $\left\{(x, y) \,\middle|\, y = -\frac{1}{3}x + \frac{2}{3}, \ x \text{ any real number}\right\}$.

43. $\begin{cases} x + y = 1 \\ 3x - 2y = \frac{4}{3} \end{cases}$

$\begin{bmatrix} 1 & 1 & | & 1 \\ 3 & -2 & | & \frac{4}{3} \end{bmatrix} \xrightarrow{R_2 = -3r_1 + r_2} \begin{bmatrix} 1 & 1 & | & 1 \\ 0 & -5 & | & -\frac{5}{3} \end{bmatrix} \xrightarrow{R_2 = -\frac{1}{5}r_2} \begin{bmatrix} 1 & 1 & | & 1 \\ 0 & 1 & | & \frac{1}{3} \end{bmatrix}$

The row echelon form of the system of equations is $\begin{cases} x = -y + 1 & (1) \\ y = \frac{1}{3} & (2) \end{cases}$

Back-substitute $\frac{1}{3}$ for y in equation (1), to get $x = -\frac{1}{3} + 1 = \frac{2}{3}$.

The solution of the system of equations is $x = \frac{2}{3}$ and $y = \frac{1}{3}$, or $\left(\frac{2}{3}, \frac{1}{3}\right)$.

45. $\begin{cases} 2x + y + z = 6 \\ x - y - z = -3 \\ 3x + y + 2z = 7 \end{cases}$

$\begin{bmatrix} 2 & 1 & 1 & | & 6 \\ 1 & -1 & -1 & | & -3 \\ 3 & 1 & 2 & | & 7 \end{bmatrix} \xrightarrow[\text{rows 1 and 2}]{\text{Interchange}} \begin{bmatrix} 1 & -1 & -1 & | & -3 \\ 2 & 1 & 1 & | & 6 \\ 3 & 1 & 2 & | & 7 \end{bmatrix} \xrightarrow[R_3 = -3r_1 + r_3]{R_2 = -2r_1 + r_2} \begin{bmatrix} 1 & -1 & -1 & | & -3 \\ 0 & 3 & 3 & | & 12 \\ 0 & 4 & 5 & | & 16 \end{bmatrix}$

$\xrightarrow{R_2 = \frac{1}{3}r_2} \begin{bmatrix} 1 & -1 & -1 & | & -3 \\ 0 & 1 & 1 & | & 4 \\ 0 & 4 & 5 & | & 16 \end{bmatrix} \xrightarrow{R_3 = -4r_2 + r_3} \begin{bmatrix} 1 & -1 & -1 & | & -3 \\ 0 & 1 & 1 & | & 4 \\ 0 & 0 & 1 & | & 0 \end{bmatrix}$

The row echelon form of the system of equations is $\begin{cases} x = y + z - 3 & (1) \\ y = -z + 4 & (2) \\ z = 0 & (3) \end{cases}$

Back-substitute 0 for z in (2) to obtain $y = 4$. Then back-substitute $z = 0$ and $y = 4$ in equation (1) to obtain $x = 4 + 0 - 3 = 1$. The solution of the system is $x = 1$, $y = 4$, $z = 0$, or (1, 4, 0).

47. $\begin{cases} 2x - 2y - z = 2 \\ 2x + 3y + z = 2 \\ 3x + 2y \quad\;\; = 0 \end{cases}$

$$\begin{bmatrix} 2 & -2 & -1 & | & 2 \\ 2 & 3 & 1 & | & 2 \\ 3 & 2 & 0 & | & 0 \end{bmatrix} \xrightarrow[R_1 = \frac{1}{2}r_1]{} \begin{bmatrix} 1 & -1 & -\frac{1}{2} & | & 1 \\ 2 & 3 & 1 & | & 2 \\ 3 & 2 & 0 & | & 0 \end{bmatrix} \xrightarrow[\substack{R_2 = -2r_1 + r_2 \\ R_3 = -3r_1 + r_3}]{} \begin{bmatrix} 1 & -1 & -\frac{1}{2} & | & 1 \\ 0 & 5 & 2 & | & 0 \\ 0 & 5 & \frac{3}{2} & | & -3 \end{bmatrix}$$

$$\xrightarrow[R_2 = \frac{1}{5}r_2]{} \begin{bmatrix} 1 & -1 & -\frac{1}{2} & | & 1 \\ 0 & 1 & \frac{2}{5} & | & 0 \\ 0 & 5 & \frac{3}{2} & | & -3 \end{bmatrix} \xrightarrow[R_3 = -5r_2 + r_3]{} \begin{bmatrix} 1 & -1 & -\frac{1}{2} & | & 1 \\ 0 & 1 & \frac{2}{5} & | & 0 \\ 0 & 0 & -\frac{1}{2} & | & -3 \end{bmatrix}$$

$$\xrightarrow[R_3 = -2r_3]{} \begin{bmatrix} 1 & -1 & -\frac{1}{2} & | & 1 \\ 0 & 1 & \frac{2}{5} & | & 0 \\ 0 & 0 & 1 & | & 6 \end{bmatrix}$$

The row echelon form of the system of equations is $\begin{cases} x = y + \dfrac{1}{2}z + 1 & (1) \\ y = -\dfrac{2}{5}z & (2) \\ z = 6 & (3) \end{cases}$.

Back-substitute 6 for z in (2) to obtain $y = -\dfrac{12}{5}$. Then back-substitute $z = 6$ and $y = -\dfrac{12}{5}$ in equation (1), to

obtain $x = -\dfrac{12}{5} + \dfrac{1}{2}(6) + 1 = \dfrac{8}{5}$. The solution of the system of equations is $x = \dfrac{8}{5}$, $y = -\dfrac{12}{5}$, $z = 6$, or

$\left(\dfrac{8}{5}, -\dfrac{12}{5}, 6 \right)$.

49. $\begin{cases} 2x + y - z = 2 \\ x + 3y + 2z = 1 \\ x + y + z = 2 \end{cases}$

$$\begin{bmatrix} 2 & 1 & -1 & | & 2 \\ 1 & 3 & 2 & | & 1 \\ 1 & 1 & 1 & | & 2 \end{bmatrix} \xrightarrow[\substack{\text{Interchange} \\ \text{rows 1 and 2}}]{} \begin{bmatrix} 1 & 3 & 2 & | & 1 \\ 2 & 1 & -1 & | & 2 \\ 1 & 1 & 1 & | & 2 \end{bmatrix} \xrightarrow[\substack{R_2 = -2r_1 + r_2 \\ R_3 = -r_1 + r_3}]{} \begin{bmatrix} 1 & 3 & 2 & | & 1 \\ 0 & -5 & -5 & | & 0 \\ 0 & -2 & -1 & | & 1 \end{bmatrix}$$

$$\xrightarrow[R_2 = -\frac{1}{5}r_2]{} \begin{bmatrix} 1 & 3 & 2 & | & 1 \\ 0 & 1 & 1 & | & 0 \\ 0 & -2 & -1 & | & 1 \end{bmatrix} \xrightarrow[R_3 = 2r_2 + r_3]{} \begin{bmatrix} 1 & 3 & 2 & | & 1 \\ 0 & 1 & 1 & | & 0 \\ 0 & 0 & 1 & | & 1 \end{bmatrix}$$

The row echelon form of the system of equations is $\begin{cases} x = -3y - 2z + 1 & (1) \\ y = -z & (2) \\ z = 1 & (3) \end{cases}$.

Back-substitute $z = 1$ in equation (2) to get $y = -1$, and then back-substitute $y = -1$ and $z = 1$ in equation (1), to get $x = -3(-1) - 2(1) + 1 = 2$. The solution of the system is $x = 2$, $y = -1$, $z = 1$, or $(2, -1, 1)$.

51. $\begin{cases} x + y - z = 0 \\ 4x + 4y - 4z = -1 \\ 2x + y + z = 2 \end{cases}$

$$\begin{bmatrix} 1 & 1 & -1 & | & 0 \\ 4 & 4 & -4 & | & -1 \\ 2 & 1 & 1 & | & 2 \end{bmatrix} \xrightarrow[R_3 = -2r_1 + r_3]{R_2 = -4r_1 + r_2} \begin{bmatrix} 1 & 1 & -1 & | & 0 \\ 0 & 0 & 0 & | & -1 \\ 0 & -1 & 3 & | & 2 \end{bmatrix}$$

The system is inconsistent because the second row of the matrix has $0 = -1$ which is contradictory.

53. $\begin{cases} 3x + y - z = \dfrac{2}{3} \\ 2x - y + z = 1 \\ 4x + 2y = \dfrac{8}{3} \end{cases}$

$$\begin{bmatrix} 3 & 1 & -1 & | & \dfrac{2}{3} \\ 2 & -1 & 1 & | & 1 \\ 4 & 2 & 0 & | & \dfrac{8}{3} \end{bmatrix}$$

$$\xrightarrow{R_1 = r_1 - r_2} \begin{bmatrix} 1 & 2 & -2 & | & -\dfrac{1}{3} \\ 2 & -1 & 1 & | & 1 \\ 4 & 2 & 0 & | & \dfrac{8}{3} \end{bmatrix}$$

$$\xrightarrow[R_3 = -4r_1 + r_3]{R_2 = -2r_1 + r_2} \begin{bmatrix} 1 & 2 & -2 & | & -\dfrac{1}{3} \\ 0 & -5 & 5 & | & \dfrac{5}{3} \\ 0 & -6 & 8 & | & 4 \end{bmatrix}$$

$$\xrightarrow{R_2 = -\frac{1}{5}r_2} \begin{bmatrix} 1 & 2 & -2 & | & -\dfrac{1}{3} \\ 0 & 1 & -1 & | & -\dfrac{1}{3} \\ 0 & -6 & 8 & | & 4 \end{bmatrix}$$

$$\xrightarrow{R_3 = 6r_2 + r_3} \begin{bmatrix} 1 & 2 & -2 & | & -\dfrac{1}{3} \\ 0 & 1 & -1 & | & -\dfrac{1}{3} \\ 0 & 0 & 2 & | & 2 \end{bmatrix}$$

$$\xrightarrow{R_3 = \frac{1}{2}r_3} \begin{bmatrix} 1 & 2 & -2 & | & -\dfrac{1}{3} \\ 0 & 1 & -1 & | & -\dfrac{1}{3} \\ 0 & 0 & 1 & | & 1 \end{bmatrix}$$

The row echelon form of the system of

equations is $\begin{cases} x = -2y + 2z - \dfrac{1}{3} & (1) \\ y = z - \dfrac{1}{3} & (2). \\ z = 1 & (3) \end{cases}$

Back-substitute $z = 1$ in equation (2) to get $y = 1 - \dfrac{1}{3} = \dfrac{2}{3}$, and then back-substitute $y = \dfrac{2}{3}$ and $z = 1$ in equation (1), to get $x = -2\left(\dfrac{2}{3}\right) + 2(1) - \dfrac{1}{3} = \dfrac{1}{3}$. The solution of the system is $x = \dfrac{1}{3}$, $y = \dfrac{2}{3}$, $z = 1$, or $\left(\dfrac{1}{3}, \dfrac{2}{3}, 1\right)$.

Exercises 55–59 were solved using a TI-84 graphing utility.

55. $\begin{cases} 2x - 2y + z = 2 \\ x - \dfrac{1}{2}y + 2z = 1 \\ 2x + \dfrac{1}{3}y - z = 0 \end{cases}$

The row-echelon form is

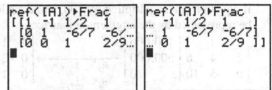

The reduced row-echelon form is

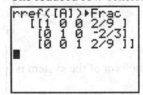

The solution to the system is $x = \dfrac{2}{9}$, $y = -\dfrac{2}{3}$, $z = \dfrac{2}{9}$, or $\left(\dfrac{2}{9}, -\dfrac{2}{3}, \dfrac{2}{9}\right)$.

57. $\begin{cases} x + y + z = 4 \\ x - y - z = 0 \\ y - z = -4 \end{cases}$

The row-echelon form is

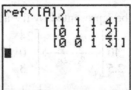

The reduced row-echelon form is

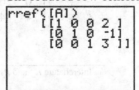

The solution to the system is $x = 2$, $y = -1$, $z = 3$, or $(2, -1, 3)$.

59.
$$\begin{cases} x_1 + x_2 + x_3 + x_4 = 20 \\ \quad\;\; x_2 + x_3 + x_4 = 0 \\ \qquad\qquad x_3 + x_4 = 13 \\ \quad\;\; x_2 - \qquad\;\; 2x_4 = -5 \end{cases}$$

The row-echelon form is

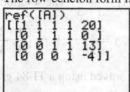

The reduced row-echelon form is

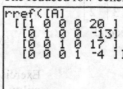

The solution to the system is $x_1 = 20$, $x_2 = -13$, $x_3 = 17$, $x_4 = -4$, or $(20, -13, 17, -4)$.

61. Let $x =$ amount invested in Treasury Bills yielding 2%, $y =$ amount invested in bank CDs yielding 4%, and $z =$ amount invested in corporate bonds yielding 6%.

The problem can be modeled with the system of equations $\begin{cases} x + y + z = 120{,}000 & (1) \\ 0.02x + 0.04y + 0.06z = 5000 & (2) \\ z = 0.5y & (3) \end{cases}$

If equation (2) is multiplied by 100, and equation (3) is multiplied by 10 and set equal to zero, the system can be represented by the augmented matrices.

$$\begin{bmatrix} 1 & 1 & 1 & | & 120{,}000 \\ 2 & 4 & 6 & | & 500{,}000 \\ 0 & -5 & 10 & | & 0 \end{bmatrix} \xrightarrow[\substack{R_2 = -2r_1 + r_2 \\ R_3 = -\frac{1}{5}r_2}]{} \begin{bmatrix} 1 & 1 & 1 & | & 120{,}000 \\ 0 & 2 & 4 & | & 260{,}000 \\ 0 & 1 & -2 & | & 0 \end{bmatrix} \xrightarrow{R_2 = \frac{1}{2}r_2} \begin{bmatrix} 1 & 1 & 1 & | & 120{,}000 \\ 0 & 1 & 2 & | & 130{,}000 \\ 0 & 1 & -2 & | & 0 \end{bmatrix}$$

$$\xrightarrow{R_3 = -r_3 + r_2} \begin{bmatrix} 1 & 1 & 1 & | & 120{,}000 \\ 0 & 1 & 2 & | & 130{,}000 \\ 0 & 0 & 4 & | & 130{,}000 \end{bmatrix} \xrightarrow{R_3 = \frac{1}{4}r_3} \begin{bmatrix} 1 & 1 & 1 & | & 120{,}000 \\ 0 & 1 & 2 & | & 130{,}000 \\ 0 & 0 & 1 & | & 32{,}500 \end{bmatrix}$$

The row-echelon form of the system is $\begin{cases} x = -y - z + 120{,}000 & (1) \\ y = -2z + 130{,}000 & (2) \\ z = 32{,}500 & (3) \end{cases}$.

Back-substituting $z = 32{,}500$ into equation (2), we have $y = 65{,}000$ Back-substituting $y = 65{,}000$ and $z = 32{,}500$ into equation (1) gives $x = 22{,}500$. Carla should invest \$22,500 in the Treasury Bills, \$65,000 in bank CDs, and \$32,500 in the corporate bonds.

63. Let $x =$ the number of servings of chicken breast, $y =$ the number of servings of potatoes, and $z =$ the number of cups of spinach needed in the diet.

The problem can be modeled with the system of equations $\begin{cases} 24x + 4y + 5z = 38 & (1) \\ 26y + 7z = 40 & (2) \\ 1.5x + 0.5z = 2.5 & (3) \end{cases}$.

The system can be represented by the augmented matrices

$$\begin{bmatrix} 24 & 4 & 5 & | & 38 \\ 0 & 26 & 7 & | & 40 \\ 1.5 & 0 & 0.5 & | & 2.5 \end{bmatrix} \xrightarrow{R_3 = 2r_3} \begin{bmatrix} 24 & 4 & 5 & | & 38 \\ 0 & 26 & 7 & | & 40 \\ 3 & 0 & 1 & | & 5 \end{bmatrix} \xrightarrow{\text{Interchange } R_1 \text{ and } R_3} \begin{bmatrix} 3 & 0 & 1 & | & 5 \\ 0 & 26 & 7 & | & 40 \\ 24 & 4 & 5 & | & 38 \end{bmatrix}$$

$$\xrightarrow{R_1 = \frac{1}{3}r_1} \begin{bmatrix} 1 & 0 & \frac{1}{3} & | & \frac{5}{3} \\ 0 & 26 & 7 & | & 40 \\ 24 & 4 & 5 & | & 38 \end{bmatrix} \xrightarrow{R_3 = -24r_1 + r_3} \begin{bmatrix} 1 & 0 & \frac{1}{3} & | & \frac{5}{3} \\ 0 & 26 & 7 & | & 40 \\ 0 & 4 & -3 & | & -2 \end{bmatrix}$$

$$\xrightarrow{\text{Interchange } R_2 \text{ and } R_3} \begin{bmatrix} 1 & 0 & \frac{1}{3} & | & \frac{5}{3} \\ 0 & 4 & -3 & | & -2 \\ 0 & 26 & 7 & | & 40 \end{bmatrix} \xrightarrow{R_2 = \frac{1}{4}r_2} \begin{bmatrix} 1 & 0 & \frac{1}{3} & | & \frac{5}{3} \\ 0 & 1 & -\frac{3}{4} & | & -\frac{1}{2} \\ 0 & 26 & 7 & | & 40 \end{bmatrix}$$

(continued on next page)

(*continued*)

$$\xrightarrow[R_3=-26r_2+r_3]{} \begin{bmatrix} 1 & 0 & \frac{1}{3} & \frac{5}{3} \\ 0 & 1 & -\frac{3}{4} & -\frac{1}{2} \\ 0 & 0 & \frac{53}{2} & 53 \end{bmatrix} \xrightarrow[R_3=\frac{2}{53}r_3]{} \begin{bmatrix} 1 & 0 & \frac{1}{3} & \frac{5}{3} \\ 0 & 1 & -\frac{3}{4} & -\frac{1}{2} \\ 0 & 0 & 1 & 2 \end{bmatrix}$$

The row-echelon form of the system is $\begin{cases} x=-\dfrac{1}{3}z+\dfrac{5}{3} & (1) \\ y=\dfrac{3}{4}z-\dfrac{1}{2} & (2) \\ z=2 & (3) \end{cases}$.

Back-substituting $z=2$ into (2) gives $y=1$. Back-substituting $z=2$ into equation (1) gives $x=1$.
The dietician should serve 1 serving of chicken, 1 potato, and 2 servings of spinach.

65. Let x = the number of cases of orange juice to be prepared, y = the number of cases of tomato juice to be prepared, and z = the number of cases of pineapple juice to be prepared.

The problem can be modeled with the system of equations $\begin{cases} 10x+12y+9z=398 & (1) \\ 4x+4y+6z=164 & (2) \\ 2x+\ y+\ z=\ 58 & (3) \end{cases}$.

The system can be represented by the augmented matrices

$$\begin{bmatrix} 10 & 12 & 9 & 398 \\ 4 & 4 & 6 & 164 \\ 2 & 1 & 1 & 58 \end{bmatrix} \xrightarrow[R_1=\frac{1}{10}r_1]{} \begin{bmatrix} 1 & 1.2 & 0.9 & 39.8 \\ 4 & 4 & 6 & 164 \\ 2 & 1 & 1 & 58 \end{bmatrix} \xrightarrow[\substack{R_2=-4r_1+r_2 \\ R_3=-2r_1+r_3}]{} \begin{bmatrix} 1 & 1.2 & 0.9 & 39.8 \\ 0 & -0.8 & 2.4 & 4.8 \\ 0 & -1.4 & -0.8 & -21.6 \end{bmatrix}$$

$$\xrightarrow[R_2=-\frac{10}{8}r_2]{} \begin{bmatrix} 1 & 1.2 & 0.9 & 39.8 \\ 0 & 1 & -3 & -6 \\ 0 & -1.4 & -0.8 & -21.6 \end{bmatrix} \xrightarrow[R_3=1.4r_2+r_3]{} \begin{bmatrix} 1 & 1.2 & 0.9 & 39.8 \\ 0 & 1 & -3 & -6 \\ 0 & 0 & -5 & -30 \end{bmatrix}$$

$$\xrightarrow[R_3=-\frac{1}{5}r_3]{} \begin{bmatrix} 1 & 1.2 & 0.9 & 39.8 \\ 0 & 1 & -3 & -6 \\ 0 & 0 & 1 & 6 \end{bmatrix}$$

The row-echelon form of the system is $\begin{cases} x=-1.2y-0.9z+39.8 & (1) \\ y=3z-6 & (2) \\ z=6 & (3) \end{cases}$.

Back-substituting $z=6$ into (2) gives $y=12$. Back-substituting $y=12$ and $z=6$ into equation (1) gives $x=20$.
The company prepared 20 cases of orange juice, 12 cases of tomato juice, and 6 cases of pineapple juice.

67. Let x = the price of a mezzanine ticket, y = the price of a lower balcony ticket, and z = the price of a middle balcony ticket.

The problem can be modeled with the system of equations $\begin{cases} 4x+6y\ \ =558 & (1) \\ 2x+7y+8z=787 & (2) \\ 3y+12z=615 & (3) \end{cases}$.

The system can be represented by the augmented matrices

$$\begin{bmatrix} 4 & 6 & 0 & 558 \\ 2 & 7 & 8 & 787 \\ 0 & 3 & 12 & 615 \end{bmatrix} \xrightarrow[R_1=0.25r_1]{} \begin{bmatrix} 1 & 1.5 & 0 & 139.5 \\ 2 & 7 & 8 & 787 \\ 0 & 3 & 12 & 615 \end{bmatrix} \xrightarrow[R_2=-2r_1+r_2]{} \begin{bmatrix} 1 & 1.5 & 0 & 139.5 \\ 0 & 4 & 8 & 508 \\ 0 & 3 & 12 & 615 \end{bmatrix}$$

$$\xrightarrow[R_2=0.25r_2]{} \begin{bmatrix} 1 & 1.5 & 0 & 139.5 \\ 0 & 1 & 2 & 127 \\ 0 & 3 & 12 & 615 \end{bmatrix} \xrightarrow[R_3=-3r_2+r_3]{} \begin{bmatrix} 1 & 1.5 & 0 & 139.5 \\ 0 & 1 & 2 & 127 \\ 0 & 0 & 6 & 234 \end{bmatrix}$$

$$\xrightarrow[R_3=\frac{1}{6}r_3]{} \begin{bmatrix} 1 & 1.5 & 0 & 139.5 \\ 0 & 1 & 2 & 127 \\ 0 & 0 & 1 & 39 \end{bmatrix}$$

(*continued on next page*)

(*continued*)

The row-echelon form of the system is $\begin{cases} x = -1.5y + 139.5 & (1) \\ y = -2z + 127 & (2) \\ z = 39 & (3) \end{cases}$.

Back-substituting $z = 39$ into (2) gives $y = 49$. Back-substituting $y = 49$ into equation (1) gives $x = 66$. So, mezzanine tickets cost \$66, lower balcony tickets cost \$49, and middle balcony tickets cost \$39.

69. Let $x =$ the number of packages of package 1 to be ordered, $y =$ the number of packages of package 2 to be ordered, and $z =$ the number of packages of package 3 to be ordered.

The problem can be modeled with the system of equations $\begin{cases} 20x + \quad\; 40z = 200 & (1) \\ 15x + 3y + 30z = 180 & (2) \\ x + y \qquad\quad = 12 & (3) \end{cases}$.

The system can be represented by the augmented matrices

$\begin{bmatrix} 20 & 0 & 40 & | & 200 \\ 15 & 3 & 30 & | & 180 \\ 1 & 1 & 0 & | & 12 \end{bmatrix} \xrightarrow[\text{rows 1 and 3}]{\text{Interchange}} \begin{bmatrix} 1 & 1 & 0 & | & 12 \\ 15 & 3 & 30 & | & 180 \\ 20 & 0 & 40 & | & 200 \end{bmatrix} \xrightarrow[R_3 = \frac{1}{20}r_3]{R_2 = \frac{1}{3}r_2} \begin{bmatrix} 1 & 1 & 0 & | & 12 \\ 5 & 1 & 10 & | & 60 \\ 1 & 0 & 2 & | & 10 \end{bmatrix}$

$\xrightarrow[R_3 = -r_1 + r_3]{R_2 = -5r_1 + r_2} \begin{bmatrix} 1 & 1 & 0 & | & 12 \\ 0 & -4 & 10 & | & 0 \\ 0 & -1 & 2 & | & -2 \end{bmatrix} \xrightarrow[\text{rows 2 and 3}]{\text{Interchange}} \begin{bmatrix} 1 & 1 & 0 & | & 12 \\ 0 & -1 & 2 & | & -2 \\ 0 & -4 & 10 & | & 0 \end{bmatrix}$

$\xrightarrow{R_2 = -r_2} \begin{bmatrix} 1 & 1 & 0 & | & 12 \\ 0 & 1 & -2 & | & 2 \\ 0 & -4 & 10 & | & 0 \end{bmatrix} \xrightarrow{R_3 = 4r_2 + r_3} \begin{bmatrix} 1 & 1 & 0 & | & 12 \\ 0 & 1 & -2 & | & 2 \\ 0 & 0 & 2 & | & 8 \end{bmatrix} \xrightarrow{R_3 = \frac{1}{2}r_3} \begin{bmatrix} 1 & 1 & 0 & | & 12 \\ 0 & 1 & -2 & | & 2 \\ 0 & 0 & 1 & | & 4 \end{bmatrix}$

The row-echelon form of the system is $\begin{cases} x = -y + 12 & (1) \\ y = 2z + 2 & (2) \\ z = 4 & (3) \end{cases}$.

Back-substituting $z = 4$ into (2) gives $y = 10$. Back-substituting $y = 10$ into equation (1) gives $x = 2$. The teacher should order 2 packages of package 1, 10 packages of package 2, and 4 packages of package 3 paper.

71. Let $x =$ the number of assorted cartons needed, $y =$ the number of mixed cartons needed, and $z =$ the number of single cartons needed to fill the recreation center's order.

The problem can be modeled with the system of equations $\begin{cases} 2x + 4y = 40 & (1) \\ 4x + 2y = 32 & (2) \\ x + 2z = 14 & (3) \end{cases}$.

The system can be represented by the augmented matrices

$\begin{bmatrix} 2 & 4 & 0 & | & 40 \\ 4 & 2 & 0 & | & 32 \\ 1 & 0 & 2 & | & 14 \end{bmatrix} \xrightarrow{R_1 = \frac{1}{2}r_1} \begin{bmatrix} 1 & 2 & 0 & | & 20 \\ 4 & 2 & 0 & | & 32 \\ 1 & 0 & 2 & | & 14 \end{bmatrix} \xrightarrow[R_3 = -r_1 + r_3]{R_2 = -4r_1 + r_2} \begin{bmatrix} 1 & 2 & 0 & | & 20 \\ 0 & -6 & 0 & | & -48 \\ 0 & -2 & 2 & | & -6 \end{bmatrix} \xrightarrow{R_2 = -\frac{1}{6}r_2} \begin{bmatrix} 1 & 2 & 0 & | & 20 \\ 0 & 1 & 0 & | & 8 \\ 0 & -2 & 2 & | & -6 \end{bmatrix}$

$\xrightarrow{R_3 = 2r_2 + r_3} \begin{bmatrix} 1 & 2 & 0 & | & 20 \\ 0 & 1 & 0 & | & 8 \\ 0 & 0 & 2 & | & 10 \end{bmatrix} \xrightarrow{R_3 = \frac{1}{2}r_3} \begin{bmatrix} 1 & 2 & 0 & | & 20 \\ 0 & 1 & 0 & | & 8 \\ 0 & 0 & 1 & | & 5 \end{bmatrix}$

The row-echelon form of the system is $\begin{cases} x = -2y + 20 & (1) \\ y = 8 & (2) \\ z = 5 & (3) \end{cases}$.

Back-substituting $y = 8$ into (2) gives $x = 4$.
The supplier should send the recreation center 4 assorted cartons, 8 mixed cartons, and 5 single cartons.

73. Let x = the number of large cans needed, y = the number mammoth cans needed, and z = the number of giant cans needed to fulfill the order.

The problem can be modeled with the system of equations $\begin{cases} y + z = 5 & (1) \\ 2x + 6y + 4z = 26 & (2) \\ x + 2y + 2z = 12 & (3) \end{cases}$.

The system can be represented by the augmented matrices

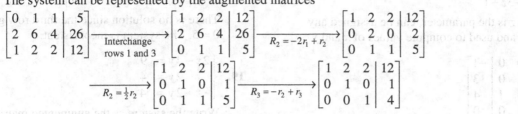

The row-echelon form of the system is $\begin{cases} x = -2y - 2z + 12 & (1) \\ y = 1 & (2) \\ z = 4 & (3) \end{cases}$.

Back-substituting $z = 4$ and $y = 1$ into (1) gives $x = 2$.
The store should use 2 large size cans, 1 mammoth size can, and 4 giant size cans to fill the order.

75. Let x = the number of shares of NCC, y = the number of shares of IBM, and z = the number of shares of ESRX.

The problem can be modeled with the system of equations $\begin{cases} 25x + 100y + 50z = 200,000 & (1) \\ 2x + 2y = 8000 & (2) \\ 0.05(100y) + 0.10(50z) = 0.04(200,000) & (3) \end{cases}$.

Note: We obtain equation (3) because growth is based on the dollar value of the investment, not on the number of shares. Equation (3) simplifies to $5y + 5z = 8000$.
The system can be represented by the augmented matrices

$$\begin{bmatrix} 25 & 100 & 50 & | & 200,000 \\ 2 & 2 & 0 & | & 8000 \\ 0 & 5 & 5 & | & 8000 \end{bmatrix} \xrightarrow[\substack{R_1 = \frac{1}{25}r_1 \\ R_2 = \frac{1}{2}r_2 \\ R_3 = \frac{1}{5}r_3}]{} \begin{bmatrix} 1 & 4 & 2 & | & 8000 \\ 1 & 1 & 0 & | & 4000 \\ 0 & 1 & 1 & | & 1600 \end{bmatrix} \xrightarrow[\text{interchange } R_3 \text{ and } R_2]{} \begin{bmatrix} 1 & 4 & 2 & | & 8000 \\ 0 & 1 & 1 & | & 1600 \\ 1 & 1 & 0 & | & 4000 \end{bmatrix}$$

$$\xrightarrow[R_3 = -r_1 + r_3]{} \begin{bmatrix} 1 & 4 & 2 & | & 8000 \\ 0 & 1 & 1 & | & 1600 \\ 0 & -3 & -2 & | & -4000 \end{bmatrix} \xrightarrow[R_3 = 3r_2 + r_3]{} \begin{bmatrix} 1 & 4 & 2 & | & 8000 \\ 0 & 1 & 1 & | & 1600 \\ 0 & 0 & 1 & | & 800 \end{bmatrix}$$

The row-echelon form of the system is $\begin{cases} x = -4y - 2z + 8000 & (1) \\ y = -z + 1600 & (2) \\ z = 800 & (3) \end{cases}$.

Back-substituting $z = 800$ into (2) gives $y = 800$. Back-substituting $y = 800$ and $z = 800$ into (1) gives $x = 3200$. Karen should invest \$3200 in NCC, \$800 in IBM, and \$800 in ESRX.

77. Answers will vary.

2.3 Systems of *m* Linear Equations Containing *n* Variables

1. False

3. $\begin{bmatrix} 1 & 2 & | & 3 \\ 0 & 0 & | & 1 \\ 0 & 0 & | & 0 \end{bmatrix}$

Yes, the matrix is in reduced row-echelon form.

5. $\begin{bmatrix} 0 & | & 1 \\ 1 & | & 0 \end{bmatrix}$

No, the matrix is not in reduced row-echelon form because the leftmost 1 in the second row is not to the right of the leftmost 1 in the first row.

7. $\begin{bmatrix} 1 & 0 & 0 & 0 & | & 0 \\ 0 & 0 & 1 & 2 & | & 0 \\ 0 & 0 & 0 & 0 & | & 1 \\ 0 & 0 & 0 & 0 & | & 0 \end{bmatrix}$

Yes, the matrix is in reduced row-echelon form.

9. $\begin{bmatrix} 1 & 0 & -2 & | & 6 \\ 0 & 1 & 3 & | & 1 \end{bmatrix}$

There are infinitely many solutions represented by the system

$$\begin{cases} x = 2z + 6 \\ y = -3z + 1 \end{cases},$$

where z is the parameter, can be assigned any value, and used to compute values of x and y.

11. $\begin{bmatrix} 1 & 0 & 0 & | & -1 \\ 0 & 1 & 0 & | & 3 \\ 0 & 0 & 1 & | & 4 \\ 0 & 0 & 0 & | & 0 \end{bmatrix}$

The system has one solution. It is $x = -1$, $y = 3$, $z = 4$, or $(-1, 3, 4)$.

13. $\begin{bmatrix} 1 & 0 & -1 & | & 1 \\ 0 & 1 & 2 & | & 1 \end{bmatrix}$

The system has infinitely many solutions. They are represented by the system

$$\begin{cases} x = z + 1 \\ y = -2z + 1 \end{cases},$$

where z is the parameter and can be assigned any real number.

15. $\begin{bmatrix} 1 & 0 & 0 & -1 & | & 4 \\ 0 & 1 & 2 & 3 & | & 0 \end{bmatrix}$

The system has infinitely many solutions. They are represented by the system

$$\begin{cases} x_1 = x_4 + 4 \\ x_2 = -2x_3 - 3x_4 \end{cases}$$

where x_3 and x_4 are the parameters and can be assigned any real numbers.

17. $\begin{cases} 3x - 3y = 12 \\ 3x + 2y = -3 \\ 2x + y = 4 \end{cases}$

Write the system as the augmented matrix,

$$\begin{bmatrix} 3 & -3 & | & 12 \\ 3 & 2 & | & -3 \\ 2 & 1 & | & 4 \end{bmatrix}.$$

Now use row operations to find the reduced row-echelon form.

$$\begin{bmatrix} 3 & -3 & | & 12 \\ 3 & 2 & | & -3 \\ 2 & 1 & | & 4 \end{bmatrix} \xrightarrow{R_1 = \frac{1}{3}r_1} \begin{bmatrix} 1 & -1 & | & 4 \\ 3 & 2 & | & -3 \\ 2 & 1 & | & 4 \end{bmatrix}$$

$$\xrightarrow[R_3 = -2r_1 + r_3]{R_2 = -3r_1 + r_2} \begin{bmatrix} 1 & -1 & | & 4 \\ 0 & 5 & | & -15 \\ 0 & 3 & | & -4 \end{bmatrix}$$

$$\xrightarrow{R_2 = \frac{1}{5}r_2} \begin{bmatrix} 1 & -1 & | & 4 \\ 0 & 1 & | & -3 \\ 0 & 3 & | & -4 \end{bmatrix}$$

$$\xrightarrow[R_3 = -3r_2 + r_3]{R_1 = r_2 + r_1} \begin{bmatrix} 1 & 0 & | & 1 \\ 0 & 1 & | & -3 \\ 0 & 0 & | & 5 \end{bmatrix}$$

There is no solution since the third row gives $0 = 5$. The system is inconsistent.

19. $\begin{cases} 2x - 4y = 9 \\ x - 2y = 4 \\ -x + 2y = -4 \end{cases}$

Write the system as the augmented matrix,

$$\begin{bmatrix} 2 & -4 & | & 8 \\ 1 & -2 & | & 4 \\ -1 & 2 & | & -4 \end{bmatrix}.$$

Now use row operations to find the reduced row-echelon form.

$$\begin{bmatrix} 2 & -4 & | & 8 \\ 1 & -2 & | & 4 \\ -1 & 2 & | & -4 \end{bmatrix} \xrightarrow[\text{rows 1 and 2}]{\text{Interchange}} \begin{bmatrix} 1 & -2 & | & 4 \\ 2 & -4 & | & 8 \\ -1 & 2 & | & -4 \end{bmatrix}$$

$$\xrightarrow[R_3 = r_1 + r_3]{R_2 = -2r_1 + r_2} \begin{bmatrix} 1 & -2 & | & 4 \\ 0 & 0 & | & 0 \\ 0 & 0 & | & 0 \end{bmatrix}$$

The system has an infinite number of solutions. They are $x = 2y + 4$, where y is the parameter and can be assigned any real number.

21. $\begin{cases} 2x + y + 3z = -1 \\ -x + y + 3z = 8 \end{cases}$

Write the system as the augmented matrix, $\begin{bmatrix} 2 & 1 & 3 & | & -1 \\ -1 & 1 & 3 & | & 8 \end{bmatrix}.$

Now use row operations to find the reduced row-echelon form.

$$\begin{bmatrix} 2 & 1 & 3 & | & -1 \\ -1 & 1 & 3 & | & 8 \end{bmatrix} \xrightarrow[\text{rows 1 and 2}]{\text{Interchange}} \begin{bmatrix} -1 & 1 & 3 & | & 8 \\ 2 & 1 & 3 & | & -1 \end{bmatrix}$$

$$\xrightarrow{R_1 = -r_1} \begin{bmatrix} 1 & -1 & -3 & | & -8 \\ 2 & 1 & 3 & | & -1 \end{bmatrix}$$

$$\xrightarrow{R_2 = -2r_1 + r_2} \begin{bmatrix} 1 & -1 & -3 & | & -8 \\ 0 & 3 & 9 & | & 15 \end{bmatrix}$$

$$\xrightarrow{R_2 = \frac{1}{3}r_2} \begin{bmatrix} 1 & -1 & -3 & | & -8 \\ 0 & 1 & 3 & | & 5 \end{bmatrix}$$

$$\xrightarrow{R_1 = r_2 + r_1} \begin{bmatrix} 1 & 0 & 0 & | & -3 \\ 0 & 1 & 3 & | & 5 \end{bmatrix}$$

There are an infinite number of solutions. They are represented by the system of equations

$$\begin{cases} x = -3 \\ y = -3z + 5 \end{cases},$$

where z is the parameter and can be assigned any real number.

23. $\begin{cases} x_1 + x_2 \qquad\qquad = 7 \\ \quad\ x_2 - x_3 + x_4 = 5 \\ x_1 - x_2 + x_3 + x_4 = 6 \\ \quad\ x_2 - \qquad x_4 = 10 \end{cases}$

Write the system as the augmented matrix, $\begin{bmatrix} 1 & 1 & 0 & 0 & 7 \\ 0 & 1 & -1 & 1 & 5 \\ 1 & -1 & 1 & 1 & 6 \\ 0 & 1 & 0 & -1 & 10 \end{bmatrix}$

Now use row operations to find the reduced row-echelon form.

$\begin{bmatrix} 1 & 1 & 0 & 0 & 7 \\ 0 & 1 & -1 & 1 & 5 \\ 1 & -1 & 1 & 1 & 6 \\ 0 & 1 & 0 & -1 & 10 \end{bmatrix} \xrightarrow{R_3 = -r_1 + r_3} \begin{bmatrix} 1 & 1 & 0 & 0 & 7 \\ 0 & 1 & -1 & 1 & 5 \\ 0 & -2 & 1 & 1 & -1 \\ 0 & 1 & 0 & -1 & 10 \end{bmatrix} \xrightarrow[\substack{R_3 = 2r_2 + r_3 \\ R_4 = -r_2 + r_4}]{R_1 = -r_2 + r_1} \begin{bmatrix} 1 & 0 & 1 & -1 & 2 \\ 0 & 1 & -1 & 1 & 5 \\ 0 & 0 & -1 & 3 & 9 \\ 0 & 0 & 1 & -2 & 5 \end{bmatrix}$

$\xrightarrow{R_3 = -r_3} \begin{bmatrix} 1 & 0 & 1 & -1 & 2 \\ 0 & 1 & -1 & 1 & 5 \\ 0 & 0 & 1 & -3 & -9 \\ 0 & 0 & 1 & -2 & 5 \end{bmatrix} \xrightarrow[\substack{R_2 = r_3 + r_2 \\ R_4 = -r_3 + r_4}]{R_1 = -r_3 + r_1} \begin{bmatrix} 1 & 0 & 0 & 2 & 11 \\ 0 & 1 & 0 & -2 & -4 \\ 0 & 0 & 1 & -3 & -9 \\ 0 & 0 & 0 & 1 & 14 \end{bmatrix}$

$\xrightarrow[\substack{R_2 = 2r_4 + r_2 \\ R_3 = 3r_4 + r_3}]{R_1 = -2r_4 + r_1} \begin{bmatrix} 1 & 0 & 0 & 0 & -17 \\ 0 & 1 & 0 & 0 & 24 \\ 0 & 0 & 1 & 0 & 33 \\ 0 & 0 & 0 & 1 & 14 \end{bmatrix}$

There is one solution, $x_1 = -17$, $x_2 = 24$, $x_3 = 33$, and $x_4 = 14$ or $(-17, 24, 33, 14)$.

25. $\begin{cases} 2x - 3y + 4z = 7 \\ x - 2y + 3z = 2 \end{cases}$

Write the system as the augmented matrix, $\begin{bmatrix} 2 & -3 & 4 & 7 \\ 1 & -2 & 3 & 2 \end{bmatrix}$.

Now use row operations to find the reduced row-echelon form.

There are an infinite number of solutions. They are represented by the system of equations $\begin{cases} x = z + 8 \\ y = 2z + 3 \end{cases}$ where

z is the parameter and can be assigned any real number.

27. $\begin{cases} x_1 + x_2 + x_3 + x_4 = 4 \\ 2x_1 - x_2 + x_3 \qquad = 0 \\ 3x_1 + 2x_2 + x_3 - x_4 = 6 \\ x_1 - 2x_2 - 2x_3 + 2x_4 = -1 \end{cases}$

Write the system as the augmented matrix, $\begin{bmatrix} 1 & 1 & 1 & 1 & 4 \\ 2 & -1 & 1 & 0 & 0 \\ 3 & 2 & 1 & -1 & 6 \\ 1 & -2 & -2 & 2 & -1 \end{bmatrix}$

Now use row operations to find the reduced row-echelon form.

$\begin{bmatrix} 1 & 1 & 1 & 1 & 4 \\ 2 & -1 & 1 & 0 & 0 \\ 3 & 2 & 1 & -1 & 6 \\ 1 & -2 & -2 & 2 & -1 \end{bmatrix} \xrightarrow[\substack{R_2 = -2r_1 + r_2 \\ R_3 = -3r_1 + r_3 \\ R_4 = -r_1 + r_4}]{} \begin{bmatrix} 1 & 1 & 1 & 1 & 4 \\ 0 & -3 & -1 & -2 & -8 \\ 0 & -1 & -2 & -4 & -6 \\ 0 & -3 & -3 & 1 & -5 \end{bmatrix} \xrightarrow[\substack{R_2 = -r_3 \\ R_3 = -r_2}]{} \begin{bmatrix} 1 & 1 & 1 & 1 & 4 \\ 0 & 1 & 2 & 4 & 6 \\ 0 & 3 & 1 & 2 & 8 \\ 0 & -3 & -3 & 1 & -5 \end{bmatrix}$

$\xrightarrow[\substack{R_1 = -r_2 + r_1 \\ R_3 = -3r_2 + r_3 \\ R_4 = 3r_2 + r_4}]{} \begin{bmatrix} 1 & 0 & -1 & -3 & -2 \\ 0 & 1 & 2 & 4 & 6 \\ 0 & 0 & -5 & -10 & -10 \\ 0 & 0 & 3 & 13 & 13 \end{bmatrix} \xrightarrow[R_3 = -\frac{1}{5}r_3]{} \begin{bmatrix} 1 & 0 & -1 & -3 & -2 \\ 0 & 1 & 2 & 4 & 6 \\ 0 & 0 & 1 & 2 & 2 \\ 0 & 0 & 3 & 13 & 13 \end{bmatrix}$

$\xrightarrow[\substack{R_1 = r_3 + r_1 \\ R_2 = -2r_3 + r_2 \\ R_4 = -3r_3 + r_4}]{} \begin{bmatrix} 1 & 0 & 0 & -1 & 0 \\ 0 & 1 & 0 & 0 & 2 \\ 0 & 0 & 1 & 2 & 2 \\ 0 & 0 & 0 & 7 & 7 \end{bmatrix} \xrightarrow[R_4 = \frac{1}{7}r_4]{} \begin{bmatrix} 1 & 0 & 0 & -1 & 0 \\ 0 & 1 & 0 & 0 & 2 \\ 0 & 0 & 1 & 2 & 2 \\ 0 & 0 & 0 & 1 & 1 \end{bmatrix} \xrightarrow[\substack{R_1 = r_4 + r_1 \\ R_3 = -2r_4 + r_3}]{} \begin{bmatrix} 1 & 0 & 0 & 0 & 1 \\ 0 & 1 & 0 & 0 & 2 \\ 0 & 0 & 1 & 0 & 0 \\ 0 & 0 & 0 & 1 & 1 \end{bmatrix}$

There is one solution, $x_1 = 1$, $x_2 = 2$, $x_3 = 0$, and $x_4 = 1$ or $(1, 2, 0, 1)$.

29. Let x be the amount of supplement 1 used, y be the amount of supplement 2 used, and z be the amount of supplement 3 used. The problem is modeled by the system of equations $\begin{cases} 0.2x + 0.4y + 0.3z = 40 & (1) \\ 0.3x + 0.2y + 0.5z = 30 & (2) \end{cases}$ and

can be represented by the augmented matrix $\begin{bmatrix} 0.2 & 0.4 & 0.3 & 40 \\ 0.3 & 0.2 & 0.5 & 30 \end{bmatrix}$. Multiply by 10 to remove the decimals:

$\begin{bmatrix} 2 & 4 & 3 & 400 \\ 3 & 2 & 5 & 300 \end{bmatrix}$.

Use the row operations to write the matrix in row reduced form.

$\begin{bmatrix} 2 & 4 & 3 & 400 \\ 3 & 2 & 5 & 300 \end{bmatrix} \xrightarrow[R_1 = \frac{1}{2}r_1]{} \begin{bmatrix} 1 & 2 & 1.5 & 200 \\ 3 & 2 & 5 & 300 \end{bmatrix} \xrightarrow[R_2 = -3r_1 + r_2]{} \begin{bmatrix} 1 & 2 & 1.5 & 200 \\ 0 & -4 & 0.5 & -300 \end{bmatrix} \xrightarrow[R_2 = -\frac{1}{4}r_2]{} \begin{bmatrix} 1 & 2 & 1.5 & 200 \\ 0 & 1 & -0.125 & 75 \end{bmatrix}$

There are an infinite number of solutions. They are $y = 0.125z + 75$ and
$x = -2y - 1.5z + 200 = -2(0.125z + 75) - 1.5z + 200 = -1.75z + 50$, where z is a nonnegative real number.
Since x and y also must be nonnegative, we find that $x = -1.75z + 50 \geq 0 \Rightarrow 50 \geq 1.75z \Rightarrow z \leq 28.57$. Now we create a table to show possible combinations.

$x = -1.75z + 50$	$y = 0.125z + 75$	z
50	75	0
41.25	75.625	5
32.5	76.25	10
23.75	76.875	15
15	77.5	20
0	78.571	28.57

31.a. Let x_1 denote the servings of turkey bologna he consumes, x_2 denote the number of bananas he eats, x_3 denote the servings of cottage cheese he eats, and x_4 denote the servings of chocolate milk he drinks. The problem is modeled by the system of equations

$$\begin{cases} 184x_1 + & 92x_2 + & 90x_3 + & 72x_4 = 700 & (1) \\ 13.2x_1 + & 0.48x_2 + 1.93x_3 + & 2x_4 = 20 & (2) \\ 4.85x_1 + & 23.43x_2 + 0.01x_3 + 10.40x_4 = 100 & (3) \end{cases}$$

b. Write the augmented matrix.

$$\begin{bmatrix} 184 & 92 & 90 & 72 & | & 700 \\ 13.2 & 0.48 & 1.93 & 2.00 & | & 20 \\ 4.85 & 23.43 & 0.01 & 10.40 & | & 100 \end{bmatrix}.$$

Because of the complexity of the arithmetic, we suggest using a graphing utility to solve the system. Below are the results using a TI-84 graphing calculator.
The augmented matrix:

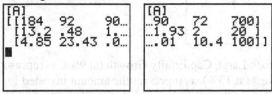

Then using Matrix Math; B: rref ([A]), we obtain

[B]
[[1 0 0 .116965...   ...1 1.187859628]
 [0 1 0 .419607...   ...9 4.0216175]
 [0 0 1 .131938...   ...2 1.238277982]]

For those who need to do it by hand, below are the steps (rounded to 5 decimal places).

$$\begin{bmatrix} 184 & 92 & 90 & 72 & | & 700 \\ 13.2 & 0.48 & 1.93 & 2 & | & 20 \\ 4.85 & 23.43 & 0.01 & 10.4 & | & 100 \end{bmatrix} \xrightarrow{R_1 = \frac{1}{184}r_1} \begin{bmatrix} 1 & 0.5 & 0.48313 & 0.39130 & | & 3.80435 \\ 13.2 & 0.48 & 1.93 & 2 & | & 20 \\ 4.85 & 23.43 & 0.01 & 10.4 & | & 100 \end{bmatrix}$$

$$\xrightarrow[R_3 = -4.85r_1 + r_3]{R_2 = -13.2r_1 + r_2} \begin{bmatrix} 1 & 0.5 & 0.48313 & 0.39130 & | & 3.80435 \\ 0 & -6.12 & -4.52652 & -3.16522 & | & -30.21739 \\ 0 & 21.005 & -2.36228 & 8.50217 & | & 81.54891 \end{bmatrix}$$

$$\xrightarrow{R_2 = -\frac{1}{6.12}r_2} \begin{bmatrix} 1 & 0.5 & 0.48313 & 0.39130 & | & 3.80435 \\ 0 & 1 & 0.73963 & 0.51719 & | & 4.93748 \\ 0 & 21.005 & -2.36228 & 8.50217 & | & 81.54891 \end{bmatrix}$$

$$\xrightarrow[R_3 = -21.005r_2 + r_3]{R_1 = -0.5r_2 + r_1} \begin{bmatrix} 1 & 0 & 0.11932 & 0.13271 & | & 1.33561 \\ 0 & 1 & 0.73963 & 0.51719 & | & 4.93748 \\ 0 & 0 & -17.89816 & -2.36145 & | & -22.16290 \end{bmatrix}$$

$$\xrightarrow{R_3 = -\frac{1}{17.89816}r_3} \begin{bmatrix} 1 & 0 & 0.11932 & 0.13271 & | & 1.33561 \\ 0 & 1 & 0.73963 & 0.51719 & | & 4.93748 \\ 0 & 0 & 1 & 0.13194 & | & 1.23828 \end{bmatrix}$$

$$\xrightarrow[R_2 = -0.73963r_3 + r_2]{R_1 = -0.11932r_3 + r_1} \begin{bmatrix} 1 & 0 & 0 & 0.11697 & | & 1.18786 \\ 0 & 1 & 0 & 0.41961 & | & 4.02161 \\ 0 & 0 & 1 & 0.13194 & | & 1.23827 \end{bmatrix}$$

Either method shows that here are multiple solutions, which can be represented by the system of equations

$$\begin{cases} x_1 = -0.117x_4 + 1.188 \\ x_2 = -0.420x_4 + 4.022, \\ x_3 = -0.132x_4 + 1.238 \end{cases}$$

where x_4 is the parameter.

c. Each of the variables must be nonnegative (you cannot eat a negative amount), so we find the least of
$x_1 = -0.117x_4 + 1.188 \geq 0 \Rightarrow x_4 \leq 10.154$, $x_2 = -0.420x_4 + 4.022 \geq 0 \Rightarrow x_2 \leq 9.576$, or
$x_3 = -0.132x_4 + 1.238 \Rightarrow x_4 \leq 9.379$
So, $0 \leq x_4 \leq 9.379$. We then can make a table.

Turkey Bologna x_1	Bananas x_2	Cottage Cheese x_3	Chocolate Milk x_4
1.2	4.0	1.2	0
1.1	3.6	1.1	1
1.0	3.2	1.0	2
0.8	2.8	0.8	3
0.7	2.3	0.7	4
0.6	1.9	0.6	5
0.5	1.5	0.4	6
0.4	1.1	0.3	7
0.3	0.7	0.2	8
0.1	0.2	0.1	9

33. a. Let x_1 represent the amount invested in John Hancock Large Cap Equity Growth (at 9%), x_2 represent the amount invested in T. Rowe Price Emerging Markets (at 13%), x_3 represent the amount invested in Templeton China World Fund (at 14%), and x_4 represent the amount invested in TCW Small Cap Growth (at 7%).

The problem is modeled by the system of equations
$$\begin{cases} x_1 + x_2 + x_3 + x_4 = 50,000 & (1) \\ 0.09x_1 + 0.13x_2 + 0.14x_3 + 0.07x_4 = 6,000 & (2) \end{cases}$$

b. The system is represented by the following augmented matrices.

$$\begin{bmatrix} 1 & 1 & 1 & 1 & | & 50,000 \\ 0.09 & 0.13 & 0.14 & 0.07 & | & 6,000 \end{bmatrix} \xrightarrow{R_2 = -0.09r_1 + r_2} \begin{bmatrix} 1 & 1 & 1 & 1 & | & 50,000 \\ 0 & 0.04 & 0.05 & -0.02 & | & 1,500 \end{bmatrix}$$

$$\xrightarrow{R_2 = 25r_2} \begin{bmatrix} 1 & 1 & 1 & 1 & | & 50,000 \\ 0 & 1 & 1.25 & -0.5 & | & 37,500 \end{bmatrix}$$

$$\xrightarrow{R_1 = -r_2 + r_1} \begin{bmatrix} 1 & 0 & -0.25 & 1.5 & | & 12,500 \\ 0 & 1 & 1.25 & -0.5 & | & 37,500 \end{bmatrix}$$

The reduced row echelon form of the augmented matrix indicates that there are an infinite number of solutions. They can be represented by the system of equations $\begin{cases} x_1 = 0.25x_3 - 1.5x_4 + 12,500 \\ x_2 = -1.25x_3 + 0.5x_4 + 37,500 \end{cases}$, where x_3 and x_4 are parameters.

c. All the variables must be nonnegative, so we begin with both parameters at zero and increase each one individually until x_1 or x_2 equal zero. A sample table follows. Your table might be different.

Amount Invested			
John Hancock Large Cap Equity Growth	T. Rowe Price Emerging Markets	Templeton China World Fund	TCW Small Cap Growth
$12,500	$37,500	$ 0	$ 0
$15,000	$25,000	$10,000	$ 0
$17,500	$12,500	$20,000	$ 0
$20,000	$ 0	$30,000	$ 0
$ 5,000	$40,000	$ 0	$ 5,000
$ 5,000	$ 5,000	$30,000	$10,000

d. Answers will vary.

35. Let x_1 represent the amount invested in the Franklin High Yield Income fund yielding 5%, x_2 represent the amount invested in the Mutual Global Discovery fund yielding 6%, x_3 represent the amount invested in the Templeton Global Bond fund yielding 8%, and x_4 represent the amount invested in the Franklin International Small Cap Growth fund which yielding 12% in interest.

a. The first couple has $20,000 to invest, and requires $2000 in income. To determine how they should allot their funds we need to solve the system of equations $\begin{cases} x_1 + x_2 + x_3 + x_4 = 20,000 & (1) \\ 0.05x_1 + 0.06x_2 + 0.08x_3 + 0.12x_4 = 2,000 & (2) \end{cases}$.

We write the system as an augmented matrix and use row operations to put it in reduced row-echelon form.

$$\begin{bmatrix} 1 & 1 & 1 & 1 & | & 20,000 \\ 0.05 & 0.06 & 0.08 & 0.12 & | & 2,000 \end{bmatrix} \xrightarrow{R_2 = -0.05r_1 + r_2} \begin{bmatrix} 1 & 1 & 1 & 1 & | & 20,000 \\ 0 & 0.01 & 0.03 & 0.07 & | & 1,000 \end{bmatrix}$$

$$\xrightarrow{R_2 = 100r_2} \begin{bmatrix} 1 & 1 & 1 & 1 & | & 20,000 \\ 0 & 1 & 3 & 7 & | & 100,000 \end{bmatrix}$$

$$\xrightarrow{R_1 = -r_2 + r_1} \begin{bmatrix} 1 & 0 & -2 & -6 & | & -80,000 \\ 0 & 1 & 3 & 7 & | & 100,000 \end{bmatrix}$$

The solutions to the system can be expressed as the system of equations $\begin{cases} x_1 = 2x_3 + 6x_4 - 80,000 \\ x_2 = -3x_3 - 7x_4 + 100,000 \end{cases}$,

where x_3 and x_4 are parameters. In this system all variables must be nonnegative, no variable can be greater than $14,286. A possible investment strategy for a couple with $20,000.00 to invest is

	Amount Invested		
Franklin High Yield Income	Mutual Global Discovery	Templeton Global Bond	Franklin International Small Cap Growth
$0	$0	$10,000	$10,000
$0	$6667	$0	$13,333
$5714	$0	$0	$14,286

b. The second couple has $25,000 to invest, and requires $2000 in income. To determine how they should allot their funds we need to solve the system of equations $\begin{cases} x_1 + x_2 + x_3 + x_4 = 25,000 & (1) \\ 0.05x_1 + 0.06x_2 + 0.08x_3 + 0.12x_4 = 2,000 & (2) \end{cases}$,

We write the system as an augmented matrix and use row operations to put it in reduced row-echelon form.

$$\begin{bmatrix} 1 & 1 & 1 & 1 & | & 25,000 \\ 0.05 & 0.06 & 0.08 & 0.12 & | & 2,000 \end{bmatrix} \xrightarrow{R_2 = -0.05r_1 + r_2} \begin{bmatrix} 1 & 1 & 1 & 1 & | & 25,000 \\ 0 & 0.01 & 0.03 & 0.07 & | & 750 \end{bmatrix}$$

$$\xrightarrow{R_2 = 100r_2} \begin{bmatrix} 1 & 1 & 1 & 1 & | & 25,000 \\ 0 & 1 & 3 & 7 & | & 75,000 \end{bmatrix}$$

$$\xrightarrow{R_1 = -r_2 + r_1} \begin{bmatrix} 1 & 0 & -2 & -6 & | & -50,000 \\ 0 & 1 & 3 & 7 & | & 75,000 \end{bmatrix}$$

The solutions to the system can be expressed as the system of equations $\begin{cases} x_1 = 2x_3 + 6x_4 - 50,000 \\ x_2 = -3x_3 - 7x_4 + 75,000 \end{cases}$, where

x_3 and x_4 are parameters. In this system all variables must be nonnegative, no variable can be greater than $25,000. A possible investment strategy for a couple with $25,000.00 to invest is

	Amount Invested		
Franklin High Yield Income	Mutual Global Discovery	Templeton Global Bond	Franklin International Small Cap Growth
$0	$0	$25,000	$0
$10,000	$5,000	$0	$10,000
$13,333	$0	$1,667	$10,000

c. The third couple has $30,000 to invest, and requires $2000 in income. To determine how they should allot their funds we need to solve the system of equations $\begin{cases} x_1 + x_2 + x_3 + x_4 = 30,000 & (1) \\ 0.05x_1 + 0.06x_2 + 0.08x_3 + 0.12x_4 = 2,000 & (2) \end{cases}$.

We write the system as an augmented matrix and use row operations to put it in reduced row-echelon form.

$$\begin{bmatrix} 1 & 1 & 1 & 1 & | & 30,000 \\ 0.05 & 0.06 & 0.08 & 0.12 & | & 2,000 \end{bmatrix} \xrightarrow{R_2 = -0.05r_1 + r_2} \begin{bmatrix} 1 & 1 & 1 & 1 & | & 30,000 \\ 0 & 0.01 & 0.03 & 0.07 & | & 500 \end{bmatrix}$$

$$\xrightarrow{R_2 = 100r_2} \begin{bmatrix} 1 & 1 & 1 & 1 & | & 30,000 \\ 0 & 1 & 3 & 7 & | & 50,000 \end{bmatrix}$$

$$\xrightarrow{R_1 = -r_2 + r_1} \begin{bmatrix} 1 & 0 & -2 & -6 & | & -20,000 \\ 0 & 1 & 3 & 7 & | & 50,000 \end{bmatrix}$$

The solutions to the system can be expressed as the system of equations $\begin{cases} x_1 = 2x_3 + 6x_4 - 20,000 \\ x_2 = -3x_3 - 7x_4 + 50,000 \end{cases}$,

where x_3 and x_4 are parameters with the restriction that all variables must be nonnegative. A possible investment strategy for a couple with $30,000.00 to invest is

Amount Invested			
Franklin High Yield Income	Mutual Global Discovery	Templeton Global Bond	Franklin International Small Cap Growth
$0	$26,667	$0	$333
$0	$20,000	$10,000	$0
$22,857	$0	$0	$7143
$13,333	$0	$16,667	$0

d. Answers will vary.

37. Let x represent the number of species 1 bacteria, y represent the number of species 2 bacteria, and z represent the number of species 3 in the test tube. To find the number of each species that can coexist in the test tube solve the system of equations

$$\begin{cases} 3x + y + 2z = 12000 \\ 2y + 4z = 12000 \\ x + 2y + 4z = 14000 \end{cases}.$$

Represent the system with the following augmented matrix, and then use row operations to write the matrix in reduced row-echelon form.

$$\begin{bmatrix} 3 & 1 & 2 & | & 12000 \\ 0 & 2 & 4 & | & 12000 \\ 1 & 2 & 4 & | & 14000 \end{bmatrix} \xrightarrow[\text{rows 1 and 3}]{\text{Interchange}} \begin{bmatrix} 1 & 2 & 4 & | & 14000 \\ 0 & 2 & 4 & | & 12000 \\ 3 & 1 & 2 & | & 12000 \end{bmatrix} \xrightarrow{R_3 = -3r_1 + r_3} \begin{bmatrix} 1 & 2 & 4 & | & 14000 \\ 0 & 2 & 4 & | & 12000 \\ 0 & -5 & -10 & | & -30000 \end{bmatrix}$$

$$\xrightarrow{R_2 = \frac{1}{2}r_3} \begin{bmatrix} 1 & 2 & 4 & | & 14000 \\ 0 & 1 & 2 & | & 6000 \\ 0 & -5 & -10 & | & -30000 \end{bmatrix} \xrightarrow[R_3 = 5r_2 + r_3]{R_1 = -2r_2 + r_1} \begin{bmatrix} 1 & 0 & 0 & | & 2000 \\ 0 & 1 & 2 & | & 6000 \\ 0 & 0 & 0 & | & 0 \end{bmatrix}$$

This system has an infinite number of solutions. They can be expressed as a system of equations $\begin{cases} x = 2000 \\ y = -2z + 6000 \end{cases}$, where z is the parameter, and all variables must be nonnegative. The nonnegativity constraint requires $z \leq 3000$. Possible combinations of the population are shown in the table. Your table might look different.

Number of Species 1	Number of Species 2	Number of Species 3
2000	6000	0
2000	4000	1000
2000	2000	2000
2000	0	3000

39. Answers will vary.

Chapter 2 Review Exercises

1. Using elimination:
$$2x - y = 5 \quad (1)$$
$$5x + 2y = 8 \quad (2)$$

Multiply equation (1) by 2, then add the resulting equation to equation (2) and solve for x.
$$4x - 2y = 10 \quad 2 \times (1)$$
$$\underline{5x + 2y = 8} \quad (2)$$
$$9x = 18 \Rightarrow x = 2$$

Substitute 2 for x in equation (1) and solve for y.
$$2(2) - y = 5 \Rightarrow y = -1$$

The solution is $x = 2$, $y = -1$, or $(2, -1)$.

3. Using elimination.
$$x - 2y - 4 = 0 \quad (1)$$
$$3x + 2y - 4 = 0 \quad (2)$$

Add the equations, then solve the resulting equation for x.
$$x - 2y - 4 = 0 \quad (1)$$
$$\underline{3x + 2y - 4 = 0} \quad (2)$$
$$4x - 8 = 0 \Rightarrow x = 2$$

Substitute 2 for x in equation (1) and solve for y.
$$2 - 2y - 4 = 0 \Rightarrow y = -1$$

The solution is $x = 2$, $y = -1$, or $(2, -1)$.

5. Using elimination.
$$3x - 2y = 8 \quad (1)$$
$$x - \frac{2}{3}y = 12 \quad (2)$$

Multiply equation (2) by 3 to eliminate the fractions.
$$3x - 2y = 8 \quad (1)$$
$$3x - 2y = 36 \quad 3 \times (2)$$

Multiply equation (2) by -1, then add the resulting equation to equation (1).
$$3x - 2y = 8 \quad (1)$$
$$\underline{-3x + 2y = -36} \quad -1 \times (2)$$
$$0 = -24$$

This is a contradiction, so the system has no solution and is inconsistent.

7.
$$x + 2y - z = 6 \quad (1)$$
$$2x - y + 3z = -13 \quad (2)$$
$$3x - 2y + 3z = -16 \quad (3)$$

Multiply equation (2) by 2, then add the resulting equation to equation (1).
$$x + 2y - z = 6 \quad (1)$$
$$\underline{4x - 2y + 6z = -26} \quad 2 \times (2)$$
$$5x + 5z = -20 \quad (4)$$

Add equations (1) and (3).
$$x + 2y - z = 6 \quad (1)$$
$$\underline{3x - 2y + 3z = -16} \quad (3)$$
$$4x + 2z = -10 \quad (5)$$

Divide equation (4) by -5 and equation (5) by 2, then add the resulting equations and solve for x.
$$-x - z = 4 \quad (4) \div (-5)$$
$$\underline{2x + z = -5} \quad (5) \div 2$$
$$x = -1$$

Substitute $x = -1$ into equation (4) and solve for z: $5(-1) + 5z = -20 \Rightarrow z = -3$

Substitute $x = -1$ and $z = -3$ into equation (1) and solve for y: $-1 + 2y - (-3) = 6 \Rightarrow y = 2$

Solution: $x = -1$, $y = 2$, $z = -3$, or $(-1, 2, -3)$

9.
$$2x - 4y + z = -15 \quad (1)$$
$$x + 2y - 4z = 27 \quad (2)$$
$$5x - 6y - 2z = -3 \quad (3)$$

Multiply equation (2) by -2, then add the resulting equation to equation (1).
$$2x - 4y + z = -15 \quad (1)$$
$$\underline{-2x - 4y + 8z = -54} \quad -2 \times (2)$$
$$-8y + 9z = -69 \quad (4)$$

Multiply equation (1) by -5 and equation (3) by 2, then add the resulting equations.
$$-10x + 20y - 5z = 75 \quad -5 \times (2)$$
$$\underline{10x - 12y - 4z = -6} \quad 2 \times (3)$$
$$8y - 9z = 69 \quad (5)$$

Add equations (4) and (5).
$$-8y + 9z = -69 \quad (4)$$
$$\underline{8y - 9z = 69} \quad (5)$$
$$0 = 0$$

The system is equivalent to a system with only two equations, so the equations are dependent and the system has infinitely many solutions. We let z be the parameter and solve for y using equation (4).
$$-8y + 9z = -69 \Rightarrow y = \frac{9}{8}z + \frac{69}{8}$$

Substitute this expression into (1) to determine x in terms of z.
$$2x - 4\left(\frac{9}{8}z + \frac{69}{8}\right) + z = -15$$
$$2x - \frac{9}{2}z - \frac{69}{2} + z = -15$$
$$x = \frac{7}{2}z + \frac{39}{4}$$

The solutions to the system can be written as the system
$$\begin{cases} x = \frac{7}{4}z + \frac{39}{4} \\ y = \frac{9}{8}z + \frac{69}{8} \end{cases}, \text{ where } z \text{ is the parameter.}$$

11. $\begin{bmatrix} 3 & 2 & | & 8 \\ 1 & 4 & | & -1 \end{bmatrix}$

The system of equations is $\begin{cases} 3x + 2y = 8 \\ x + 4y = -1 \end{cases}$.

13. $\begin{bmatrix} 1 & 0 & 0 & | & 4 \\ 0 & 1 & 0 & | & 6 \\ 0 & 0 & 1 & | & -1 \end{bmatrix}$

The system of equation is $\begin{cases} x = 4 \\ y = 6 \\ z = -1 \end{cases}$.

This system has one solution, $x = 4$, $y = 6$, $z = -1$, or $(4, 6, -1)$.

15. $\begin{cases} -5x + 2y = -2 \\ -3x + 3y = 4 \end{cases}$

Write the system as an augmented matrix and then use row operations to write it in reduced row echelon form.

$\begin{bmatrix} -5 & 2 & | & -2 \\ -3 & 3 & | & 4 \end{bmatrix} \xrightarrow{R_1 = -\frac{1}{5}r_1} \begin{bmatrix} 1 & -\frac{2}{5} & | & \frac{2}{5} \\ -3 & 3 & | & 4 \end{bmatrix} \xrightarrow{R_2 = 3r_1 + t_2} \begin{bmatrix} 1 & -\frac{2}{5} & | & \frac{2}{5} \\ 0 & \frac{9}{5} & | & \frac{26}{5} \end{bmatrix} \xrightarrow{R_2 = \frac{5}{9}r_2} \begin{bmatrix} 1 & -\frac{2}{5} & | & \frac{2}{5} \\ 0 & 1 & | & \frac{26}{9} \end{bmatrix}$

$\xrightarrow{R_1 = \frac{2}{5}r_2 + r_1} \begin{bmatrix} 1 & 0 & | & \frac{14}{9} \\ 0 & 1 & | & \frac{26}{9} \end{bmatrix}$

The solution to the system is $x = \dfrac{14}{9}$, $y = \dfrac{26}{9}$, or $\left(\dfrac{14}{9}, \dfrac{26}{9} \right)$.

17. $\begin{cases} x + 2y + 5z = 6 \\ 3x + 7y + 12z = 23 \\ x + 4y = 25 \end{cases}$

Write the system as an augmented matrix and then use row operations to write it in reduced row echelon form.

$\begin{bmatrix} 1 & 2 & 5 & | & 6 \\ 3 & 7 & 12 & | & 23 \\ 1 & 4 & 0 & | & 25 \end{bmatrix} \xrightarrow[R_3 = -r_1 + r_3]{R_2 = -3r_1 + r_2} \begin{bmatrix} 1 & 2 & 5 & | & 6 \\ 0 & 1 & -3 & | & 5 \\ 0 & 2 & -5 & | & 19 \end{bmatrix} \xrightarrow[R_3 = -2r_2 + r_3]{R_1 = -2r_2 + r_1} \begin{bmatrix} 1 & 0 & 11 & | & -4 \\ 0 & 1 & -3 & | & 5 \\ 0 & 0 & 1 & | & 9 \end{bmatrix}$

$\xrightarrow[R_2 = 3r_3 + r_2]{R_1 = -11r_3 + r_1} \begin{bmatrix} 1 & 0 & 0 & | & -103 \\ 0 & 1 & 0 & | & 32 \\ 0 & 0 & 1 & | & 9 \end{bmatrix}$

The solution to the system is $x = -103$, $y = 32$, $z = 9$, or $(-103, 32, 9)$.

19. $\begin{cases} x + 2y + 7z = 2 \\ 3x + 7y + 18z = -1 \\ x + 4y + 2z = -13 \end{cases}$

Write the system as an augmented matrix and then use row operations to write it in reduced row echelon form.

$\begin{bmatrix} 1 & 2 & 7 & | & 2 \\ 3 & 7 & 18 & | & -1 \\ 1 & 4 & 2 & | & -13 \end{bmatrix} \xrightarrow[R_3 = -r_1 + r_3]{R_2 = -3r_1 + r_2} \begin{bmatrix} 1 & 2 & 7 & | & 2 \\ 0 & 1 & -3 & | & -7 \\ 0 & 2 & -5 & | & -15 \end{bmatrix} \xrightarrow[R_3 = -2r_2 + r_3]{R_1 = -2r_2 + r_1} \begin{bmatrix} 1 & 0 & 13 & | & 16 \\ 0 & 1 & -3 & | & -7 \\ 0 & 0 & 1 & | & -1 \end{bmatrix} \xrightarrow[R_2 = 3r_3 + r_2]{R_1 = -13r_3 + r_1} \begin{bmatrix} 1 & 0 & 0 & | & 29 \\ 0 & 1 & 0 & | & -10 \\ 0 & 0 & 1 & | & -1 \end{bmatrix}$

The solution to the system is $x = 29$, $y = -10$, $z = -1$ or $(29, -10, -1)$.

21. $\begin{cases} 2x - y + z = 1 \\ x + y - z = 2 \\ 3x - y + z = 0 \end{cases}$

Write the system as an augmented matrix and then use row operations to write it in reduced row echelon form.

$$\begin{bmatrix} 2 & -1 & 1 & | & 1 \\ 1 & 1 & -1 & | & 2 \\ 3 & -1 & 1 & | & 0 \end{bmatrix} \xrightarrow[\text{rows 1 and 2}]{\text{Interchange}} \begin{bmatrix} 1 & 1 & -1 & | & 2 \\ 2 & -1 & 1 & | & 1 \\ 3 & -1 & 1 & | & 0 \end{bmatrix} \xrightarrow[R_3 = -3r_1 + r_3]{R_2 = -2r_1 + r_2} \begin{bmatrix} 1 & 1 & -1 & | & 2 \\ 0 & -3 & 3 & | & -3 \\ 0 & -4 & 4 & | & -6 \end{bmatrix}$$

$$\xrightarrow[R_2 = -\frac{1}{3}r_2]{} \begin{bmatrix} 1 & 1 & -1 & | & 2 \\ 0 & 1 & -1 & | & 1 \\ 0 & -4 & 4 & | & -6 \end{bmatrix} \xrightarrow[R_3 = 4r_2 + r_3]{} \begin{bmatrix} 1 & 1 & -1 & | & 2 \\ 0 & 1 & -1 & | & 1 \\ 0 & 0 & 0 & | & -2 \end{bmatrix}$$

Row 3 of the final matrix gives $0 = -2$, so the system has no solution and is inconsistent.

23. $\begin{cases} y - 2z = 6 \\ 3x + 2y - z = 2 \\ 4x + 3z = -1 \end{cases}$

Write the system as an augmented matrix and then use row operations to write it in reduced row echelon form.

$$\begin{bmatrix} 0 & 1 & -2 & | & 6 \\ 3 & 2 & -1 & | & 2 \\ 4 & 0 & 3 & | & -1 \end{bmatrix} \xrightarrow[\text{rows 1 and 3}]{\text{Interchange}} \begin{bmatrix} 4 & 0 & 3 & | & -1 \\ 3 & 2 & -1 & | & 2 \\ 0 & 1 & -2 & | & 6 \end{bmatrix} \xrightarrow[R_1 = r_1 - r_2]{} \begin{bmatrix} 1 & -2 & 4 & | & -3 \\ 3 & 2 & -1 & | & 2 \\ 0 & 1 & -2 & | & 6 \end{bmatrix}$$

$$\xrightarrow[R_2 = -3r_1 + r_2]{} \begin{bmatrix} 1 & -2 & 4 & | & -3 \\ 0 & 8 & -13 & | & 11 \\ 0 & 1 & -2 & | & 6 \end{bmatrix} \xrightarrow[\text{rows 2 and 3}]{\text{Interchange}} \begin{bmatrix} 1 & -2 & 4 & | & -3 \\ 0 & 1 & -2 & | & 6 \\ 0 & 8 & -13 & | & 11 \end{bmatrix}$$

$$\xrightarrow[R_3 = -8r_2 + r_3]{R_1 = 2r_2 + r_1} \begin{bmatrix} 1 & 0 & 0 & | & 9 \\ 0 & 1 & -2 & | & 6 \\ 0 & 0 & 3 & | & -37 \end{bmatrix} \xrightarrow[R_3 = \frac{1}{3}r_3]{} \begin{bmatrix} 1 & 0 & 0 & | & 9 \\ 0 & 1 & -2 & | & 6 \\ 0 & 0 & 1 & | & -\frac{37}{3} \end{bmatrix}$$

$$\xrightarrow[R_2 = 2r_3 + r_2]{} \begin{bmatrix} 1 & 0 & 0 & | & 9 \\ 0 & 1 & 0 & | & -\frac{56}{3} \\ 0 & 0 & 1 & | & -\frac{37}{3} \end{bmatrix}$$

The solution to the system is $x = 9$, $y = -\dfrac{56}{3}$, $z = -\dfrac{37}{3}$, or $\left(9, -\dfrac{56}{3}, -\dfrac{37}{3}\right)$.

25. $\begin{cases} x - 3y = 5 \\ 3y + z = 0 \\ 2x - y + 2z = 2 \end{cases}$

Write the system as an augmented matrix and then use row operations to write it in reduced row echelon form.

$$\begin{bmatrix} 1 & -3 & 0 & | & 5 \\ 0 & 3 & 1 & | & 0 \\ 2 & -1 & 2 & | & 2 \end{bmatrix} \xrightarrow[R_3 = -2r_1 + r_3]{} \begin{bmatrix} 1 & -3 & 0 & | & 5 \\ 0 & 3 & 1 & | & 0 \\ 0 & 5 & 2 & | & -8 \end{bmatrix} \xrightarrow[R_2 = \frac{1}{3}r_2]{} \begin{bmatrix} 1 & -3 & 0 & | & 5 \\ 0 & 1 & \frac{1}{3} & | & 0 \\ 0 & 5 & 2 & | & -8 \end{bmatrix}$$

$$\xrightarrow[R_3 = -5r_2 + r_3]{R_1 = 3r_2 + r_1} \begin{bmatrix} 1 & 0 & 1 & | & 5 \\ 0 & 1 & \frac{1}{3} & | & 0 \\ 0 & 0 & \frac{1}{3} & | & -8 \end{bmatrix} \xrightarrow[R_3 = 3r_3]{} \begin{bmatrix} 1 & 0 & 1 & | & 5 \\ 0 & 1 & \frac{1}{3} & | & 0 \\ 0 & 0 & 1 & | & -24 \end{bmatrix} \xrightarrow[R_2 = -\frac{1}{3}r_3 + r_2]{R_1 = -r_3 + r_1} \begin{bmatrix} 1 & 0 & 0 & | & 29 \\ 0 & 1 & 0 & | & 8 \\ 0 & 0 & 1 & | & -24 \end{bmatrix}$$

The solution to the system is $x = 29$, $y = 8$, $z = -24$ or $(29, 8, -24)$.

27. $\begin{cases} 3x + y - 2z = 3 \\ x - 2y + z = 4 \end{cases}$

Write the system as an augmented matrix and then use row operations to write it in reduced row echelon form.

$$\begin{bmatrix} 3 & 1 & -2 & | & 3 \\ 1 & -2 & 1 & | & 4 \end{bmatrix} \xrightarrow[\text{rows 1 and 2}]{\text{Interchange}} \begin{bmatrix} 1 & -2 & 1 & | & 4 \\ 3 & 1 & -2 & | & 3 \end{bmatrix} \xrightarrow{R_2 = -3r_1 + r_2} \begin{bmatrix} 1 & -2 & 1 & | & 4 \\ 0 & 7 & -5 & | & -9 \end{bmatrix}$$

$$\xrightarrow{R_2 = \frac{1}{7}r_2} \begin{bmatrix} 1 & -2 & 1 & | & 4 \\ 0 & 1 & -\frac{5}{7} & | & -\frac{9}{7} \end{bmatrix} \xrightarrow{R_1 = 2r_2 + r_1} \begin{bmatrix} 1 & 0 & -\frac{3}{7} & | & \frac{10}{7} \\ 0 & 1 & -\frac{5}{7} & | & -\frac{9}{7} \end{bmatrix}$$

The system has an infinite number of solutions. If we let z be the parameter, the solutions can be written as the

system $\begin{cases} x = \dfrac{3}{7}z + \dfrac{10}{7} \\ y = \dfrac{5}{7}z - \dfrac{9}{7} \end{cases}$. Sample solutions will vary and are found by choosing a z-value and substituting it into

the equations for x and y. Possible solutions are $\left(\dfrac{13}{7}, -\dfrac{4}{7}, 1 \right)$, $\left(\dfrac{10}{7}, -\dfrac{9}{7}, 0 \right)$, and $(1, -2, -1)$.

29. $\begin{cases} x + 2y - z = 5 \\ 2x - y + 2z = 0 \end{cases}$

Write the system as an augmented matrix and then use row operations to write it in reduced row echelon form.

$$\begin{bmatrix} 1 & 2 & -1 & | & 5 \\ 2 & -1 & 2 & | & 0 \end{bmatrix} \xrightarrow{R_2 = -2r_1 + r_2} \begin{bmatrix} 1 & 2 & -1 & | & 5 \\ 0 & -5 & 4 & | & -10 \end{bmatrix} \xrightarrow{R_2 = -\frac{1}{5}r_2} \begin{bmatrix} 1 & 2 & -1 & | & 5 \\ 0 & 1 & -\frac{4}{5} & | & 2 \end{bmatrix}$$

$$\xrightarrow{R_1 = -2r_2 + r_1} \begin{bmatrix} 1 & 0 & \frac{3}{5} & | & 1 \\ 0 & 1 & -\frac{4}{5} & | & 2 \end{bmatrix}$$

The system has an infinite number of solutions. If we let z be the parameter, the solutions are expressed as the

system $\begin{cases} x = -\dfrac{3}{5}z + 1 \\ y = \dfrac{4}{5}z + 2 \end{cases}$. Sample solutions will vary and are found by choosing a z-value and substituting it into

the equations for x and y. Sample solutions are $(-2, 6, 5)$, $(1, 2, 0)$, and $(-5, 10, 10)$.

31. $\begin{cases} 2x - y = 6 \\ x - 2y = 0 \\ 3x - y = 6 \end{cases}$

Write the system as an augmented matrix and then use row operations to write it in reduced row echelon form.

$$\begin{bmatrix} 2 & -1 & | & 6 \\ 1 & -2 & | & 0 \\ 3 & -1 & | & 6 \end{bmatrix} \xrightarrow[\text{rows 1 and 2}]{\text{Interchange}} \begin{bmatrix} 1 & -2 & | & 0 \\ 2 & -1 & | & 6 \\ 3 & -1 & | & 6 \end{bmatrix} \xrightarrow[R_3 = -3r_1 + r_3]{R_2 = -2r_1 + r_2} \begin{bmatrix} 1 & -2 & | & 0 \\ 0 & 3 & | & 6 \\ 0 & 5 & | & 6 \end{bmatrix} \xrightarrow{R_2 = \frac{1}{3}r_2} \begin{bmatrix} 1 & -2 & | & 0 \\ 0 & 1 & | & 2 \\ 0 & 5 & | & 6 \end{bmatrix}$$

$$\xrightarrow{R_3 = -5r_2 + r_3} \begin{bmatrix} 1 & -2 & | & 0 \\ 0 & 1 & | & 2 \\ 0 & 0 & | & -4 \end{bmatrix}$$

Row 3 of the final matrix gives $0 = -4$, so the system has no solution and is inconsistent.

33. $\begin{bmatrix} 1 & 4 & 3 & | & 4 \\ 0 & 1 & 0 & | & -1 \\ 0 & 0 & 1 & | & 1 \end{bmatrix}$

The augmented matrix is in row echelon form and represents the system of equations
$$\begin{cases} x+4y+3z= 4 & (1) \\ \quad\quad y \quad\quad =-1 & (2). \\ \quad\quad\quad\quad z= 1 & (3) \end{cases}$$

We can read two of the solutions, $y = -1$ and $z = 1$ directly from the system. We obtain the third variable, x, by back-substituting y and z in equation (1) and solving for x.

$x+4(-1)+3(1) = 4 \Rightarrow x-1=4 \Rightarrow x=5$

The system has one solution, $x = 5$, $y = -1$, $z = 1$, or $(5, -1, 1)$.

35. $\begin{bmatrix} 1 & 0 & 0 & 2 & | & 1 \\ 0 & 1 & 1 & 2 & | & 2 \\ 0 & 0 & 1 & 0 & | & 3 \end{bmatrix}$

The augmented matrix represents the system of equations $\begin{cases} x_1 \quad\quad\quad +2x_4 =1 & (1) \\ \quad x_2 + x_3 + 2x_4 = 2 & (2). \\ \quad\quad\quad x_3 \quad\quad = 3 & (3) \end{cases}$

If we let x_4 be the parameter, and using $x_3 = 3$, we can express the solutions by the system
$$\begin{cases} x_1 = -2x_4 +1 \\ x_2 = -2x_4 -1. \\ x_3 = 3 \end{cases}$$

37. Let x represent the number of caramels in the box and y represent the number of creams in the box. To find out how many of each kind of candy to package, we solve the system of equations $\begin{cases} x+y = 50 & (1) \\ 0.05x+0.10y = 4.00 & (2) \end{cases}$.

Solve the system using substitution. First we solve (1) for y, and substitute its value into equation (2).

$x+y = 50 \Rightarrow y = -x+50$

$0.05x+0.10(-x+50) = 4$
$0.05x-0.10x+5 = 4$
$-0.05x = -1$
$x = 20$

Sweet Delight Candies, Inc should pack 20 caramels and 30 creams in each box if they want no profit and no loss. If they want to increase profit while keeping 50 pieces of candy in the box, Sweet Delight Candies Inc. should increase the number of caramels (and decrease the number of creams) in each box.

39. Let x represent the amount of almonds needed, y represent the amount of cashews needed, and z represent the amount of peanuts needed to make the 100 bags of nuts. To determine how many pounds of each type of nut the store needs, we solve the system
$$\begin{cases} x+ y+ z =100 & (1) \\ 6x+5y+2z = 4.00(100) & (2) \end{cases}. \text{ Form an}$$
augmented matrix and use row operations to write it in reduced row echelon form.

$\begin{bmatrix} 1 & 1 & 1 & | & 100 \\ 6 & 5 & 2 & | & 400 \end{bmatrix} \xrightarrow{R_2=-6r_1+r_2} \begin{bmatrix} 1 & 1 & 1 & | & 100 \\ 0 & -1 & -4 & | & -200 \end{bmatrix}$

$\xrightarrow{R_2=-r_2} \begin{bmatrix} 1 & 1 & 1 & | & 100 \\ 0 & 1 & 4 & | & 200 \end{bmatrix}$

$\xrightarrow{R_1=-r_2+r_1} \begin{bmatrix} 1 & 0 & -3 & | & -100 \\ 0 & 1 & 4 & | & 200 \end{bmatrix}$

There are an infinite number of solutions to this system. If z is the parameter, then the solutions are represented by the system $\begin{cases} x= 3z-100 \\ y=-4z+200 \end{cases}$.

The physical constraints of this problem limit z. The three variables need to be nonnegative: $y = -4z + 200 \geq 0$ or $z \leq 50$. Possible combinations of the nuts that can be packaged include

Almonds (pounds)	Cashews (pounds)	Peanuts (pounds)
5	60	35
20	40	40
35	20	45
50	0	50

41. Let x = amount invested in Franklin Strategic Income yielding 6%, y = amount invested in Franklin Small Cap Value yielding 8%, and z = amount invested in Franklin Micro Cap Value yielding 10%. The couple invests $40,000. We find how they should allocate their funds by solving a system of equations. We solve the system by writing an augmented matrix which we put into reduced row echelon form. Before writing the augmented matrix we multiplied equation (2) by 100 to remove decimal points.

a. The couple requires $2500 in investment income.
$$\begin{cases} x+y+z = 40,000 & (1) \\ 0.06x+0.08y+0.10z = 2,500 & (2) \end{cases}$$

(*continued on next page*)

(*continued*)

$$\begin{bmatrix} 1 & 1 & 1 & | & 40{,}000 \\ 6 & 8 & 10 & | & 250{,}000 \end{bmatrix} \xrightarrow{R_2=-6r_1+r_2} \begin{bmatrix} 1 & 1 & 1 & | & 40{,}000 \\ 0 & 2 & 4 & | & 10{,}000 \end{bmatrix} \xrightarrow{R_2=\frac{1}{2}r_2} \begin{bmatrix} 1 & 1 & 1 & | & 40{,}000 \\ 0 & 1 & 2 & | & 5{,}000 \end{bmatrix}$$

$$\xrightarrow{R_1=-r_2+r_1} \begin{bmatrix} 1 & 0 & -1 & | & 35{,}000 \\ 0 & 1 & 2 & | & 5{,}000 \end{bmatrix}$$

The system of equations has infinite solutions. If we let z be the parameter, we can express the solutions as the system $\begin{cases} x = z + 35{,}000 \\ y = -2z + 5000 \end{cases}$. Since the variables must be nonnegative, $y = -2z + 5000 \geq 0 \Rightarrow z \leq 2500$

and $x = z + 35{,}000 \geq 0 \Rightarrow z \geq 0,\ x \geq 35{,}000$. Possible investments allocations available to the couple include

Strategic Income	Small Cap Value	Micro Cap Value
$35,000	$5000	0
$35,500	$4000	$500
$36,000	$3000	$1000
$36,500	$2000	$1500
$37,000	$1000	$2000
$37,500	0	$2500

b. The couple requires $3000 in investment income.

$$\begin{cases} x + y + z = 40{,}000 & (1) \\ 0.06x + 0.08y + 0.10z = 3000 & (2) \end{cases}$$

$$\begin{bmatrix} 1 & 1 & 1 & | & 40{,}000 \\ 6 & 8 & 10 & | & 300{,}000 \end{bmatrix} \xrightarrow{R_2=-6r_1+r_2} \begin{bmatrix} 1 & 1 & 1 & | & 40{,}000 \\ 0 & 2 & 4 & | & 60{,}000 \end{bmatrix} \xrightarrow{R_2=\frac{1}{2}r_2} \begin{bmatrix} 1 & 1 & 1 & | & 40{,}000 \\ 0 & 1 & 2 & | & 30{,}000 \end{bmatrix}$$

$$\xrightarrow{R_1=-r_2+r_1} \begin{bmatrix} 1 & 0 & -1 & | & 10{,}000 \\ 0 & 1 & 2 & | & 30{,}000 \end{bmatrix}$$

The system of equations has infinite solutions. If we let z be the parameter, we can express the solutions as the system $\begin{cases} x = z + 10{,}000 \\ y = -2z + 30{,}000 \end{cases}$. Since the variables must be nonnegative,

$y = -2z + 30{,}000 \geq 0 \Rightarrow z \leq 15{,}000$. Possible investments allocations available to the couple include

Strategic Income	Small Cap Value	Micro Cap Value
$10,000	$30,000	0
$15,000	$20,000	$5000
$20,000	$10,000	$10,000
$22,000	$6,000	$12,000
$25,000	0	$15,000

c. The couple requires $3500 in investment income.

$$\begin{cases} x + y + z = 40{,}000 & (1) \\ 0.06x + 0.08y + 0.10z = 3500 & (2) \end{cases}$$

$$\begin{bmatrix} 1 & 1 & 1 & | & 40{,}000 \\ 6 & 8 & 10 & | & 350{,}000 \end{bmatrix} \xrightarrow{R_2=-6r_1+r_2} \begin{bmatrix} 1 & 1 & 1 & | & 40{,}000 \\ 0 & 2 & 4 & | & 110{,}000 \end{bmatrix} \xrightarrow{R_2=\frac{1}{2}r_2} \begin{bmatrix} 1 & 1 & 1 & | & 40{,}000 \\ 0 & 1 & 2 & | & 55{,}000 \end{bmatrix}$$

$$\xrightarrow{R_1=-r_2+r_1} \begin{bmatrix} 1 & 0 & -1 & | & -15{,}000 \\ 0 & 1 & 2 & | & 55{,}000 \end{bmatrix}$$

(*continued on next page*)

(*continued*)

The system of equations has infinite solutions. If we let z be the parameter, we can express the solutions as the system $\begin{cases} x = z - 15,000 \\ y = -2z + 55,000 \end{cases}$. Since the variables must be nonnegative, $x = z - 15,000 \geq 0 \Rightarrow z \geq 15,000$ and $y = -2z + 55,000 \geq 0 \Rightarrow z \leq 27,000$. Possible investments allocations available to the couple include

Strategic Income	Small Cap Value	Micro Cap Value
0	$25,000	$15,000
$2500	$20,000	$17,500
$5000	$15,000	$20,000
$7500	$10,000	$22,500
$10,000	$5,000	$25,000
$12,500	0	$27,500

Chapter 2 Project

1. $A = 39.99 + 0.45(x - 450) + 0.02y + 1.99z$

3. Substitute $x = 600$ in the equations for A and B, then set the equations equal to each other and solve for y in terms of z.

$A = 39.99 + 0.45(600 - 450) + 0.02y + 1.99z = 107.49 + 0.02y + 1.99z$

$B = 44.99 + 0.50(600 - 500) + 0.04y + 2.29z = 94.99 + 0.04y + 2.29z$

$107.49 + 0.02y + 1.99z = 94.99 + 0.04y + 2.29z$

$-0.02y = -12.5 + 0.3z$

$y = 625 - 15z$

Since the variables must be nonnegative, $y = 625 - 15z \geq 0 \Rightarrow z \leq 41\frac{2}{3}$.

5. Substitute $y = 350$ and $z = 8$ in the equations for A and B, then solve $A \leq B$ for x.

$A = 39.99 + 0.45(x - 450) + 0.02(350) + 1.99(8) = -139.59 + 0.45x$

$B = 44.99 + 0.50(x - 500) + 0.04(350) + 2.29(8) = -172.69 + 0.50x$

$-139.59 + 0.45x \leq -172.69 + 0.5x$

$33.1 \leq 0.05x$

$662 \leq x$

Plan A will cost less than Plan B if $x \geq 662$.

7. Substitute $x = 600$ and $z = 3$ in the equations for A and B, then solve $B \leq A$ for x.

$A = 39.99 + 0.45(600 - 450) + 0.02y + 1.99(3) = 113.46 + 0.02y$

$B = 44.99 + 0.50(600 - 500) + 0.04y + 2.29(3) = 101.86 + 0.04y$

$101.86 + 0.04y \leq 113.46 + 0.02y$

$0.02y \leq 11.6$

$y \leq 580$

Plan B will cost less than Plan A if $0 \leq y \leq 580$.

9. $C = 59.99 + 0.55(x - 500) = -215.01 + 0.55x$

11. Substitute $y = 800$ and $z = 0$ in the equations for A and B, then solve $C \leq A$ and $C \leq B$ for x.

$$A = 39.99 + 0.45(x - 450) + 0.02(800) + 1.99(0)$$
$$= -146.51 + 0.45x$$
$$B = 44.99 + 0.50(x - 500) + 0.04(800) + 2.29(0)$$
$$= -173.01 + 0.5x$$
$$C \leq A$$
$$-215.01 + 0.55x \leq -146.51 + 0.45x$$
$$0.10x \leq 68.5$$
$$x \leq 685$$
$$C \leq B$$
$$-215.01 + 0.55x \leq -173.01 + 0.5x$$
$$0.05x \leq 42$$
$$x \leq 840$$

$C \leq A$ and $C \leq B$ for $500 \leq x \leq 685$.
Now solve $A \leq B$ for x.

$$A \leq B$$
$$-146.51 + 0.45x \leq -173.01 + 0.5x$$
$$26.5 \leq 0.05x$$
$$530 \leq x$$

$A \leq C$ and $A \leq B$ for $x \geq 685$.

Mathematical Questions from Professional Exams

1. Since Akron owns 80% of the capital stock of Benson and 70% of the capital stock of Cashin, the equation is $A_e = 190,000 + 0.8B_e + 0.7C_e$.
The answer is (b).

2. Since Benson owns 15% of the capital stock of Benson, the equation is $B_e = 170,000 + 0.15C_e$.
The answer is (b).

3. From the diagram, $B = 0.15C$.
The answer is (d)

4. We have $\begin{cases} A_e = 190,000 + 0.8B_e + 0.7C_e \\ B_e = 170,000 \qquad\quad + 0.15C_e \\ C_e = 230,000 + 0.25A_e \end{cases}$.

Substitute the expression for C_e into the first two equations, then solve.

$$A_e = 190,000 + 0.8B_e$$
$$+ 0.7(230,000 + 0.25A_e)$$
$$= 351,000 + 0.8B_e + 0.175A_e$$
$$0.825A_e = 351,000 + 0.8B_e \qquad (1)$$
$$B_e = 170,000 + 0.15(230,000 + 0.25A_e)$$
$$= 204,500 + 0.0375A_e \qquad (2)$$

Substitute (2) into (1) to obtain

$$0.825A_e = 351,000 + 0.8(204,500 + 0.0375A_e)$$
$$0.825A_e = 514,600 + 0.03A_e$$
$$0.795A_e = 514,600$$
$$A_e \approx 647,295.60$$

Now use (2) to solve for B_e.

$$B_e = 204,500 + 0.0375(647,295.60)$$
$$\approx 228,773.58$$

Since Akron has an 80% stake in Benson, Benson's minority interest in consolidated net income is $0.2(228,773.58) = \$45,754.72$.

The answer is (c).

Chapter 3 Matrices

3.1 Matrix Algebra

1. False. Subtraction is not commutative.

3. The dimension of the matrix $\begin{bmatrix} 1 & 4 & \sqrt{2} \\ 0 & 0 & 0 \end{bmatrix}$ is

2×3 (number of rows $\times$ number of columns).

5. True. For any two matrices A and B with the same dimensions, $A + B = B + A$.

7. $\begin{bmatrix} 3 & 2 \\ -1 & 3 \end{bmatrix}$ is a 2×2 matrix. It is square.

9. $\begin{bmatrix} 1 & 2 \\ 2 & 1 \\ 0 & -3 \end{bmatrix}$ is a 3×2 matrix.

11. $\begin{bmatrix} 1 & 4 \\ -2 & 8 \\ 0 & 0 \end{bmatrix}$ is 3×2 matrix.

13. $\begin{bmatrix} 4 \\ 1 \end{bmatrix}$ is a 2×1 column matrix.

15. False. To be equal, two matrices must have the same dimension.

17. True

19. True

21. True

23. True

25. False. $\begin{bmatrix} 8 \\ 1 \end{bmatrix} + \begin{bmatrix} 2 \\ 9 \end{bmatrix} = \begin{bmatrix} 10 \\ 10 \end{bmatrix}$

27. $\begin{bmatrix} 3 & -1 \\ 4 & 2 \end{bmatrix} + \begin{bmatrix} -2 & 2 \\ 2 & 5 \end{bmatrix} = \begin{bmatrix} 3+(-2) & (-1)+2 \\ 4+2 & 2+5 \end{bmatrix}$

$= \begin{bmatrix} 1 & 1 \\ 6 & 7 \end{bmatrix}$

29. $3 \begin{bmatrix} 2 & 6 & 0 \\ 4 & -2 & 1 \end{bmatrix} = \begin{bmatrix} 3 \cdot 2 & 3 \cdot 6 & 3 \cdot 0 \\ 3 \cdot 4 & 3 \cdot (-2) & 3 \cdot 1 \end{bmatrix}$

$= \begin{bmatrix} 6 & 18 & 0 \\ 12 & -6 & 3 \end{bmatrix}$

31. $2 \begin{bmatrix} 1 & -1 & 8 \\ 2 & 4 & 1 \end{bmatrix} - 3 \begin{bmatrix} 0 & -2 & 8 \\ 1 & 4 & 1 \end{bmatrix}$

$= \begin{bmatrix} 2 & -2 & 16 \\ 4 & 8 & 2 \end{bmatrix} - \begin{bmatrix} 0 & -6 & 24 \\ 3 & 12 & 3 \end{bmatrix}$

$= \begin{bmatrix} 2-0 & (-2)-(-6) & 16-24 \\ 4-3 & 8-12 & 2-3 \end{bmatrix}$

$= \begin{bmatrix} 2 & 4 & -8 \\ 1 & -4 & -1 \end{bmatrix}$

33. $3 \begin{bmatrix} a & 8 \\ b & 1 \\ c & -2 \end{bmatrix} + 5 \begin{bmatrix} 2a & 6 \\ -b & -2 \\ -c & 0 \end{bmatrix}$

$= \begin{bmatrix} 3a & 24 \\ 3b & 3 \\ 3c & -6 \end{bmatrix} + \begin{bmatrix} 10a & 30 \\ -5b & -10 \\ -5c & 0 \end{bmatrix}$

$= \begin{bmatrix} 3a+10a & 24+30 \\ 3b-5b & 3-10 \\ 3c-5c & -6+0 \end{bmatrix} = \begin{bmatrix} 13a & 54 \\ -2b & -7 \\ -2c & -6 \end{bmatrix}$

35. $A - B = \begin{bmatrix} 2 & -3 & 4 \\ 0 & 2 & 1 \end{bmatrix} - \begin{bmatrix} 1 & -2 & 0 \\ 5 & 1 & 2 \end{bmatrix}$

$= \begin{bmatrix} 2-1 & -3-(-2) & 4-0 \\ 0-5 & 2-1 & 1-2 \end{bmatrix}$

$= \begin{bmatrix} 1 & -1 & 4 \\ -5 & 1 & -1 \end{bmatrix}$

37. $2A - 3C = 2 \begin{bmatrix} 2 & -3 & 4 \\ 0 & 2 & 1 \end{bmatrix} - 3 \begin{bmatrix} -3 & 0 & 5 \\ 2 & 1 & 3 \end{bmatrix}$

$= \begin{bmatrix} 4 & -6 & 8 \\ 0 & 4 & 2 \end{bmatrix} - \begin{bmatrix} -9 & 0 & 15 \\ 6 & 3 & 9 \end{bmatrix}$

$= \begin{bmatrix} 4+9 & -6+0 & 8-15 \\ 0-6 & 4-3 & 2-9 \end{bmatrix}$

$= \begin{bmatrix} 13 & -6 & -7 \\ -6 & 1 & -7 \end{bmatrix}$

39. $(A + B) - 2C$

$= \left(\begin{bmatrix} 2 & -3 & 4 \\ 0 & 2 & 1 \end{bmatrix} + \begin{bmatrix} 1 & -2 & 0 \\ 5 & 1 & 2 \end{bmatrix} \right) - 2 \begin{bmatrix} -3 & 0 & 5 \\ 2 & 1 & 3 \end{bmatrix}$

$= \begin{bmatrix} 3 & -5 & 4 \\ 5 & 3 & 3 \end{bmatrix} - \begin{bmatrix} -6 & 0 & 10 \\ 4 & 2 & 6 \end{bmatrix}$

$= \begin{bmatrix} 3+6 & -5-0 & 4-10 \\ 5-4 & 3-2 & 3-6 \end{bmatrix} = \begin{bmatrix} 9 & -5 & -6 \\ 1 & 1 & -3 \end{bmatrix}$

41. $3A + 4(B + C) = 3\begin{bmatrix} 2 & -3 & 4 \\ 0 & 2 & 1 \end{bmatrix} + 4\left(\begin{bmatrix} 1 & -2 & 0 \\ 5 & 1 & 2 \end{bmatrix} + \begin{bmatrix} -3 & 0 & 5 \\ 2 & 1 & 3 \end{bmatrix}\right) = \begin{bmatrix} 6 & -9 & 12 \\ 0 & 6 & 3 \end{bmatrix} + 4\begin{bmatrix} -2 & -2 & 5 \\ 7 & 2 & 5 \end{bmatrix}$

$= \begin{bmatrix} 6 & -9 & 12 \\ 0 & 6 & 3 \end{bmatrix} + \begin{bmatrix} -8 & -8 & 20 \\ 28 & 8 & 20 \end{bmatrix} = \begin{bmatrix} 6-8 & -9-8 & 12+20 \\ 0+28 & 6+8 & 3+20 \end{bmatrix} = \begin{bmatrix} -2 & -17 & 32 \\ 28 & 14 & 23 \end{bmatrix}$

43. The commutative property for addition is $A + B = B + A$.

$A + B = \begin{bmatrix} 2 & -3 & 4 \\ 0 & 2 & 1 \end{bmatrix} + \begin{bmatrix} 1 & -2 & 0 \\ 5 & 1 & 2 \end{bmatrix} = \begin{bmatrix} 2+1 & -3-2 & 4+0 \\ 0+5 & 2+1 & 1+2 \end{bmatrix} = \begin{bmatrix} 3 & -5 & 4 \\ 5 & 3 & 3 \end{bmatrix}$

$B + A = \begin{bmatrix} 1 & -2 & 0 \\ 5 & 1 & 2 \end{bmatrix} + \begin{bmatrix} 2 & -3 & 4 \\ 0 & 2 & 1 \end{bmatrix} = \begin{bmatrix} 1+2 & -2-3 & 0+4 \\ 5+0 & 1+2 & 2+1 \end{bmatrix} = \begin{bmatrix} 3 & -5 & 4 \\ 5 & 3 & 3 \end{bmatrix} = A + B$

45. The additive inverse property states that $A + (-A) = 0$.

$A + (-A) = \begin{bmatrix} 2 & -3 & 4 \\ 0 & 2 & 1 \end{bmatrix} + \begin{bmatrix} -2 & 3 & -4 \\ 0 & -2 & -1 \end{bmatrix} = \begin{bmatrix} 2+(-2) & (-3)+3 & 4+(-4) \\ 0+0 & 2+(-2) & 1+(-1) \end{bmatrix} = \begin{bmatrix} 0 & 0 & 0 \\ 0 & 0 & 0 \end{bmatrix} = \mathbf{0}$

47. Property 3 of scalar multiplication states that $(k + h)A = kA + hA$.

$(2+3)B = 5B = 5\begin{bmatrix} 1 & -2 & 0 \\ 5 & 1 & 2 \end{bmatrix} = \begin{bmatrix} 5 & -10 & 0 \\ 25 & 5 & 10 \end{bmatrix}$

$2B + 3B = 2\begin{bmatrix} 1 & -2 & 0 \\ 5 & 1 & 2 \end{bmatrix} + 3\begin{bmatrix} 1 & -2 & 0 \\ 5 & 1 & 2 \end{bmatrix} = \begin{bmatrix} 2 & -4 & 0 \\ 10 & 2 & 4 \end{bmatrix} + \begin{bmatrix} 3 & -6 & 0 \\ 15 & 3 & 6 \end{bmatrix}$

$= \begin{bmatrix} 2+3 & -4-6 & 0+0 \\ 10+15 & 2+3 & 4+6 \end{bmatrix} = \begin{bmatrix} 5 & -10 & 0 \\ 25 & 5 & 10 \end{bmatrix} = 5B$

49. Two matrices are equal if they have the same dimension and if corresponding entries are equal. $\begin{bmatrix} x \\ 4 \end{bmatrix}$ and $\begin{bmatrix} -4 \\ z \end{bmatrix}$ have the same dimension, so the matrices are equal if $x = -4$ and $z = 4$.

51. Two matrices are equal if they have the same dimension and if corresponding entries are equal. $\begin{bmatrix} x-2y & 0 \\ -2 & 6 \end{bmatrix}$

and $\begin{bmatrix} 3 & 0 \\ -2 & x+y \end{bmatrix}$ have the same dimension so the matrices are equal if $x - 2y = 3$ and $x + y = 6$. To find the

values of x and y, solve the system $\begin{cases} x - 2y = 3 \\ x + y = 6 \end{cases}$. Writing the system as an augmented matrix and then using

row operations to put it in reduced row-echelon form, we have

$\begin{bmatrix} 1 & -2 & | & 3 \\ 1 & 1 & | & 6 \end{bmatrix} \xrightarrow{R_2 = -r_1 + r_2} \begin{bmatrix} 1 & -2 & | & 3 \\ 0 & 3 & | & 3 \end{bmatrix} \xrightarrow{R_2 = \frac{1}{3}r_2} \begin{bmatrix} 1 & -2 & | & 3 \\ 0 & 1 & | & 1 \end{bmatrix} \xrightarrow{R_1 = 2r_2 + r_1} \begin{bmatrix} 1 & 0 & | & 5 \\ 0 & 1 & | & 1 \end{bmatrix}$

The matrices are equal when $x = 5$ and $y = 1$.

53. $\begin{bmatrix} 2 & 3 & -4 \end{bmatrix} + \begin{bmatrix} x & 2y & z \end{bmatrix} = \begin{bmatrix} 2+x & 3+2y & -4+z \end{bmatrix} = \begin{bmatrix} 6 & -9 & 2 \end{bmatrix}$

Two matrices are equal if they have the same dimension and if corresponding entries are equal.
$\begin{bmatrix} 2+x & 3+2y & -4+z \end{bmatrix}$ and $\begin{bmatrix} 6 & -9 & 2 \end{bmatrix}$ have the same dimension so the matrices are equal if the

corresponding entries are equal. The values of x, y, and z, solve the system $\begin{cases} x + 2 = 6 \\ 2y + 3 = -9 \\ z - 4 = 2 \end{cases}$.

Thus, $x = 4$, $y = -6$, and $z = 6$.

55.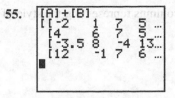

$$A + B = \begin{bmatrix} -2 & 1 & 7 & 5 \\ 4 & 6 & 7 & 5 \\ -3.5 & 8 & -4 & 13 \\ 12 & -1 & 7 & 6 \end{bmatrix}$$

57.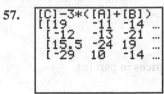

$$C - 3(A + B) = \begin{bmatrix} 19 & -11 & -14 & -15 \\ -12 & -13 & -21 & -17 \\ 15.5 & -24 & 19 & -39 \\ -29 & 10 & -14 & -11 \end{bmatrix}$$

59. The matrix will have dimension 2×3.

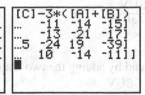

	DEMOCRATS	REPUBLICANS	INDEPENDENTS
UNDER $25,000	351	271	73
OVER $25,000	203	215	55

61. Before constructing the matrix, we need to determine how many students of each gender majored in each of the three areas.
LAS: females: (0.50)(500) = 250
males: (0.50)(500) = 250
ENG: males: (0.75)(300) = 225
females: 300 − 225 = 75
EDUC Majors: 1000 − (500 + 300) = 200
females: (0.60)(200) = 120
males: 200 − 120 = 80

	LAS	ENG	EDUC
MALE	250	225	80
FEMALE	250	75	120

63. Before constructing any matrix, we need to find the number of degrees earned by each gender.
Associate: women:
735,000 − 275,000 =460,000 degrees
Bachelor: men:
1,582,000 − 949,000 =633,000 degrees
Master: men:
693,000 − 418,000 =275,000 degrees
Doctoral: women:
54,900 − 27,600 =27,300 degrees

  a. The data can be displayed in a matrix with dimension 2×4.

	Associate	Bachelor	Master	Doctoral
Male	275,000	633,000	275,000	27,600
Female	460,000	949,000	418,000	27,300

  b. The rows represent the distribution of projected post-secondary degrees for each gender. Row 1 represents the degrees projected to be earned by males, and row 2 represents the projections for females.

The columns represent the distribution of the projected number of each post secondary degree by gender. Column 1 represents associate degrees, column 2 bachelor's degrees, column 3 master's degrees, and column 4 doctoral degrees.

c. The data can also be displayed in a matrix with dimension 4×2.

	Male	Female
Associate	275,000	460,000
Bachelor	633,000	949,000
Master	275,000	418,000
Doctoral	27,600	27,300

d. The rows of this matrix represent the numbers of various post-secondary degrees projected to be earned distributed between the two genders: male (column 1) and female (column 2). Column 1 represents the numbers of the various degrees to be earned by males, and column 2 represents the numbers of the various degrees to be earned by females.

65. Before constructing the matrix, we need to find the numbers of males and females in each type of prison.
Local: female: (0.127)(785,556) = 99,766
male: 785,556 − 99,766 = 685,790
State: male: (0.928)(1,409,166) = 1,307,706
female: 1,409,166 − 1,307,706 = 101,460
Federal: female: (0.066)(201,280) = 13,284
male: 201,280 − 13,284 = 187,996

	Local	State	Federal
Male	685,790	1,307,706	187,996
Female	99,766	101,460	13,284

67.a. A 2 × 3 matrix will have rows representing the location of the dealer and columns representing the style of the cars sold.

January Sales	Subcompacts	Intermediate	SUV
City	350	225	80
Suburban	375	200	75

February Sales	Subcompacts	Intermediate	SUV
City	300	175	40
Suburban	325	150	50

b. Combined sales from January and February are found by adding the two matrices in part (a).

$$\begin{bmatrix} 350 & 225 & 80 \\ 375 & 200 & 75 \end{bmatrix} + \begin{bmatrix} 300 & 175 & 40 \\ 325 & 150 & 50 \end{bmatrix} = \begin{array}{c} \text{Sub} \quad \text{Int} \quad \text{SUV} \\ \begin{bmatrix} 650 & 400 & 120 \\ 700 & 350 & 125 \end{bmatrix} \begin{array}{l} \text{City} \\ \text{Suburban} \end{array} \end{array}$$

c. A 3 × 2 will have the rows representing the style of car sold and the columns representing the location of the dealer.

January Sales	City	Suburban
Subcompact	350	375
Intermediate	225	200
SUV	80	75

February Sales	City	Suburban
Subcompact	300	325
Intermediate	175	150
SUV	40	50

d. Combined sales from January and February are found by adding the two matrices in part (c).

$$\begin{bmatrix} 350 & 375 \\ 225 & 200 \\ 80 & 75 \end{bmatrix} + \begin{bmatrix} 300 & 325 \\ 175 & 150 \\ 40 & 50 \end{bmatrix} = \begin{array}{c} \text{City} \qquad \text{Suburban} \\ \begin{bmatrix} 650 & 700 \\ 400 & 350 \\ 120 & 125 \end{bmatrix} \begin{array}{l} \text{Sub} \\ \text{Int} \\ \text{SUV} \end{array} \end{array}$$

69.a. Before writing any matrices, we determine the distribution of the degrees between the genders.
2008:
Social Science: 7508 degrees; 4387 to females; 7508 − 4387 = 3121 to males,
Humanities: 4721 degrees; 2256 to males; 4721 − 2256 = 2465 to females,
Education: 6577 degrees; 4414 to females; 6577 − 4414 = 2163 to males.
2007:
Social Science: 7193 degrees; 4214 to females; 7193 − 4214 = 2979 to males,
Humanities: 5106 degrees; 2588 to males; 5106 − 2588 = 2518 to females,
Education: 6443 degrees; 4340 to females; 6443 − 4340 = 2103 to males.
In 2 × 3 matrices the rows represent the genders and the columns represent the area of study.

2008	Social Sciences	Humanities	Education		2007	Social Sciences	Humanities	Education
Male	3121	2256	2163		Male	2979	2588	2103
Female	4387	2465	4414		Female	4214	2518	4340

b. The number of doctoral degrees awarded over the two year period is given by

$$\begin{bmatrix} 3121 & 2256 & 2163 \\ 4387 & 2465 & 4414 \end{bmatrix} + \begin{bmatrix} 2979 & 2588 & 2103 \\ 4214 & 2518 & 4340 \end{bmatrix} = \begin{array}{c} \text{Soc Sci} \quad \text{Hum} \quad \text{Educ} \\ \begin{bmatrix} 6100 & 4844 & 4266 \\ 8601 & 4983 & 8754 \end{bmatrix} \begin{array}{l} \text{Male} \\ \text{Female} \end{array} \end{array}$$

Over the two-year period, there were 6100 social sciences doctorates awarded to males and 8601 awarded to females, 4844 humanities doctorates awarded to males and 4983 awarded to females, and 4266 education doctoral degrees awarded to males and 8754 awarded to females.

c. The difference in doctoral degrees awarded between 2007 and 2008 is

$$\begin{bmatrix} 3121 & 2256 & 2163 \\ 4387 & 2465 & 4414 \end{bmatrix} - \begin{bmatrix} 2979 & 2588 & 2103 \\ 4214 & 2518 & 4340 \end{bmatrix} = \begin{array}{c} \begin{array}{c}\text{Social}\\\text{Sciences}\end{array} \quad \text{Humanities} \quad \text{Education} \\ \begin{bmatrix} 142 & -332 & 60 \\ 173 & -53 & 74 \end{bmatrix} \begin{array}{l} \text{Male} \\ \text{Female} \end{array} \end{array}$$

71. The rows of the matrix represent the existence of internet access; the columns represent community type.

	Rural	Suburban	Urban
Access	[0.74	0.77	0.70]

3.2 Multiplication of Matrices

1. True

3. If A is of dimension 3×5 and B is of dimension 5×7, the dimension of the matrix AB is $\underline{3 \times 7}$.

5. True. For example, $AB = \begin{bmatrix} 3 & 2 \\ 4 & 2 \end{bmatrix}\begin{bmatrix} -1 & 1 \\ 2 & -3 \end{bmatrix} = \begin{bmatrix} 1 & -3 \\ 0 & -2 \end{bmatrix}$; $BA = \begin{bmatrix} -1 & 1 \\ 2 & -3 \end{bmatrix}\begin{bmatrix} 3 & 2 \\ 4 & 2 \end{bmatrix} = \begin{bmatrix} 1 & 0 \\ -6 & -2 \end{bmatrix} \neq AB$

7. $\begin{bmatrix} 1 & 3 \end{bmatrix}\begin{bmatrix} 2 \\ 4 \end{bmatrix} = [1(2)+3(4)] = [14]$

9. $\begin{bmatrix} 1 & -2 & 3 \end{bmatrix}\begin{bmatrix} 0 \\ 1 \\ 2 \end{bmatrix} = [1\cdot0+(-2)\cdot1+3\cdot2] = [4]$

11. $\begin{bmatrix} 1 & 4 \end{bmatrix}\begin{bmatrix} 2 & 0 \\ 4 & -2 \end{bmatrix} = [1\cdot2+4\cdot4 \quad 1\cdot0+4\cdot(-2)] = [18 \quad -8]$

13. $\begin{bmatrix} 2 & 0 \\ 4 & -2 \end{bmatrix}\begin{bmatrix} 2 & 1 \\ 3 & -2 \end{bmatrix} = \begin{bmatrix} 2\cdot2+0\cdot3 & 2\cdot1+0\cdot(-2) \\ 4\cdot2+(-2)\cdot3 & 4\cdot1+(-2)(-2) \end{bmatrix} = \begin{bmatrix} 4 & 2 \\ 2 & 8 \end{bmatrix}$

15. $\begin{bmatrix} 1 & -2 & 3 \end{bmatrix}\begin{bmatrix} 0 & 1 \\ 1 & 2 \\ 2 & 3 \end{bmatrix} = [1\cdot0+(-2)\cdot1+3\cdot2 \quad 1\cdot1+(-2)\cdot2+3\cdot3] = [4 \quad 6]$

17. $\begin{bmatrix} 1 & -2 & 3 \\ 4 & 0 & 6 \end{bmatrix}\begin{bmatrix} 0 & -2 \\ 1 & 0 \\ 2 & -4 \end{bmatrix} = \begin{bmatrix} 1\cdot0+(-2)\cdot1+3\cdot2 & 1\cdot(-2)+(-2)\cdot0+3\cdot(-4) \\ 4\cdot0+0\cdot1+6\cdot2 & 4\cdot(-2)+0\cdot0+6\cdot(-4) \end{bmatrix} = \begin{bmatrix} 4 & -14 \\ 12 & -32 \end{bmatrix}$

19. $\begin{bmatrix} 2 & 0 \\ 4 & -2 \\ 6 & -1 \end{bmatrix}\begin{bmatrix} 2 & 1 \\ 3 & -2 \end{bmatrix} = \begin{bmatrix} 2\cdot2+0\cdot3 & 2\cdot1+0\cdot(-2) \\ 4\cdot2+(-2)\cdot3 & 4\cdot1+(-2)(-2) \\ 6\cdot2+(-1)\cdot3 & 6\cdot1+(-1)(-2) \end{bmatrix} = \begin{bmatrix} 4 & 2 \\ 2 & 8 \\ 9 & 8 \end{bmatrix}$

21. $\begin{bmatrix} 1 & -1 & 6 \\ 2 & 0 & -1 \\ 3 & 1 & 2 \end{bmatrix}\begin{bmatrix} 3 & 2 \\ 0 & 1 \\ 1 & 0 \end{bmatrix} = \begin{bmatrix} 1\cdot3+(-1)\cdot0+6\cdot1 & 1\cdot2+(-1)\cdot1+6\cdot0 \\ 2\cdot3+0\cdot0+(-1)\cdot1 & 2\cdot2+0\cdot1+(-1)\cdot0 \\ 3\cdot3+1\cdot0+2\cdot1 & 3\cdot2+1\cdot1+2\cdot0 \end{bmatrix} = \begin{bmatrix} 9 & 1 \\ 5 & 4 \\ 11 & 7 \end{bmatrix}$

23. BA is defined and is a 3×4 matrix.

25. AB is not defined.

27. $(BA)C$ is not defined. (BA is 3×4 and C is 2×3.)

29. $BA + A$ is defined and is a 3×4 matrix.

31. $CB - A$ is not defined. (The dimension of CB is 2×3, and A is a 3×4 matrix. To find the difference between 2 matrices, they must have the same dimension.)

33. $AB = \begin{bmatrix} 1 & 2 \\ 0 & 4 \end{bmatrix}\begin{bmatrix} 1 & 2 & 3 \\ -1 & 4 & -2 \end{bmatrix}$
$= \begin{bmatrix} 1(1)+2(-1) & 1(2)+2(4) & 1(3)+2(-2) \\ 0(1)+4(-1) & 0(2)+4(4) & 0(3)+4(-2) \end{bmatrix}$
$= \begin{bmatrix} 1-2 & 2+8 & 3-4 \\ 0-4 & 0+16 & 0-8 \end{bmatrix} = \begin{bmatrix} -1 & 10 & -1 \\ -4 & 16 & -8 \end{bmatrix}$

35. $BC = \begin{bmatrix} 1 & 2 & 3 \\ -1 & 4 & -2 \end{bmatrix}\begin{bmatrix} 3 & 1 \\ 4 & -1 \\ 0 & 2 \end{bmatrix} = \begin{bmatrix} 1(3)+2(4)+3(0) & 1(1)+2(-1)+3(2) \\ -1(3)+4(4)-2(0) & -1(1)+4(-1)-2(2) \end{bmatrix}$

$\quad\quad = \begin{bmatrix} 3+8+0 & 1-2+6 \\ -3+16+0 & -1-4-4 \end{bmatrix} = \begin{bmatrix} 11 & 5 \\ 13 & -9 \end{bmatrix}$

37. $(D+I_3)C = \left(\begin{bmatrix} 1 & 0 & 4 \\ 0 & 1 & 2 \\ 0 & -1 & 1 \end{bmatrix} + \begin{bmatrix} 1 & 0 & 0 \\ 0 & 1 & 0 \\ 0 & 0 & 1 \end{bmatrix}\right)\begin{bmatrix} 3 & 1 \\ 4 & -1 \\ 0 & 2 \end{bmatrix} = \begin{bmatrix} 2 & 0 & 4 \\ 0 & 2 & 2 \\ 0 & -1 & 2 \end{bmatrix}\begin{bmatrix} 3 & 1 \\ 4 & -1 \\ 0 & 2 \end{bmatrix}$

$\quad\quad = \begin{bmatrix} 2(3)+0(4)+4(0) & 2(1)+0(-1)+4(2) \\ 0(3)+2(4)+2(0) & 0(1)+2(-1)+2(2) \\ 0(3)-1(4)+2(0) & 0(1)-1(-1)+2(2) \end{bmatrix} = \begin{bmatrix} 6+0+0 & 2+0+8 \\ 0+8+0 & 0-2+4 \\ 0-4+0 & 0+1+4 \end{bmatrix} = \begin{bmatrix} 6 & 10 \\ 8 & 2 \\ -4 & 5 \end{bmatrix}$

39. $EI_2 = \begin{bmatrix} 3 & -1 \\ 4 & 2 \end{bmatrix}\begin{bmatrix} 1 & 0 \\ 0 & 1 \end{bmatrix} = \begin{bmatrix} 3(1)+(-1)0 & 3(0)+(-1)1 \\ 4(1)+2(0) & 4(0)+(2)1 \end{bmatrix} = \begin{bmatrix} 3 & -1 \\ 4 & 2 \end{bmatrix}$

41. $(2E)B = 2\begin{bmatrix} 3 & -1 \\ 4 & 2 \end{bmatrix}\begin{bmatrix} 1 & 2 & 3 \\ -1 & 4 & -2 \end{bmatrix} = \begin{bmatrix} 6 & -2 \\ 8 & 4 \end{bmatrix}\begin{bmatrix} 1 & 2 & 3 \\ -1 & 4 & -2 \end{bmatrix}$

$\quad\quad = \begin{bmatrix} 6(1)+-2(-1) & 6(2)+(-2)(4) & 6(3)+(-2)(-2) \\ 8(1)+4(-1) & 8(2)+4(4) & 8(3)+4(-2) \end{bmatrix} = \begin{bmatrix} 8 & 4 & 22 \\ 4 & 32 & 16 \end{bmatrix}$

43. $-5E + A = -5\begin{bmatrix} 3 & -1 \\ 4 & 2 \end{bmatrix} + \begin{bmatrix} 1 & 2 \\ 0 & 4 \end{bmatrix} = \begin{bmatrix} -15 & 5 \\ -20 & -10 \end{bmatrix} + \begin{bmatrix} 1 & 2 \\ 0 & 4 \end{bmatrix} = \begin{bmatrix} -14 & 7 \\ -20 & -6 \end{bmatrix}$

45. $3CB + 4D = 3\begin{bmatrix} 3 & 1 \\ 4 & -1 \\ 0 & 2 \end{bmatrix}\begin{bmatrix} 1 & 2 & 3 \\ -1 & 4 & -2 \end{bmatrix} + 4\begin{bmatrix} 1 & 0 & 4 \\ 0 & 1 & 2 \\ 0 & -1 & 1 \end{bmatrix} = \begin{bmatrix} 9 & 3 \\ 12 & -3 \\ 0 & 6 \end{bmatrix}\begin{bmatrix} 1 & 2 & 3 \\ -1 & 4 & -2 \end{bmatrix} + \begin{bmatrix} 4 & 0 & 16 \\ 0 & 4 & 8 \\ 0 & -4 & 4 \end{bmatrix}$

$\quad\quad = \begin{bmatrix} 9(1)+3(-1) & 9(2)+3(4) & 9(3)+3(-2) \\ 12(1)-3(-1) & 12(2)-3(4) & 12(3)-3(-2) \\ 0(1)+6(-1) & 0(2)+6(4) & 0(3)+6(-2) \end{bmatrix} = \begin{bmatrix} 6 & 30 & 21 \\ 15 & 12 & 42 \\ -6 & 24 & -12 \end{bmatrix} + \begin{bmatrix} 4 & 0 & 16 \\ 0 & 4 & 8 \\ 0 & -4 & 4 \end{bmatrix}$

$\quad\quad = \begin{bmatrix} 10 & 30 & 37 \\ 15 & 16 & 50 \\ -6 & 20 & -8 \end{bmatrix}$

47. First we will find $D(CB)$ and then $(DC)B$ and compare the results.

$CB = \begin{bmatrix} 3 & 1 \\ 4 & -1 \\ 0 & 2 \end{bmatrix}\begin{bmatrix} 1 & 2 & 3 \\ -1 & 4 & -2 \end{bmatrix} = \begin{bmatrix} 3(1)+1(-1) & 3(2)+1(4) & 3(3)+1(-2) \\ 4(1)+(-1)(-1) & 4(2)-1(4) & 4(3)+(-1)(-2) \\ 0(1)+2(-1) & 0(2)+2(4) & 0(3)+2(-2) \end{bmatrix} = \begin{bmatrix} 2 & 10 & 7 \\ 5 & 4 & 14 \\ -2 & 8 & -4 \end{bmatrix}$

$D(CB) = \begin{bmatrix} 1 & 0 & 4 \\ 0 & 1 & 2 \\ 0 & -1 & 1 \end{bmatrix}\begin{bmatrix} 2 & 10 & 7 \\ 5 & 4 & 14 \\ -2 & 8 & -4 \end{bmatrix} = \begin{bmatrix} 1\cdot2+0\cdot5+4(-2) & 1\cdot10+0\cdot4+4\cdot8 & 1\cdot7+0\cdot14+4(-4) \\ 0\cdot2+1\cdot5+2(-2) & 0\cdot10+1\cdot4+2\cdot8 & 0\cdot7+1\cdot14+2(-4) \\ 0\cdot2-1\cdot5+1(-2) & 0\cdot10-1\cdot4+1\cdot8 & 0\cdot7-1\cdot14+1(-4) \end{bmatrix}$

$\quad\quad = \begin{bmatrix} -6 & 42 & -9 \\ 1 & 20 & 6 \\ -7 & 4 & -18 \end{bmatrix}$

$DC = \begin{bmatrix} 1 & 0 & 4 \\ 0 & 1 & 2 \\ 0 & -1 & 1 \end{bmatrix}\begin{bmatrix} 3 & 1 \\ 4 & -1 \\ 0 & 2 \end{bmatrix} = \begin{bmatrix} 1\cdot3+0\cdot4+4\cdot0 & 1\cdot1+0(-1)+4\cdot2 \\ 0\cdot3+1\cdot4+2\cdot0 & 0\cdot1+1(-1)+2\cdot2 \\ 0\cdot3-1\cdot4+1\cdot0 & 0\cdot1-1(-1)+1\cdot2 \end{bmatrix} = \begin{bmatrix} 3 & 9 \\ 4 & 3 \\ -4 & 3 \end{bmatrix}$

$(DC)B = \begin{bmatrix} 3 & 9 \\ 4 & 3 \\ -4 & 3 \end{bmatrix}\begin{bmatrix} 1 & 2 & 3 \\ -1 & 4 & -2 \end{bmatrix} = \begin{bmatrix} 3\cdot1+9(-1) & 3\cdot2+9\cdot4 & 3\cdot3+9(-2) \\ 4\cdot1+3(-1) & 4\cdot2+3\cdot4 & 4\cdot3+3(-2) \\ -4\cdot1+3(-1) & -4\cdot2+3\cdot4 & -4\cdot3+3(-2) \end{bmatrix} = \begin{bmatrix} -6 & 42 & -9 \\ 1 & 20 & 6 \\ -7 & 4 & -18 \end{bmatrix} = D(CB)$

49. $AB = \begin{bmatrix} 1 & -1 \\ 2 & 0 \end{bmatrix}\begin{bmatrix} 3 & 2 \\ -2 & 4 \end{bmatrix} = \begin{bmatrix} 1\cdot3+(-1)\cdot(-2) & 1\cdot2+(-1)\cdot4 \\ 2\cdot3+0\cdot(-2) & 2\cdot2+0\cdot4 \end{bmatrix} = \begin{bmatrix} 5 & -2 \\ 6 & 4 \end{bmatrix}$

$BA = \begin{bmatrix} 3 & 2 \\ -2 & 4 \end{bmatrix}\begin{bmatrix} 1 & -1 \\ 2 & 0 \end{bmatrix} = \begin{bmatrix} 3\cdot1+2\cdot2 & 3\cdot(-1)+2\cdot0 \\ (-2)\cdot1+4\cdot2 & (-2)\cdot(-1)+4\cdot0 \end{bmatrix} = \begin{bmatrix} 7 & -3 \\ 6 & 2 \end{bmatrix} \neq AB$

51. $AB = \begin{bmatrix} 0.5 & 16 & -30 & 25 \\ 21 & 14 & 28 & 64 \\ 19.5 & 8 & -23 & 33 \\ -9.5 & 45 & -9 & 83 \end{bmatrix}$

53. $(AB)C = \begin{bmatrix} 31.5 & 251 & -31.5 & 143 \\ 861 & 350 & 791 & 420 \\ 369.5 & 115 & 206.5 & 215 \\ 412.5 & 882 & 451.5 & 491 \end{bmatrix}$

55. $B(A+C) = \begin{bmatrix} 66 & 74 & 94 & 38 \\ 71 & 13 & 106 & 28 \\ 165 & 124.5 & 79 & 52 \\ 158 & -3 & 152 & 46 \end{bmatrix}$

57. $A(2B-3C) = \begin{bmatrix} -5 & 23 & -102 & 44 \\ -108 & -56 & -70 & 122 \\ 108 & -152 & -67 & 36 \\ -346 & 279 & -249 & 187 \end{bmatrix}$

59. a. Matrix A represents the price at the close of trading on July 7, 2009 of a share of Chevron Corporation, ConocoPhillips, and Exxon Mobil Corporation respectively. The dimension of A is 1×3.

b. $AB = \begin{bmatrix} 62.70 & 39.99 & 66.56 \end{bmatrix}\begin{bmatrix} 100 & 415 & 175 \\ 250 & 90 & 200 \\ 325 & 225 & 130 \end{bmatrix} = \begin{bmatrix} 37,899.50 & 44,595.60 & 27,623.30 \end{bmatrix}$

Matrix AB represents the total amount each of the three people spent on the three stocks on July 7, 2009 at their closing price.

c. $C = \begin{bmatrix} 1 \\ 1 \\ 1 \end{bmatrix}$

$BC = \begin{bmatrix} 100 & 415 & 175 \\ 250 & 90 & 200 \\ 325 & 225 & 130 \end{bmatrix}\begin{bmatrix} 1 \\ 1 \\ 1 \end{bmatrix} = \begin{bmatrix} 100+415+175 \\ 250+90+200 \\ 325+225+130 \end{bmatrix} = \begin{bmatrix} 690 \\ 540 \\ 680 \end{bmatrix}$

Matrix BC represents the total number of shares of Chevron Corporation, ConocoPhillips, and Exxon Mobil Corporation, respectively, purchased by the three funds at the close of trading on July 7, 2009.

61. a.

	Credit Hours JJC CSU			Cost per Credit
$A =$ John	12	4	$B =$ JJC	103
Marsha	8	8	CSU	249

b. A is 2×2.

c. B is 2×1.

d. $AB = \begin{bmatrix} 12 & 4 \\ 8 & 8 \end{bmatrix} \begin{bmatrix} 103 \\ 249 \end{bmatrix} = \begin{bmatrix} 2232 \\ 2816 \end{bmatrix}$

e. AB has dimension 2×1. Its rows are called John and Marsha, and its column is called total cost of tuition.

f. The first entry in AB is the total cost, $2232, John spends on tuition at the two schools, and the second entry is the total cost, $2816, Marsha spends on tuition.

63. a.

	GE	TM	JNJ
Oct 2008	22	73	64
$A =$ Apr 2009	11	78	53
Oct 2009	17	79	61

	Bill	Dan
GE	50	40
$B =$ TM	30	60
JNJ	20	30

b. A has dimension 3×3.

c. B has dimension 3×2.

d. $AB = \begin{bmatrix} 22 & 73 & 64 \\ 11 & 78 & 53 \\ 17 & 79 & 61 \end{bmatrix} \begin{bmatrix} 50 & 40 \\ 30 & 60 \\ 20 & 30 \end{bmatrix} = \begin{bmatrix} 4570 & 7180 \\ 3950 & 6710 \\ 4440 & 7250 \end{bmatrix}$

e. AB has dimension 3×2. The rows are the dates. The columns are named Bill and Dan.

f. The entries in AB represent the value of each man's total investment on each of the particular dates. For example, 4570 tells us that Bill's total investment in October 2008 was worth $4570.

65.

	Number of pants bought	Number of shirts bought	Number of jackets bought
Lee	6	8	2
Chan	2	5	3

	Cost per item
Pants	$25.00
Shirts	$18.00
Jackets	$39.00

$\begin{bmatrix} 6 & 8 & 2 \\ 2 & 5 & 3 \end{bmatrix} \begin{bmatrix} 25 \\ 18 \\ 39 \end{bmatrix} = \begin{bmatrix} (6)(25)+(8)(18)+(2)(39) \\ (2)(25)+(5)(18)+(3)(39) \end{bmatrix} = \begin{bmatrix} 372 \\ 257 \end{bmatrix}$

Lee spent $372 and Chan spent $257.

67. To find the value of the variable, x, first multiply the matrices.

$$\begin{bmatrix} x & 4 & 1 \end{bmatrix} \begin{bmatrix} 2 & 1 & 0 \\ 1 & 0 & 2 \\ 0 & 2 & 4 \end{bmatrix} \begin{bmatrix} x \\ -7 \\ 5/4 \end{bmatrix} = \begin{bmatrix} 2x+4 & x+2 & 12 \end{bmatrix} \begin{bmatrix} x \\ -7 \\ 5/4 \end{bmatrix}$$

$$= (2x+4)x + (x+2)(-7) + 15 = 2x^2 - 3x + 1$$

Since the product of the three matrices is $\mathbf{0}$, set $2x^2 - 3x + 1$ equal to 0 and solve the quadratic equation.

$2x^2 - 3x + 1 = 0$

$(2x-1)(x-1) = 0$

$2x - 1 = 0 \Rightarrow x = \dfrac{1}{2} \quad \bigg| \quad x - 1 = 0 \Rightarrow x = 1$

69. When $A = \begin{bmatrix} a & b \\ c & d \end{bmatrix}$ and $B = \begin{bmatrix} 1 & 1 \\ -1 & 1 \end{bmatrix}$, then $AB = \begin{bmatrix} a-b & a+b \\ c-d & c+d \end{bmatrix}$ and $BA = \begin{bmatrix} a+c & b+d \\ -a+c & -b+d \end{bmatrix}$.

Since $AB = BA$, $a - b = a + c \Rightarrow -b = c$ and $c - d = -a + c \Rightarrow -d = -a \Rightarrow d = a$.

71. $A^2 = A \cdot A = \begin{bmatrix} a & 1-a \\ 1+a & -a \end{bmatrix}\begin{bmatrix} a & 1-a \\ 1+a & -a \end{bmatrix}$

$= \begin{bmatrix} a^2+(1-a)(1+a) & a(1-a)+(1-a)(-a) \\ (1+a)a-a(1+a) & (1+a)(1-a)-a(-a) \end{bmatrix}$

$= \begin{bmatrix} a^2+1-a^2 & a-a^2-a+a^2 \\ a+a^2-a-a^2 & 1-a^2+a^2 \end{bmatrix}$

$= \begin{bmatrix} 1 & 0 \\ 0 & 1 \end{bmatrix} = I_2$

73. $AB = \begin{bmatrix} a & b \\ -b & a \end{bmatrix}\begin{bmatrix} c & d \\ -d & c \end{bmatrix} = \begin{bmatrix} ac-bd & ad+bc \\ -bc-ad & -bd+ac \end{bmatrix}$

$BA = \begin{bmatrix} c & d \\ -d & c \end{bmatrix}\begin{bmatrix} a & b \\ -b & a \end{bmatrix} = \begin{bmatrix} ca-db & cb+da \\ -da-cb & -db+ca \end{bmatrix}$

$= \begin{bmatrix} ac-bd & ad+bc \\ -bc-ad & ac-bd \end{bmatrix} = AB$

Since the entries *a*, *b*, *c*, and *d* represent numbers, the commutative properties of arithmetic can be used in the last matrix of product *BA*.

75. $A^2 = A \cdot A = \begin{bmatrix} 3 & 1 \\ -2 & -1 \end{bmatrix}\begin{bmatrix} 3 & 1 \\ -2 & -1 \end{bmatrix} = \begin{bmatrix} 7 & 2 \\ -4 & -1 \end{bmatrix}$

$A^3 = A \cdot A^2 = \begin{bmatrix} 3 & 1 \\ -2 & -1 \end{bmatrix}\begin{bmatrix} 7 & 2 \\ -4 & -1 \end{bmatrix} = \begin{bmatrix} 17 & 5 \\ -10 & -3 \end{bmatrix}$

$A^4 = A \cdot A^3 = \begin{bmatrix} 3 & 1 \\ -2 & -1 \end{bmatrix}\begin{bmatrix} 17 & 5 \\ -10 & -3 \end{bmatrix} = \begin{bmatrix} 41 & 12 \\ -24 & -7 \end{bmatrix}$

77. $A^2 = A \cdot A = \begin{bmatrix} \frac{1}{2} & \frac{1}{2} \\ \frac{1}{4} & \frac{3}{4} \end{bmatrix}\begin{bmatrix} \frac{1}{2} & \frac{1}{2} \\ \frac{1}{4} & \frac{3}{4} \end{bmatrix} = \begin{bmatrix} \frac{3}{8} & \frac{5}{8} \\ \frac{5}{16} & \frac{11}{16} \end{bmatrix}$

$A^3 = A \cdot A^2 = \begin{bmatrix} \frac{1}{2} & \frac{1}{2} \\ \frac{1}{4} & \frac{3}{4} \end{bmatrix}\begin{bmatrix} \frac{3}{8} & \frac{5}{8} \\ \frac{5}{16} & \frac{11}{16} \end{bmatrix} = \begin{bmatrix} \frac{11}{32} & \frac{21}{32} \\ \frac{21}{64} & \frac{43}{64} \end{bmatrix}$

$A^4 = A \cdot A^3 = \begin{bmatrix} \frac{1}{2} & \frac{1}{2} \\ \frac{1}{4} & \frac{3}{4} \end{bmatrix}\begin{bmatrix} \frac{11}{32} & \frac{21}{32} \\ \frac{21}{64} & \frac{43}{64} \end{bmatrix}$

$= \begin{bmatrix} \frac{43}{128} & \frac{85}{128} \\ \frac{85}{256} & \frac{171}{256} \end{bmatrix}$

79. Answers will vary. However, $A^n = \begin{bmatrix} \frac{1}{3} & \frac{2}{3} \\ \frac{1}{3} & \frac{2}{3} \end{bmatrix}$.

81. $B^2 = \begin{bmatrix} 0.674 & 0.043 & 0.230 & 0.054 \\ 0.382 & 0.096 & 0.212 & 0.310 \\ 0.500 & 0.070 & 0.550 & 0.210 \\ 0.130 & 0.110 & 0.200 & 0.560 \end{bmatrix}$

$B^{10} = \begin{bmatrix} 0.514 & 0.065 & 0.221 & 0.201 \\ 0.491 & 0.068 & 0.220 & 0.222 \\ 0.500 & 0.067 & 0.220 & 0.213 \\ 0.470 & 0.070 & 0.218 & 0.241 \end{bmatrix}$

$B^{15} = \begin{bmatrix} 0.503 & 0.066 & 0.220 & 0.211 \\ 0.498 & 0.067 & 0.220 & 0.215 \\ 0.500 & 0.067 & 0.220 & 0.213 \\ 0.494 & 0.067 & 0.220 & 0.219 \end{bmatrix}$

```
[B]²                    [B]²
[[.674 .043 .23...      ...043 .230 .054]
 [.382 .096 .21...      ...096 .212 .310]
 [.500 .070 .22...      ...070 .220 .210]
 [.130 .110 .20...      ...110 .200 .560]]

[B]^10                  [B]^10
[[.514 .065 .22...      ...065 .221 .201]
 [.491 .068 .22...      ...068 .220 .222]
 [.500 .067 .22...      ...067 .220 .213]
 [.470 .070 .21...      ...070 .218 .241]]

[B]^15                  [B]^15
[[.503 .066 .22...      ...066 .220 .211]
 [.498 .067 .22...      ...067 .220 .215]
 [.500 .067 .22...      ...067 .220 .213]
 [.494 .067 .22...      ...067 .220 .219]]
```

83. Answers will vary. See exercise 49 for an example.

3.3 The Inverse of a Matrix

1. True

3. If *A* and *B* are matrices of dimension $n \times n$, and if $AB = BA = I_n$, then *B* is called the <u>inverse</u> of *A*.

5. Since the product

$\begin{bmatrix} 1 & 2 \\ 2 & 3 \end{bmatrix}\begin{bmatrix} -3 & 2 \\ 2 & -1 \end{bmatrix} = \begin{bmatrix} -3+4 & 2-2 \\ -6+6 & 4-3 \end{bmatrix}$

$= \begin{bmatrix} 1 & 0 \\ 0 & 1 \end{bmatrix} = I_2$

and the product

$\begin{bmatrix} -3 & 2 \\ 2 & -1 \end{bmatrix}\begin{bmatrix} 1 & 2 \\ 2 & 3 \end{bmatrix} = \begin{bmatrix} -3+4 & -6+6 \\ 2-2 & 4-3 \end{bmatrix}$

$= \begin{bmatrix} 1 & 0 \\ 0 & 1 \end{bmatrix} = I_2,$

the matrices are inverses of each other.

7. Since the product $\begin{bmatrix} -1 & -2 \\ 3 & 4 \end{bmatrix} \begin{bmatrix} 2 & 1 \\ -\dfrac{3}{2} & -\dfrac{1}{2} \end{bmatrix} = \begin{bmatrix} -2+3 & -1+1 \\ 6-6 & 3-2 \end{bmatrix} = \begin{bmatrix} 1 & 0 \\ 0 & 1 \end{bmatrix} = I_2$, and the product

$\begin{bmatrix} 2 & 1 \\ -\dfrac{3}{2} & -\dfrac{1}{2} \end{bmatrix} \begin{bmatrix} -1 & -2 \\ 3 & 4 \end{bmatrix} = \begin{bmatrix} 1 & 0 \\ 0 & 1 \end{bmatrix} = I_2$, the matrices are inverses of each other.

9. Since the product $\begin{bmatrix} 1 & 2 & 3 \\ 2 & 3 & 4 \\ 1 & 2 & 1 \end{bmatrix} \begin{bmatrix} -\dfrac{5}{2} & 2 & -\dfrac{1}{2} \\ 1 & -1 & 1 \\ \dfrac{1}{2} & 0 & -\dfrac{1}{2} \end{bmatrix} = \begin{bmatrix} -\dfrac{5}{2}+\dfrac{4}{2}+\dfrac{3}{2} & 2-2+0 & -\dfrac{1}{2}+\dfrac{4}{2}-\dfrac{3}{2} \\ -5+3+2 & 4-3+0 & -1+3-2 \\ -\dfrac{5}{2}+\dfrac{4}{2}+\dfrac{1}{2} & 2-2+0 & -\dfrac{1}{2}+\dfrac{4}{2}-\dfrac{1}{2} \end{bmatrix} = \begin{bmatrix} 1 & 0 & 0 \\ 0 & 1 & 0 \\ 0 & 0 & 1 \end{bmatrix} = I_3$

and the product $\begin{bmatrix} -\dfrac{5}{2} & 2 & -\dfrac{1}{2} \\ 1 & -1 & 1 \\ \dfrac{1}{2} & 0 & -\dfrac{1}{2} \end{bmatrix} \begin{bmatrix} 1 & 2 & 3 \\ 2 & 3 & 4 \\ 1 & 2 & 1 \end{bmatrix} = \begin{bmatrix} 1 & 0 & 0 \\ 0 & 1 & 0 \\ 0 & 0 & 1 \end{bmatrix} = I_3$, the matrices are inverses of each other.

For problems 11–15, first augment the matrix with I_2, then use row operations to obtain the reduced row echelon form of the matrix.

11. $\left[\begin{array}{cc|cc} 3 & 7 & 1 & 0 \\ 2 & 5 & 0 & 1 \end{array}\right] \xrightarrow{R_1=r_1-r_2} \left[\begin{array}{cc|cc} 1 & 2 & 1 & -1 \\ 2 & 5 & 0 & 1 \end{array}\right] \xrightarrow{R_2=-2r_1+r_2} \left[\begin{array}{cc|cc} 1 & 2 & 1 & -1 \\ 0 & 1 & -2 & 3 \end{array}\right] \xrightarrow{R_1=-2r_2+r_1} \left[\begin{array}{cc|cc} 1 & 0 & 5 & -7 \\ 0 & 1 & -2 & 3 \end{array}\right]$

    Since the identity matrix I_2 is on the left side, the matrix on the right is the inverse.

    $\begin{bmatrix} 3 & 7 \\ 2 & 5 \end{bmatrix}^{-1} = \begin{bmatrix} 5 & -7 \\ -2 & 3 \end{bmatrix}$

13. $\left[\begin{array}{cc|cc} 1 & -1 & 1 & 0 \\ 3 & -4 & 0 & 1 \end{array}\right] \xrightarrow{R_2=-3r_1+r_2} \left[\begin{array}{cc|cc} 1 & -1 & 1 & 0 \\ 0 & -1 & -3 & 1 \end{array}\right] \xrightarrow{R_2=-r_2} \left[\begin{array}{cc|cc} 1 & -1 & 1 & 0 \\ 0 & 1 & 3 & -1 \end{array}\right] \xrightarrow{R_1=r_2+r_1} \left[\begin{array}{cc|cc} 1 & 0 & 4 & -1 \\ 0 & 1 & 3 & -1 \end{array}\right]$

    Since the identity matrix I_2 is on the left side, the matrix on the right is the inverse.

    $\begin{bmatrix} 1 & -1 \\ 3 & -4 \end{bmatrix}^{-1} = \begin{bmatrix} 4 & -1 \\ 3 & -1 \end{bmatrix}$

15. $\left[\begin{array}{cc|cc} 2 & 1 & 1 & 0 \\ 4 & 3 & 0 & 1 \end{array}\right] \xrightarrow{R_1=\frac{1}{2}r_1} \left[\begin{array}{cc|cc} 1 & \dfrac{1}{2} & \dfrac{1}{2} & 0 \\ 4 & 3 & 0 & 1 \end{array}\right] \xrightarrow{R_2=-4r_1+r_2} \left[\begin{array}{cc|cc} 1 & \dfrac{1}{2} & \dfrac{1}{2} & 0 \\ 0 & 1 & -2 & 1 \end{array}\right] \xrightarrow{R_1=-\frac{1}{2}r_2+r_1} \left[\begin{array}{cc|cc} 1 & 0 & \dfrac{3}{2} & -\dfrac{1}{2} \\ 0 & 1 & -2 & 1 \end{array}\right]$

    Since the identity matrix I_2 is on the left side, the matrix on the right is the inverse.

    $\begin{bmatrix} 2 & 1 \\ 4 & 3 \end{bmatrix}^{-1} = \begin{bmatrix} \dfrac{3}{2} & -\dfrac{1}{2} \\ -2 & 1 \end{bmatrix}$

For problems 17–21, first augment the matrix with I_3, then use row operations to obtain the reduced row echelon form of the matrix.

17. $\left[\begin{array}{ccc|ccc} 0 & 0 & 1 & 1 & 0 & 0 \\ 0 & 1 & 0 & 0 & 1 & 0 \\ 1 & 0 & 0 & 0 & 0 & 1 \end{array}\right] \xrightarrow[\text{rows 1 and 3}]{\text{Interchange}} \left[\begin{array}{ccc|ccc} 1 & 0 & 0 & 0 & 0 & 1 \\ 0 & 1 & 0 & 0 & 1 & 0 \\ 0 & 0 & 1 & 1 & 0 & 0 \end{array}\right]$

    Since the identity matrix I_3 is on the left side, the matrix on the right is the inverse.

    $\begin{bmatrix} 0 & 0 & 1 \\ 0 & 1 & 0 \\ 1 & 0 & 0 \end{bmatrix}^{-1} = \begin{bmatrix} 0 & 0 & 1 \\ 0 & 1 & 0 \\ 1 & 0 & 0 \end{bmatrix}$

19. $\left[\begin{array}{ccc|ccc} 1 & 1 & -1 & 1 & 0 & 0 \\ 3 & -1 & 0 & 0 & 1 & 0 \\ 2 & -3 & 4 & 0 & 0 & 1 \end{array}\right] \xrightarrow[R_3=-2r_1+r_3]{R_2=-3r_1+r_2} \left[\begin{array}{ccc|ccc} 1 & 1 & -1 & 1 & 0 & 0 \\ 0 & -4 & 3 & -3 & 1 & 0 \\ 0 & -5 & 6 & -2 & 0 & 1 \end{array}\right] \xrightarrow{R_2=-\frac{1}{4}r_2} \left[\begin{array}{ccc|ccc} 1 & 1 & -1 & 1 & 0 & 0 \\ 0 & 1 & -\dfrac{3}{4} & \dfrac{3}{4} & -\dfrac{1}{4} & 0 \\ 0 & -5 & 6 & -2 & 0 & 1 \end{array}\right]$

$\xrightarrow[R_3=5r_2+r_3]{R_1=-r_2+r_1} \left[\begin{array}{ccc|ccc} 1 & 0 & -\dfrac{1}{4} & \dfrac{1}{4} & \dfrac{1}{4} & 0 \\ 0 & 1 & -\dfrac{3}{4} & \dfrac{3}{4} & -\dfrac{1}{4} & 0 \\ 0 & 0 & \dfrac{9}{4} & \dfrac{7}{4} & -\dfrac{5}{4} & 1 \end{array}\right] \xrightarrow{R_3=\frac{4}{9}r_3} \left[\begin{array}{ccc|ccc} 1 & 0 & -\dfrac{1}{4} & \dfrac{1}{4} & \dfrac{1}{4} & 0 \\ 0 & 1 & -\dfrac{3}{4} & \dfrac{3}{4} & -\dfrac{1}{4} & 0 \\ 0 & 0 & 1 & \dfrac{7}{9} & -\dfrac{5}{9} & \dfrac{4}{9} \end{array}\right]$

$\xrightarrow[R_2=\frac{3}{4}r_3+r_2]{R_1=\frac{1}{4}r_3+r_1} \left[\begin{array}{ccc|ccc} 1 & 0 & 0 & \dfrac{4}{9} & \dfrac{1}{9} & \dfrac{1}{9} \\ 0 & 1 & 0 & \dfrac{4}{3} & -\dfrac{2}{3} & \dfrac{1}{3} \\ 0 & 0 & 1 & \dfrac{7}{9} & -\dfrac{5}{9} & \dfrac{4}{9} \end{array}\right]$

Since the identity matrix I_3 is on the left side, the matrix on the right is the inverse.

$$\left[\begin{array}{ccc} 1 & 1 & -1 \\ 3 & -1 & 0 \\ 2 & -3 & 4 \end{array}\right]^{-1} = \left[\begin{array}{ccc} \dfrac{4}{9} & \dfrac{1}{9} & \dfrac{1}{9} \\ \dfrac{4}{3} & -\dfrac{2}{3} & \dfrac{1}{3} \\ \dfrac{7}{9} & -\dfrac{5}{9} & \dfrac{4}{9} \end{array}\right]$$

21. $\left[\begin{array}{ccc|ccc} 1 & 1 & -1 & 1 & 0 & 0 \\ 2 & 1 & 1 & 0 & 1 & 0 \\ 1 & 0 & 1 & 0 & 0 & 1 \end{array}\right] \xrightarrow[R_3=-r_1+r_3]{R_2=-2r_1+r_2} \left[\begin{array}{ccc|ccc} 1 & 1 & -1 & 1 & 0 & 0 \\ 0 & -1 & 3 & -2 & 1 & 0 \\ 0 & -1 & 2 & -1 & 0 & 1 \end{array}\right] \xrightarrow{R_2=-r_2} \left[\begin{array}{ccc|ccc} 1 & 1 & -1 & 1 & 0 & 0 \\ 0 & 1 & -3 & 2 & -1 & 0 \\ 0 & -1 & 2 & -1 & 0 & 1 \end{array}\right]$

$\xrightarrow[R_3=r_2+r_3]{R_1=-r_2+r_1} \left[\begin{array}{ccc|ccc} 1 & 0 & 2 & -1 & 1 & 0 \\ 0 & 1 & -3 & 2 & -1 & 0 \\ 0 & 0 & -1 & 1 & -1 & 1 \end{array}\right] \xrightarrow{R_3=-r_3} \left[\begin{array}{ccc|ccc} 1 & 0 & 2 & -1 & 1 & 0 \\ 0 & 1 & -3 & 2 & -1 & 0 \\ 0 & 0 & 1 & -1 & 1 & -1 \end{array}\right]$

$\xrightarrow[R_2=3r_3+r_2]{R_1=-2r_3+r_1} \left[\begin{array}{ccc|ccc} 1 & 0 & 0 & 1 & -1 & 2 \\ 0 & 1 & 0 & -1 & 2 & -3 \\ 0 & 0 & 1 & -1 & 1 & -1 \end{array}\right]$

Since the identity matrix I_3 is on the left side, the matrix on the right is the inverse.

$$\left[\begin{array}{ccc} 1 & 1 & -1 \\ 2 & 1 & 1 \\ 1 & 0 & 1 \end{array}\right]^{-1} = \left[\begin{array}{ccc} 1 & -1 & 2 \\ -1 & 2 & -3 \\ -1 & 1 & -1 \end{array}\right]$$

23. First, we augment the matrix with I_4, and then we use row operations to obtain the reduced row echelon form of the matrix.

$\left[\begin{array}{cccc|cccc} 1 & 1 & 0 & 0 & 1 & 0 & 0 & 0 \\ 0 & 1 & -1 & 1 & 0 & 1 & 0 & 0 \\ 1 & -1 & 1 & 1 & 0 & 0 & 1 & 0 \\ 0 & 1 & 0 & -1 & 0 & 0 & 0 & 1 \end{array}\right] \xrightarrow{R_3=-r_1+r_3} \left[\begin{array}{cccc|cccc} 1 & 1 & 0 & 0 & 1 & 0 & 0 & 0 \\ 0 & 1 & -1 & 1 & 0 & 1 & 0 & 0 \\ 0 & -2 & 1 & 1 & -1 & 0 & 1 & 0 \\ 0 & 1 & 0 & -1 & 0 & 0 & 0 & 1 \end{array}\right]$

$\xrightarrow[\substack{R_3=2r_2+r_3 \\ R_4=-r_2+r_4}]{R_1=-r_2+r_1} \left[\begin{array}{cccc|cccc} 1 & 0 & 1 & -1 & 1 & -1 & 0 & 0 \\ 0 & 1 & -1 & 1 & 0 & 1 & 0 & 0 \\ 0 & 0 & -1 & 3 & -1 & 2 & 1 & 0 \\ 0 & 0 & 1 & -2 & 0 & -1 & 0 & 1 \end{array}\right]$

(continued on next page)

(continued)

$$\xrightarrow{R_3=-r_3} \begin{bmatrix} 1 & 0 & 1 & -1 & 1 & -1 & 0 & 0 \\ 0 & 1 & -1 & 1 & 0 & 1 & 0 & 0 \\ 0 & 0 & 1 & -3 & 1 & -2 & -1 & 0 \\ 0 & 0 & 1 & -2 & 0 & -1 & 0 & 1 \end{bmatrix}$$

$$\xrightarrow[\substack{R_1=-r_3+r_1 \\ R_2=r_3+r_2 \\ R_4=-r_3+r_4}]{} \begin{bmatrix} 1 & 0 & 0 & 2 & 0 & 1 & 1 & 0 \\ 0 & 1 & 0 & -2 & 1 & -1 & -1 & 0 \\ 0 & 0 & 1 & -3 & 1 & -2 & -1 & 0 \\ 0 & 0 & 0 & 1 & -1 & 1 & 1 & 1 \end{bmatrix}$$

$$\xrightarrow[\substack{R_1=-2r_4+r_1 \\ R_2=2r_4+r_2 \\ R_3=3r_4+r_3}]{} \begin{bmatrix} 1 & 0 & 0 & 0 & 2 & -1 & -1 & -2 \\ 0 & 1 & 0 & 0 & -1 & 1 & 1 & 2 \\ 0 & 0 & 1 & 0 & -2 & 1 & 2 & 3 \\ 0 & 0 & 0 & 1 & -1 & 1 & 1 & 1 \end{bmatrix}$$

Since the identity matrix I_4 is on the left side, the matrix on the right is the inverse.

$$\begin{bmatrix} 1 & 1 & 0 & 0 \\ 0 & 1 & -1 & 1 \\ 1 & -1 & 1 & 1 \\ 0 & 1 & 0 & -1 \end{bmatrix}^{-1} = \begin{bmatrix} 2 & -1 & -1 & -2 \\ -1 & 1 & 1 & 2 \\ -2 & 1 & 2 & 3 \\ -1 & 1 & 1 & 1 \end{bmatrix}$$

25. First, we augment the matrix with I_2; then we use row operations to obtain the reduced row echelon form of the matrix.

$$\begin{bmatrix} 4 & 6 & 1 & 0 \\ 2 & 3 & 0 & 1 \end{bmatrix} \xrightarrow{R_1=\frac{1}{4}r_1} \begin{bmatrix} 1 & \frac{3}{2} & \frac{1}{4} & 0 \\ 2 & 3 & 0 & 1 \end{bmatrix}$$

$$\xrightarrow{R_2=-2r_1+r_2} \begin{bmatrix} 1 & \frac{3}{2} & \frac{1}{4} & 0 \\ 0 & 0 & -\frac{1}{2} & 1 \end{bmatrix}$$

The 0s in row 2 indicate that we cannot get the identity matrix. This tells us that the original matrix has no inverse.

27. First, we augment the matrix with I_2; then we use row operations to obtain the reduced row echelon form of the matrix.

$$\begin{bmatrix} -8 & 4 & 1 & 0 \\ -4 & 2 & 0 & 1 \end{bmatrix} \xrightarrow{R_1=-\frac{1}{8}r_1} \begin{bmatrix} 1 & -\frac{1}{2} & -\frac{1}{8} & 0 \\ -4 & 2 & 0 & 1 \end{bmatrix}$$

$$\xrightarrow{R_2=4r_1+r_2} \begin{bmatrix} 1 & -\frac{1}{2} & -\frac{1}{8} & 0 \\ 0 & 0 & -\frac{1}{2} & 1 \end{bmatrix}$$

The 0s in row 2 indicate that we cannot get the identity matrix. This tells us that the original matrix has no inverse.

29. Since the original matrix has a row of zeros, it has no inverse.

31. To find the inverse we augment the matrix with I_2, and then use row operations to obtain the reduced row echelon form.

$$\begin{bmatrix} 1 & 1 & 1 & 0 \\ 1 & 2 & 0 & 1 \end{bmatrix} \xrightarrow{R_2=-r_1+r_2} \begin{bmatrix} 1 & 1 & 1 & 0 \\ 0 & 1 & -1 & 1 \end{bmatrix}$$

$$\xrightarrow{R_1=-r_2+r_1} \begin{bmatrix} 1 & 0 & 2 & -1 \\ 0 & 1 & -1 & 1 \end{bmatrix}$$

Since the identity matrix I_2 is on the left side, $\begin{bmatrix} 1 & 1 \\ 1 & 2 \end{bmatrix}^{-1} = \begin{bmatrix} 2 & -1 \\ -1 & 1 \end{bmatrix}$.

33. To find the inverse we augment the matrix with I_2, and then use row operations to obtain the reduced row echelon form.

$$\begin{bmatrix} 3 & 2 & 1 & 0 \\ 6 & 4 & 0 & 1 \end{bmatrix} \xrightarrow{R_1=\frac{1}{3}r_1} \begin{bmatrix} 1 & \frac{2}{3} & \frac{1}{3} & 0 \\ 6 & 4 & 0 & 1 \end{bmatrix}$$

$$\xrightarrow{R_2=-6r_1+r_2} \begin{bmatrix} 1 & \frac{2}{3} & \frac{1}{3} & 0 \\ 0 & 0 & -2 & 1 \end{bmatrix}$$

The 0s in the row 2 tell us we cannot get the identity matrix. This indicates that the original matrix has no inverse.

35. To find A^{-1} we use row operations to obtain the reduced row echelon form of the matrix $[A \mid I_2]$.

$$\begin{bmatrix} 1 & 2 & 1 & 0 \\ 2 & -1 & 0 & 1 \end{bmatrix} \xrightarrow{R_2=-2r_1+r_2} \begin{bmatrix} 1 & 2 & 1 & 0 \\ 0 & -5 & -2 & 1 \end{bmatrix}$$

$$\xrightarrow{R_2=-\frac{1}{5}r_2} \begin{bmatrix} 1 & 2 & 1 & 0 \\ 0 & 1 & \frac{2}{5} & -\frac{1}{5} \end{bmatrix}$$

$$\xrightarrow{R_1=-2r_2+r_1} \begin{bmatrix} 1 & 0 & \frac{1}{5} & \frac{2}{5} \\ 0 & 1 & \frac{2}{5} & -\frac{1}{5} \end{bmatrix}$$

So, $A^{-1} = \begin{bmatrix} \frac{1}{5} & \frac{2}{5} \\ \frac{2}{5} & -\frac{1}{5} \end{bmatrix}$.

To find B^{-1} we use row operations to obtain the reduced row echelon form of the matrix $[B \mid I_2]$.

$$\begin{bmatrix} 1 & 3 & 1 & 0 \\ 2 & 1 & 0 & 1 \end{bmatrix} \xrightarrow{R_2=-2r_1+r_2} \begin{bmatrix} 1 & 3 & 1 & 0 \\ 0 & -5 & -2 & 1 \end{bmatrix}$$

$$\xrightarrow{R_2=-\frac{1}{5}r_2} \begin{bmatrix} 1 & 3 & 1 & 0 \\ 0 & 1 & \frac{2}{5} & -\frac{1}{5} \end{bmatrix}$$

$$\xrightarrow{R_1=-3r_2+r_1} \begin{bmatrix} 1 & 0 & -\frac{1}{5} & \frac{3}{5} \\ 0 & 1 & \frac{2}{5} & -\frac{1}{5} \end{bmatrix}$$

(continued on next page)

(*continued*)

So, $B^{-1} = \begin{bmatrix} -\dfrac{1}{5} & \dfrac{3}{5} \\ \dfrac{2}{5} & -\dfrac{1}{5} \end{bmatrix}$.

$A^{-1} - B^{-1} = \begin{bmatrix} \dfrac{1}{5} & \dfrac{2}{5} \\ \dfrac{2}{5} & -\dfrac{1}{5} \end{bmatrix} - \begin{bmatrix} -\dfrac{1}{5} & \dfrac{3}{5} \\ \dfrac{2}{5} & -\dfrac{1}{5} \end{bmatrix}$

$= \begin{bmatrix} \dfrac{1}{5}+\dfrac{1}{5} & \dfrac{2}{5}-\dfrac{3}{5} \\ \dfrac{2}{5}-\dfrac{2}{5} & -\dfrac{1}{5}+\dfrac{1}{5} \end{bmatrix} = \begin{bmatrix} \dfrac{2}{5} & -\dfrac{1}{5} \\ 0 & 0 \end{bmatrix}$

For exercises 37–49, we used a TI-84 calculator to perform the computations.

37. To solve the system $\begin{cases} x+3y+2z = 2 \\ 2x+7y+3z = 1 \\ x \qquad +6z = 3 \end{cases}$, we

define $A = \begin{bmatrix} 1 & 3 & 2 \\ 2 & 7 & 3 \\ 1 & 0 & 6 \end{bmatrix}$, $X = \begin{bmatrix} x \\ y \\ z \end{bmatrix}$, and $B = \begin{bmatrix} 2 \\ 1 \\ 3 \end{bmatrix}$.

The solution to the system is $X = A^{-1}B$.

$\begin{bmatrix} x \\ y \\ z \end{bmatrix} = \begin{bmatrix} 1 & 3 & 2 \\ 2 & 7 & 3 \\ 1 & 0 & 6 \end{bmatrix}^{-1} \begin{bmatrix} 2 \\ 1 \\ 3 \end{bmatrix}$

$= \begin{bmatrix} 42 & -18 & -5 \\ -9 & 4 & 1 \\ -7 & 3 & 1 \end{bmatrix}\begin{bmatrix} 2 \\ 1 \\ 3 \end{bmatrix} = \begin{bmatrix} 51 \\ -11 \\ -8 \end{bmatrix}$

The solution is $x = 51$, $y = -11$, $z = -8$, or (51, −11, −8).

For Problems 39–43, we use $A = \begin{bmatrix} 3 & 7 \\ 2 & 5 \end{bmatrix}$ *and*

$A^{-1} = \begin{bmatrix} 3 & 7 \\ 2 & 5 \end{bmatrix}^{-1} = \begin{bmatrix} 5 & -7 \\ -2 & 3 \end{bmatrix}$ *from problem 11.*

39. To solve the system $\begin{cases} 3x+7y = 10 \\ 2x+5y = 2 \end{cases}$ we define

$A = \begin{bmatrix} 3 & 7 \\ 2 & 5 \end{bmatrix}$, $X = \begin{bmatrix} x \\ y \end{bmatrix}$, and $B = \begin{bmatrix} 10 \\ 2 \end{bmatrix}$.

The solution to the system is $X = A^{-1}B$.

$\begin{bmatrix} x \\ y \end{bmatrix} = \begin{bmatrix} 3 & 7 \\ 2 & 5 \end{bmatrix}^{-1}\begin{bmatrix} 10 \\ 2 \end{bmatrix} = \begin{bmatrix} 5 & -7 \\ -2 & 3 \end{bmatrix}\begin{bmatrix} 10 \\ 2 \end{bmatrix} = \begin{bmatrix} 36 \\ -14 \end{bmatrix}$

The solution is $x = 36$, $y = -14$, or (36, −14).

41. To solve the system $\begin{cases} 3x+7y = 13 \\ 2x+5y = 9 \end{cases}$ we define

$A = \begin{bmatrix} 3 & 7 \\ 2 & 5 \end{bmatrix}$, $X = \begin{bmatrix} x \\ y \end{bmatrix}$, and $B = \begin{bmatrix} 13 \\ 9 \end{bmatrix}$.

The solution to the system is $X = A^{-1}B$.

$\begin{bmatrix} x \\ y \end{bmatrix} = \begin{bmatrix} 3 & 7 \\ 2 & 5 \end{bmatrix}^{-1}\begin{bmatrix} 13 \\ 9 \end{bmatrix} = \begin{bmatrix} 5 & -7 \\ -2 & 3 \end{bmatrix}\begin{bmatrix} 13 \\ 9 \end{bmatrix} = \begin{bmatrix} 2 \\ 1 \end{bmatrix}$

The solution is $x = 2$, $y = 1$, or (2, 1).

43. To solve the system $\begin{cases} 3x+7y = 12 \\ 2x+5y = -4 \end{cases}$ we define

$A = \begin{bmatrix} 3 & 7 \\ 2 & 5 \end{bmatrix}$, $X = \begin{bmatrix} x \\ y \end{bmatrix}$, and $B = \begin{bmatrix} 12 \\ -4 \end{bmatrix}$.

The solution to the system is $X = A^{-1}B$.

$\begin{bmatrix} x \\ y \end{bmatrix} = \begin{bmatrix} 3 & 7 \\ 2 & 5 \end{bmatrix}^{-1}\begin{bmatrix} 12 \\ -4 \end{bmatrix} = \begin{bmatrix} 5 & -7 \\ -2 & 3 \end{bmatrix}\begin{bmatrix} 12 \\ -4 \end{bmatrix} = \begin{bmatrix} 88 \\ -36 \end{bmatrix}$

The solution is $x = 88$, $y = -36$, or (88, −36).

For Problems 45–49, we use $A = \begin{bmatrix} 1 & 1 & -1 \\ 3 & -1 & 0 \\ 2 & -3 & 4 \end{bmatrix}$

and $A^{-1} = \begin{bmatrix} 1 & 1 & -1 \\ 3 & -1 & 0 \\ 2 & -3 & 4 \end{bmatrix}^{-1} = \begin{bmatrix} \dfrac{4}{9} & \dfrac{1}{9} & \dfrac{1}{9} \\ \dfrac{4}{3} & -\dfrac{2}{3} & \dfrac{1}{3} \\ \dfrac{7}{9} & -\dfrac{5}{9} & \dfrac{4}{9} \end{bmatrix}$

from problem 19.

45. To solve $\begin{cases} x+y-z = 3 \\ 3x-y = -4 \\ 2x-3y+4z = 6 \end{cases}$ we define

$A = \begin{bmatrix} 1 & 1 & -1 \\ 3 & -1 & 0 \\ 2 & -3 & 4 \end{bmatrix}$, $X = \begin{bmatrix} x \\ y \\ z \end{bmatrix}$, and $B = \begin{bmatrix} 3 \\ -4 \\ 6 \end{bmatrix}$.

The solution to the system is $X = A^{-1}B$.

$\begin{bmatrix} x \\ y \\ z \end{bmatrix} = \begin{bmatrix} 1 & 1 & -1 \\ 3 & -1 & 0 \\ 2 & -3 & 4 \end{bmatrix}^{-1}\begin{bmatrix} 3 \\ -4 \\ 6 \end{bmatrix}$

$= \begin{bmatrix} \dfrac{4}{9} & \dfrac{1}{9} & \dfrac{1}{9} \\ \dfrac{4}{3} & -\dfrac{2}{3} & \dfrac{1}{3} \\ \dfrac{7}{9} & -\dfrac{5}{9} & \dfrac{4}{9} \end{bmatrix}\begin{bmatrix} 3 \\ -4 \\ 6 \end{bmatrix} = \begin{bmatrix} \dfrac{14}{9} \\ \dfrac{26}{3} \\ \dfrac{65}{9} \end{bmatrix}$

The solution is $x = \dfrac{14}{9}$, $y = \dfrac{26}{3}$, $z = \dfrac{65}{9}$, or $\left(\dfrac{14}{9}, \dfrac{26}{3}, \dfrac{65}{9}\right)$.

47. To solve $\begin{cases} x+ \ y- \ z= \ 12 \\ 3x- \ y \ \ \ \ = -4 \\ 2x-3y+4z= \ 16 \end{cases}$ we define

$$A = \begin{bmatrix} 1 & 1 & -1 \\ 3 & -1 & 0 \\ 2 & -3 & 4 \end{bmatrix}, \ X = \begin{bmatrix} x \\ y \\ z \end{bmatrix}, \text{ and } B = \begin{bmatrix} 12 \\ -4 \\ 16 \end{bmatrix}.$$

The solution to the system is $X = A^{-1}B$.

$$\begin{bmatrix} x \\ y \\ z \end{bmatrix}\begin{bmatrix} 1 & 1 & -1 \\ 3 & -1 & 0 \\ 2 & -3 & 4 \end{bmatrix}^{-1}\begin{bmatrix} 12 \\ -4 \\ 16 \end{bmatrix}$$

$$= \begin{bmatrix} \dfrac{4}{9} & \dfrac{1}{9} & \dfrac{1}{9} \\ \dfrac{4}{3} & -\dfrac{2}{3} & \dfrac{1}{3} \\ \dfrac{7}{9} & -\dfrac{5}{9} & \dfrac{4}{9} \end{bmatrix}\begin{bmatrix} 12 \\ -4 \\ 16 \end{bmatrix} = \begin{bmatrix} \dfrac{60}{9} \\ \dfrac{72}{3} \\ \dfrac{168}{9} \end{bmatrix} = \begin{bmatrix} \dfrac{20}{3} \\ 24 \\ \dfrac{56}{3} \end{bmatrix}$$

The solution is $x = \dfrac{20}{3}$, $y = 24$, $z = \dfrac{56}{3}$, or

$\left(\dfrac{20}{3}, 24, \dfrac{56}{3} \right)$.

49. To solve $\begin{cases} x+ \ y- \ z= \ 0 \\ 3x- \ y \ \ \ \ = -8 \\ 2x-3y+4z= -6 \end{cases}$ we define

$$A = \begin{bmatrix} 1 & 1 & -1 \\ 3 & -1 & 0 \\ 2 & -3 & 4 \end{bmatrix}, \ X = \begin{bmatrix} x \\ y \\ z \end{bmatrix}, \text{ and } B = \begin{bmatrix} 0 \\ -8 \\ -6 \end{bmatrix}.$$

The solution to the system is $X = A^{-1}B$.

$$\begin{bmatrix} x \\ y \\ z \end{bmatrix}\begin{bmatrix} 1 & 1 & -1 \\ 3 & -1 & 0 \\ 2 & -3 & 4 \end{bmatrix}^{-1}\begin{bmatrix} 0 \\ -8 \\ -6 \end{bmatrix}$$

$$= \begin{bmatrix} \dfrac{4}{9} & \dfrac{1}{9} & \dfrac{1}{9} \\ \dfrac{4}{3} & -\dfrac{2}{3} & \dfrac{1}{3} \\ \dfrac{7}{9} & -\dfrac{5}{9} & \dfrac{4}{9} \end{bmatrix}\begin{bmatrix} 0 \\ -8 \\ -6 \end{bmatrix} = \begin{bmatrix} -\dfrac{14}{9} \\ \dfrac{10}{3} \\ \dfrac{16}{9} \end{bmatrix}$$

The solution is $x = -\dfrac{14}{9}$, $y = \dfrac{10}{3}$, $z = \dfrac{16}{9}$, or

$\left(-\dfrac{14}{9}, \dfrac{10}{3}, \dfrac{16}{9} \right)$.

For problems 51–56, we used a TI-84 Plus graphing calculator.

51.

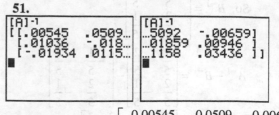

The inverse is $\begin{bmatrix} 0.00545 & 0.0509 & -0.0066 \\ 0.01036 & -0.0186 & 0.0095 \\ -0.0193 & 0.0116 & 0.0344 \end{bmatrix}$

53.

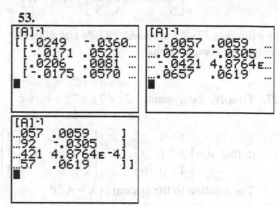

The inverse is

$$\begin{bmatrix} 0.0249 & -0.0360 & -0.0057 & 0.0059 \\ -0.0171 & 0.0521 & 0.0292 & -0.0305 \\ 0.0206 & 0.0081 & -0.0421 & 0.0005 \\ -0.0175 & 0.0570 & 0.0657 & 0.0619 \end{bmatrix}$$

55.

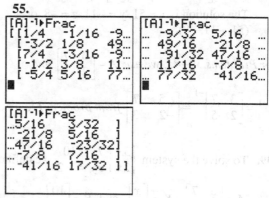

The inverse is $\begin{bmatrix} \dfrac{1}{4} & -\dfrac{1}{16} & -\dfrac{9}{32} & \dfrac{5}{16} & \dfrac{3}{32} \\ -\dfrac{3}{2} & \dfrac{1}{8} & \dfrac{49}{16} & -\dfrac{21}{8} & \dfrac{5}{16} \\ \dfrac{7}{4} & -\dfrac{3}{16} & -\dfrac{91}{32} & \dfrac{47}{16} & -\dfrac{23}{32} \\ -\dfrac{1}{2} & \dfrac{3}{8} & \dfrac{11}{16} & -\dfrac{7}{8} & \dfrac{7}{16} \\ -\dfrac{5}{4} & \dfrac{5}{16} & \dfrac{77}{32} & -\dfrac{41}{16} & \dfrac{17}{32} \end{bmatrix}$

57. To solve the system of equations using a graphing utility (or Excel) first we write the system in the form $AX = B$, where

$$A = \begin{bmatrix} 25 & 61 & -12 \\ 18 & -12 & 7 \\ 3 & 4 & -1 \end{bmatrix}, X = \begin{bmatrix} x \\ y \\ z \end{bmatrix}, \text{ and }$$

$$B = \begin{bmatrix} 10 \\ -9 \\ 12 \end{bmatrix}.$$ Enter matrices A and B into the

graphing utility's matrix editor. Then calculate $A^{-1}B$.

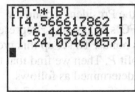

```
[A]⁻¹*[B]
[[4.566617862 ]
 [-6.44363104 ]
 [-24.07467057]]
■
```

So, $x = 4.567$, $y = -6.444$, and $z = -24.075$, or $(4.567, -6.444, -24.075)$.

59. To solve the system of equations using a graphing utility (or Excel) first we write the system in the form $AX = D$, where

$$A = \begin{bmatrix} 25 & 61 & -12 \\ 18 & -12 & 7 \\ 3 & 4 & -1 \end{bmatrix}, X = \begin{bmatrix} x \\ y \\ z \end{bmatrix}, \text{ and }$$

$$D = \begin{bmatrix} 21 \\ 7 \\ -2 \end{bmatrix}.$$ Enter matrices A and D into the

graphing utility's matrix editor. Then calculate $A^{-1}D$.

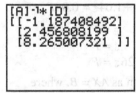

```
[A]⁻¹*[D]
[[-1.187408492]
 [2.456808199 ]
 [8.265007321 ]]
```

So, $x = -1.187$, $y = 2.457$, and $z = 8.265$, or $(-1.187, 2.457, 8.265)$.

61. For each part of the problem, let x denote the amount invested in the Mid Cap Growth Fund at 4.57% and let y denote the amount invested in the bond fund at 6.39%. Then solve the system of equations $\begin{cases} x + y = 3000 \\ 0.0457x + 0.0639y = 3000i \end{cases}$ where i is the average return desired. Therefore, we can write the system as $AX = B$ or

$$\begin{bmatrix} 1 & 1 \\ 0.0457 & 0.0639 \end{bmatrix}\begin{bmatrix} x \\ y \end{bmatrix} = \begin{bmatrix} 3000 \\ 3000i \end{bmatrix} \text{ and } X = A^{-1}B$$

or $\begin{bmatrix} x \\ y \end{bmatrix} = \begin{bmatrix} 1 & 1 \\ 0.0457 & 0.0639 \end{bmatrix}^{-1}\begin{bmatrix} 3000 \\ 3000i \end{bmatrix}.$

Using a TI-84 Plus graphing calculator, we find that $A^{-1} = \begin{bmatrix} 3.5110 & -54.9451 \\ -2.5110 & 54.9451 \end{bmatrix}.$

a. When the average rate of return is 5%, (3000)(0.05) = 150.00 will be earned, and

$$X = \begin{bmatrix} x \\ y \end{bmatrix} = \begin{bmatrix} 3.5110 & -54.9451 \\ -2.5110 & 54.9451 \end{bmatrix}\begin{bmatrix} 3000 \\ 150 \end{bmatrix}$$

$$= \begin{bmatrix} 2291.21 \\ 708.79 \end{bmatrix}$$

You should invest $2291.21 in the Mid Cap fund and $708.79 in the Bond fund.

b. When the average rate of return is 5.5%, (3000)(0.055) = 165.00 will be earned, and

$$X = \begin{bmatrix} x \\ y \end{bmatrix} = \begin{bmatrix} 3.5110 & -54.9451 \\ -2.5110 & 54.9451 \end{bmatrix}\begin{bmatrix} 3000 \\ 165 \end{bmatrix}$$

$$= \begin{bmatrix} 1467.03 \\ 1532.97 \end{bmatrix}$$

You should invest $1467.03 in the Mid Cap fund and $1532.97 in the Bond fund.

c. When the average rate of return is 6%, (3000)(0.05) = 180.00 will be earned, and

$$X = \begin{bmatrix} x \\ y \end{bmatrix} = \begin{bmatrix} 3.5110 & -54.9451 \\ -2.5110 & 54.9451 \end{bmatrix}\begin{bmatrix} 3000 \\ 180 \end{bmatrix}$$

$$= \begin{bmatrix} 642.86 \\ 2357.14 \end{bmatrix}$$

You should invest $642.86 in the Mid Cap fund and $2357.14 in the Bond fund.

63. For each part of the problem, let x denote the amount invested in the Small Cap Growth Fund at 2.56%, let y denote the amount invested in the Large Cap Value Fund at 4.79%, and let z denote the amount invested in the International Fund at 8.73%. Then solve the system of equations

$$\begin{cases} x + y + z = 8000 \\ x = y + 1000 \\ 0.0256x + 0.0479y + 0.0873z = 8000i \end{cases}$$

where i is the average return desired.
We will write the system as $AX = B$, where

$$A = \begin{bmatrix} 1 & 1 & 1 \\ 1 & -1 & 0 \\ 0.0256 & 0.0479 & 0.0873 \end{bmatrix}, X = \begin{bmatrix} x \\ y \\ z \end{bmatrix},$$

and $B = \begin{bmatrix} 8000 \\ 1000 \\ 8000i \end{bmatrix}$.

The solution will be given by $X = A^{-1}B$ or

$$\begin{bmatrix} x \\ y \\ z \end{bmatrix} = \begin{bmatrix} 1 & 1 & 1 \\ 1 & -1 & 0 \\ 0.0256 & 0.0479 & 0.0873 \end{bmatrix}^{-1} \begin{bmatrix} 8000 \\ 1000 \\ 8000i \end{bmatrix}.$$

Using a TI-84 Plus graphing calculator, we find
that $A^{-1} = \begin{bmatrix} 0.8635 & 0.3897 & -9.8912 \\ 0.8635 & -0.6103 & -9.8912 \\ -0.7270 & 0.2206 & 19.7824 \end{bmatrix}$.

a. When the average rate of return is 4%, ($8000)(0.04) = \$320.00$ will be earned, and

$$X = \begin{bmatrix} 0.8635 & 0.3897 & -9.8912 \\ 0.8635 & -0.6103 & -9.8912 \\ -0.7270 & 0.2206 & 19.7824 \end{bmatrix} \begin{bmatrix} 8000 \\ 1000 \\ 320 \end{bmatrix}$$

$$= \begin{bmatrix} 4132.54 \\ 3132.54 \\ 734.92 \end{bmatrix}$$

You should invest $4132.54 in the Small Cap fund, $3132.54 in the Large Cap fund, and $734.92 in the International fund.

b. When the average rate of return is 4.5%, ($8000)(0.045) = \$360.00$ will be earned, and

$$X = \begin{bmatrix} 0.8635 & 0.3897 & -9.8912 \\ 0.8635 & -0.6103 & -9.8912 \\ -0.7270 & 0.2206 & 19.7824 \end{bmatrix} \begin{bmatrix} 8000 \\ 1000 \\ 360 \end{bmatrix}$$

$$= \begin{bmatrix} 3736.89 \\ 2736.89 \\ 1526.21 \end{bmatrix}$$

You should invest $3736.89 in the Small Cap fund, $2736.89 in the Large Cap fund, and $1526.21 in the International fund.

c. When the average rate of return is 5%, ($8000)(0.05) = \$400.00$ will be earned, and

$$X = \begin{bmatrix} 0.8635 & 0.3897 & -9.8912 \\ 0.8635 & -0.6103 & -9.8912 \\ -0.7270 & 0.2206 & 19.7824 \end{bmatrix} \begin{bmatrix} 8000 \\ 1000 \\ 400 \end{bmatrix}$$

$$= \begin{bmatrix} 3341.25 \\ 2341.25 \\ 2317.51 \end{bmatrix}$$

You should invest $3341.25 in the Small Cap fund, $2341.25 in the Large Cap fund, and $2317.51 in the International fund.

65. Let x represent the President's bonus, y represent the CFO's bonus, and z represent the vice-president's bonus. To set up this problem, let's call the profit P. Then we find that the bonuses can be determined as follows.

President: $x = 0.08\left[P - (x + y + z)\right] \Rightarrow$
$x = 0.08P - 0.08x - 0.08y - 0.08z \Rightarrow$
$1.08x + 0.08y + 0.08z = 0.08P$

CFO: $y = 0.05\left[P - (x + y + z)\right] \Rightarrow$
$y = 0.05P - 0.05x - 0.05y - 0.05z \Rightarrow$
$0.05x + 1.05y + 0.05z = 0.05P$

Vice President: $z = 0.04\left[P - (x + y + z)\right] \Rightarrow$
$z = 0.04P - 0.04x - 0.04y - 0.04z \Rightarrow$
$0.04x + 0.04y + 1.04z = 0.04P$

This gives the system of equations

$$\begin{cases} 1.08x + 0.08y + 0.08z = 0.08P \\ 0.05x + 1.05y + 0.05z = 0.05P \text{ or} \\ 0.04x + 0.04y + 1.04z = 0.04P \end{cases}$$

$$\begin{cases} 13.5x + y + z = P \\ x + 21y + z = P \\ x + y + 26z = P \end{cases}$$

Write the system as $AX = B$, where

$$A = \begin{bmatrix} 13.5 & 1 & 1 \\ 1 & 21 & 1 \\ 1 & 1 & 26 \end{bmatrix}, X = \begin{bmatrix} x \\ y \\ z \end{bmatrix}, \text{ and}$$

$$B = \begin{bmatrix} P \\ P \\ P \end{bmatrix} \text{ for each value of } P. \text{ The solution is}$$

given by $X = A^{-1}B$. Using a TI-84 Plus graphing calculator, we find that

$$A^{-1} = \begin{bmatrix} 0.0745 & -0.0034 & -0.0027 \\ -0.0034 & 0.0479 & -0.0017 \\ -0.0027 & -0.0017 & 0.0386 \end{bmatrix}$$

a. If the profits are $300,000, then $P = 300,000$, and the bonuses are given by

$$X = \begin{bmatrix} 0.0745 & -0.0034 & -0.0027 \\ -0.0034 & 0.0479 & -0.0017 \\ -0.0027 & -0.0017 & 0.0386 \end{bmatrix} \begin{bmatrix} 300,000 \\ 300,000 \\ 300,000 \end{bmatrix}$$

$$= \begin{bmatrix} 20,512.82 \\ 12,820.51 \\ 10,256.41 \end{bmatrix}$$

So, when profits equal $300.000, the president's bonus is $20,512.82, the CFO's bonus is $12,820.51, and the vice-president's bonus is $10,256.41.

b. If the profits are $500,000, then $P = 500,000$, and the bonuses are given by

$$X = \begin{bmatrix} 0.0745 & -0.0034 & -0.0027 \\ -0.0034 & 0.0479 & -0.0017 \\ -0.0027 & -0.0017 & 0.0386 \end{bmatrix} \begin{bmatrix} 500,000 \\ 500,000 \\ 500,000 \end{bmatrix}$$

$$= \begin{bmatrix} 34,188.03 \\ 21,367.52 \\ 17,094.02 \end{bmatrix}$$

So when profits equal $500.000, the president's bonus is $34,188.03, the CFO's bonus is $21,367.52, and the vice-president's bonus is $17,094.02.

c. If the profits are $750,000, then $P = 750,000$, and the bonuses are given by

$$X = \begin{bmatrix} 0.0745 & -0.0034 & -0.0027 \\ -0.0034 & 0.0479 & -0.0017 \\ -0.0027 & -0.0017 & 0.0386 \end{bmatrix} \begin{bmatrix} 750,000 \\ 750,000 \\ 750,000 \end{bmatrix}$$

$$= \begin{bmatrix} 51,282.05 \\ 32,051.28 \\ 25,641.03 \end{bmatrix}$$

So when profits equal $750,000, the president's bonus is $51,282.05, the CFO's bonus is $32,051.28, and the vice-president's bonus is $25,641.03.

67. If $A = \begin{bmatrix} 1 & 0 & 0 \\ 3 & 1 & 5 \\ -2 & 0 & 1 \end{bmatrix}$, then $A^{-1} = \begin{bmatrix} 1 & 0 & 0 \\ -13 & 1 & -5 \\ 2 & 0 & 1 \end{bmatrix}$.

To decode the message, we form 3×1 column matrices and multiply on the left by A^{-1}.

$$\begin{bmatrix} 1 & 0 & 0 \\ -13 & 1 & -5 \\ 2 & 0 & 1 \end{bmatrix} \begin{bmatrix} 4 \\ 152 \\ 17 \end{bmatrix} = \begin{bmatrix} 4 \\ 15 \\ 25 \end{bmatrix} \begin{matrix} D \\ O \\ Y \end{matrix}$$

$$\begin{bmatrix} 1 & 0 & 0 \\ -13 & 1 & -5 \\ 2 & 0 & 1 \end{bmatrix} \begin{bmatrix} 15 \\ 106 \\ -22 \end{bmatrix} = \begin{bmatrix} 15 \\ 21 \\ 8 \end{bmatrix} \begin{matrix} O \\ U \\ H \end{matrix}$$

$$\begin{bmatrix} 1 & 0 & 0 \\ -13 & 1 & -5 \\ 2 & 0 & 1 \end{bmatrix} \begin{bmatrix} 1 \\ 50 \\ 3 \end{bmatrix} = \begin{bmatrix} 1 \\ 22 \\ 5 \end{bmatrix} \begin{matrix} A \\ V \\ E \end{matrix}$$

$$\begin{bmatrix} 1 & 0 & 0 \\ -13 & 1 & -5 \\ 2 & 0 & 1 \end{bmatrix} \begin{bmatrix} 1 \\ 52 \\ 7 \end{bmatrix} = \begin{bmatrix} 1 \\ 4 \\ 9 \end{bmatrix} \begin{matrix} A \\ D \\ I \end{matrix}$$

$$\begin{bmatrix} 1 & 0 & 0 \\ -13 & 1 & -5 \\ 2 & 0 & 1 \end{bmatrix} \begin{bmatrix} 19 \\ 195 \\ -12 \end{bmatrix} = \begin{bmatrix} 19 \\ 8 \\ 26 \end{bmatrix} \begin{matrix} S \\ H \\ Z \end{matrix}$$

The message is: Do you have a dish

69. To encode a message using A, assign each letter in the message a number according to the scheme, then group the numbers into 3×1 column matrices, and multiply each on the left by matrix A.

I	H	A	V	E	T	I	C	K	E	T	S	T
9	8	1	22	5	20	9	3	11	5	20	19	20

O	T	H	E	F	I	N	A	L	F	O	U	R
15	20	8	5	6	9	14	1	12	6	15	21	18	26

$$\begin{bmatrix} 1 & 0 & 0 \\ 3 & 1 & 5 \\ -2 & 0 & 1 \end{bmatrix} \begin{bmatrix} 9 \\ 8 \\ 1 \end{bmatrix} = \begin{bmatrix} 9 \\ 40 \\ -17 \end{bmatrix}$$

$$\begin{bmatrix} 1 & 0 & 0 \\ 3 & 1 & 5 \\ -2 & 0 & 1 \end{bmatrix} \begin{bmatrix} 22 \\ 5 \\ 20 \end{bmatrix} = \begin{bmatrix} 22 \\ 171 \\ -24 \end{bmatrix}$$

$$\begin{bmatrix} 1 & 0 & 0 \\ 3 & 1 & 5 \\ -2 & 0 & 1 \end{bmatrix} \begin{bmatrix} 9 \\ 3 \\ 11 \end{bmatrix} = \begin{bmatrix} 9 \\ 85 \\ -7 \end{bmatrix}$$

$$\begin{bmatrix} 1 & 0 & 0 \\ 3 & 1 & 5 \\ -2 & 0 & 1 \end{bmatrix} \begin{bmatrix} 5 \\ 20 \\ 19 \end{bmatrix} = \begin{bmatrix} 5 \\ 130 \\ 9 \end{bmatrix}$$

$$\begin{bmatrix} 1 & 0 & 0 \\ 3 & 1 & 5 \\ -2 & 0 & 1 \end{bmatrix} \begin{bmatrix} 20 \\ 15 \\ 20 \end{bmatrix} = \begin{bmatrix} 20 \\ 175 \\ -20 \end{bmatrix}$$

$$\begin{bmatrix} 1 & 0 & 0 \\ 3 & 1 & 5 \\ -2 & 0 & 1 \end{bmatrix} \begin{bmatrix} 8 \\ 5 \\ 6 \end{bmatrix} = \begin{bmatrix} 8 \\ 59 \\ -10 \end{bmatrix}$$

$$\begin{bmatrix} 1 & 0 & 0 \\ 3 & 1 & 5 \\ -2 & 0 & 1 \end{bmatrix} \begin{bmatrix} 9 \\ 14 \\ 1 \end{bmatrix} = \begin{bmatrix} 9 \\ 46 \\ -17 \end{bmatrix}$$

$$\begin{bmatrix} 1 & 0 & 0 \\ 3 & 1 & 5 \\ -2 & 0 & 1 \end{bmatrix} \begin{bmatrix} 12 \\ 6 \\ 15 \end{bmatrix} = \begin{bmatrix} 12 \\ 117 \\ -9 \end{bmatrix}$$

$$\begin{bmatrix} 1 & 0 & 0 \\ 3 & 1 & 5 \\ -2 & 0 & 1 \end{bmatrix} \begin{bmatrix} 21 \\ 18 \\ 26 \end{bmatrix} = \begin{bmatrix} 21 \\ 211 \\ -16 \end{bmatrix}$$

The encoded message is; 9 40 −17 22 171 −24 9 85 −7 5 130 9 20 175 −20 8 59 −10 9 46 −17 12 117 −9 21 211 −16.

71. To encode a message using A, assign each letter in the message a number according to the scheme, then group the numbers into 3×1 column matrices, and multiply each on the left by matrix A.

a.

I	A	M	G	O	I	N	G
9	1	13	7	15	9	14	7

T	O	D	I	S	N	E	Y
20	15	4	9	19	14	5	25

$$\begin{bmatrix} 1 & 0 & 0 \\ 3 & 1 & 5 \\ -2 & 0 & 1 \end{bmatrix}\begin{bmatrix} 9 \\ 1 \\ 13 \end{bmatrix} = \begin{bmatrix} 9 \\ 93 \\ -5 \end{bmatrix}$$

$$\begin{bmatrix} 1 & 0 & 0 \\ 3 & 1 & 5 \\ -2 & 0 & 1 \end{bmatrix}\begin{bmatrix} 7 \\ 15 \\ 9 \end{bmatrix} = \begin{bmatrix} 7 \\ 81 \\ -5 \end{bmatrix}$$

$$\begin{bmatrix} 1 & 0 & 0 \\ 3 & 1 & 5 \\ -2 & 0 & 1 \end{bmatrix}\begin{bmatrix} 14 \\ 7 \\ 20 \end{bmatrix} = \begin{bmatrix} 14 \\ 149 \\ -8 \end{bmatrix}$$

$$\begin{bmatrix} 1 & 0 & 0 \\ 3 & 1 & 5 \\ -2 & 0 & 1 \end{bmatrix}\begin{bmatrix} 15 \\ 4 \\ 9 \end{bmatrix} = \begin{bmatrix} 15 \\ 94 \\ -21 \end{bmatrix}$$

$$\begin{bmatrix} 1 & 0 & 0 \\ 3 & 1 & 5 \\ -2 & 0 & 1 \end{bmatrix}\begin{bmatrix} 19 \\ 14 \\ 5 \end{bmatrix} = \begin{bmatrix} 19 \\ 96 \\ -33 \end{bmatrix}$$

$$\begin{bmatrix} 1 & 0 & 0 \\ 3 & 1 & 5 \\ -2 & 0 & 1 \end{bmatrix}\begin{bmatrix} 25 \\ 25 \\ 26 \end{bmatrix} = \begin{bmatrix} 25 \\ 230 \\ -24 \end{bmatrix}$$

The encoded message is: 9 93 −5 7 81 −5 14 149 −8 15 94 −21 19 96 −33 25 230 −24.

b.

W	E	S	U	R	F	T	H	E	N	E	T
23	5	19	21	18	6	20	8	5	14	5	20

$$\begin{bmatrix} 1 & 0 & 0 \\ 3 & 1 & 5 \\ -2 & 0 & 1 \end{bmatrix}\begin{bmatrix} 23 \\ 5 \\ 19 \end{bmatrix} = \begin{bmatrix} 23 \\ 169 \\ -27 \end{bmatrix}$$

$$\begin{bmatrix} 1 & 0 & 0 \\ 3 & 1 & 5 \\ -2 & 0 & 1 \end{bmatrix}\begin{bmatrix} 21 \\ 18 \\ 6 \end{bmatrix} = \begin{bmatrix} 21 \\ 111 \\ -36 \end{bmatrix}$$

$$\begin{bmatrix} 1 & 0 & 0 \\ 3 & 1 & 5 \\ -2 & 0 & 1 \end{bmatrix}\begin{bmatrix} 20 \\ 8 \\ 5 \end{bmatrix} = \begin{bmatrix} 20 \\ 93 \\ -35 \end{bmatrix}$$

$$\begin{bmatrix} 1 & 0 & 0 \\ 3 & 1 & 5 \\ -2 & 0 & 1 \end{bmatrix}\begin{bmatrix} 14 \\ 5 \\ 20 \end{bmatrix} = \begin{bmatrix} 14 \\ 147 \\ -8 \end{bmatrix}$$

The encoded message is: 23 169 −27 21 111 −36 20 93 −35 14 147 −8.

c.

L	E	T	S	G	O
12	5	20	19	7	15

C	L	U	B	B	I	N	G
3	12	21	2	2	9	14	7

$$\begin{bmatrix} 1 & 0 & 0 \\ 3 & 1 & 5 \\ -2 & 0 & 1 \end{bmatrix}\begin{bmatrix} 12 \\ 5 \\ 20 \end{bmatrix} = \begin{bmatrix} 12 \\ 141 \\ -4 \end{bmatrix}$$

$$\begin{bmatrix} 1 & 0 & 0 \\ 3 & 1 & 5 \\ -2 & 0 & 1 \end{bmatrix}\begin{bmatrix} 19 \\ 7 \\ 15 \end{bmatrix} = \begin{bmatrix} 19 \\ 139 \\ -23 \end{bmatrix}$$

$$\begin{bmatrix} 1 & 0 & 0 \\ 3 & 1 & 5 \\ -2 & 0 & 1 \end{bmatrix}\begin{bmatrix} 3 \\ 12 \\ 21 \end{bmatrix} = \begin{bmatrix} 3 \\ 126 \\ 15 \end{bmatrix}$$

$$\begin{bmatrix} 1 & 0 & 0 \\ 3 & 1 & 5 \\ -2 & 0 & 1 \end{bmatrix}\begin{bmatrix} 2 \\ 2 \\ 9 \end{bmatrix} = \begin{bmatrix} 2 \\ 53 \\ 5 \end{bmatrix}$$

$$\begin{bmatrix} 1 & 0 & 0 \\ 3 & 1 & 5 \\ -2 & 0 & 1 \end{bmatrix}\begin{bmatrix} 14 \\ 7 \\ 26 \end{bmatrix} = \begin{bmatrix} 14 \\ 179 \\ -2 \end{bmatrix}$$

The encoded message is: 12 141 −4 19 139 −23 3 126 15 2 53 5 14 179 −2.

73. a. First, we write the student ID number as a

4×3 matrix $D = \begin{bmatrix} 8 & 3 & 0 \\ 1 & 9 & 6 \\ 8 & 7 & 0 \\ 2 & 4 & 9 \end{bmatrix}$, then we use K

to encrypt D.

$$E = D \cdot K = \begin{bmatrix} 8 & 3 & 0 \\ 1 & 9 & 6 \\ 8 & 7 & 0 \\ 2 & 4 & 9 \end{bmatrix}\begin{bmatrix} 0 & 0 & 1 \\ 1 & 1 & 1 \\ 0 & 1 & 0 \end{bmatrix}$$

$$= \begin{bmatrix} 0+3+0 & 0+3+0 & 8+3+0 \\ 0+9+0 & 0+9+6 & 1+9+0 \\ 0+7+0 & 0+7+0 & 8+7+0 \\ 0+4+0 & 0+4+9 & 2+4+0 \end{bmatrix}$$

$$= \begin{bmatrix} 3 & 3 & 11 \\ 9 & 15 & 10 \\ 7 & 7 & 15 \\ 4 & 13 & 6 \end{bmatrix}$$

or 3, 3, 11, 9, 15, 10, 7, 7, 15, 4, 13, 6.

b. The encrypted ID, written as a 4×3 matrix

$$E = \begin{bmatrix} 6 & 14 & 10 \\ 2 & 6 & 2 \\ 6 & 7 & 15 \\ 9 & 14 & 17 \end{bmatrix}, \text{ then the original ID}$$

number is found by multiplying E and

$$K^{-1} = \begin{bmatrix} -1 & 1 & -1 \\ 0 & 0 & 1 \\ 1 & 0 & 0 \end{bmatrix}.$$

$$D = E \cdot K^{-1}$$

$$= \begin{bmatrix} 6 & 14 & 10 \\ 2 & 6 & 2 \\ 6 & 7 & 15 \\ 9 & 14 & 17 \end{bmatrix} \begin{bmatrix} -1 & 1 & -1 \\ 0 & 0 & 1 \\ 1 & 0 & 0 \end{bmatrix}$$

$$= \begin{bmatrix} 4 & 6 & 8 \\ 0 & 2 & 4 \\ 9 & 6 & 1 \\ 8 & 9 & 5 \end{bmatrix}$$

or 4, 6, 8, 0, 2, 4, 9, 6, 1, 8, 9, 5.

75. To find the inverse of $\begin{bmatrix} 1 & 2 \\ 2 & 3 \end{bmatrix}$ we first find

$\Delta = ad - bc = 1 \cdot 3 - 2 \cdot 2 = -1$, and then write

the inverse $\begin{bmatrix} 1 & 2 \\ 2 & 3 \end{bmatrix}^{-1} = \begin{bmatrix} \frac{3}{-1} & \frac{-2}{-1} \\ \frac{-2}{-1} & \frac{1}{-1} \end{bmatrix} = \begin{bmatrix} -3 & 2 \\ 2 & -1 \end{bmatrix}.$

77. To find the inverse of $\begin{bmatrix} -1 & -2 \\ 3 & 4 \end{bmatrix}$ we first find

$\Delta = ad - bc = -1 \cdot 4 - (-2) \cdot 3 = 2$, and then

write the inverse

$\begin{bmatrix} -1 & -2 \\ 3 & 4 \end{bmatrix}^{-1} = \begin{bmatrix} \frac{4}{2} & \frac{2}{2} \\ \frac{-3}{2} & \frac{-1}{2} \end{bmatrix} = \begin{bmatrix} 2 & 1 \\ -\frac{3}{2} & -\frac{1}{2} \end{bmatrix}.$

79. Answers will vary.

3.4 Application in Economics (the Leontief Model)

1. We let x represent A's wages, y represent B's wages, and z represent C's wages, which we are told equal \$30,000. The amount paid out by each A, B, and C equals the amount each

receives, so $\begin{bmatrix} x \\ y \\ z \end{bmatrix} = \begin{bmatrix} \frac{1}{2} & \frac{1}{3} & \frac{1}{4} \\ \frac{1}{4} & \frac{1}{3} & \frac{1}{4} \\ \frac{1}{4} & \frac{1}{3} & \frac{1}{2} \end{bmatrix} \begin{bmatrix} x \\ y \\ z \end{bmatrix}$ which

gives the system $\begin{cases} x = \dfrac{1}{2}x + \dfrac{1}{3}y + \dfrac{1}{4}z \\ y = \dfrac{1}{4}x + \dfrac{1}{3}y + \dfrac{1}{4}z \\ z = \dfrac{1}{4}x + \dfrac{1}{3}y + \dfrac{1}{2}z \end{cases}$ or

$$\begin{cases} \dfrac{1}{2}x - \dfrac{1}{3}y - \dfrac{1}{4}z = 0 \\ -\dfrac{1}{4}x + \dfrac{2}{3}y - \dfrac{1}{4}z = 0. \\ -\dfrac{1}{4}x - \dfrac{1}{3}y + \dfrac{1}{2}z = 0 \end{cases}$$

Solving the equations for x, y, and z, we find

$$\begin{bmatrix} \dfrac{1}{2} & -\dfrac{1}{3} & -\dfrac{1}{4} & \Big| & 0 \\ -\dfrac{1}{4} & \dfrac{2}{3} & -\dfrac{1}{4} & \Big| & 0 \\ -\dfrac{1}{4} & -\dfrac{1}{3} & \dfrac{1}{2} & \Big| & 0 \end{bmatrix}$$

$$\xrightarrow[\substack{R_1 = 2r_1 \\ R_2 = \frac{1}{4}r_1 + r_2 \\ R_3 = \frac{1}{4}r_1 + r_3}]{} \begin{bmatrix} 1 & -\dfrac{2}{3} & -\dfrac{1}{2} & \Big| & 0 \\ 0 & \dfrac{1}{2} & -\dfrac{3}{8} & \Big| & 0 \\ 0 & -\dfrac{1}{2} & \dfrac{3}{8} & \Big| & 0 \end{bmatrix}$$

$$\xrightarrow[\substack{R_2 = 2r_2 \\ R_1 = \frac{2}{3}r_2 + r_1 \\ R_3 = \frac{1}{2}r_2 + r_3}]{} \begin{bmatrix} 1 & 0 & -1 & \Big| & 0 \\ 0 & 1 & -\dfrac{3}{4} & \Big| & 0 \\ 0 & 0 & 0 & \Big| & 0 \end{bmatrix}$$

or $x = z$ and $y = \dfrac{3}{4}z$, where z is the parameter.

Since we are told C's wages are \$30,000, we know A's wages are also \$30,000, and B's

wages are $\dfrac{3}{4}(30,000) = \$22,500$.

3. We let x represent A's wages, y represent B's wages, and z represent C's wages, which we are told equal $30,000. The amount paid out by each A, B, and C equals the amount each receives, so

$$\begin{bmatrix} x \\ y \\ z \end{bmatrix} = \begin{bmatrix} 0.2 & 0.3 & 0.1 \\ 0.6 & 0.4 & 0.2 \\ 0.2 & 0.3 & 0.7 \end{bmatrix} \begin{bmatrix} x \\ y \\ z \end{bmatrix}$$ which gives the system $\begin{cases} x = 0.2x + 0.3y + 0.1z \\ y = 0.6x + 0.4y + 0.2z \\ z = 0.2x + 0.3y + 0.7z \end{cases}$ or $\begin{cases} 0.8x - 0.3y - 0.1z = 0 \\ -0.6x + 0.6y - 0.2z = 0 \\ -0.2x - 0.3y + 0.3z = 0 \end{cases}$ or

$\begin{cases} 8x - 3y - 1z = 0 \\ -6x + 6y - 2z = 0 \\ -2x - 3y + 3z = 0 \end{cases}$ Solving the equations for x, y, and z, we find

$$\begin{bmatrix} 8 & -3 & -1 & | & 0 \\ -6 & 6 & -2 & | & 0 \\ -2 & -3 & 3 & | & 0 \end{bmatrix} \xrightarrow[\substack{R_1 = \frac{1}{8}r_1 \\ R_2 = 6r_1 + r_2 \\ R_3 = 2r_1 + r_3}]{} \begin{bmatrix} 1 & -\frac{3}{8} & -\frac{1}{8} & | & 0 \\ 0 & \frac{15}{4} & -\frac{11}{4} & | & 0 \\ 0 & -\frac{15}{4} & \frac{11}{4} & | & 0 \end{bmatrix} \xrightarrow[\substack{R_2 = \frac{4}{15}r_2 \\ R_1 = \frac{3}{8}r_2 + r_1 \\ R_3 = \frac{15}{4}r_2 + r_3}]{} \begin{bmatrix} 1 & 0 & -\frac{2}{5} & | & 0 \\ 0 & 1 & -\frac{11}{15} & | & 0 \\ 0 & 0 & 0 & | & 0 \end{bmatrix}$$

or $x = \dfrac{2}{5}z$ and $y = \dfrac{11}{15}z$, where z is the parameter. Since we are told C's wages are $30,000, we find A's

wages are $\dfrac{2}{5}(30,000) = \$12,000$ and B's wages are $\dfrac{11}{15}(30,000) = \$22,000$.

5. The total output vector X for an open Leontief model is from the system $X = \left[I_3 - A \right]^{-1} D$ where D represents

future demand for the goods produced in the system. If $D_2 = \begin{bmatrix} 80 \\ 90 \\ 60 \end{bmatrix}$, then using the information from Table 2,

we get $X = \begin{bmatrix} 1.6048 & 0.3568 & 0.7131 \\ 0.2946 & 1.3363 & 0.3857 \\ 0.3660 & 0.2721 & 1.4013 \end{bmatrix} \begin{bmatrix} 80 \\ 90 \\ 60 \end{bmatrix} = \begin{bmatrix} 203.28 \\ 166.98 \\ 137.85 \end{bmatrix}$.

The total output of R, S, and T required for the forecast demand D_2 is to produce 203.28 units of product R, 166.98 units of product S, and 137.85 units of product T.

7. We place the information in the 4×4 input/output matrix, A,

$$A: \quad \begin{array}{l} \text{farmer} \\ \text{builder} \\ \text{tailor} \\ \text{rancher} \end{array} \begin{array}{cccc} \text{farmer} & \text{builder} & \text{tailor} & \text{rancher} \\ \begin{bmatrix} 0.3 & 0.3 & 0.3 & 0.2 \\ 0.2 & 0.3 & 0.3 & 0.2 \\ 0.2 & 0.1 & 0.1 & 0.2 \\ 0.3 & 0.3 & 0.3 & 0.4 \end{bmatrix} \end{array}$$

and define the variables x_1 as the farmer's income, x_2 as the builder's income, x_3 as the tailor's income, and x_4 as the rancher's income (which we are told is $25,000). Since in a closed Leontief model the amount paid equals the amount received for each member, we form the system of equations $X = AX$ or $\left(I_4 - A \right) X = 0$.

$\begin{cases} 0.3x_1 + 0.3x_2 + 0.3x_3 + 0.2x_4 = x_1 \\ 0.2x_1 + 0.3x_2 + 0.3x_3 + 0.2x_4 = x_2 \\ 0.2x_1 + 0.1x_2 + 0.1x_3 + 0.2x_4 = x_3 \\ 0.3x_1 + 0.3x_2 + 0.3x_3 + 0.4x_4 = x_4 \end{cases}$ or $\begin{cases} 0.7x_1 - 0.3x_2 - 0.3x_3 - 0.2x_4 = 0 \\ -0.2x_1 + 0.7x_2 - 0.3x_3 - 0.2x_4 = 0 \\ -0.2x_1 - 0.1x_2 + 0.9x_3 - 0.2x_4 = 0 \\ -0.3x_1 - 0.3x_2 - 0.3x_3 + 0.6x_4 = 0 \end{cases}$

(continued on next page)

(continued)

We solve the system of equations for *X*.

$$\begin{bmatrix} 7 & -3 & -3 & -2 & | & 0 \\ -2 & 7 & -3 & -2 & | & 0 \\ -2 & -1 & 9 & -2 & | & 0 \\ -3 & -3 & -3 & 6 & | & 0 \end{bmatrix} \xrightarrow[\substack{R_1=\frac{1}{7}r_1 \\ R_2=2r_1+r_2 \\ R_3=2r_1+r_3 \\ R_4=3r_1+r_4}]{} \begin{bmatrix} 1 & -\frac{3}{7} & -\frac{3}{7} & -\frac{2}{7} & | & 0 \\ 0 & \frac{43}{7} & -\frac{27}{7} & -\frac{18}{7} & | & 0 \\ 0 & -\frac{13}{7} & \frac{57}{7} & -\frac{18}{7} & | & 0 \\ 0 & -\frac{30}{7} & -\frac{30}{7} & \frac{36}{7} & | & 0 \end{bmatrix} \xrightarrow[\substack{R_2=\frac{7}{43}r_2 \\ R_1=\frac{3}{7}r_2+r_1 \\ R_3=\frac{13}{7}r_2+r_3 \\ R_4=\frac{30}{7}r_2+r_4}]{} \begin{bmatrix} 1 & 0 & -\frac{30}{43} & -\frac{20}{43} & | & 0 \\ 0 & 1 & -\frac{27}{43} & -\frac{18}{43} & | & 0 \\ 0 & 0 & \frac{300}{43} & -\frac{144}{43} & | & 0 \\ 0 & 0 & -\frac{300}{43} & \frac{144}{43} & | & 0 \end{bmatrix}$$

$$\xrightarrow[\substack{R_3=\frac{43}{300}r_3 \\ R_1=\frac{30}{43}r_3+r_1 \\ R_2=\frac{27}{43}r_3+r_2 \\ R_4=\frac{300}{43}r_3+r_4}]{} \begin{bmatrix} 1 & 0 & 0 & -\frac{4}{5} & | & 0 \\ 0 & 1 & 0 & -\frac{18}{25} & | & 0 \\ 0 & 0 & 1 & -\frac{12}{25} & | & 0 \\ 0 & 0 & 0 & 0 & | & 0 \end{bmatrix}$$

The solutions to the system are $x_1 = \frac{4}{5}x_4$, $x_2 = \frac{18}{25}x_4$, and $x_3 = \frac{12}{25}x_4$, where x_4 is the parameter. Since the

rancher's income was \$25,000, we get the farmer's income was $\frac{4}{5}(25,000) = \$20,000$; the builder's income

was $\frac{18}{25}(25,000) = \$18,000$; and the tailor's income was $\frac{12}{25}(25,000) = \$12,000$.

9. In an open Leontief model total output must satisfy producer needs and consumer demand. First we construct
matrix *A* which specifies the material needed per unit of production.

$$A = \begin{array}{c} \\ R \\ S \end{array} \overset{\begin{array}{cc} R & S \end{array}}{\begin{bmatrix} \frac{3}{13} & \frac{4}{7} \\ \frac{2}{13} & \frac{1}{7} \end{bmatrix}}$$

We define *X* as the total output of *R* and *S* needed to meet future demand. *AX* + *D* = *X* or

$$X = [I - A]^{-1} D = \begin{bmatrix} \frac{10}{13} & -\frac{4}{7} \\ -\frac{2}{13} & \frac{6}{7} \end{bmatrix}^{-1} \begin{bmatrix} 80 \\ 40 \end{bmatrix} = \begin{bmatrix} \frac{3}{2} & 1 \\ \frac{7}{26} & \frac{35}{26} \end{bmatrix} \begin{bmatrix} 80 \\ 40 \end{bmatrix} = \begin{bmatrix} 160 \\ \frac{980}{13} \end{bmatrix}.$$

To meet future demand, produce 160 units of product *R* and 75.385 units of product *S*. (We get $[I - A]^{-1}$ using
the formula from Section 3.3 Problem 74.)

$$\Delta = \frac{10}{13} \cdot \frac{6}{7} - \left(-\frac{4}{7}\right)\left(-\frac{2}{13}\right) = \frac{60-8}{7 \cdot 13} = \frac{4}{7} \text{ and } (I_2 - A)^{-1} = \begin{bmatrix} \frac{6}{7} \cdot \frac{7}{4} & \frac{4}{7} \cdot \frac{7}{4} \\ \frac{2}{13} \cdot \frac{7}{4} & \frac{10}{13} \cdot \frac{7}{4} \end{bmatrix} = \begin{bmatrix} \frac{3}{2} & 1 \\ \frac{7}{26} & \frac{35}{26} \end{bmatrix}.$$

11. We place the information in the 3×3 input/output matrix, A, in which the columns represent the work done by each of the members and the rows represent the work done for each of the members.

	Physician	Attorney	Financial Planner
Physician	0.3	0.5	0.2
Attorney	0.4	0.2	0.2
Financial Planner	0.3	0.3	0.6

and define the variables x: the physician's income; y: the attorney's income; and z: the financial planner's income. Since in a closed Leontief model the amount paid equals the amount received by each member, we

form the system of equations $\begin{cases} 0.3x + 0.5y + 0.2z = x \\ 0.4x + 0.2y + 0.2z = y \\ 0.3x + 0.3y + 0.6z = z \end{cases}$ and represent it by $X = AX$ or $(I_3 - A)X = 0$ where

$X = \begin{bmatrix} x \\ y \\ z \end{bmatrix}$. We solve the system of equations for X and get

$$\begin{bmatrix} 0.7 & -0.5 & -0.2 & | & 0 \\ -0.4 & 0.8 & -0.2 & | & 0 \\ -0.3 & -0.3 & 0.4 & | & 0 \end{bmatrix} \xrightarrow{A=10A} \begin{bmatrix} 7 & -5 & -2 & | & 0 \\ -4 & 8 & -2 & | & 0 \\ -3 & -3 & 4 & | & 0 \end{bmatrix} \xrightarrow[\text{rows 1 and 3}]{\text{Interchange}} \begin{bmatrix} -3 & -3 & 4 & | & 0 \\ -4 & 8 & -2 & | & 0 \\ 7 & -5 & -2 & | & 0 \end{bmatrix}$$

$$\xrightarrow{R_1 = r_1 - r_2} \begin{bmatrix} 1 & -11 & 6 & | & 0 \\ -4 & 8 & -2 & | & 0 \\ 7 & -5 & -2 & | & 0 \end{bmatrix} \xrightarrow[R_3 = -7r_1 + r_3]{R_2 = 4r_1 + r_2} \begin{bmatrix} 1 & -11 & 6 & | & 0 \\ 0 & -36 & 22 & | & 0 \\ 0 & 72 & -44 & | & 0 \end{bmatrix}$$

$$\xrightarrow{R_2 = -\frac{1}{36}r_2} \begin{bmatrix} 1 & -11 & 6 & | & 0 \\ 0 & 1 & -\frac{11}{18} & | & 0 \\ 0 & 72 & -44 & | & 0 \end{bmatrix} \xrightarrow[R_3 = -72r_2 + r_3]{R_1 = 11r_2 + r_1} \begin{bmatrix} 1 & 0 & -\frac{13}{18} & | & 0 \\ 0 & 1 & -\frac{11}{18} & | & 0 \\ 0 & 0 & 0 & | & 0 \end{bmatrix}$$

The solutions to the system are $x = \frac{13}{18}z$ and $y = \frac{11}{18}z$, where z is the parameter. Since it was decided that

each person's work was valued at approximately \$20,000, if we let the financial planner earn \$27,000, then the physician will earn \$19,500, and the attorney will earn \$16,500. (You can choose a different amount for the financial planner's income and recalculate the incomes of the other two members of the model.)

3.4 Application in Accounting

1. We first find the total costs for the 2 service departments by solving the system $\begin{cases} x_1 = 2000 + \frac{1}{9}x_1 + \frac{3}{9}x_2 \\ x_2 = 1000 + \frac{3}{9}x_1 + \frac{1}{9}x_2 \end{cases}$.

$$X = \begin{bmatrix} x_1 \\ x_2 \end{bmatrix}, \quad C = \begin{bmatrix} \frac{1}{9} & \frac{3}{9} \\ \frac{3}{9} & \frac{1}{9} \end{bmatrix}, \quad D = \begin{bmatrix} 2000 \\ 1000 \end{bmatrix}$$

The system can now be represented by $X = D + CX$, which means the total costs of the two service departments can be obtained by solving $X = \left[I_2 - C \right]^{-1} D$.

$$\left[I_2 - C \right]^{-1} = \begin{bmatrix} \frac{8}{9} & -\frac{3}{9} \\ -\frac{3}{9} & \frac{8}{9} \end{bmatrix}^{-1} = \begin{bmatrix} \frac{72}{55} & \frac{27}{55} \\ \frac{27}{55} & \frac{72}{55} \end{bmatrix}, \text{ so the problem has a solution because } \left[I_2 - C \right]^{-1} \text{ exists and}$$

because both $\left[I_2 - C \right]^{-1}$ and D have only nonnegative entries.

(continued on next page)

(*continued*)

$$X = \begin{bmatrix} \dfrac{72}{55} & \dfrac{27}{55} \\ \dfrac{27}{55} & \dfrac{72}{55} \end{bmatrix} \begin{bmatrix} 2000 \\ 1000 \end{bmatrix} = \begin{bmatrix} 3109.09 \\ 2290.91 \end{bmatrix}.$$

By substituting $x_1 = \$3109.09$ and $x_2 = \$2290.91$ into the table we find all direct and indirect costs.

			Indirect Costs	
Dept.	Total Costs	Direct Costs	S_1	S_2
S_1	$3109.09	$2000	$345.45	$763.64
S_2	$2290.91	$1000	$1036.36	$254.55
P_1	$3354.54	$2500	$345.45	$509.09
P_2	$2790.91	$1500	$1036.36	$254.55
P_3	$3854.54	$3000	$345.45	$509.09
Totals:	$15,399.99	$10,000	$3109.07	$2290.92

Finally, we show that the total of service charges allocated to production departments P_1, P_2, and P_3 equals the sum of the direct costs of the service departments S_1 and S_2.

Service charges allocated to P_1, P_2, P_3:

$345.45 + \$509.09 + \$1036.36 + \$254.55 + \$345.45 + \$509.09 = \$2999.99

Direct costs of S_1 and S_2: $2000 + \$1000 = \$3000

3. The total costs for the 2 service departments come from the two first lines of the table. They are given by the system of equations $\begin{cases} x_1 = \quad 800 + 0.2x_1 + 0.1x_2 \\ x_2 = 4000 + 0.1x_1 + 0.3x_2 \end{cases}$.

$$X = \begin{bmatrix} x_1 \\ x_2 \end{bmatrix}, \quad C = \begin{bmatrix} 0.2 & 0.1 \\ 0.1 & 0.3 \end{bmatrix}, \quad D = \begin{bmatrix} 800 \\ 4000 \end{bmatrix}$$

The system can now be represented by $X = D + CX$, which means the total costs of the two service departments can be obtained by solving $X = \left[I_2 - C \right]^{-1} D$.

$$\left[I_2 - C \right]^{-1} = \begin{bmatrix} 0.8 & -0.1 \\ -0.1 & 0.7 \end{bmatrix}^{-1} = \begin{bmatrix} \dfrac{70}{55} & \dfrac{10}{55} \\ \dfrac{10}{55} & \dfrac{80}{55} \end{bmatrix},$$ so the problem has a solution because $\left[I_2 - C \right]^{-1}$ exists and

because both $\left[I_2 - C \right]^{-1}$ and D have only nonnegative entries.

$$X = \begin{bmatrix} \dfrac{70}{55} & \dfrac{10}{55} \\ \dfrac{10}{55} & \dfrac{80}{55} \end{bmatrix} \begin{bmatrix} 800 \\ 4000 \end{bmatrix} = \begin{bmatrix} 1745.4545 \\ 5963.6364 \end{bmatrix}.$$

(*continued on next page*)

(*continued*)

By substituting $x_1 = \$1745.45$ and $x_2 = \$5963.64$ into the table we find all direct and indirect costs.

Dept.	Total Costs	Direct Costs	Indirect Costs S_1	S_2
S_1	$1745.45	$800	$349.09	$596.36
S_2	$5963.64	$4000	$174.55	$1789.09
P_1	$2445.45	$1500	$349.09	$596.36
P_2	$2216.37	$500	$523.64	$1192.73
P_3	$3338.18	$1200	$349.09	$1789.09
Totals:	$15709.09	$8000	$1745.46	$5963.63

Finally, we show that the total of service charges allocated to production departments P_1, P_2, and P_3 equals the sum of the direct costs of the service departments S_1 and S_2.

Service charges allocated to P_1, P_2, P_3:

$349.09 + \$596.36 + \$523.64 + \$1192.73 + \$349.09 + \$1789.09 = \$4800$

Direct costs of S_1 and S_2: $800 + \$4000 = \$4800$

3.4 Application in Statistics: The Method of Least Squares

In problems 1–5, to find the transpose A^T of matrix A, interchange the rows and the columns of A.

1. $A = \begin{bmatrix} 4 & 1 & 2 \\ 3 & 1 & 0 \end{bmatrix} \Rightarrow A^T = \begin{bmatrix} 4 & 3 \\ 1 & 1 \\ 2 & 0 \end{bmatrix}$

3. $A = \begin{bmatrix} 1 & 11 \\ 0 & 12 \\ 1 & 4 \end{bmatrix} \Rightarrow A^T = \begin{bmatrix} 1 & 0 & 1 \\ 11 & 12 & 4 \end{bmatrix}$

5. $A = \begin{bmatrix} 8 \\ 6 \\ 3 \end{bmatrix} \Rightarrow A^T = \begin{bmatrix} 8 & 6 & 3 \end{bmatrix}$

7.a. $\begin{bmatrix} 1 & 1 & 2 \\ 1 & 0 & 1 \\ 3 & 2 & 3 \end{bmatrix}^T = \begin{bmatrix} 1 & 1 & 3 \\ 1 & 0 & 2 \\ 2 & 1 & 3 \end{bmatrix}$

Since the matrix does not equal its transpose, the matrix is not symmetric.

b. $\begin{bmatrix} 0 & 1 & 3 \\ 1 & 4 & 7 \\ 3 & 7 & 5 \end{bmatrix}^T = \begin{bmatrix} 0 & 1 & 3 \\ 1 & 4 & 7 \\ 3 & 7 & 5 \end{bmatrix}$

Since the matrix equals its transpose, the matrix is symmetric.

c. $\begin{bmatrix} 1 & 2 & 3 & 0 \\ 2 & 4 & 5 & 0 \\ 3 & 5 & 1 & 0 \end{bmatrix}^T = \begin{bmatrix} 1 & 2 & 3 \\ 2 & 4 & 5 \\ 3 & 5 & 1 \\ 0 & 0 & 0 \end{bmatrix}$

Since the matrix does not equal its transpose, the matrix is not symmetric. A symmetric matrix must be square. If A has dimension $n \times m$, then A^T has dimension $m \times n$. If A is symmetric, then $A = A^T$. For two matrices to be equal they must have the same dimensions, so $n = m$.

9. a. To find the least squares line, we define $A = \begin{bmatrix} 3 & 1 \\ 5 & 1 \\ 6 & 1 \\ 7 & 1 \end{bmatrix}$, $Y = \begin{bmatrix} 10 \\ 13 \\ 15 \\ 16 \end{bmatrix}$, and $X = \begin{bmatrix} m \\ b \end{bmatrix}$ and use these matrices to form

the equation $A^TAX = A^TY$.

$$\begin{bmatrix} 3 & 5 & 6 & 7 \\ 1 & 1 & 1 & 1 \end{bmatrix} \begin{bmatrix} 3 & 1 \\ 5 & 1 \\ 6 & 1 \\ 7 & 1 \end{bmatrix} \begin{bmatrix} m \\ b \end{bmatrix} = \begin{bmatrix} 3 & 5 & 6 & 7 \\ 1 & 1 & 1 & 1 \end{bmatrix} \begin{bmatrix} 10 \\ 13 \\ 15 \\ 16 \end{bmatrix}$$

This simplifies to $\begin{bmatrix} 119 & 21 \\ 21 & 4 \end{bmatrix} \begin{bmatrix} m \\ b \end{bmatrix} = \begin{bmatrix} 297 \\ 54 \end{bmatrix}$ which represents the system $\begin{cases} 119m + 21b = 297 \\ 21m + 4b = \ 54 \end{cases}$.

We solve the system of equations for m and b.

$$\begin{bmatrix} 119 & 21 & | & 297 \\ 21 & 4 & | & 54 \end{bmatrix} \xrightarrow[\substack{R_1 = \frac{1}{119}r_1 \\ R_2 = -21r_1 + r_2}]{} \begin{bmatrix} 1 & \frac{3}{17} & | & \frac{297}{119} \\ 0 & \frac{5}{17} & | & \frac{27}{17} \end{bmatrix} \xrightarrow[\substack{R_2 = \frac{17}{5}r_2 \\ R_1 = -\frac{3}{17}r_2 + r_1}]{} \begin{bmatrix} 1 & 0 & | & \frac{54}{35} \\ 0 & 1 & | & \frac{27}{5} \end{bmatrix}$$

Thus $m = \dfrac{54}{35}$ and $b = \dfrac{27}{5}$. The least squares line of best fit is $y = \dfrac{54}{35}x + \dfrac{27}{5}$.

b. The predicted supply when the price \$8.00 per item is $y = \dfrac{54}{35}(8) + \dfrac{27}{5} = 17.7429$. That is, when the price is

\$8, the supply is predicted to be about 17,743 units.

11. To find the least squares line, we define $A = \begin{bmatrix} 10 & 1 \\ 17 & 1 \\ 11 & 1 \\ 18 & 1 \\ 21 & 1 \end{bmatrix}$, $Y = \begin{bmatrix} 50 \\ 61 \\ 55 \\ 60 \\ 70 \end{bmatrix}$, and $X = \begin{bmatrix} m \\ b \end{bmatrix}$ and use these matrices to form

the equation $A^TAX = A^TY$.

$$\begin{bmatrix} 10 & 17 & 11 & 18 & 21 \\ 1 & 1 & 1 & 1 & 1 \end{bmatrix} \begin{bmatrix} 10 & 1 \\ 17 & 1 \\ 11 & 1 \\ 18 & 1 \\ 21 & 1 \end{bmatrix} \begin{bmatrix} m \\ b \end{bmatrix} = \begin{bmatrix} 10 & 17 & 11 & 18 & 21 \\ 1 & 1 & 1 & 1 & 1 \end{bmatrix} \begin{bmatrix} 50 \\ 61 \\ 55 \\ 60 \\ 70 \end{bmatrix}.$$ This simplifies to

$\begin{bmatrix} 1275 & 77 \\ 77 & 5 \end{bmatrix} \begin{bmatrix} m \\ b \end{bmatrix} = \begin{bmatrix} 4692 \\ 296 \end{bmatrix}$ which represents the system $\begin{cases} 1275m + 77b = 4692 \\ 77m + \ 5b = \ 296 \end{cases}$.

Solve the system of equations for m and b.

$$\begin{bmatrix} 1275 & 77 & | & 4692 \\ 77 & 5 & | & 296 \end{bmatrix} \xrightarrow[R_1 = \frac{1}{1275}r_1]{} \begin{bmatrix} 1 & \frac{77}{1275} & | & \frac{92}{25} \\ 77 & 5 & | & 296 \end{bmatrix} \xrightarrow[R_2 = -77r_1 + r_2]{} \begin{bmatrix} 1 & \frac{77}{1275} & | & \frac{92}{25} \\ 0 & \frac{446}{1275} & | & \frac{316}{25} \end{bmatrix}$$

$$\xrightarrow[R_2 = \frac{1275}{446}r_2]{} \begin{bmatrix} 1 & \frac{77}{1275} & | & \frac{92}{25} \\ 0 & 1 & | & \frac{8058}{223} \end{bmatrix} \xrightarrow[R_1 = -\frac{77}{1275}r_2 + r_1]{} \begin{bmatrix} 1 & 0 & | & \frac{334}{223} \\ 0 & 1 & | & \frac{8058}{223} \end{bmatrix}$$

Thus $m = \dfrac{334}{223}$ and $b = \dfrac{8058}{223}$. The least squares line of best fit is $y = \dfrac{334}{223}x + \dfrac{8058}{223} = 1.498x + 36.135$.

13. Define $A = \begin{bmatrix} 0 & 1 \\ 1 & 1 \\ 2 & 1 \\ 3 & 1 \\ 4 & 1 \\ 5 & 1 \\ 6 & 1 \\ 7 & 1 \\ 8 & 1 \\ 9 & 1 \\ 10 & 1 \\ 11 & 1 \\ 12 & 1 \\ 13 & 1 \\ 14 & 1 \end{bmatrix}$, $X = \begin{bmatrix} m \\ b \end{bmatrix}$, $Y = \begin{bmatrix} 8.1 \\ 8.0 \\ 8.8 \\ 9.4 \\ 10.5 \\ 11.1 \\ 12.0 \\ 12.9 \\ 13.6 \\ 14.3 \\ 15.2 \\ 16.3 \\ 16.8 \\ 17.9 \\ 24.0 \end{bmatrix}$

Use these matrices to form the equation $A^TAX = A^TY$. Using a TI-84 Plus graphing calculator, we find that this simplifies to

$$\begin{bmatrix} 1015 & 105 \\ 105 & 15 \end{bmatrix}\begin{bmatrix} m \\ b \end{bmatrix} = \begin{bmatrix} 1652.7 \\ 198.9 \end{bmatrix}$$ which

represents the system

$$\begin{cases} 1015m + 105b = 1652.7 & (1) \\ 105m + 15b = 198.9 & (2) \end{cases}.$$

Using the method of elimination, we have

$$\begin{aligned} 1015m + 105b &= 1652.7 \quad (1) \\ -735m - 105b &= -1392.3 \quad -7 \times (2) \\ \hline 280m &= 260.4 \\ m &= 0.93 \end{aligned}$$

$$105(0.93) + 15b = 198.9$$
$$97.65 + 15b = 198.9$$
$$15b = 101.25$$
$$b = 6.75$$

The least squares line of best fit is $y = 0.93x + 6.75$. Using this line and $t = 17$ to represent the year 2011, we estimate that $(0.93)(17) + 6.75 = 22.56$ million people will be diagnosed with diabetes in 2011.

Chapter 3 Review Exercises

1. $A + C = \begin{bmatrix} 1 & 0 \\ 2 & 4 \\ -1 & 2 \end{bmatrix} + \begin{bmatrix} 3 & -4 \\ 1 & 5 \\ 5 & -2 \end{bmatrix} = \begin{bmatrix} 4 & -4 \\ 3 & 9 \\ 4 & 0 \end{bmatrix}$

The sum has dimension 3×2.

3. $6A = 6 \cdot \begin{bmatrix} 1 & 0 \\ 2 & 4 \\ -1 & 2 \end{bmatrix} = \begin{bmatrix} 6 & 0 \\ 12 & 24 \\ -6 & 12 \end{bmatrix}$

The scalar product has dimension 3×2.

5. $AB = \begin{bmatrix} 1 & 0 \\ 2 & 4 \\ -1 & 2 \end{bmatrix}\begin{bmatrix} 4 & -3 & 0 \\ 1 & 1 & -2 \end{bmatrix}$

$$= \begin{bmatrix} 4+0 & -3+0 & 0+0 \\ 8+4 & -6+4 & 0-8 \\ -4+2 & 3+2 & 0-4 \end{bmatrix}$$

$$= \begin{bmatrix} 4 & -3 & 0 \\ 12 & -2 & -8 \\ -2 & 5 & -4 \end{bmatrix}$$

The matrix product has dimension 3×3.

7. $CB = \begin{bmatrix} 3 & -4 \\ 1 & 5 \\ 5 & -2 \end{bmatrix}\begin{bmatrix} 4 & -3 & 0 \\ 1 & 1 & -2 \end{bmatrix}$

$$= \begin{bmatrix} 12-4 & -9-4 & 0+8 \\ 4+5 & -3+5 & 0-10 \\ 20-2 & -15-2 & 0+4 \end{bmatrix}$$

$$= \begin{bmatrix} 8 & -13 & 8 \\ 9 & 2 & -10 \\ 18 & -17 & 4 \end{bmatrix}$$

The matrix product has dimension 3×3.

9. $(A+C)B = \left(\begin{bmatrix} 1 & 0 \\ 2 & 4 \\ -1 & 2 \end{bmatrix} + \begin{bmatrix} 3 & -4 \\ 1 & 5 \\ 5 & -2 \end{bmatrix} \right)\begin{bmatrix} 4 & -3 & 0 \\ 1 & 1 & -2 \end{bmatrix}$

$$= \begin{bmatrix} 4 & -4 \\ 3 & 9 \\ 4 & 0 \end{bmatrix}\begin{bmatrix} 4 & -3 & 0 \\ 1 & 1 & -2 \end{bmatrix}$$

$$= \begin{bmatrix} 16-4 & -12-4 & 0+8 \\ 12+9 & -9+9 & 0-18 \\ 16+0 & -12+0 & 0+0 \end{bmatrix}$$

$$= \begin{bmatrix} 12 & -16 & 8 \\ 21 & 0 & -18 \\ 16 & -12 & 0 \end{bmatrix}$$

The resulting matrix has dimension 3×3.

11. $3A + 2C = 3\begin{bmatrix} 1 & 0 \\ 2 & 4 \\ -1 & 2 \end{bmatrix} + 2\begin{bmatrix} 3 & -4 \\ 1 & 5 \\ 5 & -2 \end{bmatrix}$

$$= \begin{bmatrix} 3 & 0 \\ 6 & 12 \\ -3 & 6 \end{bmatrix} + \begin{bmatrix} 6 & -8 \\ 2 & 10 \\ 10 & -4 \end{bmatrix} = \begin{bmatrix} 9 & -8 \\ 8 & 22 \\ 7 & 2 \end{bmatrix}$$

The resulting matrix has dimension 3×2.

13. $C = 0 = \begin{bmatrix} 3 & -4 \\ 1 & 5 \\ 5 & -2 \end{bmatrix} + \begin{bmatrix} 0 & 0 \\ 0 & 0 \\ 0 & 0 \end{bmatrix} = \begin{bmatrix} 3 & -4 \\ 1 & 5 \\ 5 & -2 \end{bmatrix} = C$

The resulting matrix has dimension 3×2.

15. $AB = \begin{bmatrix} 1 & 0 \\ 2 & 4 \\ -1 & 2 \end{bmatrix} \begin{bmatrix} 4 & -3 & 0 \\ 1 & 1 & -2 \end{bmatrix}$

$= \begin{bmatrix} 1(4)+0(1) & 1(-3)+0(1) & 1(0)+0(-2) \\ 2(4)+4(1) & 2(-3)+4(1) & 2(0)+4(-2) \\ -1(4)+2(1) & -1(-3)+2(1) & -1(0)+2(-2) \end{bmatrix}$

$= \begin{bmatrix} 4 & -3 & 0 \\ 12 & -2 & -8 \\ -2 & 5 & -4 \end{bmatrix}$

$CB = \begin{bmatrix} 3 & -4 \\ 1 & 5 \\ 5 & -2 \end{bmatrix} \begin{bmatrix} 4 & -3 & 0 \\ 1 & 1 & -2 \end{bmatrix}$

$= \begin{bmatrix} 3(4)-4(1) & 3(-3)-4(1) & 3(0)-4(-2) \\ 1(4)+5(1) & 1(-3)+5(1) & 1(0)+5(-2) \\ 5(4)-2(1) & 5(-3)-2(1) & 5(0)-2(-2) \end{bmatrix}$

$= \begin{bmatrix} 8 & -13 & 8 \\ 9 & 2 & -10 \\ 18 & -17 & 4 \end{bmatrix}$

$AB + BC = \begin{bmatrix} 4 & -3 & 0 \\ 12 & -2 & -8 \\ -2 & 5 & -4 \end{bmatrix} + \begin{bmatrix} 8 & -13 & 8 \\ 9 & 2 & -10 \\ 18 & -17 & 4 \end{bmatrix}$

$= \begin{bmatrix} 12 & -16 & 8 \\ 21 & 0 & -18 \\ 16 & -12 & 0 \end{bmatrix}$

The resulting matrix has dimension 3 × 3.

For problems 17–19, we use the method described in problem 74 in section 3.3.

$\begin{bmatrix} a & b \\ c & d \end{bmatrix}^{-1} = \begin{bmatrix} \dfrac{d}{ad-bc} & -\dfrac{b}{ad-bc} \\ -\dfrac{c}{ad-bc} & \dfrac{a}{ad-bc} \end{bmatrix}$

provided $ad - bc \neq 0$.

17. $ad - bc = 3(1) - 0(-2) = 3.$

Thus, $\begin{bmatrix} 3 & 0 \\ -2 & 1 \end{bmatrix}^{-1} = \begin{bmatrix} \dfrac{1}{3} & 0 \\ \dfrac{2}{3} & 1 \end{bmatrix}.$

19. $ad - bc = 4(3) - 2(6) = 0.$

Therefore, the inverse does not exist.

21. Augment the matrix with I_3, then use row operations to write the matrix in reduced row echelon form.

$\begin{bmatrix} 4 & 3 & -1 & | & 1 & 0 & 0 \\ 0 & 2 & 2 & | & 0 & 1 & 0 \\ 3 & -1 & 0 & | & 0 & 0 & 1 \end{bmatrix}$

$\xrightarrow{R_1 = r_1 - r_3} \begin{bmatrix} 1 & 4 & -1 & | & 1 & 0 & -1 \\ 0 & 2 & 2 & | & 0 & 1 & 0 \\ 3 & -1 & 0 & | & 0 & 0 & 1 \end{bmatrix}$

$\xrightarrow{R_3 = -3r_1 + r_3} \begin{bmatrix} 1 & 4 & -1 & | & 1 & 0 & -1 \\ 0 & 2 & 2 & | & 0 & 1 & 0 \\ 0 & -13 & 3 & | & -3 & 0 & 4 \end{bmatrix}$

$\xrightarrow[\substack{R_1 = -4r_2 + r_1 \\ R_3 = 13r_2 + r_3}]{R_2 = \frac{1}{2}r_2} \begin{bmatrix} 1 & 0 & -5 & | & 1 & -2 & -1 \\ 0 & 1 & 1 & | & 0 & \frac{1}{2} & 0 \\ 0 & 0 & 16 & | & -3 & \frac{13}{2} & 4 \end{bmatrix}$

$\xrightarrow{R_3 = \frac{1}{16}r_3} \begin{bmatrix} 1 & 0 & -5 & | & 1 & -2 & -1 \\ 0 & 1 & 1 & | & 0 & \frac{1}{2} & 0 \\ 0 & 0 & 1 & | & -\frac{3}{16} & \frac{13}{32} & \frac{1}{4} \end{bmatrix}$

$\xrightarrow[\substack{R_1 = 5r_3 + r_1 \\ R_2 = -r_3 + r_2}]{} \begin{bmatrix} 1 & 0 & 0 & | & \frac{1}{16} & \frac{1}{32} & \frac{1}{4} \\ 0 & 1 & 0 & | & \frac{3}{16} & \frac{3}{32} & -\frac{1}{4} \\ 0 & 0 & 1 & | & -\frac{3}{16} & \frac{13}{32} & \frac{1}{4} \end{bmatrix}$

Since the left side is now I_3, the right side of the matrix is the inverse.

$\begin{bmatrix} 4 & 3 & -1 \\ 0 & 2 & 2 \\ 3 & -1 & 0 \end{bmatrix}^{-1} = \begin{bmatrix} \frac{1}{16} & \frac{1}{32} & \frac{1}{4} \\ \frac{3}{16} & \frac{3}{32} & -\frac{1}{4} \\ -\frac{3}{16} & \frac{13}{32} & \frac{1}{4} \end{bmatrix}$

23. We want $AB = BA$ when $A = \begin{bmatrix} x & y \\ z & w \end{bmatrix}$ and

$B = \begin{bmatrix} 1 & 1 \\ -1 & 1 \end{bmatrix}.$

$AB = \begin{bmatrix} x & y \\ z & w \end{bmatrix} \begin{bmatrix} 1 & 1 \\ -1 & 1 \end{bmatrix} = \begin{bmatrix} x-y & x+y \\ z-w & z+w \end{bmatrix}$ and

$BA = \begin{bmatrix} 1 & 1 \\ -1 & 1 \end{bmatrix} \begin{bmatrix} x & y \\ z & w \end{bmatrix} = \begin{bmatrix} x+z & y+w \\ -x+z & -y+w \end{bmatrix}.$

If the two products are equal then

$\begin{bmatrix} x-y & x+y \\ z-w & z+w \end{bmatrix}$ and $\begin{bmatrix} x+z & y+w \\ -x+z & -y+w \end{bmatrix}$ must be

equal. These 2 matrices are equal when their corresponding entries are equal. So,

$\begin{array}{lll} x-y=x+z & x+y=y+w & z-w=-x+z \\ -y=z & x=w & w=x \end{array}$

$z+w=-y+w$

$z=-y$

Thus, for $AB = BA$, $x = w$ and $y = -z$.

25.a. To encode the message, first we assign a number to each letter.

H	E	Y	D	U	D	E
19	22	2	23	6	23	22

W	H	A	T	S	U	P
4	19	26	7	8	6	11

To use matrix *A* to encode the message, group the numbers into 2 × 1 column matrices and multiply each column matrix on the left by *A*.

$$\begin{bmatrix} 3 & 1 \\ 2 & 2 \end{bmatrix}\begin{bmatrix} 19 \\ 22 \end{bmatrix} = \begin{bmatrix} 79 \\ 82 \end{bmatrix}; \begin{bmatrix} 3 & 1 \\ 2 & 2 \end{bmatrix}\begin{bmatrix} 2 \\ 23 \end{bmatrix} = \begin{bmatrix} 29 \\ 50 \end{bmatrix}$$

$$\begin{bmatrix} 3 & 1 \\ 2 & 2 \end{bmatrix}\begin{bmatrix} 6 \\ 23 \end{bmatrix} = \begin{bmatrix} 41 \\ 58 \end{bmatrix}; \begin{bmatrix} 3 & 1 \\ 2 & 2 \end{bmatrix}\begin{bmatrix} 22 \\ 4 \end{bmatrix} = \begin{bmatrix} 70 \\ 52 \end{bmatrix}$$

$$\begin{bmatrix} 3 & 1 \\ 2 & 2 \end{bmatrix}\begin{bmatrix} 19 \\ 26 \end{bmatrix} = \begin{bmatrix} 83 \\ 90 \end{bmatrix}; \begin{bmatrix} 3 & 1 \\ 2 & 2 \end{bmatrix}\begin{bmatrix} 7 \\ 8 \end{bmatrix} = \begin{bmatrix} 29 \\ 30 \end{bmatrix}$$

$$\begin{bmatrix} 3 & 1 \\ 2 & 2 \end{bmatrix}\begin{bmatrix} 6 \\ 11 \end{bmatrix} = \begin{bmatrix} 29 \\ 34 \end{bmatrix}$$

The encoded message is 79 82 29 50 41 58 70 52 83 90 29 30 29 34.

To use matrix *B* to encode the message, group the numbers into 2 × 1 column matrices and multiply each column matrix on the left by *B*.

$$\begin{bmatrix} 1 & 4 & 2 \\ 2 & 0 & 2 \\ 0 & 0 & 4 \end{bmatrix}\begin{bmatrix} 19 \\ 22 \\ 2 \end{bmatrix} = \begin{bmatrix} 111 \\ 42 \\ 8 \end{bmatrix}$$

$$\begin{bmatrix} 1 & 4 & 2 \\ 2 & 0 & 2 \\ 0 & 0 & 4 \end{bmatrix}\begin{bmatrix} 23 \\ 6 \\ 23 \end{bmatrix} = \begin{bmatrix} 93 \\ 92 \\ 92 \end{bmatrix}$$

$$\begin{bmatrix} 1 & 4 & 2 \\ 2 & 0 & 2 \\ 0 & 0 & 4 \end{bmatrix}\begin{bmatrix} 22 \\ 4 \\ 19 \end{bmatrix} = \begin{bmatrix} 76 \\ 82 \\ 76 \end{bmatrix}$$

$$\begin{bmatrix} 1 & 4 & 2 \\ 2 & 0 & 2 \\ 0 & 0 & 4 \end{bmatrix}\begin{bmatrix} 26 \\ 7 \\ 8 \end{bmatrix} = \begin{bmatrix} 70 \\ 68 \\ 32 \end{bmatrix}$$

$$\begin{bmatrix} 1 & 4 & 2 \\ 2 & 0 & 2 \\ 0 & 0 & 4 \end{bmatrix}\begin{bmatrix} 6 \\ 11 \\ 1 \end{bmatrix} = \begin{bmatrix} 52 \\ 14 \\ 4 \end{bmatrix}$$

The encoded message is 111 42 8 93 92 92 76 82 76 70 68 32 52 14 4.

b. To encode the message, first we assign a number to each letter.

C	A	L	L	M	E	O	N
24	26	15	15	14	22	12	13

M	Y	C	E	L	L
14	2	24	22	15	15

To use matrix *A* to encode the message, group the numbers into 2 × 1 column matrices and multiply each column matrix on the left by *A*.

$$\begin{bmatrix} 3 & 1 \\ 2 & 2 \end{bmatrix}\begin{bmatrix} 24 \\ 26 \end{bmatrix} = \begin{bmatrix} 98 \\ 100 \end{bmatrix}; \begin{bmatrix} 3 & 1 \\ 2 & 2 \end{bmatrix}\begin{bmatrix} 15 \\ 15 \end{bmatrix} = \begin{bmatrix} 60 \\ 60 \end{bmatrix}$$

$$\begin{bmatrix} 3 & 1 \\ 2 & 2 \end{bmatrix}\begin{bmatrix} 14 \\ 22 \end{bmatrix} = \begin{bmatrix} 64 \\ 72 \end{bmatrix}; \begin{bmatrix} 3 & 1 \\ 2 & 2 \end{bmatrix}\begin{bmatrix} 12 \\ 13 \end{bmatrix} = \begin{bmatrix} 49 \\ 50 \end{bmatrix}$$

$$\begin{bmatrix} 3 & 1 \\ 2 & 2 \end{bmatrix}\begin{bmatrix} 14 \\ 2 \end{bmatrix} = \begin{bmatrix} 44 \\ 32 \end{bmatrix}; \begin{bmatrix} 3 & 1 \\ 2 & 2 \end{bmatrix}\begin{bmatrix} 24 \\ 22 \end{bmatrix} = \begin{bmatrix} 94 \\ 92 \end{bmatrix}$$

$$\begin{bmatrix} 3 & 1 \\ 2 & 2 \end{bmatrix}\begin{bmatrix} 15 \\ 15 \end{bmatrix} = \begin{bmatrix} 60 \\ 60 \end{bmatrix}$$

The encoded message is 98 100 60 60 64 72 49 50 44 32 94 92 60 60.

To use matrix *B* to encode the message, group the numbers into 2 × 1 column matrices and multiply each column matrix on the left by *B*.

$$\begin{bmatrix} 1 & 4 & 2 \\ 2 & 0 & 2 \\ 0 & 0 & 4 \end{bmatrix}\begin{bmatrix} 24 \\ 26 \\ 15 \end{bmatrix} = \begin{bmatrix} 158 \\ 78 \\ 60 \end{bmatrix}$$

$$\begin{bmatrix} 1 & 4 & 2 \\ 2 & 0 & 2 \\ 0 & 0 & 4 \end{bmatrix}\begin{bmatrix} 15 \\ 14 \\ 22 \end{bmatrix} = \begin{bmatrix} 115 \\ 74 \\ 88 \end{bmatrix}$$

$$\begin{bmatrix} 1 & 4 & 2 \\ 2 & 0 & 2 \\ 0 & 0 & 4 \end{bmatrix}\begin{bmatrix} 12 \\ 13 \\ 14 \end{bmatrix} = \begin{bmatrix} 92 \\ 52 \\ 56 \end{bmatrix}$$

$$\begin{bmatrix} 1 & 4 & 2 \\ 2 & 0 & 2 \\ 0 & 0 & 4 \end{bmatrix}\begin{bmatrix} 2 \\ 24 \\ 22 \end{bmatrix} = \begin{bmatrix} 142 \\ 48 \\ 88 \end{bmatrix}$$

$$\begin{bmatrix} 1 & 4 & 2 \\ 2 & 0 & 2 \\ 0 & 0 & 4 \end{bmatrix}\begin{bmatrix} 15 \\ 15 \\ 1 \end{bmatrix} = \begin{bmatrix} 77 \\ 32 \\ 4 \end{bmatrix}$$

The encoded message is 158 78 60 115 74 88 92 52 56 142 48 88 77 32 4.

c. To decode the message: 75 78 45 42 72 64 35 42 17 26 67 62 using A we group the numbers into 2×1 column matrices and multiply each column matrix on the left by A^{-1}.

$$A^{-1} = \begin{bmatrix} 3 & 1 \\ 2 & 2 \end{bmatrix}^{-1} = \begin{bmatrix} \dfrac{1}{2} & -\dfrac{1}{4} \\ -\dfrac{1}{2} & \dfrac{3}{4} \end{bmatrix}$$

$$\begin{bmatrix} \dfrac{1}{2} & -\dfrac{1}{4} \\ -\dfrac{1}{2} & \dfrac{3}{4} \end{bmatrix} \begin{bmatrix} 75 \\ 78 \end{bmatrix} = \begin{bmatrix} 18 \\ 21 \end{bmatrix} \qquad \begin{bmatrix} \dfrac{1}{2} & -\dfrac{1}{4} \\ -\dfrac{1}{2} & \dfrac{3}{4} \end{bmatrix} \begin{bmatrix} 45 \\ 42 \end{bmatrix} = \begin{bmatrix} 12 \\ 9 \end{bmatrix} \qquad \begin{bmatrix} \dfrac{1}{2} & -\dfrac{1}{4} \\ -\dfrac{1}{2} & \dfrac{3}{4} \end{bmatrix} \begin{bmatrix} 72 \\ 64 \end{bmatrix} = \begin{bmatrix} 20 \\ 12 \end{bmatrix}$$

$$\begin{bmatrix} \dfrac{1}{2} & -\dfrac{1}{4} \\ -\dfrac{1}{2} & \dfrac{3}{4} \end{bmatrix} \begin{bmatrix} 35 \\ 42 \end{bmatrix} = \begin{bmatrix} 7 \\ 14 \end{bmatrix} \qquad \begin{bmatrix} \dfrac{1}{2} & -\dfrac{1}{4} \\ -\dfrac{1}{2} & \dfrac{3}{4} \end{bmatrix} \begin{bmatrix} 17 \\ 26 \end{bmatrix} = \begin{bmatrix} 2 \\ 11 \end{bmatrix} \qquad \begin{bmatrix} \dfrac{1}{2} & -\dfrac{1}{4} \\ -\dfrac{1}{2} & \dfrac{3}{4} \end{bmatrix} \begin{bmatrix} 67 \\ 62 \end{bmatrix} = \begin{bmatrix} 18 \\ 13 \end{bmatrix}$$

The decoded numbers are 18 21 12 9 20 12 7 14 2 11 18 13. Replacing the numbers with the appropriate letters gives the message:
I F O R G O T M Y P I N

27. a. On February 3, 2010 the three stocks cost

$$\begin{array}{ccc} \text{Microsoft} & \text{Intel} & \text{Dell} \\ A = [28.50 & 19.50 & 13.30] \end{array}$$

b.

$$\begin{array}{c} \\ B = \begin{array}{c} \text{Microsoft} \\ \text{Intel} \\ \text{Dell} \end{array} \end{array} \begin{array}{c} \text{Juanita} \quad \text{Debbie} \quad \text{Dawn} \\ \begin{bmatrix} 20 & 15 & 10 \\ 30 & 25 & 20 \\ 10 & 20 & 25 \end{bmatrix} \end{array}$$

c. The entries of matrix AB is represents each woman's cost of purchasing the three stocks.

$$AB = [28.50 \quad 19.50 \quad 13.30] \begin{bmatrix} 20 & 15 & 10 \\ 30 & 25 & 20 \\ 10 & 20 & 25 \end{bmatrix}$$

$$= [570 + 585 + 133 \quad 427.50 + 487.50 + 266 \quad 285 + 390 + 332.50]$$

$$= [1288 \quad 1181 \quad 1007.50]$$

The entries of matrix AB represent the cost of Juanita's, Debbie's, and Dawn's purchases, respectively, in dollars.

Chapter 3 Project

1. Matrix A represents the amounts of materials needed to produce 1 unit of a product. The entries in the matrix are quotients. Column 1 is found by dividing each entry in the agriculture column of the table by total gross output of agriculture. Column 2 entries are the quotients found by dividing each entry in the manufacturing column by total gross output of manufacturing, and column 3 entries are the quotients found by dividing entries by total gross outcome of services. D is the consumer demand. It is labeled Open Sector in Table 1 in the text.

$$A = \begin{bmatrix} 0.410 & 0.030 & 0.026 \\ 0.062 & 0.378 & 0.105 \\ 0.124 & 0.159 & 0.192 \end{bmatrix}, \quad D = \begin{bmatrix} 39.24 \\ 60.02 \\ 130.65 \end{bmatrix}$$

3. To calculate the new X_1, solve the equation $X_1 = [I_3 - A]^{-1} D_1$.

$$X_1 = \begin{bmatrix} 1.720 & 0.100 & 0.068 \\ 0.223 & 1.676 & 0.225 \\ 0.308 & 0.345 & 1.292 \end{bmatrix} \begin{bmatrix} 40.24 \\ 60.02 \\ 130.65 \end{bmatrix} = \begin{bmatrix} 84.099 \\ 138.963 \\ 201.901 \end{bmatrix}$$

Note that answers may vary slightly due to rounding.

5. Using the interpretation from Problem 4, we find that service production would need to increase by 0.308 unit for a one unit increase in agriculture.

Use Excel to solve problems 7–11.

7.

$$A = \begin{bmatrix} 0.21387 & 0.00000 & 0.00469 & 0.53058 & 0.00015 & 0.00935 & 0.00714 & 0.05473 & 0.00608 \\ 0.00106 & 0.13709 & 0.01894 & 0.86578 & 0.25893 & 0.00025 & 0.00436 & 0.00123 & 0.03622 \\ 0.00083 & 0.00009 & 0.00099 & 0.00644 & 0.00836 & 0.00676 & 0.03234 & 0.02738 & 0.05148 \\ 0.01180 & 0.00961 & 0.06151 & 0.32111 & 0.04528 & 0.02868 & 0.02062 & 0.09088 & 0.06326 \\ 0.00551 & 0.00544 & 0.01464 & 0.09620 & 0.17368 & 0.04597 & 0.06231 & 0.13175 & 0.07116 \\ 0.00545 & 0.00287 & 0.04604 & 0.12726 & 0.01805 & 0.02474 & 0.01203 & 0.04016 & 0.01516 \\ 0.00535 & 0.00679 & 0.00946 & 0.02687 & 0.03227 & 0.03957 & 0.19974 & 0.11416 & 0.02227 \\ 0.00116 & 0.00600 & 0.01757 & 0.06712 & 0.04734 & 0.05200 & 0.06444 & 0.13262 & 0.05954 \\ 0.00004 & 0.00003 & 0.00060 & 0.00134 & 0.00323 & 0.00434 & 0.00462 & 0.01736 & 0.00419 \end{bmatrix}$$

$$D_0 = \begin{bmatrix} 64,207.5 \\ -146,059 \\ 1,185,993 \\ 1,740,597 \\ 1,003,964 \\ 1,759,363 \\ 2,454,395 \\ 3,719,337 \\ 2,278,087 \end{bmatrix}$$

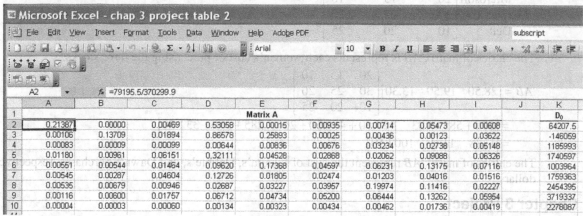

9. In the original matrix, demand for transportation, communication, and utilities = $3,393,307.434 million.

$$X_1 = \begin{bmatrix} I_9 - A \end{bmatrix} D_1 = \begin{bmatrix} 3477656.333 \\ 5342855.827 \\ 1687390.369 \\ 4304942.495 \\ 3393307.478 \\ 2864558.874 \\ 4438166.491 \\ 5550845.314 \\ 2435608.149 \end{bmatrix} \begin{matrix} \text{Agriculture} \\ \text{Mining} \\ \text{Construction} \\ \text{Manufacturing} \\ \text{Transportation/Communication/Utilities} \\ \text{Trade} \\ \text{Finance/Insurance/Real Estate} \\ \text{Services} \\ \text{Other} \end{matrix}$$

D_1	$(I_9 - A)^{-1}D_1$
64207.5	3477656.333
-146059	5342855.827
1185994	1687390.369
1740597	4304942.495
1003964	3393307.478
1759363	2864558.874
2454395	4438166.491
3719337	5550845.314
2278087	2435608.149

Demand for transportation, communication, and utilities = $1,687,390.369 million. If demand for construction increases by 1 million dollars, demand for transportation, communication, and utilities increases by $1,687,390.369 − $1,003,964 = $683,426.369 million.

11. An increase of $1 million in demand for construction produces an composite increase of $1.5 million in the economy. This is obtained by adding the entries in the third column of the $(I - A)^{-1}$ matrix.

Chapter 4 Linear Programming with Two Variables

4.1 Systems of Linear Inequalities

1. a. False   **b.** False

3.

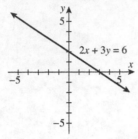

5. False. The slope of the line $2x + y = 6$ is -2, while the slope of the line $y = 2x + 2$ is 2. Since the slopes are not equal, the lines are not parallel.

7. True

9. $x \geq 0$

The corresponding linear equation is $x = 0$. We graph a solid line, since the inequality is nonstrict. The test point $(1, 1)$ satisfies the inequality, so we shade the region containing it.

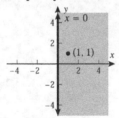

11. $x < 4$

The corresponding linear equation is $x = 4$. We graph a dashed line, since the inequality is strict. The test point $(0, 0)$ satisfies the inequality, so we shade the region containing it.

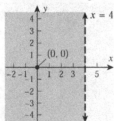

13. $y \geq 1$

The corresponding linear equation is $y = 1$. We graph a solid line, since the inequality is non-strict. The test point $(0, 0)$ does not satisfy the inequality, so we shade the region opposite it.

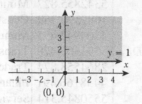

15. $2x + y \leq 4$

The corresponding linear equation is $2x + y = 4$. We graph a solid line, since the inequality is nonstrict. The test point $(0, 0)$ satisfies the inequality, so we shade the region containing it.

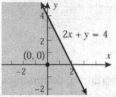

17. $5x + y \geq 10$

The corresponding linear equation is $5x + y = 10$. We graph a solid line, since the inequality is nonstrict. The test point $(0, 0)$ does not satisfy the inequality, so we shade the region opposite it.

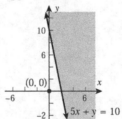

19. $x + 5y < 5$

The corresponding linear equation is $x + 5y = 5$. We graph a dashed line, since the inequality is strict. The test point $(0, 0)$ satisfies the inequality, so we shade the region containing it.

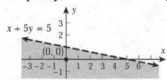

In exercises 21–24, to determine which points are part of the graph of the system, check each point to see if it satisfies both of the inequalities.

21.

Point	Inequality 1 $x + 3y \geq 0$	Inequality 2 $-3x + 2y \geq 0$	Conclusion
$P_1 = (3, 8)$	Is $3 + 3 \cdot 8 \geq 0$? $27 > 0$	Is $-3 \cdot 3 + 2 \cdot 8 \geq 0$? $7 > 0$	Both inequalities are satisfied; (3, 8) is part of the graph.
$P_2 = (12, 9)$	Is $12 + 3 \cdot 9 \geq 0$? $39 > 0$	Is $-3 \cdot 12 + 2 \cdot 9 \geq 0$? $-18 < 0$	Since the second inequality is not satisfied, (12, 9) is not part of the graph.
$P_3 = (5, 1)$	Is $5 + 3 \cdot 1 \geq 0$? $8 > 0$	Is $-3 \cdot 5 + 2 \cdot 1 \geq 0$? $-13 < 0$	Since the second inequality is not satisfied, (5, 1) is not part of the graph.

23.

Point	Inequality 1 $3x + 2y \geq 0$	Inequality 2 $x + y \leq 15$	Conclusion
$P_1 = (2, 3)$	Is $3 \cdot 2 + 2 \cdot 3 \geq 0$? $12 > 0$	Is $2 + 3 \leq 15$? $5 < 15$	Both inequalities are satisfied; (2, 3) is part of the graph.
$P_2 = (10, 10)$	Is $3 \cdot 10 + 2 \cdot 10 \geq 0$? $50 > 0$	Is $10 + 10 \leq 15$? $20 > 15$	The second inequality is not satisfied; (10, 10) is not part of the graph.
$P_3 = (5, 1)$	Is $3 \cdot 5 + 2 \cdot 1 \geq 0$? $17 > 0$	Is $5 + 1 \leq 15$? $6 < 15$	Both inequalities are satisfied; (5, 1) is part of the graph.

25. The region that represents the graph of the system is the set of points common to the solutions of each individual inequality. We use the test point (0, 0).

$$5x - 4y \leq 8 \qquad 2x + 5y \leq 23$$
$$5 \cdot 0 - 4 \cdot 0 = 0 \quad 2 \cdot 0 + 5 \cdot 0 = 0$$
$$0 < 8 \qquad\qquad 0 < 23$$

The test point (0, 0) satisfies both inequalities, so the region (**b**) represents the graph of the system.

27. The region that represents the graph of the system is the set of points common to the solutions of each individual inequality. We use the test point (0, 0).

$$2x - 3y \geq -3 \qquad 2x + 3y \leq 16$$
$$2 \cdot 0 - 3 \cdot 0 = 0 \quad 2 \cdot 0 + 6 \cdot 0 = 0$$
$$0 > -3 \qquad\qquad 0 < 16$$

The test point (0, 0) satisfies both inequalities, so the region (**c**) represents the graph of the system.

29. The region that represents the graph of the system is the set of points common to the solutions of each individual inequality. We use the test point (0, 0).

$$5x - 3y \geq 3 \qquad 2x + 6y \geq 30$$
$$5 \cdot 0 - 3 \cdot 0 = 0 \quad 2 \cdot 0 + 6 \cdot 0 = 0$$
$$0 < 3 \qquad\qquad 0 < 30$$

The test point (0, 0) satisfies neither inequality, so the region representing each inequality is on the opposite side of the line from (0, 0). This means (**d**) represents the graph of the system.

For exercises 31–37, first graph each inequality separately. Then graph the lines with the points the separate graphs have in common. The solution to the system is the set of all points common to both graphs.

31.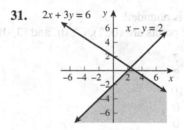

33.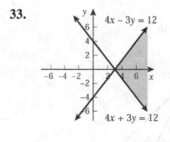

35.

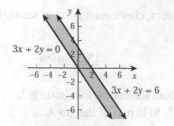

37.

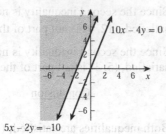

39. a. $\begin{cases} x + y \geq 2 \\ x \geq 0 \\ y \geq 0 \end{cases}$

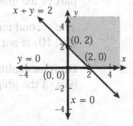

b. The graph is unbounded.
The corner points are $(0, 2)$ and $(2, 0)$.

41. a. $\begin{cases} x + y \geq 2 \\ 2x + 3y \leq 6 \\ x \geq 0 \\ y \geq 0 \end{cases}$

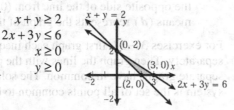

b. The graph is bounded.
The corner points are $(0, 2)$, $(2, 0)$, and $(3, 0)$.

43. a. $\begin{cases} x + y \geq 2 \\ x + y \leq 8 \\ 2x + y \leq 10 \\ x \geq 0 \\ y \geq 0 \end{cases}$

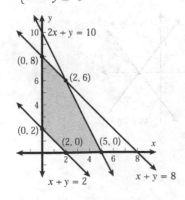

We find the point $(2, 6)$ by solving the system formed by the 2nd and 3rd equations.

$$\begin{cases} x + y = 8 & (2) \\ 2x + y = 10 & (3) \end{cases}$$

Solving (2) for y, we get $y = 8 - x$ (2).
Back-substituting y into (3), gives

$$2x + (8 - x) = 10$$
$$x = 2$$

Then using x in (2), we solve for y.
$y = 8 - 2 = 6$.

b. The graph is bounded. The corner points are
$(0, 8)$, $(0, 2)$, $(2, 0)$, $(5, 0)$, $(2, 6)$.

45. a. $\begin{cases} x + y \geq 2 \\ 2x + 3y \leq 12 \\ 3x + y \leq 12 \\ x \geq 0 \\ y \geq 0 \end{cases}$

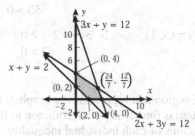

To find the point $\left(\dfrac{24}{7}, \dfrac{12}{7} \right)$ solve the system:

$$\begin{cases} 2x + 3y = 12 & (2) \\ 3x + y = 12 & (3) \end{cases}$$

Multiply equation (3) by -3, then add the resulting equation to equation (2) and solve for x.

$$\begin{array}{r} 2x + 3y = 12 \quad (2) \\ -9x - 3y = -36 \quad -3 \times (3) \\ \hline -7x \quad\quad\ = -24 \end{array}$$

$$x = \frac{24}{7}$$

Back-substituting $x = \dfrac{24}{7}$ into equation (3)

gives $y = -3\left(\dfrac{24}{7} \right) + 12 = \dfrac{12}{7}$.

b. The graph is bounded. The corner points are
$(0, 4)$, $(0, 2)$, $(2, 0)$, $(4, 0)$ and $\left(\dfrac{24}{7}, \dfrac{12}{7} \right)$.

47. a. $\begin{cases} x+2y \geq 1 \\ \quad y \leq 4 \\ \quad x \geq 0 \\ \quad y \geq 0 \end{cases}$

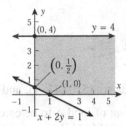

b. The graph is unbounded. The corner points

are: $(1, 0)$, $\left(0, \dfrac{1}{2}\right)$, and $(0, 4)$.

49. a. $\begin{cases} x+2y \geq 2 \\ x+\ y \leq 4 \\ 3x+\ y \leq 3 \\ \quad x \geq 0 \\ \quad y \geq 0 \end{cases}$

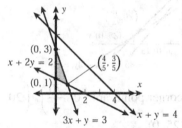

To find the corner point $\left(\dfrac{4}{5}, \dfrac{3}{5}\right)$, we find the

intersection of the lines $\begin{cases} x+2y=2 & (1) \\ 3x+y=3 & (3) \end{cases}$.

Using the substitution method, we solve (1)
for x and substitute the result in (3).

$x = 2-2y$

$3(2-2y)+y=3$

$-5y=-3$

$y = \dfrac{3}{5}$

Back-substituting into (1), we find

$x = 2-2\left(\dfrac{3}{5}\right) = 2-\dfrac{6}{5} = \dfrac{4}{5}$

b. The graph is bounded. The corner points are

$(0, 3)$, $(0, 1)$, and $\left(\dfrac{4}{5}, \dfrac{3}{5}\right)$.

51. a. We let $x =$ the number of pound packages of
low-grade mix, and $y =$ the number of
packages of high-grade mix to be prepared.
There are 60 pounds (960 ounces) of almonds
and 90 pounds (1440 ounces) of peanuts
available. The system of inequalities
representing the system is

$\begin{cases} \dfrac{1}{4}x + \dfrac{1}{2}y \leq 60 & (1) \\ \dfrac{3}{4}x + \dfrac{1}{2}y \leq 90 & (2) \\ \quad x \geq \ 0 & (3) \\ \quad y \geq \ 0 & (4) \end{cases}$ or

$\begin{cases} x+2y \leq 240 & (1) \\ 3x+2y \leq 360 & (2) \\ \quad x \geq \ 0 & (3) \\ \quad y \geq \ 0 & (4) \end{cases}$.

b.

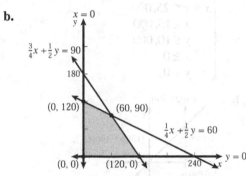

From the graph we see that three of the corner
points are $(0, 0)$, $(120, 0)$, and $(0, 120)$. The
fourth corner $(60, 90)$, is found by solving the
system of equations $\begin{cases} x+2y=240 & (1) \\ 3x+2y=360 & (2) \end{cases}$.

Using elimination, we subtract (1) from (2)
and get $2x = 120$ or $x = 60$. Back-substituting
into (1) we find

$60+2y=240$ (1)

$2y=180$

$y=90$

53. a. First we note that meaningful values for x nd y
are nonnegative values. Next we note that
there are only limited numbers of grinding and
finishing hours. There are $80 = 2 \cdot 40$ grinding
hours available, and $120 = 3 \cdot 40$ finishing
hours available. Using these constraints, the
system is

$\begin{cases} 3x+2y \leq \ 80 & (1) \\ 4x+3y \leq 120 & (2) \\ \quad x \geq \ \ 0 & (3) \\ \quad y \geq \ \ 0 & (4) \end{cases}$.

b.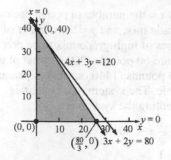

The corner points are (0, 40), (0, 0) and $\left(\dfrac{80}{3}, 0\right)$.

55.a. First we note that meaningful values of x and y are nonnegative. Then we see that there is at most \$25,000 to be invested. Using these constraints, the system is
$$\begin{cases} x + y \le 25{,}000 \\ x \ge 15{,}000 \\ y \le 10{,}000 \\ x \ge 0 \\ y \ge 0 \end{cases}$$

b.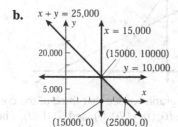

The corner points are (15000, 10000), (15000, 0), and (25000, 0).

c. The x-value of each corner point represents the amount in dollars to be invested in the Dividend Equity Fund and the y-value is the amount in dollars to be invested in the Global Bond Fund.

57.a. Let x = the number of units of grain 1 and let y = the number of units of grain 2.
$$\begin{cases} x + 2y \ge 5 \\ 5x + y \ge 16 \\ x \ge 0 \\ y \ge 0 \end{cases}$$

b.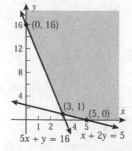

The corner points are (5, 0), (3, 1), and (0, 16).

59.a. Let x = the amount of food A to be eaten and let y = the amount of food B to be eaten.
$$\begin{cases} 5x + 4y \ge 85 \\ 3x + 3y \ge 70 \\ 2x + 3y \ge 50 \\ x \ge 0 \\ y \ge 0 \end{cases}$$

b.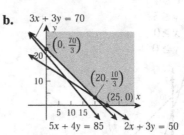

The corner points are $\left(0, \dfrac{70}{3}\right)$, $\left(20, \dfrac{10}{3}\right)$, and (25, 0).

61.a.
$$\begin{cases} x + y \le 30{,}000 & (1) \\ 0.0435x + 0.0505y \ge 500 & (2) \\ x \ge 3{,}000 & (3) \\ y \ge 5{,}000 & (4) \end{cases}$$

b.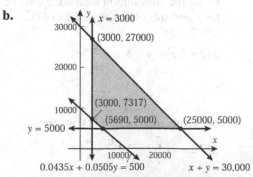

The corner points are (5690, 5000), (25000, 5000), (3000, 7317), and (3000, 27000).

c. Each corner point represents an amount x deposited in the ICON Bond fund and an amount y deposited in ING Mid Cap Opportunites fund that satisfy the conditions given.

63. a. $\begin{cases} x+y \le 200{,}000 & (1) \\ 3x-y \le \quad 0 & (2) \\ x \ge \quad 20{,}000 & (3) \\ y \ge \quad 0 & (4) \end{cases}$

b.

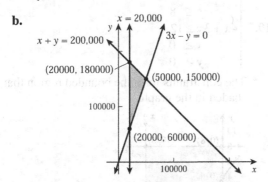

Two of the corner points lie along the line $x = 20{,}000$. So they have values $(20000, 60000)$ and $(20000, 180000)$. The third corner point is found by solving the system of equations formed by (1) and (2).

$$\begin{array}{rl} x+y = & 200{,}000 \\ 3x-y = & 0 \\ \hline 4x \quad = & 200{,}000 \\ x = & 50{,}000 \end{array}$$

Add.

Back-substitute: $y = 3 \cdot 50{,}000 = 150{,}000$

The third corner point is $(50000, 150000)$.

c. The yields are found by evaluation the expression $0.0997x + 0.2365y$ at each of the corner points.

$(20000, 60000)$:
$(0.0997)(20{,}000) + (0.2365)(60{,}000)$
$\qquad\qquad\qquad = \$16{,}184$

$(20000, 180000)$:
$(0.0997)(20{,}000) + (0.2365)(180{,}000)$
$\qquad\qquad\qquad = \$44{,}564$

$(50000, 150000)$:
$(0.0997)(50{,}000) + (0.2365)(150{,}000)$
$\qquad\qquad\qquad = \$40{,}460$

65. Answers will vary. Sample answer:
$\begin{cases} x-y \le -2 \\ x-y \ge \ 2 \end{cases}$

67. Answers will vary. Sample graph:

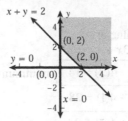

4.2 A Geometric Approach to Linear Programming Problems with Two Variables

1. In a linear programming problem the function to be maximized (or minimized) is called the <u>objective function</u>.

3. False

In exercises 5–13, evaluate the objective function at each corner point, then choose the maximum and minimum values.

5. $z = 2x + 3y$

Corner point	Value of $z = 2x + 3y$	
$(2, 2)$	$z = 2(2) + 3(2) = 10$	minimum
$(8, 1)$	$z = 2(8) + 3(1) = 19$	
$(2, 7)$	$z = 2(2) + 3(7) = 25$	
$(7, 8)$	$z = 2(7) + 3(8) = 38$	maximum

7. $z = x + y$

Corner point	Value of $z = x + y$	
$(2, 2)$	$z = 2 + 2 = 4$	minimum
$(8, 1)$	$z = 8 + 1 = 9$	
$(2, 7)$	$z = 2 + 7 = 9$	
$(7, 8)$	$z = 7 + 8 = 15$	maximum

9. $z = x + 6y$

Corner point	Value of $z = x + 6y$	
$(2, 2)$	$z = 2 + 6(2) = 14$	minimum
$(8, 1)$	$z = 8 + 6(1) = 14$	minimum
$(2, 7)$	$z = 2 + 6(7) = 44$	
$(7, 8)$	$z = 7 + 6(8) = 55$	maximum

Note that all feasible points on the line containing the points $(2, 2)$ and $(8, 1)$ will give the minimum value of z.

11. $z = 3x + 4y$

Corner point	Value of $z = 3x + 4y$	
(2, 2)	$z = 3(2) + 4(2) = 14$	minimum
(8, 1)	$z = 3(8) + 4(1) = 28$	
(2, 7)	$z = 3(2) + 4(7) = 34$	
(7, 8)	$z = 3(7) + 4(8) = 53$	maximum

13. $z = 10x + y$

Corner point	Value of $z = 10x + y$	
(2, 2)	$z = 10(2) + 2 = 22$	minimum
(8, 1)	$z = 10(8) + 1 = 81$	maximum
(2, 7)	$z = 10(2) + 7 = 27$	
(7, 8)	$z = 10(7) + 8 = 78$	

15. $\begin{cases} x \le 13 \\ 4x + 3y \ge 12 \\ x \ge 0 \\ y \ge 0 \end{cases}$

The constraints form the unbounded region that is shaded in the graph.

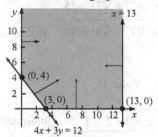

The corner points are: (0, 4), (3, 0), and (13, 0).

17. $\begin{cases} y \le 10 \\ x + y \le 15 \\ x \ge 0 \\ y \ge 0 \end{cases}$

The constraints form the bounded region that is shaded in the graph.

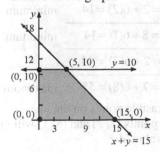

Three of the corner points, (0, 0), (15, 0), and (0, 10) are easy to identify. The fourth corner, (5, 10) is found by solving

$$\begin{cases} y = 10 \\ x + y = 15 \end{cases}.$$

Substituting $y = 10$ into the second equation, we find $x = 5$. The corner points are: (0, 0), (15, 0), (0, 10), and (5, 10).

19. $\begin{cases} x \le 10 \\ y \le 8 \\ 4x + 3y \ge 12 \\ x \ge 0 \\ y \ge 0 \end{cases}$

The constraints form the bounded region that is shaded in the graph.

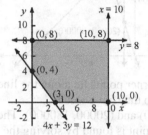

Four of the corner points (3, 0), (10, 0), (0, 4), and (0, 8) are easy to identify. The fifth corner, (10, 8) is found at the intersection of the lines $x = 10$ and $y = 8$. The corner points are (3, 0), (10, 0), (0, 4), (0, 8), and (10, 8).

In exercises 21–27, to maximize $z = 5x + 7y$, graph the system of linear inequalities, shade the set of feasible points, and locate the corner points. Then evaluate the objective function at each corner point.

21. $\begin{cases} x + y \le 2 \\ y \ge 1 \\ x \ge 0 \\ y \ge 0 \end{cases}$

The corner points are (0, 1), (0, 2) and (1, 1). The third point was obtained by solving

$$\begin{cases} x + y = 2 \\ y = 1 \end{cases}.$$

Substituting $y = 1$ into the first equation, we find $x = 1$.

(continued on next page)

(continued)

Corner point	Value of $z = 5x + 7y$	
(0, 1)	$z = 5(0) + 7(1) = 7$	
(0, 2)	$z = 5(0) + 7(2) = 14$	maximum
(1, 1)	$z = 5(1) + 7(1) = 12$	

The maximum value of 14 occurs at (0, 2).

23. $\begin{cases} x + y \geq 2 \\ 2x + 3y \leq 6 \\ x \geq 0 \\ y \geq 0 \end{cases}$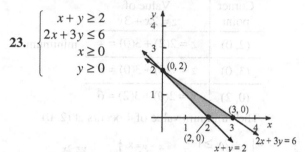

The corner points are (2, 0), (0, 2) and (3, 0).

Corner point	Value of $z = 5x + 7y$	
(2, 0)	$z = 5(2) + 7(0) = 10$	
(0, 2)	$z = 5(0) + 7(2) = 14$	
(3, 0)	$z = 5(3) + 7(0) = 15$	maximum

The maximum value of 15 occurs at (3, 0).

25. $\begin{cases} x + y \geq 2 \\ x + y \leq 8 \\ 2x + y \leq 10 \\ x \geq 0 \\ y \geq 0 \end{cases}$

The corner points are (2, 0), (5, 0), (0, 2), (0, 8), and (2, 6). The fifth point was obtained by solving
$\begin{cases} x + y = 8 \\ 2x + y = 10 \end{cases}$.

Subtracting the first equation from the second equation, we find $x = 2$. Back substituting into equation 1, we find $y = 6$.

Corner point	Value of $z = 5x + 7y$	
(2, 0)	$z = 5(2) + 7(0) = 10$	
(5, 0)	$z = 5(5) + 7(0) = 25$	
(0, 2)	$z = 5(0) + 7(2) = 14$	
(0, 8)	$z = 5(0) + 7(8) = 56$	maximum
(2, 6)	$z = 5(2) + 7(6) = 52$	

The maximum value of 56 occurs at (0, 8).

27. $\begin{cases} x + y \leq 10 \\ x \geq 6 \\ x \geq 0 \\ y \geq 0 \end{cases}$

The corner points are (6, 0), (10, 0), and (6, 4). The fifth point was obtained by solving
$\begin{cases} x = 6 \\ x + y = 10 \end{cases}$.

Substituting $x = 6$ into equation 2, we find $y = 4$.

Corner point	Value of $z = 5x + 7y$	
(6, 0)	$z = 5(6) + 7(0) = 30$	
(10, 0)	$z = 5(10) + 7(0) = 50$	
(6, 4)	$z = 5(6) + 7(4) = 58$	maximum

The maximum value of 58 occurs at (6, 4).

In exercises 29–36, to minimize $z = 2x + 3y$, graph the system of linear inequalities, shade the set of feasible points, and locate the corner points. Then evaluate the objective function at each corner point.

29. $\begin{cases} x + y \leq 2 \\ y \leq x \\ x \geq 0 \\ y \geq 0 \end{cases}$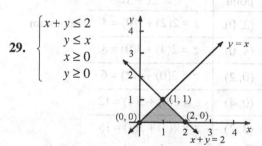

The corner points are (0, 0), (2, 0) and (1, 1). The third point was obtained by solving
$\begin{cases} x + y = 2 \\ y = x \end{cases}$.

(continued on next page)

(continued)

Substituting x for y into the first equation, we find $x = 1$. Back-substituting in the second equation gives $y = 1$.

Corner point	Value of $z = 2x + 3y$	
$(0, 0)$	$z = 2(0) + 3(0) = 0$	minimum
$(2, 0)$	$z = 2(2) + 3(0) = 4$	
$(1, 1)$	$z = 2(1) + 3(1) = 5$	

The minimum value of 0 occurs at $(0, 0)$.

31. $\begin{cases} x + y \geq 2 \\ x + 3y \leq 12 \\ 3x + y \leq 12 \\ x \geq 0 \\ y \geq 0 \end{cases}$

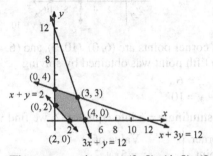

The corner points are $(2, 0)$, $(4, 0)$, $(0, 2)$, $(0, 4)$, and $(3, 3)$. The third point was obtained by solving

$\begin{cases} 3x + y = 12 \\ x + 3y = 12 \end{cases}$.

Subtracting the first equation from three times the second equation gives $8y = 24$ or $y = 3$. Back-substituting 3 for y in the first equation gives $3x + 3 = 12$ or $x = 3$.

Corner point	Value of $z = 2x + 3y$	
$(2, 0)$	$z = 2(2) + 3(0) = 4$	minimum
$(4, 0)$	$z = 2(4) + 3(0) = 8$	
$(0, 2)$	$z = 2(0) + 3(2) = 6$	
$(0, 4)$	$z = 2(0) + 3(4) = 12$	
$(3, 3)$	$z = 2(3) + 3(3) = 15$	

The minimum value of 4 occurs at $(2, 0)$.

33. $\begin{cases} x + y \geq 2 \\ x + y \leq 10 \\ 2x + 3y \leq 6 \\ x \geq 0 \\ y \geq 0 \end{cases}$

The corner points are $(2, 0)$, $(3, 0)$, and $(0, 2)$.

Corner point	Value of $z = 2x + 3y$	
$(2, 0)$	$z = 2(2) + 3(0) = 4$	minimum
$(3, 0)$	$z = 2(3) + 3(0) = 6$	
$(0, 2)$	$z = 2(0) + 3(2) = 6$	

The minimum value of 4 occurs at $(2, 0)$.

35. $\begin{cases} x + 2y \geq 1 \\ x + 2y \leq 10 \\ y \geq 2x \\ x + y \leq 8 \\ x \geq 0 \\ y \geq 0 \end{cases}$

The corner points are $(0, 5)$, $(2, 4)$, $\left(0, \dfrac{1}{2} \right)$, and $\left(\dfrac{1}{5}, \dfrac{2}{5} \right)$. The point $(2, 4)$ is found by solving the system of equations $\begin{cases} x + 2y = 10 \\ y = 2x \end{cases}$ and the point $\left(\dfrac{1}{5}, \dfrac{2}{5} \right)$ is found by solving the system of equations $\begin{cases} x + 2y = 1 \\ y = 2x \end{cases}$.

Corner point	Value of $z = 2x + 3y$	
$(0, 5)$	$z = 2(0) + 3(5) = 15$	
$(2, 4)$	$z = 2(2) + 3(4) = 16$	
$\left(0, \dfrac{1}{2} \right)$	$z = 2(0) + 3\left(\dfrac{1}{2} \right) = \dfrac{3}{2}$	minimum
$\left(\dfrac{1}{5}, \dfrac{2}{5} \right)$	$z = 2\left(\dfrac{1}{5} \right) + 3\left(\dfrac{2}{5} \right) = \dfrac{8}{5}$	

The minimum value of $\dfrac{3}{2}$ occurs at $\left(0, \dfrac{1}{2} \right)$.

Use the graph below and its corner points to find the maximum and minimum values of z in exercises 37–43. Then evaluate the objective function at z, to identify the maximum and minimum values of z.

$$\begin{cases} x+y\le10 \\ 2x+y\ge10 \\ x+2y\ge10 \\ x\ge0 \\ y\ge0 \end{cases}$$

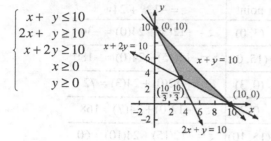

The corner points are $(10, 0)$, $(0, 10)$, and $\left(\dfrac{10}{3}, \dfrac{10}{3}\right)$.

The third point was found by solving $\begin{cases} 2x+y=10 \\ x+2y=10 \end{cases}$.

37.

Corner point	Value of $z = x + y$	
$(10, 0)$	$z = 10 + 0 = 10$	minimum
$(0, 10)$	$z = 0 + 10 = 10$	minimum
$\left(\dfrac{10}{3}, \dfrac{10}{3}\right)$	$z = \dfrac{10}{3} + \dfrac{10}{3} = \dfrac{20}{3}$	maximum

The maximum value of z is 10 at any point on the line $x + y = 10$ between $(10, 0)$ and $(0, 10)$.

The minimum value of z is $\dfrac{20}{3}$ at $\left(\dfrac{10}{3}, \dfrac{10}{3}\right)$.

39.

Corner point	Value of $z = 5x + 2y$	
$(10, 0)$	$z = 5(10) + 2(0) = 50$	maximum
$(0, 10)$	$z = 5(0) + 2(10) = 20$	minimum
$\left(\dfrac{10}{3}, \dfrac{10}{3}\right)$	$z = 5\left(\dfrac{10}{3}\right) + 2\left(\dfrac{10}{3}\right) = \dfrac{70}{3}$	

The maximum value of z is 50 at $(10, 0)$.
The minimum value of z is 20 at $(0, 10)$.

41.

Corner point	Value of $z = 3x + 4y$	
$(10, 0)$	$z = 3(10) + 4(0) = 30$	
$(0, 10)$	$z = 3(0) + 4(10) = 40$	maximum
$\left(\dfrac{10}{3}, \dfrac{10}{3}\right)$	$z = 3\left(\dfrac{10}{3}\right) + 4\left(\dfrac{10}{3}\right) = 30$	minimum

The maximum value of z is 40 at $(0, 10)$. The minimum value of z is $\dfrac{70}{3}$ at $\left(\dfrac{10}{3}, \dfrac{10}{3}\right)$.

43.

Corner point	Value of $z = 10x + y$	
$(10, 0)$	$z = 10(10) + 0 = 100$	maximum
$(0, 10)$	$z = 10(0) + 10 = 10$	minimum
$\left(\dfrac{10}{3}, \dfrac{10}{3}\right)$	$z = 10\left(\dfrac{10}{3}\right) + \dfrac{10}{3} = \dfrac{110}{3}$	

The maximum value of z is 100 at $(10, 0)$
The minimum value of z is 10 at $(0, 10)$.

In exercises 45–47, to find the maximum and minimum values of z, we graph the constraints, shade the set of feasible points, find the corner points, and evaluate the objective function, z, at each corner point.

45. $\begin{cases} 3x+3y\ge9 \\ -x+4y\le12 \\ 4x-y\le12 \\ x\ge0 \\ y\ge0 \end{cases}$

The corner points are $(3, 0)$, $(0, 3)$, and $(4, 4)$.
We find $(4, 4)$ by solving $\begin{cases} -x+4y=12 & (1) \\ 4x-y=12 & (2) \end{cases}$.

Multiplying (1) by 4 and adding it to (2) gives $15y = 60$ or $y = 4$. Back-substituting into (2) gives $4x = 16$ or $x = 4$.

Corner point	Value of $z = 18x + 30y$	
$(3, 0)$	$z = 18(3) + 30(0) = 54$	minimum
$(0, 3)$	$z = 18(0) + 30(3) = 90$	
$(4, 4)$	$z = 18(4) + 30(4) = 192$	maximum

The maximum value of 192 occurs at $(4, 4)$.
The minimum value of 54 occurs at $(3, 0)$.

47. $\begin{cases} 2x+3y\ge6 \\ -3x+4y\le8 \\ 5x-y\le15 \\ x\ge0 \\ y\ge0 \end{cases}$

The corner points are $(3, 0)$, $(0, 2)$, and $(4, 5)$.
We find $(4, 5)$ by solving $\begin{cases} -3x+4y=8 & (2) \\ 5x-y=15 & (3) \end{cases}$.

Multiplying (3) by 4 and adding it to (2) gives $17x = 68$ or $x = 4$. Back-substituting into (3) gives $y = 5$.

(continued on next page)

(*continued*)

Corner point	Value of $z = 7x + 6y$	
(3, 0)	$z = 7(3) + 6(0) = 21$	
(0, 2)	$z = 7(0) + 6(2) = 12$	minimum
(4, 5)	$z = 7(4) + 6(5) = 58$	maximum

The maximum value of 58 occurs at (4, 5).
The minimum value of 12 occurs at (0, 2).

In exercises 49–52, to find the maximum and minimum values of z, we graph the constraints, shade the set of feasible points, find the corner points, and evaluate the objective function, z, at each corner point.

49. $\begin{cases} x \le 15 \\ y \le 10 \\ 5x + 3y \ge 15 \\ -3x + 3y \le 21 \\ x \ge 0 \\ y \ge 0 \end{cases}$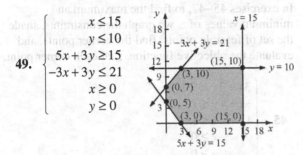

The corner points are (3, 0), (15, 0) (0, 5), (0, 7), (15, 10), and (3, 10). We find (3, 10) by evaluating $-3x + 3y = 21$ when $y = 10$.

Corner point	Value of $z = -20x + 30y$	
(3, 0)	$z = -20(3) + 30(0) = -60$	
(15, 0)	$z = -20(15) + 30(0) = -300$	
(0, 5)	$z = -20(0) + 30(5) = 150$	
(0, 7)	$z = -20(0) + 30(7) = 210$	
(15, 10)	$z = -20(15) + 30(10) = 0$	
(3, 10)	$z = -20(3) + 30(10) = 240$	maximum

The maximum value of 240 occurs at (3, 10).

51. $\begin{cases} x \le 15 \\ y \le 10 \\ 3x + 3y \ge 9 \\ -3x + 2y \le 14 \\ x \ge 0 \\ y \ge 0 \end{cases}$

The corner points are (3, 0), (15, 0) (0, 3), (0, 7), (15, 10), and (2, 10). We find (2, 10) by evaluating $-3x + 2y = 14$ when $y = 10$.

Corner point	Value of $z = -12x + 24y$	
(3, 0)	$z = -12(3) + 24(0) = -36$	
(15, 0)	$z = -12(15) + 24(0) = -180$	
(0, 3)	$z = -12(0) + 24(3) = 72$	
(0, 7)	$z = -12(0) + 24(7) = 168$	
(15, 10)	$z = -12(15) + 24(10) = 60$	
(2, 10)	$z = -12(2) + 24(10) = 216$	maximum

The maximum value of 216 occurs at (2, 10).

53. Answers will vary.

4.3 Models Utilizing Linear Programming with Two Variables

1. To answer parts (a), (b), and (c) we set up a linear programming problem and analyze its solution. We begin by naming the variables. Let x denote the number of acres of soybeans planted, and y denote the number of acres of wheat planted. We want to maximize $P = 180x + 100y$ subject to the constraints

$\begin{cases} x + y \le 70 \\ 60x + 30y \le 1800 \\ 3x + 4y \le 120 \\ x \ge 0 \\ y \ge 0 \end{cases}$ or $\begin{cases} x + y \le 70 \\ 2x + y \le 60 \\ 3x + 4y \le 120 \\ x \ge 0 \\ y \ge 0 \end{cases}$.

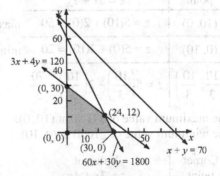

a. Evaluate the objective function at each corner.

Corner point	$P = 180x + 100y$	
(0, 0)	$180 \cdot 0 + 100 \cdot 0 = 0$	
(30, 0)	$180 \cdot 30 + 100 \cdot 0 = 5400$	
(24, 12)	$180 \cdot 24 + 100 \cdot 12 = 5520$	maximum
(0, 30)	$180 \cdot 0 + 100 \cdot 30 = 3000$	

The farmer should plant 24 acres of soybeans and 12 acres of wheat.

b. The maximum profit is $5520.

c. To find the maximum if the preparation constraint is raised to $2400, we rewrite the system of inequalities and re-evaluate the objective function at the corner points.

$$\begin{cases} x+ \ y \le \ 70 \\ 60x+30y \le 2400 \\ 3x+4y \le \ 120 \\ x \ge \ 0 \\ y \ge \ 0 \end{cases} \text{ or } \begin{cases} x+ \ y \le \ 70 \\ 2x+ \ y \le \ 80 \\ 3x+4y \le 120 \\ x \ge \ 0 \\ y \ge \ 0 \end{cases}$$

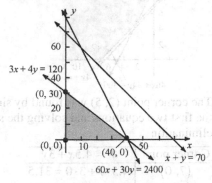

The graph shows that there is only 1 new corner point (40, 0), and that point (24, 12) is no longer a corner. So we evaluate the objective function at (40, 0) and compare its value to that at the other point.
$$P = 180 \cdot 40 + 100 \cdot 0 = 7200$$
Thus, the maximum profit is $7200 if the preparation constraint is raised to $2400.

3. Let x = amount invested in the Franklin Natural Resources fund bonds, and let y = the amount invested in the Oppenheimer Developing Markets fund. We want to maximize the return on the investment. That is, maximize
$P = 0.10x + 0.15y$ subject to the constraints

$$\begin{cases} x+y \le 20,000 \\ x \ge \ 5000 \\ y \le \ 8000 \ . \\ x \ge \ \ 0 \\ y \ge \ \ 0 \end{cases}$$

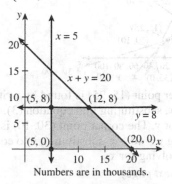

Numbers are in thousands.

The constraints are graphed in thousands of dollars, and the feasible points are shaded. We now evaluate the objective function at each corner point to find the amount that should be invested in each type of bond to maximize her return.

Corner point	$P = 0.10x + 0.15y$
(5000, 0)	$P = 0.10(5000) + 0.15(0) = 500$
(20000, 0)	$P = 0.10(20000) + 0.15(0) = 2000$
(5000, 8000)	$P = 0.10(5000) + 0.15(8000) = 1700$
(12000, 8000)	$P = 0.10(12000) + 0.15(8000) = 2400$

The broker should invest $12,000 in the Franklin Natural Resources fund bonds and $8000 in the Oppenheimer Developing Markets fund for a maximum return is $2400.

5. Let x = number (in thousands) of high potency vitamins produced, and let y = number (in thousands) of calcium-enriched vitamins produced. We want to maximize the profit, P, which is given by: $P = [0.10x + 0.05y] \cdot 1000$.

The limited amount of vitamin C and calcium available put the following constraints on the problem.

$$\begin{cases} 500x+100y \le 300,000 & (1) \\ 40x+400y \le 220,000 & (2) \\ x \ge 0 & (3) \\ y \ge 0 & (4) \end{cases} \text{ or}$$

$$\begin{cases} 5x+y \le 3000 \\ x+10y \le 5500 \\ x \ge 0 \\ y \ge 0 \end{cases}$$

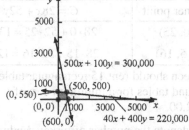

The corner point (500, 500) is found by solving equations (1) and (2). The objective function P is evaluated at the corner points as shown in the table.

(continued on next page)

(*continued*)

Since x and y are in thousands, the profit corresponding to each corner point is:

Corner point	$P = (0.10x + 0.05y)1000$
(0, 0)	$P = (0.10(0) + 0.05(0))1000 = 0$
(600, 0)	$P = (0.10(600) + 0.05(0))1000 = 60,000$
(0, 550)	$P = (0.10(0) + 0.05(550))1000 = 27,500$
(500, 500)	$P = (0.10(500) + 0.05(500))1000 = 75,000$

The maximum profit of $75,000 is made when 500,000 high-potency vitamins and 500,000 calcium enriched vitamins are produced.

7. Let x denote the number of rectangular tables rented, and let y denote the number of round tables rented. Kathleen wants to minimize her cost, $C = 28x + 52y$, subject to the constraints

$$\begin{cases} 6x + 10y = 250 \\ x + y \le 35 \\ x \le 15 \\ x \ge 0 \\ y \ge 0 \end{cases}$$

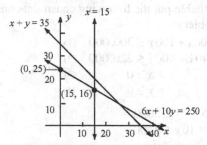

The objective function is evaluated at the corner points and the minimum value is found.

Corner point	$C = 28x + 52y$
(0, 25)	$28 \cdot 0 + 52 \cdot 25 = 1300$
(15, 16)	$28 \cdot 15 + 52 \cdot 16 = 1252$

Kathleen should rent 15 rectangular tables and 16 round tables for a minimum cost of $1252.00.

9. Let x denote the number of turkey sandwiches, and y denote the number of veggie sandwiches that Eric consumes. Eric wants to minimize his fat intake while assuring that he has sufficient protein, fiber, and carbohydrates.

He wants to minimize his fat intake, given by $F = 4.5x + 3y$, subject to the constraints

$$\begin{cases} 18x + 9y \ge 81 \\ 4x + 4y \ge 28 \\ 46x + 44y \ge 300 \\ x \ge 0 \\ y \ge 0 \end{cases}$$

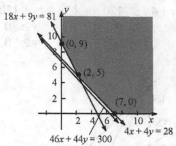

The corner point (2, 5) was found by simplifying the first two equations and solving the system by elimination.

Corner point	$F = 4.5x + 3y$
(7, 0)	$4.5 \cdot 7 + 3 \cdot 0 = 31.5$
(0, 9)	$4.5 \cdot 0 + 3 \cdot 9 = 27$
(2, 5)	$4.5 \cdot 2 + 3 \cdot 5 = 24$ minimum

Eric should eat 2 turkey and 5 veggie sandwiches to meet his dietary requirements and to minimize his fat. He will eat 24 grams of fat.

11. Let $x =$ units of the first product produced, and $y =$ units of the second product produced. The objective is to maximize the profit, $P = 40x + 60y$. The hours that each machine is available form the constraints on the problem.

$$\begin{cases} 2x + y \le 70 & \text{hrs. available machine 1} & (1) \\ x + y \le 40 & \text{hrs. available machine 2} & (2) \\ x + 3y \le 90 & \text{hrs. available machine 3} & (3) \\ x \ge 0 & & (4) \\ y \ge 0 & & (5) \end{cases}$$

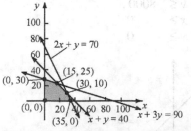

The corner point (15, 25) is found by writing $x = 40 - y$, substituting into equation (3), and solving for y. The corner point (30, 10) is found by writing $y = 40 - x$, substituting into equation (1), and solving for x.

(*continued on next page*)

(*continued*)

Corner Point	$P = 40x + 60y$
(0, 0)	$P = 40(0) + 60(0) = 0$
(35, 0)	$P = 40(35) + 60(0) = 1400$
(0, 30)	$P = 40(0) + 60(30) = 1800$
(30, 10)	$P = 40(30) + 60(10) = 1800$
(15, 25)	$P = 40(15) + 60(25) = 2100$ max

The factory should manufacture 15 units of the first product and 25 units of the second product, maximizing the profit at $2100.

13. Let x denote the units of Supplement A, and y denote the units of Supplement B in the diet. We want to minimize $C = 1.5x + y$ subject to the constraints

$$\begin{cases} 5x + 2y \ge 60 & (1) \\ 3x + 2y \ge 45 & (2) \\ 4x - y \ge 30 & (3) \\ x \ge 0 & (4) \\ y \ge 0 & (5) \end{cases}$$

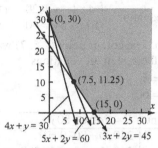

Corner point (7.5, 11.25) was found by solving the system $\begin{cases} 5x + 2y = 60 \\ 3x + 2y = 45 \end{cases}$ using substitution.

$2y = 60 - 5x$ and $2y = 45 - 3x$.

$60 - 5x = 45 - 3x$

$\qquad 15 = 2x$

$\qquad x = 7.5$

Back-substituting, $y = \dfrac{(60 - 5 \cdot 7.5)}{2} = 11.25$

Corner point	$C = 1.5x + y$
(15, 0)	$1.5 \cdot 15 + 0 = 22.5$
(0, 30)	$1.5 \cdot 0 + 30 = 30$
(7.5, 11.25)	$1.5 \cdot 7.5 + 11.25 = 22.5$

One should take any combination of supplements A and B that satisfy the equation $3x + 2y = 45$ between the points (15, 0) and (7.5, 11.25). These combinations minimize the cost at $22.50.

15. Let x denote the amount allocated to the 30 year loans, and y denote the amount allocated to the 15 year loans. We want to maximize

$$P = \frac{0.05125}{12}x + \frac{0.04375}{12}y \text{ subject to the}$$

constraints

$$\begin{cases} x + y \le 72 \\ x \le 2y \\ 0 \le x \le 15 \\ 0 \le y \le 15 \end{cases}$$

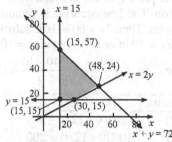

The point (48, 24) is found by substitution. $x = 2y$ is substituted into the first equation $2y + y = 72$ to obtain $y = 24$. Back-substituting, we find $x = 2 \cdot 24 = 48$.

Corner point	$P = \dfrac{0.05125}{12}x + \dfrac{0.04375}{12}y$
(30, 15)	$\dfrac{0.05125}{12} \cdot 30 + \dfrac{0.04375}{12} \cdot 15 = 0.182813$
(48, 24)	$\dfrac{0.05125}{12} \cdot 48 + \dfrac{0.04375}{12} \cdot 24 = 0.2925$
(15, 57)	$\dfrac{0.05125}{12} \cdot 15 + \dfrac{0.04375}{12} \cdot 57 = 0.271875$
(15, 15)	$\dfrac{0.05125}{12} \cdot 15 + \dfrac{0.04375}{12} \cdot 15 = 0.11875$

Remembering the amounts were given in millions, we conclude that Fremont Bank should allocate $48 million to the 30 year loans and $24 million to 15 year loans. The bank will receive $292,500 in interest.

17. Let x denote the number of racing skates and let y denote the number of figure skates to be manufactured. We want to maximize $P = 10x + 12y$ subject to the constraints

$$\begin{cases} 6x + 4y \le 120 \\ x + 2y \le 40 \\ x \ge 0 \\ y \ge 0 \end{cases}$$

(*continued on next page*)

(*continued*)

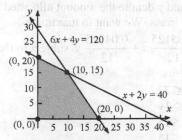

The corner point (10, 15) is found by substitution. The first equation is simplified to $3x + 2y = 60$. Then $2y = 60 - 3x$ is substituted into $x + 2y = 40$ in order to obtain $x = 10$. Back-substitute x, so $y = \dfrac{40 - x}{2} = \dfrac{40 - 10}{2} = 15$.

Corner point	$P = 10x + 12y$
(0, 0)	$10 \cdot 0 + 12 \cdot 0 = 0$
(20, 0)	$10 \cdot 20 + 12 \cdot 0 = 200$
(0, 20)	$10 \cdot 0 + 12 \cdot 20 = 240$
(10, 15)	$10 \cdot 10 + 12 \cdot 15 = 280$

The factory should manufacture 10 pairs of racing skates and 15 pairs of figure skates to obtain a maximum profit of $280.

19. Let x denote the number of items of A produced and y denote the number of items of B produced. We want to maximize $P = 1.5x + y$ subject to the constraints

$$\begin{cases} 2x + 4y \le 3000 & \text{carbon monoxide limits} \\ 6x + 3y \le 5400 & \text{sulfur dioxide limits} \\ x \ge 0 \\ y \ge 0 \end{cases}$$

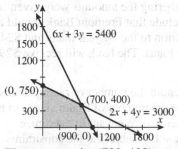

The corner point (700, 400) was found by solving the simplified system of equations

$$\begin{cases} x + 2y = 1500 \\ 2x + y = 1800 \end{cases}.$$

Corner point	$P = 1.5x + y$
(0, 0)	$1.5(0) + 0 = 0$
(900, 0)	$1.5(900) + 0 = 1350$
(0, 750)	$1.5(0) + 750 = 750$
(700, 400)	$1.5(700) + 400 = 1450$

The maximum profit is $1450, and it is obtained when 700 units of type A compound and 400 units of type B compound are produced.

21. Let x denote the number of tons of Type 1 steel produced and y denote the number of tons of Type 2 steel produced. We want to maximize $P = 240x + 80y$ subject to the constraints

$$\begin{cases} 2x + 5y \le 40 & \text{melting} & (1) \\ 4x + y \le 20 & \text{cutting} & (2) \\ 10x + 5y \le 60 & \text{rolling} & (3) \\ x \ge 0 & & (4) \\ y \ge 0 & & (5) \end{cases}$$

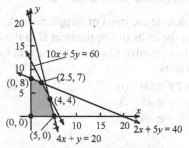

The corner point (4, 4) was found by solving the system of equations $\begin{cases} 4x + y = 20 & (2) \\ 10x + 5y = 60 & (3) \end{cases}$ by substitution. The corner point (2.5, 7) was found by solving the system of equations

$$\begin{cases} 2x + 5y = 40 & (1) \\ 10x + 5y = 60 & (3) \end{cases}.$$

Corner point	$P = 240x + 80y$
(0, 0)	$240 \cdot 0 + 80 \cdot 0 = 0$
(5, 0)	$240 \cdot 5 + 80 \cdot 0 = 1200$
(0, 8)	$240 \cdot 0 + 80 \cdot 8 = 640$
$\left(\dfrac{5}{2}, 7\right)$	$240 \cdot \dfrac{5}{2} + 80 \cdot 7 = 1160$
(4, 4)	$240 \cdot 4 + 80 \cdot 4 = 1280$

The maximum profit is $1280, and it is obtained when 4 tons of Type 1 steel and 4 tons of Type 2 steel are produced.

23. Let x denote the number of ounces of Supplement I added and y denote the number of Supplement II added. Danny wants to minimize his cost, C, while maintaining a healthy diet for his chickens. He is minimizing $C = 0.03x + 0.04y$ subject to the constraints

$$\begin{cases} 5x + 25y \ge 50 & \text{vitamin 1} & (1) \\ 25x + 10y \ge 100 & \text{vitamin 2} & (2) \\ 10x + 10y \ge 60 & \text{vitamin 3} & (3) \\ 35x + 20y \ge 180 & \text{vitamin 4} & (4) \\ x \ge 0 & & (5) \\ y \ge 0 & & (6) \end{cases}$$

(*continued on next page*)

(*continued*)

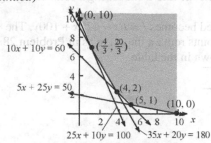

The corner point (5, 1) was found by solving the

system of equations $\begin{cases} 5x+25y=50 & (1) \\ 10x+10y=60 & (3) \end{cases}$.

The corner point (4, 2) was found by solving the

system of equations $\begin{cases} 10x+10y=60 & (3) \\ 35x+20y=180 & (4) \end{cases}$.

The corner point $\left(\dfrac{4}{3}, \dfrac{20}{3}\right)$ is the solution to the

system of equations $\begin{cases} 25x+10y=100 & (2) \\ 35x+20y=180 & (4) \end{cases}$.

Corner point	$P=0.03x+0.04y$
(10, 0)	$0.03(10)+0.04(0)=0.3$
(0, 10)	$0.03(0)+0.04(10)=0.4$
(5, 1)	$0.03(5)+0.04(1)=0.19$
$\left(\dfrac{4}{3}, \dfrac{20}{3}\right)$	$0.03\left(\dfrac{4}{3}\right)+0.04\left(\dfrac{20}{3}\right)=0.31$
(4, 2)	$0.03(4)+0.04(2)=0.20$

The minimum cost is $0.19 when Danny adds 5 ounces of Supplement I and 1 ounce of Supplement II to every 100 ounces of feed.

25. Let x denote the amount invested in the ICON Bond fund and let y denote the amount invested in ING Mid Cap Opportunities. The couple wants to maximize $I = 0.0435x + 0.0505y$ subject to the constraints
$$\begin{cases} x+y \le 45000 \\ x \ge 10000 \\ y \ge 10000 \\ x \le 15000 \\ y \le 2x \end{cases}$$

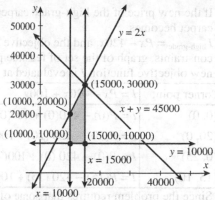

Corner point	$I = 0.0435x + 0.0505y$
(15000, 10000)	$I = 1157.50$
(15000, 30000)	$I = 2167.50$
(10000, 20000)	$I = 1445$
(10000, 10000)	$I = 940$

The couple should invest $15,000 in the ICON Bond fund and $30,000 in ING Mid Cap Opportunities for a maximum annual yield of $2167.50.

27. Let x denote the number of sales reps from NYC attending the meeting, and let y denote the number of sales reps from Chicago attending the meeting. The company wants to minimize travel costs $C = 343x + 329y$ subject to the constraints
$$\begin{cases} x+y \ge 40 \\ x \le 28 \\ y \le 22 \\ x \ge 16 \\ y \ge 12 \end{cases}$$

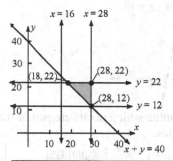

Corner points	$C = 343x + 329y$
(28, 12)	$343 \cdot 28 + 329 \cdot 12 = 13,552$
(28, 22)	$343 \cdot 28 + 329 \cdot 22 = 16,842$
(18, 22)	$343 \cdot 18 + 329 \cdot 22 = 13,412$

Eighteen sales representatives from New York City and 22 from Chicago should be sent to the meeting to minimize the total airfare. The minimum airfare is $13,412.00.

29. If the new price of the high-grade carpet is denoted by P, then the income from selling x rolls of high-grade carpet becomes
$I_{\text{high-grade}} = Px - 420x$, and the objective function to be maximized becomes $I = Px - 420x + 100y$. The constraints, graph of the set of feasible points, and the corner points remain the same as in Problem 28. So the new objective function I is evaluated at the corner points as shown in the table

Corner point	$I = Px - 420x + 100y$
(0, 0)	$I = P(0) - 420(0) + 100(0) = 0$
(20, 0)	$I = P(20) - 420\,(20) + 100(0) = 20P - 8400$
(0, 25)	$I = P(0) - 420\,(0) + 100(25) = 2500$
(15, 10)	$I = P(15) - 420\,(15) + 100(10) = 15P - 5300$

Since the problem requires that some of each grade carpet be manufactured, the corner point associated with the maximum income must be (15, 10). This means that both of the following inequalities must be true.
$15P - 5300 \geq 2500 \Rightarrow 15P \geq 7800 \Rightarrow P \geq 520$ and $15P - 5300 \geq 20P - 8400 \Rightarrow -5P \geq -3100 \Rightarrow P \leq 620$
So if high-grade carpet is priced between \$520 and \$620 per roll, some of each type of carpeting will be manufactured to maximize income.

Chapter 4 Review Exercises

1. The inequality is nonstrict, so the line is solid. The point $(-2, 0)$ satisfies the inequality, so we shade the side of the graph containing $(-2, 0)$.

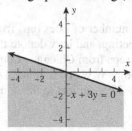

3. The inequality is strict, so the line is dashed. The point $(4, 0)$ satisfies the inequality, so we shade the side of the graph containing $(4, 0)$.

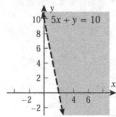

5. To determine which points are part of the graph of the system, check each point to see if it satisfies both of the inequalities.

Point	Inequality 1 $x + 2y \leq 8$	Inequality 2 $2x - y \geq 4$	Conclusion
$P_1 = (4, -3)$	Is $4 + 2 \cdot (-3) \leq 8$? $-2 < 8$	Is $2 \cdot 4 - (-3) \geq 4$? $11 > 4$	Both inequalities are satisfied $(4, -3)$ is part of the graph.
$P_2 = (2, -6)$	Is $2 + 2 \cdot (-6) \leq 8$? $-10 < 8$	Is $2 \cdot 2 - (-6) \geq 4$? $10 > 4$	Both inequalities are satisfied $(2, -6)$ is part of the graph.
$P_3 = (8, -3)$	Is $8 + 2 \cdot (-3) \leq 8$? $2 < 8$	Is $2 \cdot 8 - (-3) \geq 4$? $19 > 4$	Both inequalities are satisfied $(8, -3)$ is part of the graph.

7. The region that represents the graph of the system is the set of points common to the solutions of each individual inequality. Use the test point (0, 0).

$$6x - 4y \le 12 \qquad 3x + 2y \le 18$$
$$6 \cdot 0 - 4 \cdot 0 = 0 \qquad 3 \cdot 0 + 2 \cdot 0 = 0$$
$$0 < 12 \qquad\qquad 0 < 18$$

The test point (0, 0) satisfies both inequalities, so the region (**a**) represents the graph of the system.

9.

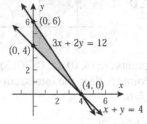

The graph is bounded. The corner points are (4, 0), (0, 4), and (0, 6).

11.

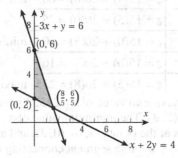

The graph is bounded. The corner points are (0, 2), (0, 6), and $\left(\dfrac{8}{5}, \dfrac{6}{5}\right)$. To find the third corner point, we solved the system by elimination.

$$\begin{cases} x + 2y = 4 & (1) \\ 3x + y = 6 & (2) \end{cases}$$

Subtracting (1) from twice (2) gives $5x = 8$ or $x = \dfrac{8}{5}$. Back-substituting in (1) we get $y = \dfrac{6}{5}$.

13.

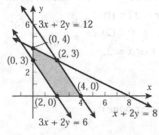

The graph is bounded. The corner points are (2, 0), (4, 0), (0, 3), (0, 4), and (2, 3).

To find the (2, 3), we solved the system by elimination.

$$\begin{cases} x + 2y = 8 & (1) \\ 3x + 2y = 12 & (2) \end{cases}$$

Subtracting (1) from (2) gives $2x = 4$ or $x = 2$. Back-substituting in (1) we get $2y = 6$ or $y = 3$.

For exercises 15–21, use the graph below and its corner points to find the maximum and minimum values of z.

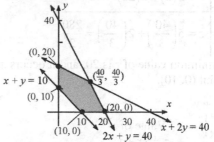

The corner points are (10, 0), (20, 0), (0, 10), (0, 20), and $\left(\dfrac{40}{3}, \dfrac{40}{3}\right)$. We found $\left(\dfrac{40}{3}, \dfrac{40}{3}\right)$ by solving the system of equations

$$\begin{cases} x + 2y = 40 & (1) \\ 2x + y = 40 & (2) \end{cases}.$$

Using elimination, we subtract (2) from two times (1) and find $3y = 40$ or $y = \dfrac{40}{3}$. Back-substituting y in

(2) gives $2x + \dfrac{40}{3} = 40$ or $x = \dfrac{40}{3}$.

To find the maximum and minimum values of z, we construct a table and evaluate the objective function at the corner points.

15. $z = x + y$

Corner point	Value of $z = x + y$	
(10, 0)	$z = 10 + 0 = 10$	
(20, 0)	$z = 20 + 0 = 20$	
(0, 10)	$z = 0 + 10 = 10$	
(0, 20)	$z = 0 + 20 = 20$	
$\left(\dfrac{40}{3}, \dfrac{40}{3}\right)$	$z = \dfrac{40}{3} + \dfrac{40}{3} = \dfrac{80}{3}$	maximum

The maximum value of z is $\dfrac{80}{3} \approx 26.67$, and it

occurs at the point $\left(\dfrac{40}{3}, \dfrac{40}{3}\right)$.

17. $z = 5x + 2y$

Corner point	Value of $z = 5x + 2y$	
(10, 0)	$z = 5(10) + 2(0) = 50$	
(20, 0)	$z = 5(20) + 2(0) = 100$	
(0, 10)	$z = 5(0) + 2(10) = 20$	minimum
(0, 20)	$z = 5(0) + 2(20) = 40$	
$\left(\dfrac{40}{3}, \dfrac{40}{3}\right)$	$z = 5\left(\dfrac{40}{3}\right) + 2\left(\dfrac{40}{3}\right) = \dfrac{280}{3}$	

The minimum value of z is 20, and it occurs at the point (0, 10).

19. $z = 2x + y$

Corner point	Value of $z = 2x + y$	
(10, 0)	$z = 2(10) + 0 = 20$	
(20, 0)	$z = 2(20) + 0 = 40$	maximum
(0, 10)	$z = 2(0) + 10 = 10$	
(0, 20)	$z = 2(0) + 20 = 20$	
$\left(\dfrac{40}{3}, \dfrac{40}{3}\right)$	$z = 2\left(\dfrac{40}{3}\right) + \dfrac{40}{3} = 40$	maximum

The maximum value of z is 40, and it occurs at the points (20, 0) and $\left(\dfrac{40}{3}, \dfrac{40}{3}\right)$ and at all points on the line segment connecting them.

21. $z = 2x + 5y$

Corner point	Value of $z = 2x + 5y$	
(10, 0)	$z = 2(10) + 5(0) = 20$	minimum
(20, 0)	$z = 2(20) + 5(0) = 40$	
(0, 10)	$z = 2(0) + 5(10) = 50$	
(0, 20)	$z = 2(0) + 5(20) = 100$	
$\left(\dfrac{40}{3}, \dfrac{40}{3}\right)$	$z = 2\left(\dfrac{40}{3}\right) + 5\left(\dfrac{40}{3}\right) = \dfrac{280}{3}$	

The minimum value of z is 20, and it occurs at the point (10, 0).

For exercises 23–25, to find the maximum and minimum values of z, we will graph the constraints, shade the set of feasible points, find the corner points, and evaluate the objective function, z, at each corner point.

23. $\begin{cases} x \le 5 \\ x \le 8 \\ 3x + 4y \ge 12 \\ x \ge 0 \\ y \ge 0 \end{cases}$

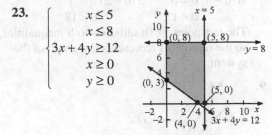

The corner points are (4, 0), (5, 0), (0, 3), (0, 8), and (5, 8). Point (5, 8) is the intersection of lines $x = 5$ and $y = 8$.

Corner Point	$z = 15x + 20y$	
(4, 0)	$z = 15(4) + 20(0) = 60$	minimum
(5, 0)	$z = 15(5) + 20(0) = 75$	
(0, 3)	$z = 15(0) + 20(3) = 60$	minimum
(0, 8)	$z = 15(0) + 20(8) = 160$	
(5, 8)	$z = 15(5) + 20(8) = 235$	maximum

The maximum value of $z = 235$; it occurs at the point (5, 8). The minimum value of $z = 60$; it occurs at the points (4, 0), (0, 3), and at all points on the line segment connecting them.

25. $\begin{cases} 2x + 3y \le 22 \\ x \le 5 \\ y \le 6 \\ x \ge 0 \\ y \ge 0 \end{cases}$

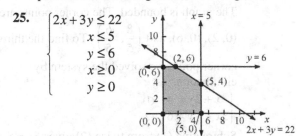

The corner points are (0, 0), (5, 0), (0, 6), (5, 4), and (2, 6). To find (5, 4) solve
$\begin{cases} 2x + 3y = 22 & (1) \\ x = 5 & (2) \end{cases}$. Substitute 5 for x in (1) to get $3y = 12$ or $y = 4$. To find (2, 6) solve
$\begin{cases} 2x + 3y = 22 & (1) \\ y = 6 & (2) \end{cases}$. Substitute 6 for y in (1) to get $2x = 4$ or $x = 2$.

(continued on next page)

(*continued*)

Corner Point	$z = 15x + 20y$	
(0, 0)	$z = 15(0) + 20(0) = 0$	minimum
(5, 0)	$z = 15(5) + 20(0) = 75$	
(0, 6)	$z = 15(0) + 20(6) = 120$	
(5, 4)	$z = 15(5) + 20(4) = 155$	maximum
(2, 6)	$z = 15(2) + 20(6) = 150$	

The maximum value of $z = 155$; it occurs at the point (5, 4). The minimum value of $z = 0$ occurs at the point (0, 0).

27. $\begin{cases} x \le 9 \\ y \le 8 \\ x + y \ge 3 \\ x \ge 0 \\ y \ge 0 \end{cases}$

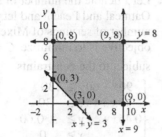

The corner points are (3, 0), (9, 0), (0, 3), (0, 8), and (9, 8).

Corner point	Value of $z = 2x + 3y$	
(3, 0)	$z = 2(3) + 3(0) = 6$	
(9, 0)	$z = 2(9) + 3(0) = 18$	
(0, 3)	$z = 2(0) + 3(3) = 9$	
(0, 8)	$z = 2(0) + 3(8) = 24$	
(9, 8)	$z = 2(9) + 3(8) = 42$	maximum

The maximum value of $z = 42$; it occurs at the point (9, 8).

29. $\begin{cases} x + y \ge 1 \\ y \le 2x \\ x \le 8 \\ y \le 8 \\ x \ge 0 \\ y \ge 0 \end{cases}$

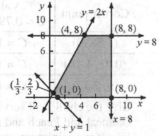

The corner points are (1, 0), (8, 0), (4, 8), $\left(\dfrac{1}{3}, \dfrac{2}{3}\right)$, and (8, 8). $\left(\dfrac{1}{3}, \dfrac{2}{3}\right)$ is the solution of

$\begin{cases} x + y = 1 & (1) \\ y = 2x & (2) \end{cases}$.

By substituting $2x$ for y in (1), we get $x = \dfrac{1}{3}$,

then back-substituting into (2) gives $y = \dfrac{2}{3}$.

Point (4, 8) is the intersection of the lines $y = 8$ and $y = 2x$.

Corner point	Value of $z = x + 2y$	
(1, 0)	$z = 1 + 2(0) = 1$	
(8, 0)	$z = 8 + 2(0) = 8$	
(4, 8)	$z = 4 + 2(8) = 20$	
$\left(\dfrac{1}{3}, \dfrac{2}{3}\right)$	$z = \dfrac{1}{3} + 2\left(\dfrac{2}{3}\right) = \dfrac{5}{3}$	
(8, 8)	$z = 8 + 2(8) = 24$	maximum

The maximum value of z is 24, and it occurs at the point (8, 8).

31. $\begin{cases} x + 2y \ge 8 \\ 3x + y \ge 6 \\ x \le 8 \\ x \ge 0 \\ y \ge 0 \end{cases}$

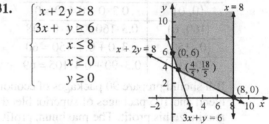

The corner points are (8, 0), (0, 6), and $\left(\dfrac{4}{5}, \dfrac{18}{5}\right)$. Solve $\begin{cases} x + 2y = 8 & (1) \\ 3x + y = 6 & (2) \end{cases}$ to find $\left(\dfrac{4}{5}, \dfrac{18}{5}\right)$.

Corner point	Value of $z = 3x + 2y$	
(8, 0)	$z = 3(8) + 2(0) = 24$	
(0, 6)	$z = 3(0) + 2(6) = 12$	
$\left(\dfrac{4}{5}, \dfrac{18}{5}\right)$	$z = 3\left(\dfrac{4}{5}\right) + 2\left(\dfrac{18}{5}\right) = \dfrac{48}{5}$	minimum

The minimum value of z is $\dfrac{48}{5}$, and it occurs at the point $\left(\dfrac{4}{5}, \dfrac{18}{5}\right)$.

33. a.
$$\begin{cases} \frac{1}{4}x+\frac{1}{2}y \le 75 & (1) \\ \frac{3}{4}x+\frac{1}{2}y \le 120 & (2) \end{cases} \text{ or simplified}$$
$$\begin{aligned} x \ge\ 0 & \quad (3) \\ y \ge\ 0 & \quad (4) \end{aligned}$$

$$\begin{cases} x+2y \le 300 & (1) \\ 3x+2y \le 480 & (2) \\ x \ge\ 0 & (3) \\ y \ge\ 0 & (4) \end{cases}$$

b.

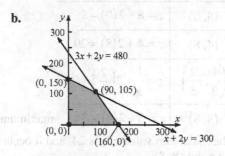

c.

Corner Point	$P = 0.3x + 0.4y$
$(0, 0)$	$0.3 \cdot 0 + 0.4 \cdot 0 = 0$
$(160, 0)$	$0.3 \cdot 160 + 0.4 \cdot 0 = 48$
$(0, 150)$	$0.3 \cdot 0 + 0.4 \cdot 150 = 60$
$(90, 105)$	$0.3 \cdot 90 + 0.4 \cdot 105 = 69$

Bill should prepare 90 packages of economy blend and 105 packages of superior blend to maximize his profit. The maximum profit will be $69.

35. Let x denote the number of downhill skis, and let y denote the number of cross-country skis. The objective is to maximize $P = 70x + 50y$ subject to the constraints
$$\begin{cases} 2x+y \le 40 & (1) \\ x+y \le 32 & (2) \\ x \ge\ 0 & (3) \\ y \ge\ 0 & (4) \end{cases}.$$

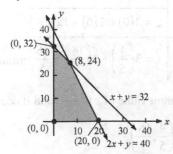

The corner points are $(20, 0)$, $(8, 24)$, $(0, 32)$, and $(0, 0)$. The corner point $(8, 24)$ is found by solving the system of equations
$$\begin{cases} 2x+y = 40 & (1) \\ x+y = 32 & (2) \end{cases}.$$

Corner point	Value of $P = 70x + 50y$	
$(0, 0)$	$P = 70(0) + 50(0) = 0$	
$(20, 0)$	$P = 70(20) + 50(0) = 1400$	
$(0, 32)$	$P = 70(0) + 50(32) = 1600$	
$(8, 24)$	$P = 70(8) + 50(24) = 1760$	maximum

The manufacturer should produce 8 pairs of downhill skis and 24 of cross-country skis for a maximum profit of $1760.

37. Let x denote the number of servings of Banana Oatmeal and Peach, and let y denote the number of servings of Mixed Fruit Juice. The objective is to minimize $C = 0.79x + 0.65y$ subject to the constraints
$$\begin{cases} 90x+60y \ge\ 130 & (1) \\ 19x+15y \ge\ 30 & (2) \\ 0.45x+\ y \ge\ 0.60 & (3) \\ x \ge\ 0 & (4) \\ y \ge\ 0 & (5) \end{cases}$$

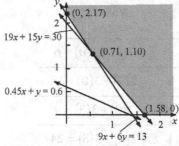

The corner point $(0.71, 1.10)$ is found by solving the system of equations
$$\begin{cases} 90x+60y = 130 & (1) \\ 19x+15y = 30 & (2) \end{cases}.$$

Corner point	Value of $C = 0.79x + 0.65y$	
$(1.58, 0)$	$C = 1.248$	minimum
$(0.71, 1.10)$	$C = 1.276$	
$(0, 2.17)$	$C = 1.411$	

Give the child 1.58 servings of Gerber Banana Oatmeal and Peach and no Gerber Mixed Fruit Juice for a minimum cost of $1.25.

39. Let x denote the number (in thousands) of high-potency vitamins to be produced and let y denote the number (in thousands) of calcium-enriched vitamins to be produced. The amount of available ingredients is given in kilograms. 1kg = 1,000,000 mg. The objective is to maximize

$$P = 0.10(1000)x + 0.05(1000)y = 100x + 50y$$

subject to the constraints

$$\begin{cases} 500,000x + 100,000y \leq 300,000,000 \\ 40,000x + 400,000y \leq 122,000,000 \\ 100,000x + 40,000y \leq 65,000,000 \\ \qquad\qquad\qquad x \geq \qquad 0 \\ \qquad\qquad\qquad y \geq \qquad 0 \end{cases}$$

or $\begin{cases} 5x + \quad y \leq 3000 \\ 4x + 40y \leq 12200 \\ 10x + 4y \leq 6500. \\ \quad x \geq 0 \\ \quad y \geq 0 \end{cases}$

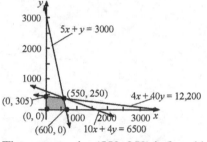

The corner point (550, 250) is found by solving the system of equations formed by any two of the first three constraints.

Corner point	Value of $P = 100x + 50y$	
(0, 0)	$P = 100(0) + 50(0) = 0$	
(600, 0)	$P = 100(600) + 50(0) = 60000$	
(0, 305)	$P = 100(0) + 50(305) = 15250$	
(550, 250)	$P = 100(550) + 50(250) = 67500$	max

The maximum profit is $67,500. It is attained when 550,000 high-potency vitamins and 250,000 calcium-enriched vitamins are produced.

41. Let x denote the number of shares of Boeing and let y denote the number of shares of Honeywell. The investor wants to maximize the quarterly yield, $Y = 0.42x + 0.30y$ subject to the constraints

$$\begin{cases} 64x + 40y \leq 500,000 & (1) \\ \qquad\quad x \geq \quad 2,000 & (2) \\ \qquad\quad y \geq \quad 2,000 & (3) \\ 64x \leq 300,000 & (4) \\ 40y \leq 300,000 & (5) \end{cases}$$

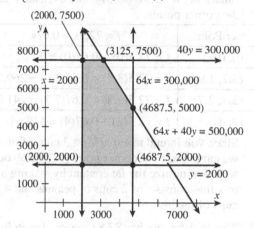

Corner point	Value of $Y = 0.42x + 0.30y$	
(2000, 2000)	$Y = 1440$	
(4687.5, 2000)	$Y = 2568.75$	
(4687.5, 5000)	$Y = 3468.75$	
(3125, 7500)	$Y = 3562.50$	max
(2000, 7500)	$Y = 3090$	

To maximize the projected dividend, the fund should purchase 3125 shares of Boeing and 7500 shares of Honeywell. The projected dividend is $3562.50.

Chapter 4 Project

1. Objective function: $c = 31.4x + 114.74y$

3.

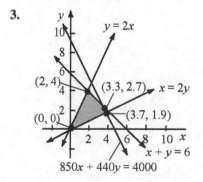

5. The total mix contains 6 cups, so there are $521.76 \div 6 = 86.96$ grams of carbohydrates per cup. Since each cup of raisins contain 440 calories and each cup of peanuts have 850 calories, the calorie count per cup is $(4 \cdot 440 + 2 \cdot 850) \div 6 = 576.67$ calories per cup of mix.

7. To determine the recipe needed to minimize the fat in the mix, construct a new objective function, $f = 72.5x + 0.67y$, and evaluate it at the corner points.

Corner Point	$f = 72.5x + 0.67y$
(0, 0)	$f = 72.5(0) + 0.67(0) = 0$
(3.7, 1.9)	$f = 72.5(3.7) + 0.67(1.9) = 269.52$
(3.3, 2.7)	$f = 72.5(3.3) + 0.67(2.7) = 241.06$
(2, 4)	$f = 72.5(2) + 0.67(4) = 147.68$

Since you intend to eat at least 3 cups of mix, we cannot use the corner point (0, 0). So you would minimize the fat content by making a mix that consists of 2 cups of peanuts and 4 cups of raisins.

9. This low-fat mix has 87.82 grams of protein. See exercise 6.

Mathematical Questions from Professional Exams

1. b

3. c

5. Profit is the difference between revenue and cost, so profit for product $Q = 20 - 12 = 8$, and profit for product $P = 17 - 13 = 4$. The objective function is choice d.

7. c

9. b

11. b

13. c

Chapter 5 Linear Programming: Simplex Method

5.1 The Simplex Tableau; Pivoting

1.
$$\begin{bmatrix} 0 & 3 & 2 & 1 & 0 & 200 \\ 0 & 1 & 3 & 0 & 1 & 150 \\ 1 & -2 & -3 & 0 & 0 & 0 \end{bmatrix} \xrightarrow{R_1 = \frac{1}{3}r_1}$$

$$\begin{bmatrix} 0 & 1 & \frac{2}{3} & \frac{1}{3} & 0 & \frac{200}{3} \\ 0 & 1 & 3 & 0 & 1 & 150 \\ 1 & -2 & -3 & 0 & 0 & 0 \end{bmatrix}$$

3. True

5. False. Slack variables are always nonnegative.

For exercises 7–15, a maximum problem is in standard form if both of the following conditions are met.

Condition 1 All the variables are nonnegative.
Condition 2 Every other constraint is written as a linear expression that is less than or equal to a positive constant.

7. The problem is in standard form since both conditions are met.

9. The problem is not in standard form since variables x_2 and x_3 are not given as nonnegative.

11. The problem is not in standard form. The constraint $2x_1 + x_2 + 4x_3 \geq 6$ is not a linear expression that is less than or equal to a positive constant.

13. The problem is not in standard form. The constraint $x_1 + x_2 \geq -6$ is not a linear expression that is less than or equal to a positive constant.

15. The problem is in standard form since both conditions are met.

17. The maximum problem cannot be modified to be in standard form.

19. The maximum problem cannot be modified to be in standard form.

21. The problem can be modified so it is in standard form. Multiply the first two constraints by −1. The problem becomes maximize
$P = 2x_1 + x_2 + 3x_3$ subject to the constraints
$x_1 - x_2 - x_3 \leq 6, -2x_1 + 3x_2 \leq 12,$
$x_3 \leq 2, x_1 \geq 0, x_2 \geq 0, x_3 \geq 0.$

23. We write the objective function as
$P - 2x_1 - x_2 - 3x_3 = 0$ subject to the constraints
$$\begin{aligned} 5x_1 + 2x_2 + \ x_3 + s_1 & = 20 \\ 6x_1 - \ x_2 + 4x_3 \quad + s_2 & = 24 \\ x_1 + \ x_2 + 4x_3 \quad\quad + s_3 & = 16 \end{aligned}$$
$x_1 \geq 0, x_2 \geq 0, x_3 \geq 0,$
$s_1 \geq 0, \ s_2 \geq 0, \ s_3 \geq 0$
The initial tableau is

BV	P	x_1	x_2	x_3	s_1	s_2	s_3	RHS
s_1	0	5	2	1	1	0	0	20
s_2	0	6	−1	4	0	1	0	24
s_3	0	1	1	4	0	0	1	16
P	1	−2	−1	−3	0	0	0	0

25. We write the objective function as
$P - 3x_1 - 5x_2 = 0$ subject to the constraints
$$\begin{aligned} 2.2x_1 - 1.8x_2 + s_1 & = 5 \\ 0.8x_1 + 1.2x_2 \quad + s_2 & = 2.5 \\ x_1 + \ x_2 \quad\quad + s_3 & = 0.1 \end{aligned}$$
$x_1 \geq 0, x_2 \geq 0,$
$s_1 \geq 0, \ s_2 \geq 0, s_3 \geq 0$
The initial tableau is

BV	P	x_1	x_2	s_1	s_2	s_3	RHS
s_1	0	2.2	−1.8	1	0	0	5
s_2	0	0.8	1.2	0	1	0	2.5
s_3	0	1	1	0	0	1	0.1
P	1	−3	−1	0	0	0	0

27. We write the objective function as
$P - 2x_1 - 3x_2 - x_3 = 0$ subject to the constraints
$$\begin{aligned} x_1 + \ x_2 + x_3 + s_1 & = 50 \\ 3x_1 + 2x_2 + x_3 \quad + s_2 & = 10 \end{aligned}$$
$x_1 \geq 0, x_2 \geq 0, x_3 \geq 0,$
$s_1 \geq 0, \ s_2 \geq 0$
The initial tableau is

BV	P	x_1	x_2	x_3	s_1	s_2	RHS
s_1	0	1	1	1	1	0	50
s_2	0	3	2	1	0	1	10
P	1	−2	−3	−1	0	0	0

29. We write the objective function as
$P - 3x_1 - 4x_2 - 2x_3 = 0$ subject to the
constraints

$$3x_1 + x_2 + 4x_3 + s_1 \qquad = 5$$
$$x_1 - x_2 \qquad + s_2 \qquad = 5$$
$$2x_1 - x_2 + x_3 \qquad + s_3 = 6$$
$$x_1 \geq 0,\ x_2 \geq 0,\ x_3 \geq 0,$$
$$s_1 \geq 0,\ s_2 \geq 0,\ s_3 \geq 0$$

The initial tableau is

BV	P	x_1	x_2	x_3	s_1	s_2	s_3	RHS
s_1	0	3	1	4	1	0	0	5
s_2	0	1	−1	0	0	1	0	5
s_3	0	2	−1	1	0	0	1	6
P	1	−3	−4	−2	0	0	0	0

31. We modify the problem by writing each linear
constraint as an inequality less than or equal to a
positive constant. The modified maximum
problem in standard form is maximize
$P = x_1 + 2x_2 + 5x_3$ subject to the constraints

$$x_1 - 2x_2 - 3x_3 \leq 10$$
$$-3x_1 - x_2 + x_3 \leq 12$$
$$x_1 \geq 0,\ x_2 \geq 0,\ x_3 \geq 0$$

We then introduce slack variables and set up the
initial simplex tableau.

$$P - x_1 - 2x_2 - 5x_3 \qquad = 0$$
$$x_1 - 2x_2 - 3x_3 + s_1 \qquad = 10$$
$$-3x_1 - x_2 + x_3 \qquad + s_2 = 12$$
$$x_1 \geq 0,\ x_2 \geq 0,\ x_3 \geq 0,\ s_1 \geq 0,\ s_2 \geq 0$$

The initial tableau is

BV	P	x_1	x_2	x_3	s_1	s_2	RHS
s_1	0	1	−2	−3	1	0	10
s_2	0	−3	−1	1	0	1	12
P	1	−1	−2	−5	0	0	0

33. We modify the problem by writing each linear
constraint as an inequality less than or equal to a
positive constant. The modified maximum
problem in standard form is maximize
$P = 2x_1 + 3x_2 + x_3 + 6x_4$ subject to the
constraints

$$-x_1 + x_2 + 2x_3 + x_4 \leq 10$$
$$-x_1 + x_2 - x_3 + x_4 \leq 8$$
$$x_1 + x_2 + x_3 + x_4 \leq 9$$
$$x_1 \geq 0\quad x_2 \geq 0\quad x_3 \geq 0\quad x_4 \geq 0$$

We then introduce slack variables and set up the
initial simplex tableau.

$$P - 2x_1 - 3x_2 - x_3 - 6x_4 \qquad = 0$$
$$-x_1 + x_2 + 2x_3 + x_4 + s_1 \qquad = 10$$
$$-x_1 + x_2 - x_3 + x_4 \qquad + s_2 \qquad = 8$$
$$x_1 + x_2 + x_3 + x_4 \qquad + s_3 = 9$$
$$x_1 \geq 0,\ x_2 \geq 0,\ x_3 \geq 0,\ x_4 \geq 0$$
$$s_1 \geq 0,\ s_2 \geq 0,\ s_3 \geq 0$$

The initial tableau is

BV	P	x_1	x_2	x_3	x_4	s_1	s_2	s_3	RHS
s_1	0	−1	1	2	1	1	0	0	10
s_2	0	−1	1	−1	1	0	1	0	8
s_3	0	1	1	1	1	0	0	1	9
P	1	−2	−3	−1	−6	0	0	0	0

35.a. Step 1: Divide each entry in the *pivot row* by
the *pivot element*.

BV	P	x_1	x_2	s_1	s_2	RHS
s_1	0	1	2	1	0	300
s_2	0	**3**	2	0	1	480
P	1	−1	−2	0	0	0

BV	P	x_1	x_2	s_1	s_2	RHS
s_1	0	1	2	1	0	300
$\to x_1$	0	1	$\frac{2}{3}$	0	$\frac{1}{3}$	160
P	1	−1	−2	0	0	0

Step 2: Obtain 0s elsewhere in the *pivot
column* by performing row operations using
the revised *pivot row*.

BV	P	x_1	x_2	s_1	s_2	RHS
s_1	0	0	$\frac{4}{3}$	1	$-\frac{1}{3}$	140
x_1	0	1	$\frac{2}{3}$	0	$\frac{1}{3}$	160
P	1	0	$-\frac{4}{3}$	0	$\frac{1}{3}$	160

with $R_1 = -R_2 + r_1$, $R_3 = R_2 + r_3$.

b. The system of equations corresponding to the
new tableau is

$$\begin{cases} s_1 = -\dfrac{4}{3}x_2 + \dfrac{1}{3}s_2 + 140 \\[2mm] x_1 = -\dfrac{2}{3}x_2 - \dfrac{1}{3}s_2 + 160 \\[2mm] P = \dfrac{4}{3}x_2 - \dfrac{1}{3}s_2 + 160 \end{cases}$$

c. The current values are $P = 160$, $x_1 = 160$, and
$s_1 = 140$.

37. a. Step 1: Divide each entry in the *pivot row* by the *pivot element*.

BV	P	x_1	x_2	x_3	s_1	s_2	s_3	RHS
s_1	0	1	2	4	1	0	0	24
s_2	0	2	-1	1	0	1	0	32
s_3	0	3	2	4	0	0	1	18
P	1	-1	-2	-3	0	0	0	0

$\rightarrow$

BV	P	x_1	x_2	x_3	s_1	s_2	s_3	RHS
s_1	0	1	2	4	1	0	0	24
s_2	0	2	-1	1	0	1	0	32
x_1	0	1	$\frac{2}{3}$	$\frac{4}{3}$	0	0	$\frac{1}{3}$	6
P	1	-1	-2	-3	0	0	0	0

Step 2: Obtain 0s elsewhere in the *pivot column* by performing row operations using the revised *pivot row*.

$\begin{array}{c} \\ \xrightarrow{\hspace{1cm}} \\ R_1 = -R_3 + r_1 \\ R_2 = -2R_3 + r_2 \\ R_4 = R_3 + r_4 \end{array}$

BV	P	x_1	x_2	x_3	s_1	s_2	s_3	RHS
s_1	0	0	$\frac{4}{3}$	$\frac{8}{3}$	1	0	$-\frac{1}{3}$	18
s_2	0	0	$-\frac{7}{3}$	$-\frac{5}{3}$	0	1	$-\frac{2}{3}$	20
x_1	0	1	$\frac{2}{3}$	$\frac{4}{3}$	0	0	$\frac{1}{3}$	6
P	1	0	$-\frac{4}{3}$	$-\frac{5}{3}$	0	0	$\frac{1}{3}$	6

b. The system of equations corresponding to the new tableau is

$$\begin{cases} s_1 = -\dfrac{4}{3}x_2 - \dfrac{8}{3}x_3 + \dfrac{1}{3}s_3 + 18 \\[2mm] s_2 = \dfrac{7}{3}x_2 + \dfrac{5}{3}x_3 + \dfrac{2}{3}s_3 + 20 \\[2mm] x_1 = -\dfrac{2}{3}x_2 - \dfrac{4}{3}x_3 - \dfrac{1}{3}s_3 + 6 \\[2mm] P = \dfrac{4}{3}x_2 + \dfrac{5}{3}x_3 - \dfrac{1}{3}s_3 + 6 \end{cases}$$

c. The current values are $P = 6$, $x_1 = 6$, $s_1 = 18$, and $s_2 = 20$.

39. a. Step 1: Divide each entry in the *pivot row* by the *pivot element*.

BV	P	x_1	x_2	x_3	x_4	s_1	s_2	s_3	s_4	RHS
s_1	0	-3	0	1	0	1	0	0	0	20
s_2	0	2	0	0	1	0	1	0	0	24
s_3	0	0	-3	1	0	0	0	1	0	28
s_4	0	0	-3	0	1	0	0	0	1	24
P	1	-1	-2	-3	-4	0	0	0	0	0

$\rightarrow$

BV	P	x_1	x_2	x_3	x_4	s_1	s_2	s_3	s_4	RHS
s_1	0	-3	0	1	0	1	0	0	0	20
x_1	0	1	0	0	$\frac{1}{2}$	0	$\frac{1}{2}$	0	0	12
s_3	0	0	-3	1	0	0	0	1	0	28
s_4	0	0	-3	0	1	0	0	0	1	24
P	1	-1	-2	-3	-4	0	0	0	0	0

Step 2: Obtain 0s elsewhere in the *pivot column* by performing row operations using the revised *pivot row*.

$\begin{array}{c} \\ \xrightarrow{\hspace{1cm}} \\ R_1 = 3R_2 + r_1 \\ R_5 = R_2 + r_5 \end{array}$

BV	P	x_1	x_2	x_3	x_4	s_1	s_2	s_3	s_4	RHS
s_1	0	0	0	1	$\frac{3}{2}$	1	$\frac{3}{2}$	0	0	56
x_1	0	1	0	0	$\frac{1}{2}$	0	$\frac{1}{2}$	0	0	12
s_3	0	0	-3	1	0	0	0	1	0	28
s_4	0	0	-3	0	1	0	0	0	1	24
P	1	0	-2	-3	$-\frac{7}{2}$	0	$\frac{1}{2}$	0	0	12

b. The system of equations corresponding to the new tableau is

$$\begin{cases} s_1 = -x_3 - \dfrac{3}{2}x_4 - \dfrac{3}{2}s_2 + 56 \\[2mm] x_1 = -\dfrac{1}{2}x_4 - \dfrac{1}{2}s_2 + 12 \\[2mm] s_3 = 3x_2 - x_3 + 28 \\[1mm] s_4 = 3x_2 - x_4 + 24 \\[1mm] P = 2x_2 + 2x_3 + \dfrac{7}{2}x_4 - \dfrac{1}{2}s_2 + 12 \end{cases}$$

c. The current values are $P = 12$, $x_1 = 12$, $s_1 = 56$, $s_3 = 28$, and $s_4 = 24$.

41.a. P is the profit, x_1 is the number of sleeping dolls, x_2 is the number of talking dolls, and x_3 is the number of walking dolls.

b. To find the profit, we subtract the cost of each doll from its selling price.
Sleeping baby dolls: $12 − $6 = $6; talking baby dolls: $13.50 − $7.50 = $6
Walking baby dolls: $17 − $9 = $8.
We use these values in the objective function.
Maximize $P = 6x_1 + 6x_2 + 8x_3$ subject to the constraints

$$x_1 + x_2 + x_3 \le 60$$
$$6x_1 + 7.5x_2 + 9x_3 \le 405$$
$$x_1 \ge 0 \quad x_2 \ge 0 \quad x_3 \ge 0$$

c. System with slack variables added

$$x_1 + x_2 + x_3 + s_1 = 60$$
$$6x_1 + 7.5x_2 + 9x_3 + s_2 = 405$$
$$P - 6x_1 - 6x_2 - 8x_3 = 0$$
$$x_1 \ge 0 \quad x_2 \ge 0 \quad x_3 \ge 0 \quad s_1 \ge 0 \quad s_2 \ge 0$$

d. The initial tableau is

BV	P	x_1	x_2	x_3	s_1	s_2	RHS
s_1	0	1	1	1	1	0	60
s_2	0	6	7.5	9	0	1	405
P	1	−6	−6	−8	0	0	0

43.a. P is the daily profit, x_1 is the number of shirts produced, x_2 is the number of jackets produced, and x_3 is the number of pairs of pants produced each day.

b. Maximize $P = 19x_1 + 34x_2 + 15x_3$ subject to the constraints

$$10x_1 + 20x_2 + 20x_3 \le 500$$
$$20x_1 + 40x_2 + 30x_3 \le 900$$
$$x_1 + x_2 + x_3 \le 30$$
$$x_1 \ge 0 \quad x_2 \ge 0 \quad x_3 \ge 0$$

c. System with slack variables added

$$10x_1 + 20x_2 + 20x_3 + s_1 = 500$$
$$20x_1 + 40x_2 + 30x_3 + s_2 = 900$$
$$x_1 + x_2 + x_3 + s_3 = 30$$
$$P - 19x_1 - 34x_2 - 15x_3 = 0$$
$$x_1 \ge 0 \quad x_2 \ge 0 \quad x_3 \ge 0 \quad s_1 \ge 0 \quad s_2 \ge 0 \quad s_3 \ge 0$$

d. The initial tableau is

BV	P	x_1	x_2	x_3	s_1	s_2	s_3	RHS
s_1	0	10	20	20	1	0	0	500
s_2	0	20	40	30	0	1	0	900
s_3	0	1	1	1	0	0	1	30
P	1	−19	−34	−15	0	0	0	0

45. a. P is the return on her investments, x_1 is the amount invested in the money market, x_2 is the amount invested in the mutual fund, and x_3 is the amount invested in the CD.

b. Maximize $P = 0.0105x_1 + 0.05x_2 + 0.0166x_3$ subject to the constraints

$$x_1 + x_2 + x_3 \le 25{,}000$$
$$x_2 + x_3 \le 10{,}000$$
$$-x_1 \quad\;\; + x_3 \le \;\; 1{,}500$$
$$x_1 \ge 0 \;\; x_2 \ge 0 \;\; x_3 \ge 0$$

c. System with slack variables added

$$x_1 + \quad x_2 + \quad x_3 + s_1 \qquad\qquad = 25{,}000$$
$$x_2 + \quad x_3 \quad\;\; + s_2 \qquad = 10{,}000$$
$$-x_1 \quad + \quad x_3 \qquad\quad + s_3 = \;\; 1{,}500$$
$$P - 0.0105x_1 - 0.05x_2 - 0.0166x_3 \qquad\qquad = 0$$
$$x_1 \ge 0 \;\; x_2 \ge 0 \;\; x_3 \ge 0 \;\; s_1 \ge 0 \;\; s_2 \ge 0 \;\; s_3 \ge 0$$

d. The initial tableau is

BV	P	x_1	x_2	x_3	s_1	s_2	s_3	RHS
s_1	0	1	1	1	1	0	0	25,000
s_2	0	0	1	1	0	1	0	10,000
s_3	0	−1	0	1	0	0	1	1,500
P	1	−0.0105	−0.05	−0.0166	0	0	0	0

5.2 The Simplex Method: Solving Maximum Problems in Standard Form

1. False. The pivot element is never in the objective row. The pivot element is chosen from the positive entries above the objective row in the pivot column.

3. False. The maximum problem is unbounded and has no solution only if all the entries in the pivot column are zero or negative.

5. BV P x_1 x_2 s_1 s_2 RHS

BV	P	x_1	x_2	s_1	s_2	RHS
s_1	0	1	0	1	$-\frac{1}{2}$	20
x_2	0	$\frac{1}{2}$	1	0	$\frac{1}{4}$	30
P	1	−1	0	1	1	120

(b) This requires further pivoting. Since there is still a negative entry in the objective row, the tableau needs further pivoting. We find the pivot element by selecting the first column containing a negative entry in the objective row. Here it is column x_1. We select the pivot row by computing the quotients formed by dividing the entry in the right hand side by the corresponding positive entry of the pivot column. The pivot row has the smallest nonnegative quotient.

$$20 \div 1 = 20 \text{ and } 30 \div \frac{1}{2} = 60$$

The pivot element is 1 in row s_1, column x_1.

7. BV P x_1 x_2 s_1 s_2 RHS

BV	P	x_1	x_2	s_1	s_2	RHS
s_1	0	0	$\frac{1}{14}$	1	$-\frac{1}{7}$	$\frac{186}{21}$
x_1	0	1	$\frac{12}{7}$	0	$\frac{4}{7}$	$\frac{32}{7}$
P	1	0	$\frac{12}{7}$	1	$\frac{32}{7}$	$\frac{256}{7}$

(a) This is the final tableau since there are no negative entries in the objective row. The maximum value is $P = \frac{256}{7}$. It occurs when $x_1 = \frac{32}{7}$ and $x_2 = 0$.

9. BV P x_1 x_2 s_1 s_2 RHS

BV	P	x_1	x_2	s_1	s_2	RHS
x_1	0	1	−2	0	4	24
s_1	0	0	−2	1	4	36
P	1	5	−10	12	4	20

(c) There is no solution to this problem. Although there is a negative entry in the objective row, all entries in the pivot column are negative, so the problem is unbounded and has no solution.

11.

BV	P	x_1	x_2	s_1	s_2	s_3	RHS
x_2	0	0	1	-2	0	1	6
s_2	0	0	0	1	1	4	6
x_1	0	1	0	1	0	-1	1
P	1	0	0	-10	0	-5	110

(b) This tableau requires further pivoting since there are still negative entries in the objective row. We find the pivot element by selecting the first column containing a negative entry in the objective row. Here it is column s_1. We select the pivot row by computing the quotients formed by dividing the entry in the right hand side by the corresponding positive entry of the pivot column. The pivot row has the smallest nonnegative quotient. $6 \div 1 = 6$ and $1 \div 1 = 1$, thus, the new pivot element is 1 in row x_1, column s_1.

13. To solve the problem using the simplex method, we first must introduce slack variables and construct the initial tableau. Maximize $P - 5x_1 - 7x_2 = 0$ subject to the constraints

$$2x_1 + 3x_2 + s_1 \quad\quad = 12$$
$$3x_1 + x_2 \quad\quad + s_2 = 12$$
$$x_1 \geq 0 \quad x_2 \geq 0 \quad s_1 \geq 0 \quad s_2 \geq 0$$

The initial simplex tableau with the pivot element marked is below. The first negative entry in the objective row identifies the pivot column. The smallest nonnegative quotient formed by the RHS and positive entries in the pivot column identifies the pivot row. Here the pivot element is in row s_2, column x_1.

BV	P	x_1	x_2	s_1	s_2	RHS
s_1	0	2	3	1	0	12
s_2	0	$\boxed{3}$	1	0	1	12
P	1	-5	-7	0	0	0

We pivot using row operations

$\xrightarrow{R_2 = \frac{1}{3} r_2}$

BV	P	x_1	x_2	s_1	s_2	RHS
s_1	0	2	3	1	0	12
x_1	0	1	$\frac{1}{3}$	0	$\frac{1}{3}$	4
P	1	-5	-7	0	0	0

$\xrightarrow[R_3 = 5R_2 + r_3]{R_1 = -2R_2 + r_1}$

BV	P	x_1	x_2	s_1	s_2	RHS
s_1	0	0	$\boxed{\frac{7}{3}}$	1	$-\frac{2}{3}$	4
x_1	0	1	$\frac{1}{3}$	0	$\frac{1}{3}$	4
P	1	0	$-\frac{16}{3}$	0	$\frac{5}{3}$	20

Since there is still a negative entry in the objective row, the problem needs further pivoting. We choose the pivot element in row s_1, column x_2, and pivot using row operations.

$\xrightarrow{R_1 = \frac{3}{7} r_1}$

BV	P	x_1	x_2	s_1	s_2	RHS
x_2	0	0	1	$\frac{3}{7}$	$-\frac{2}{7}$	$\frac{12}{7}$
x_1	0	1	$\frac{1}{3}$	0	$\frac{1}{3}$	4
P	1	0	$-\frac{16}{3}$	0	$\frac{5}{3}$	20

$\xrightarrow[R_3 = \frac{16}{3} R_1 + r_3]{R_2 = -\frac{1}{3} R_1 + r_2}$

BV	P	x_1	x_2	s_1	s_2	RHS
x_2	0	0	1	$\frac{3}{7}$	$-\frac{2}{7}$	$\frac{12}{7}$
x_1	0	1	0	$-\frac{1}{7}$	$\frac{3}{7}$	$\frac{24}{7}$
P	1	0	0	$\frac{16}{7}$	$\frac{1}{7}$	$\frac{204}{7}$

This is the final tableau since all entries in the objective row are positive. The maximum is

$$P = \frac{204}{7}, \text{ obtained when } x_1 = \frac{24}{7} \text{ and}$$

$$x_2 = \frac{12}{7}.$$

15. To solve the problem using the simplex method, we first must introduce slack variables and construct the initial tableau. Maximize $P - 5x_1 - 7x_2 = 0$ subject to the constraints

$$x_1 + 2x_2 + s_1 \quad\quad = 2$$
$$2x_1 + x_2 \quad\quad + s_2 = 2$$
$$x_1 \geq 0 \quad x_2 \geq 0 \quad s_1 \geq 0 \quad s_2 \geq 0$$

The initial simplex tableau with the pivot element marked is below. The first negative entry in the objective row identifies the pivot column. The smallest nonnegative quotient formed by the RHS and positive entries in the pivot column identifies the pivot row. Here the pivot element is in row s_2, column x_1. We then use row operations to make the pivot element 1 and all the other entries in the pivot column 0.

BV	P	x_1	x_2	s_1	s_2	RHS
s_1	0	1	2	1	0	2
s_2	0	$\boxed{2}$	1	0	1	2
P	1	-5	-7	0	0	0

$\xrightarrow[R_3 = 5R_2 + r_3]{\substack{R_2 = \frac{1}{2} r_2 \\ R_1 = -R_2 + r_1}}$

BV	P	x_1	x_2	s_1	s_2	RHS
s_1	0	0	$\boxed{\frac{3}{2}}$	1	$-\frac{1}{2}$	1
x_1	0	1	$\frac{1}{2}$	0	$\frac{1}{2}$	1
P	0	0	$-\frac{9}{2}$	0	$\frac{5}{2}$	5

(*continued on next page*)

(*continued*)

Since there is still a negative entry in the objective row, we need to pivot again. We choose the pivot entry as before. The pivot column will be x_2, the pivot row will be s_1 because $\dfrac{\frac{1}{3}}{\frac{3}{2}} = \dfrac{2}{3} < \dfrac{\frac{1}{2}}{\frac{1}{2}} = 2$.

$$
\begin{array}{c}
\\
\\
R_1 = \frac{2}{3} r_1 \\
R_2 = -\frac{1}{2} R_1 + r_2 \\
R_3 = \frac{9}{2} R_1 + r_3
\end{array}
\quad
\begin{array}{c|cccccc|c}
\text{BV} & P & x_1 & x_2 & s_1 & s_2 & \text{RHS} \\
\hline
x_2 & 0 & 0 & 1 & \frac{2}{3} & -\frac{1}{3} & \frac{2}{3} \\
x_1 & 0 & 1 & 0 & -\frac{1}{3} & \frac{2}{3} & \frac{2}{3} \\
P & 0 & 0 & 0 & 3 & 1 & 8
\end{array}
$$

This is the final tableau since all entries in the objective row are nonnegative. The maximum is $P = 8$, obtained when $x_1 = \dfrac{2}{3}$ and $x_2 = \dfrac{2}{3}$.

17. To solve the problem using the simplex method, we first must introduce slack variables and construct the initial tableau. Maximize $P - 3x_1 - x_2 = 0$ subject to the constraints

$$
\begin{aligned}
x_1 + \ x_2 + s_1 \qquad\qquad &= 2 \\
2x_1 + 3x_2 \qquad + s_2 \qquad &= 12 \\
3x_1 + \ x_2 \qquad\qquad + s_3 &= 12 \\
x_1 \geq 0 \quad x_2 \geq 0 \quad s_1 \geq 0 \quad s_2 \geq 0 \quad s_3 \geq 0
\end{aligned}
$$

The initial simplex tableau with the pivot element marked is below. The first negative entry in the objective row identifies the pivot column. The smallest nonnegative quotient formed by the RHS and positive entries in the pivot column identifies the pivot row. Here the pivot element is in row s_1, column x_1. We then use row operations to make all entries in the pivot column other than the pivot element 0.

$$
\begin{array}{c|cccccc|c}
\text{BV} & P & x_1 & x_2 & s_1 & s_2 & s_3 & \text{RHS} \\
\hline
s_1 & 0 & \boxed{1} & 1 & 1 & 0 & 0 & 2 \\
s_2 & 0 & 2 & 3 & 0 & 1 & 0 & 12 \\
s_3 & 0 & 3 & 1 & 0 & 0 & 1 & 12 \\
P & 1 & -3 & -1 & 0 & 0 & 0 & 0
\end{array}
\quad
\begin{array}{c}
\\
\\
R_2 = -2R_1 + r_2 \\
R_3 = -3R_1 + r_3 \\
R_4 = 3R_1 + r_4
\end{array}
\;\longrightarrow\;
\begin{array}{c|cccccc|c}
\text{BV} & P & x_1 & x_2 & s_1 & s_2 & s_3 & \text{RHS} \\
\hline
x_1 & 0 & 1 & 1 & 1 & 0 & 0 & 2 \\
s_2 & 0 & 0 & 1 & -2 & 1 & 0 & 8 \\
s_3 & 0 & 0 & -2 & -3 & 0 & 1 & 6 \\
P & 1 & 0 & 2 & 3 & 0 & 0 & 6
\end{array}
$$

This is the final tableau since all entries in the objective row are nonnegative. The maximum is $P = 6$, obtained when $x_1 = 2$ and $x_2 = 0$.

19. To solve the problem using the simplex method, we first must introduce slack variables and construct the initial tableau. Maximize $P - 2x_1 - x_2 - x_3 = 0$ subject to the constraints

$$
\begin{aligned}
-2x_1 + \ x_2 - 2x_3 + s_1 \qquad\ &= 4 \\
x_1 - 2x_2 + \ x_3 \qquad + s_2 &= 2 \\
x_1 \geq 0 \quad x_2 \geq 0 \quad x_3 \geq 0 \quad s_1 \geq 0 \quad s_2 \geq 0
\end{aligned}
$$

The initial simplex tableau with the pivot element marked is below. The first negative entry in the objective row identifies the pivot column. The smallest nonnegative quotient formed by the RHS and positive entries in the pivot column identifies the pivot row. Here the pivot element is in row s_2, column x_1. We then use row operations to make all entries in the pivot column other than the pivot element 0.

$$
\begin{array}{c|cccccc|c}
\text{BV} & P & x_1 & x_2 & x_3 & s_1 & s_2 & \text{RHS} \\
\hline
s_1 & 0 & -2 & 1 & -2 & 1 & 0 & 4 \\
s_2 & 0 & \boxed{1} & -2 & 1 & 0 & 1 & 2 \\
P & 1 & -2 & -1 & -1 & 0 & 0 & 0
\end{array}
\quad
\begin{array}{c}
\\
R_1 = 2R_2 + r_1 \\
R_3 = 2R_2 + r_3
\end{array}
\;\longrightarrow\;
\begin{array}{c|cccccc|c}
\text{BV} & P & x_1 & x_2 & x_3 & s_1 & s_2 & \text{RHS} \\
\hline
s_1 & 0 & 0 & -3 & 0 & 1 & 2 & 8 \\
x_1 & 0 & 1 & -2 & 1 & 0 & 1 & 2 \\
P & 1 & 0 & -5 & 1 & 0 & 2 & 4
\end{array}
$$

The new pivot column is x_2, since -5 is the only negative entry in the objective row. But all the entries in the pivot column are negative. This problem is unbounded and has no maximum.

21. To solve the problem using the simplex method, we first must introduce slack variables and construct the initial tableau. Maximize $P - 2x_1 - x_2 - 3x_3 = 0$ subject to the constraints

$$x_1 + 2x_2 + x_3 + s_1 \qquad = 25$$
$$3x_1 + 2x_2 + 3x_3 \qquad + s_2 = 30$$
$$x_1 \geq 0 \quad x_2 \geq 0 \quad x_3 \geq 0 \quad s_1 \geq 0 \quad s_2 \geq 0$$

The initial simplex tableau with the pivot element marked is below. The first negative entry in the objective row identifies the pivot column. The smallest nonnegative quotient formed by the RHS and positive entries in the pivot column identifies the pivot row. Here the pivot element is in row s_2, column x_1. We then use row operations tomake the pivot element 1 and all other entries in the pivot column 0.

BV	P	x_1	x_2	x_3	s_1	s_2	RHS
s_1	0	1	2	1	1	0	25
s_2	0	[3]	2	3	0	1	30
P	1	-2	-1	-3	0	0	0

$$\xrightarrow[\substack{R_1 = -R_2 + r_1 \\ R_3 = 2R_2 + r_3}]{R_2 = \frac{1}{3}r_2}$$

BV	P	x_1	x_2	x_3	s_1	s_2	RHS
s_1	0	0	$\frac{4}{3}$	0	1	$-\frac{1}{3}$	15
x_1	0	1	$\frac{2}{3}$	[1]	0	$\frac{1}{3}$	10
P	1	0	$\frac{1}{3}$	-1	0	$\frac{2}{3}$	20

Since there is still a negative entry in the objective row, we need to pivot again. We choose the pivot entry as before. The pivot column will be x_3, the pivot row will be x_1.

$$\xrightarrow{R_3 = R_2 + r_3}$$

BV	P	x_1	x_2	x_3	s_1	s_2	RHS
s_1	0	0	$\frac{4}{3}$	0	1	$-\frac{1}{3}$	15
x_3	0	1	$\frac{2}{3}$	1	0	$\frac{1}{3}$	10
P	1	1	1	0	0	1	30

Since all the entries in the objective row are nonnegative, this is the final tableau. The maximum is $P = 30$, obtained when $x_1 = 0$, $x_2 = 0$, and $x_3 = 10$.

23. To solve the problem using the simplex method, we first must introduce slack variables and construct the initial tableau. Maximize $P - 2x_1 - 4x_2 - x_3 - x_4 = 0$ subject to the constraints

$$2x_1 + x_2 + 2x_3 + 3x_4 + s_1 \qquad = 12$$
$$2x_2 + x_3 + 2x_4 \qquad + s_2 \qquad = 20$$
$$2x_1 + x_2 + 4x_3 \qquad + s_3 = 16$$
$$x_1 \geq 0 \quad x_2 \geq 0 \quad x_3 \geq 0 \quad x_4 \geq 0 \quad s_1 \geq 0 \quad s_2 \geq 0 \quad s_3 \geq 0$$

The initial tableau is below. The pivot element is 2 (row s_1, column x_1), making x_1 a basic variable.

BV	P	x_1	x_2	x_3	x_4	s_1	s_2	s_3	RHS
s_1	0	[2]	1	2	3	1	0	0	12
s_2	0	0	2	1	2	0	1	0	20
s_3	0	2	1	4	0	0	0	1	16
P	1	-2	-4	-1	-1	0	0	0	0

$$\xrightarrow[\substack{R_3 = -2R_1 + r_3 \\ R_4 = 2R_1 + r_4}]{R_1 = \frac{1}{2}r_1}$$

BV	P	x_1	x_2	x_3	x_4	s_1	s_2	s_3	RHS
x_1	0	1	$\frac{1}{2}$	1	$\frac{3}{2}$	$\frac{1}{2}$	0	0	6
s_2	0	0	2	[2]	1	2	0	1	0	20
s_3	0	0	0	2	-3	-1	0	1	4
P	1	0	-3	1	2	1	0	0	12

The new tableau needs further pivoting. The new pivot column is x_2 and the new pivot row is s_2 making 2 the pivot element.

$$\xrightarrow[\substack{R_1 = -\frac{1}{2}R_2 + r_1 \\ R_4 = 3R_2 + r_4}]{R_2 = \frac{1}{2}r_2}$$

BV	P	x_1	x_2	x_3	x_4	s_1	s_2	s_3	RHS
x_1	0	1	0	$\frac{3}{4}$	1	$\frac{1}{2}$	$-\frac{1}{4}$	0	1
x_2	0	0	1	$\frac{1}{2}$	1	0	$\frac{1}{2}$	0	10
s_3	0	0	0	2	-3	-1	0	1	4
P	1	0	0	$\frac{5}{2}$	5	1	$\frac{3}{2}$	0	42

This is the final tableau since all the entries in the objective row are nonnegative. The maximum is $P = 42$, obtained when $x_1 = 1$, $x_2 = 10$, $x_3 = 0$, and $x_4 = 0$.

25. To solve the problem using the simplex method, we first must introduce slack variables and construct the initial tableau. Maximize $P - 2x_1 - x_2 - x_3 = 0$ subject to the constraints

$$x_1 + 2x_2 + 4x_3 + s_1 \qquad\qquad = 20$$
$$2x_1 + 4x_2 + 4x_3 \qquad + s_2 \qquad = 60$$
$$3x_1 + 4x_2 + \ x_3 \qquad\qquad + s_3 = 90$$
$$x_1 \ge 0 \quad x_2 \ge 0 \quad x_3 \ge 0 \quad s_1 \ge 0 \quad s_2 \ge 0 \quad s_3 \ge 0$$

The initial pivot is below. The pivot element is 1; the entering variable is x_1.

BV	P	x_1	x_2	x_3	s_1	s_2	s_3	RHS
s_1	0	1	2	4	1	0	0	20
s_2	0	2	4	4	0	1	0	60
s_3	0	3	4	1	0	0	1	90
P	1	−2	−1	−1	0	0	0	0

$R_2 = -2r_1 + r_2$
$R_3 = -3r_1 + r_3$
$R_4 = 2r_1 + r_4$

BV	P	x_1	x_2	x_3	s_1	s_2	s_3	RHS
x_1	0	1	2	4	1	0	0	20
s_2	0	0	0	−4	−2	1	0	20
s_3	0	0	−2	−11	−3	0	1	30
P	1	0	3	7	2	0	0	40

This is the final tableau since all the entries in the objective row are nonnegative. The maximum is $P = 40$, obtained when $x_1 = 20$, $x_2 = 0$, and $x_3 = 0$.

27. To solve the problem using the simplex method, we first must introduce slack variables and construct the initial tableau. Maximize $P - x_1 - 2x_2 - 4x_3 + x_4 = 0$ subject to the constraints

$$5x_1 \qquad\quad + 4x_3 + 6x_4 + s_1 \qquad = 20$$
$$4x_1 + 2x_2 + 2x_3 + 8x_4 \qquad + s_2 = 40$$
$$x_1 \ge 0 \quad x_2 \ge 0 \quad x_3 \ge 0 \quad x_4 \ge 0 \quad s_1 \ge 0 \quad s_2 \ge 0$$

The initial tableau is below. The pivot element is 5 and the entering variable is x_1.

BV	P	x_1	x_2	x_3	x_4	s_1	s_2	RHS
s_1	0	5	0	4	6	1	0	20
s_2	0	4	2	2	8	0	1	40
P	1	−1	−2	−4	1	0	0	0

$R_1 = \frac{1}{5}r_1$
$R_2 = -4R_1 + r_2$
$R_3 = R_1 + r_3$

BV	P	x_1	x_2	x_3	x_4	s_1	s_2	RHS
x_1	0	1	0	$\frac{4}{5}$	$\frac{6}{5}$	$\frac{1}{5}$	0	4
s_2	0	0	2	$-\frac{6}{5}$	$\frac{16}{5}$	$-\frac{4}{5}$	1	24
P	1	0	−2	$\frac{16}{5}$	$\frac{11}{5}$	$\frac{1}{5}$	0	4

Since there is still a negative entry in the objective row, this tableau needs further pivoting. The new pivot will be 2, x_2 will be the entering basic variable.

BV	P	x_1	x_2	x_3	x_4	s_1	s_2	RHS
x_1	0	1	0	$\frac{4}{5}$	$\frac{6}{5}$	$\frac{1}{5}$	0	4
x_2	0	0	1	$-\frac{3}{5}$	$\frac{8}{5}$	$-\frac{2}{5}$	$\frac{1}{2}$	12
P	1	0	0	$-\frac{22}{5}$	$\frac{27}{5}$	$-\frac{3}{5}$	1	28

$R_2 = \frac{1}{2}r_2$
$R_3 = 2R_2 + r_3$

$R_1 = \frac{5}{4}r_1$
$R_2 = \frac{3}{5}R_1 + r_2$
$R_3 = \frac{22}{5}R_1 + r_3$

BV	P	x_1	x_2	x_3	x_4	s_1	s_2	RHS
x_3	0	$\frac{5}{4}$	0	1	$\frac{3}{2}$	$\frac{1}{4}$	0	5
x_2	0	$\frac{3}{4}$	1	0	$\frac{5}{2}$	$-\frac{1}{4}$	$\frac{1}{2}$	15
P	1	$\frac{11}{2}$	0	0	12	$\frac{1}{2}$	1	50

This is the final tableau. The maximum $P = 50$, is obtained when $x_1 = 0$, $x_2 = 15$, $x_3 = 5$, and $x_4 = 0$.

29. a. We let x_1, x_2, and x_3 represent the number of type I jeans, type II jeans and type III jeans produced. We want to maximize $P = 4x_1 + 4.50x_2 + 6x_3$ subject to (manufacturing) constraints

$$8x_1 + 12x_2 + 18x_3 \le 5200 \qquad \text{Cutting constraint}$$
$$12x_1 + 18x_2 + 24x_3 \le 6000 \qquad \text{Sewing constraint}$$
$$4x_1 + \ 8x_2 + 12x_3 \le 2200 \qquad \text{Finishing constraint}$$
$$x_1 \ge 0, \quad x_2 \ge 0, \quad x_3 \ge 0$$

b. Slack variables are added and the initial tableau is constructed and shown below.

BV	P	x_1	x_2	x_3	s_1	s_2	s_3	RHS
s_1	0	8	12	18	1	0	0	5200
s_2	0	12	18	24	0	1	0	6000
s_3	0	4	8	12	0	0	1	2200
P	1	−4	−4.5	−6	0	0	0	0

(*continued on next page*)

(continued)

Using the standard pivoting procedure, the pivot element is chosen from column x_1. The quotients of the RHS and the entries in column x_1 are computed and compared.

$$\frac{5200}{8} = 650, \quad \frac{6000}{12} = 500, \quad \text{and} \quad \frac{2200}{4} = 550$$

The smallest quotient, 500, indicates that the pivot element is 12. So x_1 will enter the basis and s_2 will exit the basis.

$$
\begin{array}{c}
\\
\\
\xrightarrow{\begin{array}{l} R_2 = \frac{1}{12} r_2 \\ R_1 = -8R_2 + r_1 \\ R_3 = -4R_2 + r_3 \\ R_4 = 4R_2 + r_4 \end{array}}
\end{array}
\begin{array}{c|ccccccc|c}
\text{BV} & P & x_1 & x_2 & x_3 & s_1 & s_2 & s_3 & \text{RHS} \\
\hline
s_1 & 0 & 0 & 0 & 2 & 1 & -\frac{2}{3} & 0 & 1200 \\
x_1 & 0 & 1 & \frac{3}{2} & 2 & 0 & \frac{1}{12} & 0 & 500 \\
s_3 & 0 & 0 & 2 & 4 & 0 & -\frac{1}{3} & 1 & 200 \\
\hline
P & 1 & 0 & \frac{3}{2} & 2 & 0 & \frac{1}{3} & 0 & 2000
\end{array}
$$

This is the final tableau. The maximum is $P = 2000$, obtained when $x_1 = 500$, $x_2 = 0$, and $x_3 = 0$.

   c. The manufacturer should produce 500 pairs of type I jeans and no pairs of type II or type III jeans for a maximum profit of \$2000.

31. a. Let P be the weekly profit, x_1 be the number of rolls of birthday wrapping paper sold per week, x_2 be the number of rolls of holiday wrapping paper sold weekly, and x_3 be the number of rolls of wedding wrapping paper sold per week. We want to maximize $P = x_1 + x_2 + 2x_3$ subject to the constraints

$$
\begin{aligned}
3x_1 + 5x_2 + 4x_3 &\le 500 \\
10x_1 + 15x_2 + 12x_3 &\le 1800 \\
0.50x_1 + x_3 &\le 75 \\
x_1 \ge 0 \quad x_2 \ge 0 \quad x_3 &\ge 0
\end{aligned}
$$

   b. Slack variables are added and the initial tableau is constructed and shown below. The pivot element is 1/2. Pivoting gives the tableau on the right.

$$
\begin{array}{c|ccccccc|c}
\text{BV} & P & x_1 & x_2 & x_3 & s_1 & s_2 & s_3 & \text{RHS} \\
\hline
s_1 & 0 & 3 & 5 & 4 & 1 & 0 & 0 & 500 \\
s_2 & 0 & 10 & 15 & 12 & 0 & 1 & 0 & 1800 \\
s_3 & 0 & 0.5 & 0 & 1 & 0 & 0 & 1 & 75 \\
\hline
P & 1 & -1 & -1 & -2 & 0 & 0 & 0 & 0
\end{array}
$$

$$
\xrightarrow{\begin{array}{l} R_3 = 2r_3 \\ R_1 = -3R_3 + r_1 \\ R_2 = -10R_3 + r_2 \\ R_4 = R_3 + r_4 \end{array}}
\begin{array}{c|ccccccc|c}
\text{BV} & P & x_1 & x_2 & x_3 & s_1 & s_2 & s_3 & \text{RHS} \\
\hline
s_1 & 0 & 0 & 5 & -2 & 1 & 0 & -6 & 50 \\
s_2 & 0 & 0 & 15 & -8 & 0 & 1 & -20 & 300 \\
x_1 & 0 & 1 & 0 & 2 & 0 & 0 & 2 & 150 \\
\hline
P & 1 & 0 & -1 & 0 & 0 & 0 & 2 & 150
\end{array}
$$

Since there is still a negative entry in the objective row, the new tableau requires further pivoting. The pivot column is x_2. The pivot row is x_1.

$$
\xrightarrow{\begin{array}{l} R_1 = \frac{1}{5} r_1 \\ R_2 = -15R_1 + r_2 \\ R_4 = R_1 + r_4 \end{array}}
\begin{array}{c|ccccccc|c}
\text{BV} & P & x_1 & x_2 & x_3 & s_1 & s_2 & s_3 & \text{RHS} \\
\hline
x_2 & 0 & 0 & 1 & -\frac{2}{5} & \frac{1}{5} & 0 & -\frac{6}{5} & 10 \\
s_2 & 0 & 0 & 0 & -2 & -3 & 1 & -2 & 150 \\
x_1 & 0 & 1 & 0 & 2 & 0 & 0 & 2 & 150 \\
\hline
P & 1 & 0 & 0 & -\frac{2}{5} & \frac{1}{5} & 0 & \frac{4}{5} & 160
\end{array}
$$

Since there is still a negative entry in the objective row, the new tableau requires further pivoting. The pivot column is x_3. The pivot row is x_1.

$$
\xrightarrow{\begin{array}{l} R_3 = \frac{1}{2} r_3 \\ R_1 = \frac{2}{5} R_3 + r_1 \\ R_2 = 2R_3 + r_2 \\ R_4 = \frac{2}{5} R_1 + r_4 \end{array}}
\begin{array}{c|ccccccc|c}
\text{BV} & P & x_1 & x_2 & x_3 & s_1 & s_2 & s_3 & \text{RHS} \\
\hline
x_2 & 0 & \frac{1}{5} & 1 & 0 & \frac{1}{5} & 0 & -\frac{4}{5} & 40 \\
s_2 & 0 & 1 & 0 & 0 & -3 & 1 & 0 & 300 \\
x_3 & 0 & \frac{1}{2} & 0 & 1 & 0 & 0 & 1 & 75 \\
\hline
P & 1 & \frac{1}{5} & 0 & 0 & \frac{1}{5} & 0 & \frac{6}{5} & 190
\end{array}
$$

Since all the entries in the objective row are nonnegative, this is the final tableau. The maximum is $P = 190$, obtained when $x_1 = 0$, $x_2 = 40$, and $x_3 = 75$.

   c. The lacrosse team makes a maximum profit of $190 a week when constrained as in the problem by selling no rolls of birthday wrapping paper, 40 rolls of holiday wrapping paper, and 75 rolls of wedding wrapping paper.

33. a. Let P denote the revenue, x_1 denote the number of Can I nuts packaged, x_2 denote the number of Can II nuts packaged, and x_3 denote the number of Can III nuts packaged. We must maximize $P = 28x_1 + 24x_2 + 21x_3$ subject to the constraints

$$3x_1 + 4x_2 + 5x_3 \le 500$$
$$x_1 + \tfrac{1}{2}x_2 \qquad \le 100$$
$$x_1 + \tfrac{1}{2}x_2 \qquad \le \ 50$$
$$x_1 \ge 0 \quad x_2 \ge 0 \quad x_3 \ge 0$$

  b. Slack variables are added and the initial tableau is constructed and shown below.

BV	P	x_1	x_2	x_3	s_1	s_2	s_3	RHS
s_1	0	3	4	5	1	0	0	500
s_2	0	1	$\tfrac{1}{2}$	0	0	1	0	100
s_3	0	$\boxed{1}$	$\tfrac{1}{2}$	0	0	0	1	50
P	1	−28	−24	−21	0	0	0	0

$R_1 = -3R_3 + r_1$
$R_2 = -R_3 + r_2$
$R_4 = 28R_3 + r_4$

BV	P	x_1	x_2	x_3	s_1	s_2	s_3	RHS
s_1	0	0	$\tfrac{5}{2}$	5	1	0	−3	350
s_2	0	0	0	0	0	1	−1	50
x_1	0	1	$\boxed{\tfrac{1}{2}}$	0	0	0	1	50
P	1	0	−10	−21	0	0	28	1400

$R_3 = 2r_3$
$R_1 = -\tfrac{5}{2}R_3 + r_1$
$R_4 = 10R_3 + r_4$

BV	P	x_1	x_2	x_3	s_1	s_2	s_3	RHS
s_1	0	−5	0	$\boxed{5}$	1	0	−8	100
s_2	0	0	0	0	0	1	−1	50
x_2	0	2	1	0	0	0	2	100
P	1	20	0	−21	0	0	48	2400

$R_1 = \tfrac{1}{5}r_1$
$R_4 = 21R_1 + r_4$

BV	P	x_1	x_2	x_3	s_1	s_2	s_3	RHS
x_3	0	−1	0	1	$\tfrac{1}{5}$	0	$-\tfrac{8}{5}$	20
s_2	0	0	0	0	0	1	−1	50
x_2	0	$\boxed{2}$	1	0	0	0	2	100
P	1	−1	0	0	$\tfrac{21}{5}$	0	$\tfrac{72}{5}$	2820

$R_3 = \tfrac{1}{2}r_3$
$R_1 = R_3 + r_1$
$R_4 = R_3 + r_4$

BV	P	x_1	x_2	x_3	s_1	s_2	s_3	RHS
x_3	0	0	$\tfrac{1}{2}$	1	$\tfrac{1}{5}$	0	$-\tfrac{8}{5}$	70
s_2	0	0	0	0	0	1	−1	50
x_1	0	1	$\tfrac{1}{2}$	0	0	0	1	50
P	1	0	$\tfrac{1}{2}$	0	$\tfrac{21}{5}$	0	$\tfrac{77}{5}$	2870

Since all the entries in the objective row are nonnegative, this is the final tableau. The maximum $P = 2870$, obtained when $x_1 = 50$, $x_2 = 0$, and $x_3 = 70$.

  c. Revenue is maximized at $2870.00 when 50 packages of Can I nuts, 70 packages of Can III nuts, and no packages of Can II nuts are produced.

35.a. P is the profit, x_1 is the number of sleeping dolls, x_2 is the number of talking dolls, and x_3 is the number of walking dolls. We want to maximize $P = 6x_1 + 6x_2 + 8x_3$ subject to the constraints

$$x_1 + x_2 + x_3 \le 60$$
$$6x_1 + 7.5x_2 + 9x_3 \le 405$$
$$x_1 \ge 0 \quad x_2 \ge 0 \quad x_3 \ge 0$$

We begin the simplex method using the initial tableau from Problem 41, Section 5.1. The first pivot element is in row s_1, column x_1.

BV	P	x_1	x_2	x_3	s_1	s_2	RHS
s_1	0	[1]	1	1	1	0	60
s_2	0	6	7.5	9	0	1	405
P	1	−6	−6	−8	0	0	0

$\xrightarrow{\substack{R_2=-6R_1+r_2\\R_3=6R_1+r_3}}$

BV	P	x_1	x_2	x_3	s_1	s_2	RHS
x_1	0	1	1	1	1	0	60
s_2	0	0	1.5	[3]	−6	1	45
P	1	0	0	−2	6	0	360

$\xrightarrow{\substack{R_2=\frac{1}{3}r_2\\R_1=-R_2+r_1\\R_3=2R_2+r_3}}$

BV	P	x_1	x_2	x_3	s_1	s_2	RHS
x_1	0	1	$\frac{1}{2}$	0	3	$-\frac{1}{3}$	45
x_3	0	0	$\frac{1}{2}$	1	−2	$\frac{1}{3}$	15
P	1	0	1	0	2	$\frac{2}{3}$	390

The store should order 45 sleeping dolls, 15 walking dolls and no talking dolls to maximize profit.

b. The maximum profit is $390.

37.a. We begin the simplex method using the initial tableau from Problem 43, Section 5.1. The first pivot element is marked in the initial tableau

BV	P	x_1	x_2	x_3	s_1	s_2	s_3	RHS
s_1	0	10	20	20	1	0	0	500
s_2	0	20	40	30	0	1	0	900
s_3	0	[1]	1	1	0	0	1	30
P	1	−19	−34	−15	0	0	0	0

$\xrightarrow{\substack{R_1=-10R_3+r_1\\R_2=-20R_3+r_2\\R_4=19R_3+r_4}}$

BV	P	x_1	x_2	x_3	s_1	s_2	s_3	RHS
s_1	0	0	10	10	1	0	−10	200
s_2	0	0	[20]	10	0	1	−20	300
x_1	0	1	1	1	0	0	1	30
P	1	0	−15	4	0	0	19	570

$\xrightarrow{\substack{R_2=\frac{1}{20}r_2\\R_1=-10R_2+r_1\\R_3=-R_2+r_3\\R_4=15R_2+r_4}}$

BV	P	x_1	x_2	x_3	s_1	s_2	s_3	RHS
s_1	0	0	0	5	1	−0.5	0	50
x_2	0	0	1	0.5	0	0.05	−1	15
x_1	0	1	0	0.5	0	−0.05	2	15
P	1	0	0	11.5	0	0.75	4	795

There are no longer any negative entries in the objective row indicating that we have found the maximum P. Steven should produce 15 shirts, 15 jackets, and no pants to maximize his profit.

b. The maximum profit is $795.

39. a. We let P denote the company's revenue and we let x_1, x_2, and x_3 represent the number of gallons of regular, premium, and super premium gasoline, respectively to be refined. The company wants to refine amounts that will maximize its revenue $P = 2.80x_1 + 2.97x_2 + 3.08x_3$ subject to its available resources

$$0.6x_1 + 0.7x_2 + 0.8x_3 \le 140,000$$
$$0.4x_1 + 0.3x_2 + 0.2x_3 \le 120,000$$
$$x_1 + x_2 + x_3 \le 225,000$$
$$x_1 \ge 0 \quad x_2 \ge 0 \quad x_3 \ge 0$$

b.

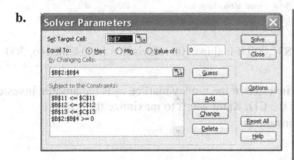

	B7	▾	f_x =2.8*B2+2.97*B3+3.08*B4	
	A	B	C	D
1	Variables			
2	gallons of regular, x_1	175000		
3	gallons of premium, x_2	50000		
4	gallons of super premium, x_3	0		
5				
6	Objective			
7	Maximize revenue	638500		
8				
9				
10	Constraints	Amount Used	Maximum	
11	amount of high octane	140000	140000	
12	amount of low octane	85000	120000	
13	total number of gallons to be mixed	225000	225000	

c. To achieve a maximize revenue of $638,500, the company should refine 175,000 gallons of regular gasoline, 50,000 gallons of premium gasoline, and no super premium gasoline.

41. a. Let P be the total yield, x_1 be the amount invested in stocks, x_2 be the amount invested in corporate bonds, and x_3 be the amount invested in municipal bonds. The consultant wants to maximize the yield $P = 0.08x_1 + 0.05x_2 + 0.03x_3$ subject to the constraints

$$x_1 + x_2 + x_3 \le 90,000$$
$$x_1 \le 45,000$$
$$x_2 - x_3 \le 18,000$$
$$x_1 \ge 0 \quad x_2 \ge 0 \quad x_3 \ge 0$$

b.

	B7	▾	f_x =0.08*B2+0.05*B3+0.03*B4
	A	B	C
1	Variables		
2	amount in stocks, x_1	45000	
3	amount in corporate bonds, x_2	31500	
4	amount in municipal bonds, x_3	13500	
5			
6	Objective		
7	Maximize revenue	5580	
8			
9			
10	Constraints	Amount Used	Maximum
11	total invested	90000	90000
12	amount in stocks	45000	45000
13	difference between amount in corporate bonds and minicipal bonds	18000	18000

c. The financial planner can maximize her client's investment income, while maintaining her investment strategy by investing $45,000 in stocks, $31,500 in corporate bonds, and $13,500 in municipal bonds. The maximum return will be $5580.

43. a. Let P denote the profit, x_1 denote the number of acres of soybeans planted, x_2 denote the number of acres of corn planted, and x_3 denote the number of acres of wheat planted. The farmer wants to maximize the profit $P = 70x_1 + 90x_2 + 50x_3$ subject to the constraints

$$x_1 + x_2 + x_3 \le 200$$
$$40x_1 + 50x_2 + 30x_3 \le 18,000$$
$$20x_1 + 30x_2 + 15x_3 \le 4,200$$
$$x_1 \ge 0 \quad x_2 \ge 0 \quad x_3 \ge 0$$

b.

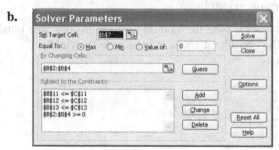

c. The farmer realizes a maximum profit of $14,400 when planting 180 acres of soybeans, 20 acres of corn, and no wheat.

45.a. Let P denote the profit, x_1 denote the number televisions shipped from Chicago, x_2 denote the number shipped from New York, and x_3 denote the number shipped from Denver. We want to maximize the profit $P = 70x_1 + 80x_2 + 40x_3$ subject to the constraints

$$
\begin{aligned}
x_1 + x_2 + x_3 &\le 400 \\
50x_1 + 40x_2 + 80x_3 &\le 20{,}000 \\
6x_1 + 8x_2 + 4x_3 &\le 3{,}000 \\
x_1 \ge 0 \quad x_2 \ge 0 \quad x_3 &\ge 0
\end{aligned}
$$

b.

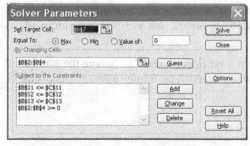

c. The manufacturer will obtain a maximum profit of $31,000 if it ships 100 televisions from Chicago, 300 televisions from New York and none from Denver.

47.a. P is the return on her investments, x_1 is the amount invested in the money market, x_2 is the amount invested in the mutual fund, and x_3 is the amount invested in the CD. Kami wants to maximize the return $P = 0.0105x_1 + 0.05x_2 + 0.0166x_3$ subject to the constraints

$$
\begin{aligned}
x_1 + x_2 + x_3 &\le 25{,}000 \\
x_2 + x_3 &\le 10{,}000 \\
-x_1 + x_3 &\le 1{,}500 \\
x_1 \ge 0 \quad x_2 \ge 0 \quad x_3 &\ge 0
\end{aligned}
$$

Kami should invest $15,000 in the money market, $10,000 in the mutual fund and nothing in the CD.

b. The maximum return on Kami's investments is $657.50.

5.3 Solving Minimum Problems Using the Duality Principle

1. 2×4

3. True

5. True

7. The <u>Duality Principle</u> states that the solution of a minimum problem, if it exists, is the same as the solution of the maximum problem, which is its dual.

For exercises 9–13, a minimum problem is in standard form if the following conditions are met.

Condition 1   All the variables are nonnegative.
Condition 2   All other constraints are written as linear expressions that are greater than or equal to a positive constant.
Condition 3   The objective function is a linear expression with nonnegative coefficients.

9. The problem is in standard form since all conditions are met.

11. The problem is not in standard form. Condition 3 is not met since the objective function $C = 2x_1 - x_2$ has a negative coefficient.

13. The problem is not in standard form. Condition 2 is not met since the constraint $x_1 + x_3 \leq 6$ is not written as a linear expression that is greater than or equal to a positive constant.

In exercises 15–19 use the following 4 steps to write the Dual Problem.

STEP 1   Write the minimum problem in standard form.
STEP 2   Construct a matrix that represents the constraints and the objective function, placing the objective function in the bottom row.
STEP 3   Interchange the rows and columns to form the matrix of the dual problem.
STEP 4   Translate this matrix into a maximum problem in standard form.

15. **STEP 1**: The problem is in standard form since both conditions are met.
STEP 2: The matrix with the objective function in the bottom row.

$$\begin{array}{cc} x_1 & x_2 \end{array}$$
$$\begin{bmatrix} 1 & 1 & | & 2 \\ 2 & 3 & | & 6 \\ 2 & 3 & | & 0 \end{bmatrix}$$

STEP 3: The matrix with the rows and columns interchanged.

$$\begin{bmatrix} 1 & 2 & | & 2 \\ 1 & 3 & | & 3 \\ 2 & 6 & | & 0 \end{bmatrix}$$

STEP 4: The corresponding maximum problem in standard form. Maximize $P = 2y_1 + 6y_2$ subject to the constraints

$$y_1 + 2y_2 \leq 2$$
$$y_1 + 3y_2 \leq 3$$
$$y_1 \geq 0 \qquad y_2 \geq 0$$

This maximum problem is the dual of the minimum problem.

17. **STEP 1**: The problem is in standard form since both conditions are met.
STEP 2: The matrix with the objective function in the bottom row.

$$\begin{array}{ccc} x_1 & x_2 & x_3 \end{array}$$
$$\begin{bmatrix} 1 & 1 & 1 & | & 5 \\ 2 & 1 & 0 & | & 4 \\ 3 & 1 & 1 & | & 0 \end{bmatrix}$$

STEP 3: The matrix with the rows and columns interchanged.

$$\begin{bmatrix} 1 & 2 & | & 3 \\ 1 & 1 & | & 1 \\ 1 & 0 & | & 1 \\ 5 & 4 & | & 0 \end{bmatrix}$$

STEP 4: The corresponding maximum problem in standard form. Maximize $P = 5y_1 + 4y_2$ subject to the constraints

$$y_1 + 2y_2 \leq 3$$
$$y_1 + y_2 \leq 1$$
$$y_1 \leq 1$$
$$y_1 \geq 0 \qquad y_2 \geq 0$$

This maximum problem is the dual of the minimum problem.

19. STEP 1: The problem is in standard form since both conditions are met.

STEP 2: The matrix with the objective function in the bottom row.

$$
\begin{array}{cccc}
x_1 & x_2 & x_3 & x_4
\end{array}
$$
$$
\left[\begin{array}{cccc|c}
1 & 1 & 1 & 2 & 60 \\
3 & 2 & 1 & 2 & 90 \\
3 & 4 & 1 & 2 & 0
\end{array}\right]
$$

STEP 3: The matrix with the rows and columns interchanged.

$$
\left[\begin{array}{cc|c}
1 & 3 & 3 \\
1 & 2 & 4 \\
1 & 1 & 1 \\
2 & 2 & 2 \\
60 & 90 & 0
\end{array}\right]
$$

STEP 4: The corresponding maximum problem in standard form. Maximize $P = 60y_1 + 90y_2$ subject to the constraints

$$
\begin{aligned}
y_1 + 3y_2 &\le 3 \\
y_1 + 2y_2 &\le 4 \\
y_1 + y_2 &\le 1 \\
2y_1 + 2y_2 &\le 2 \\
y_1 \ge 0 \qquad y_2 &\ge 0
\end{aligned}
$$

This maximum problem is the dual of the minimum problem.

21. STEP 1: Write the dual problem.

$$
\begin{array}{cc}
x_1 & x_2
\end{array}
$$

The matrix:
$$
\left[\begin{array}{cc|c}
1 & 1 & 2 \\
2 & 6 & 6 \\
6 & 3 & 0
\end{array}\right]
$$

Its transpose:
$$
\left[\begin{array}{cc|c}
1 & 2 & 6 \\
1 & 6 & 3 \\
2 & 6 & 0
\end{array}\right]
$$

Maximize $P = 2y_1 + 6y_2$ subject to the constraints

$$
\begin{aligned}
y_1 + 2y_2 &\le 6 \\
y_1 + 6y_2 &\le 3 \\
y_1 \ge 0 \qquad y_2 &\ge 0
\end{aligned}
$$

STEP 2: Set up the initial tableau and use the simplex method to solve the dual problem.

BV	P	y_1	y_2	s_1	s_2	RHS
s_1	0	1	2	1	0	6
s_2	0	**1**	6	0	1	3
P	1	−2	−6	0	0	0

$\xrightarrow[\substack{R_1 = -R_2 + r_1 \\ R_3 = 2R_2 + r_3}]{}$

BV	P	y_1	y_2	s_1	s_2	RHS
s_1	0	0	−4	1	−1	3
y_1	0	1	6	0	1	3
P	1	0	6	0	2	6

STEP 3: This is the final tableau. From it we read that the minimum cost $C = 6$ is obtained when $x_1 = 0$ and $x_2 = 2$.

23. STEP 1: Write the dual problem.

$$
\begin{array}{cc}
x_1 & x_2
\end{array}
$$

The matrix:
$$
\left[\begin{array}{cc|c}
1 & 1 & 4 \\
3 & 4 & 12 \\
6 & 3 & 0
\end{array}\right]
$$
Its transpose:
$$
\left[\begin{array}{cc|c}
1 & 3 & 6 \\
1 & 4 & 3 \\
4 & 12 & 0
\end{array}\right]
$$

Maximize $P = 4y_1 + 12y_2$ subject to the constraints

$$
\begin{aligned}
y_1 + 3y_2 &\le 6 \\
y_1 + 4y_2 &\le 3 \\
y_1 \ge 0 \qquad y_2 &\ge 0
\end{aligned}
$$

STEP 2: Set up the initial tableau and use the simplex method to solve the dual problem.

BV	P	y_1	y_2	s_1	s_2	RHS
s_1	0	1	3	1	0	6
s_2	0	**1**	4	0	1	3
P	1	−4	−12	0	0	0

$\xrightarrow[\substack{R_1 = -R_2 + r_1 \\ R_3 = 4R_2 + r_3}]{}$

BV	P	y_1	y_2	s_1	s_2	RHS
s_1	0	0	−1	1	−1	3
t_1	0	1	4	0	1	3
P	1	0	4	0	4	12

STEP 3: This is the final tableau. From it we read that the minimum cost $C = 12$ is obtained when $x_1 = 0$ and $x_2 = 4$.

25. STEP 1: Write the dual problem.

$$
\begin{array}{ccc}
x_1 & x_2 & x_3
\end{array}
$$

The matrix:
$$
\left[\begin{array}{ccc|c}
1 & -3 & 4 & 12 \\
3 & 1 & 2 & 10 \\
1 & -1 & -1 & -8 \\
1 & 2 & 1 & 0
\end{array}\right]
$$
; its transpose:

$$
\left[\begin{array}{cccc|c}
1 & 3 & 1 & 1 \\
-3 & 1 & -1 & 2 \\
4 & 2 & -1 & 1 \\
12 & 10 & -8 & 0
\end{array}\right]
$$

Maximize $P = 12y_1 + 10y_2 - 8y_3$ subject to the constraints

$$
\begin{aligned}
y_1 + 3y_2 + y_3 &\le 1 \\
-3y_1 + y_2 - y_3 &\le 2 \\
4y_1 + 2y_2 - y_3 &\le 1 \\
y_1 \ge 0 \qquad y_2 \ge 0 \qquad y_3 &\ge 0
\end{aligned}
$$

(continued on next page)

(*continued*)

STEP 2: Set up the initial tableau and use the simplex method to solve the dual problem.

BV	P	y_1	y_2	y_3	s_1	s_2	s_3	RHS
s_1	0	1	3	1	1	0	0	1
s_2	0	−3	1	−1	0	1	0	2
s_3	0	[4]	2	−1	0	0	1	1
P	1	−12	−10	8	0	0	0	0

$R_3 = \frac{1}{4}r_3$
$R_1 = -R_3 + r_1$
$R_2 = 3R_3 + r_2$
$R_4 = 12R_3 + r_4$

BV	P	y_1	y_2	y_3	s_1	s_2	s_3	RHS
s_1	0	0	$\frac{5}{2}$	$\frac{5}{4}$	1	0	$-\frac{1}{4}$	$\frac{3}{4}$
s_2	0	0	$\frac{5}{2}$	$-\frac{7}{4}$	0	1	$\frac{3}{4}$	$\frac{11}{4}$
y_1	0	1	$\frac{1}{2}$	$-\frac{1}{4}$	0	0	$\frac{1}{4}$	$\frac{1}{4}$
P	1	0	−4	5	0	0	3	3

$R_1 = \frac{2}{5}r_1$
$R_2 = -\frac{5}{2}R_1 + r_2$
$R_3 = -\frac{1}{2}R_1 + r_3$
$R_4 = 4R_1 + r_4$

BV	P	y_1	y_2	y_3	s_1	s_2	s_3	RHS
y_2	0	0	1	$\frac{1}{2}$	$\frac{2}{5}$	0	$-\frac{1}{10}$	$\frac{3}{10}$
s_2	0	0	0	−3	−1	1	1	2
y_1	0	1	0	$-\frac{1}{2}$	$-\frac{1}{5}$	0	$\frac{3}{10}$	$\frac{1}{10}$
P	1	0	0	7	$\frac{8}{5}$	0	$\frac{13}{5}$	$\frac{21}{5}$

STEP 3: This is the final tableau. From it we read that the minimum cost $C = \dfrac{21}{5} = 4.20$ is obtained when

$x_1 = \dfrac{8}{5}$, $x_2 = 0$, and $x_3 = \dfrac{13}{5}$.

27. STEP 1: Write the dual problem.

The matrix:
$$
\begin{array}{cccc}
x_1 & x_2 & x_3 & x_4
\end{array}
\begin{bmatrix}
1 & 0 & 1 & 0 & 1 \\
0 & 1 & 0 & 1 & 1 \\
-1 & -1 & -1 & -1 & -3 \\
1 & 4 & 2 & 4 & 0
\end{bmatrix}
$$
; its transpose:
$$
\begin{bmatrix}
1 & 0 & -1 & 1 \\
0 & 1 & -1 & 4 \\
1 & 0 & -1 & 2 \\
0 & 1 & -1 & 4 \\
1 & 1 & -3 & 0
\end{bmatrix}
$$

Maximize $P = y_1 + y_2 - 3y_3$ subject to the constraints

$y_1 \qquad - y_3 \leq 1$
$\qquad y_2 - y_3 \leq 4$
$y_1 \qquad - y_3 \leq 2$
$\qquad y_2 - y_3 \leq 4$
$y_1 \geq 0 \qquad y_2 \geq 0 \qquad y_3 \geq 0$

STEP 2: Set up the initial tableau and use the simplex method to solve the dual problem.

BV	P	y_1	y_2	y_3	s_1	s_2	s_3	s_4	RHS
s_1	0	[1]	0	−1	1	0	0	0	1
s_2	0	0	1	−1	0	1	0	0	4
s_3	0	1	0	−1	0	0	1	0	2
s_4	0	0	1	−1	0	0	0	1	4
P	1	−1	−1	3	0	0	0	0	0

$R_3 = -R_1 + r_3$
$R_5 = R_1 + r_5$

BV	P	y_1	y_2	y_3	s_1	s_2	s_3	s_4	RHS
y_1	0	1	0	−1	1	0	0	0	1
s_2	0	0	0	[1]	−1	0	1	0	0	4
s_3	0	0	0	0	−1	0	1	0	1
s_4	0	0	1	−1	0	0	0	1	4
P	1	0	−1	2	1	0	0	0	1

$R_4 = -R_2 + r_4$
$R_5 = R_2 + r_5$

BV	P	y_1	y_2	y_3	s_1	s_2	s_3	s_4	RHS
y_1	0	1	0	−1	1	0	0	0	1
y_2	0	0	1	−1	0	1	0	0	4
s_3	0	0	0	0	−1	0	1	0	1
s_4	0	0	0	0	0	−1	0	1	0
P	1	0	0	1	1	1	0	0	5

STEP 3: This is the final tableau. From it we read that the minimum cost $C = 5$ is obtained when
$x_1 = 1$, $x_2 = 1$, $x_3 = 0$, and $x_4 = 0$.

29. Let C represent the weekly payroll, and let x_1 denote the number of employees scheduled to work on Friday, x_2 denote the number of employees scheduled to work on Saturday, and x_3 denote the number of employees Sunday. Since each employee is scheduled to work exactly 2 days, then $x_1 + x_2 + x_3$ must be an even integer. We want to minimize $C = 100x_1 + 150x_2 + 150x_3$ subject to the constraints

$x_1 \geq 6, \ x_2 \geq 15, \ x_3 \geq 8$.

The matrix (on the left) representing the system and its transpose (on the right) are

$$
\begin{array}{ccc}
x_1 & x_2 & x_3
\end{array}
\begin{bmatrix}
1 & 0 & 0 & | & 6 \\
0 & 1 & 0 & | & 15 \\
0 & 0 & 1 & | & 8 \\
100 & 150 & 150 & | & 0
\end{bmatrix}
\text{ and }
\begin{array}{ccc}
y_1 & y_2 & y_3
\end{array}
\begin{bmatrix}
1 & 0 & 0 & | & 100 \\
0 & 1 & 0 & | & 150 \\
0 & 0 & 1 & | & 150 \\
6 & 15 & 8 & | & 0
\end{bmatrix}
$$

The dual of the problem is maximize $P = 6y_1 + 25y_2 + 8y_3$ subject to the constraints $y_1 \geq 100$, $y_2 \geq 150, \ y_3 \geq 150$.

We set up the initial simplex tableau and solve the maximum problem.

BV	P	y_1	y_2	y_3	s_1	s_2	s_3	RHS
s_1	0	1	0	0	1	0	0	100
s_2	0	0	1	0	0	1	0	150
s_3	0	0	0	1	0	0	1	150
P	1	−6	−15	−8	0	0	0	0

$\xrightarrow{R_4 = 6r_1 + r_4}$

BV	P	y_1	y_2	y_3	s_1	s_2	s_3	RHS
y_1	0	1	0	0	1	0	0	100
s_2	0	0	1	0	0	1	0	150
s_3	0	0	0	1	0	0	1	150
P	1	0	−15	−8	6	0	0	600

$\xrightarrow{R_4 = 15r_2 + r_4}$

BV	P	y_1	y_2	y_3	s_1	s_2	s_3	RHS
y_1	0	1	0	0	1	0	0	100
y_2	0	0	1	0	0	1	0	150
s_3	0	0	0	1	0	0	1	150
P	1	0	0	−8	6	15	0	2850

$\xrightarrow{R_4 = 8r_3 + r_4}$

BV	P	y_1	y_2	y_3	s_1	s_2	s_3	RHS
y_1	0	1	0	0	1	0	0	100
y_2	0	0	1	0	0	1	0	150
y_3	0	0	0	1	0	0	1	150
P	1	0	0	0	6	15	8	4050

C is minimized when $x_1 = 6$, $x_2 = 15$, and $x_3 = 8$. The minimum payroll would be \$4050. However, since each employee works exactly 2 days, $x_1 + x_2 + x_3$ must be an even integer. In order to satisfy this condition, we must add one more employee. Since Friday wages are the least, add one employee on Friday. Therefore, 7 employees will be scheduled on Friday, 15 on Saturday, and 8 on Sunday, making the payroll \$4150 for the 3-day weekend. If there are 15 employees, then 7 employees will work a Friday/Saturday schedule, and 8 employees will work a Saturday/Sunday schedule. No employees will work a Friday/Sunday schedule.

31.a. Let C represent the cost of the order, and let x_1, x_2, and x_3 represent the number of lunch #1, lunch #2, and lunch #3 respectively that Mrs. Mintz purchases. She and her friends want to order the foods they need at minimum cost. They want to minimize $C = 6.20x_1 + 7.40x_2 + 9.10x_3$ subject to the constraints

$$
\begin{aligned}
x_1 & \geq 4 \\
x_1 + x_2 + x_3 & \geq 9 \\
x_1 + x_3 & \geq 6 \\
x_2 + x_3 & \geq 5 \\
x_1 \geq 0 \quad x_2 \geq 0 \quad x_3 & \geq 0
\end{aligned}
$$

b. The matrix (on the left) representing the system and its transpose (on the right) are

$$
\begin{array}{ccc}
x_1 & x_2 & x_3
\end{array}
\begin{bmatrix}
1 & 0 & 0 & | & 4 \\
1 & 1 & 1 & | & 9 \\
1 & 0 & 1 & | & 6 \\
0 & 1 & 1 & | & 5 \\
6.2 & 7.4 & 9.1 & | & 0
\end{bmatrix}
\text{ and }
\begin{array}{cccc}
y_1 & y_2 & y_3 & y_4
\end{array}
\begin{bmatrix}
1 & 1 & 1 & 0 & | & 6.2 \\
0 & 1 & 0 & 1 & | & 7.4 \\
0 & 1 & 1 & 1 & | & 9.1 \\
4 & 9 & 6 & 5 & | & 0
\end{bmatrix}
$$

(continued on next page)

(continued)

The dual of the problem is maximize $P = 4y_1 + 9y_2 + 6y_3 + 5y_4$ subject to the constraints

$$y_1 + y_2 + y_3 \qquad \leq 6.2$$
$$y_2 + \qquad y_4 \leq 7.4$$
$$y_2 + y_3 + y_4 \leq 9.1$$
$$y_1 \geq 0 \qquad y_2 \geq 0 \qquad y_3 \geq 0 \qquad y_4 \geq 0$$

We set up the initial simplex tableau and solve the maximum problem.

BV	P	y_1	y_2	y_3	y_4	s_1	s_2	s_3	RHS
s_1	0	[1]	1	1	0	1	0	0	6.2
s_2	0	0	1	0	1	0	1	0	7.4
s_3	0	0	1	1	1	0	0	1	9.1
P	1	−4	−9	−6	−5	0	0	0	0

$R_4 = 4R_1 + r_4 \longrightarrow$

BV	P	y_1	y_2	y_3	y_4	s_1	s_2	s_3	RHS
y_1	0	1	[1]	1	0	1	0	0	6.2
s_2	0	0	1	0	1	0	1	0	7.4
s_3	0	0	1	1	1	0	0	1	9.1
P	1	0	−5	−2	−5	4	0	0	24.80

$\begin{array}{l} R_2 = -R_1 + r_2 \\ R_3 = -R_1 + r_3 \\ R_4 = 5R_1 + r_4 \end{array} \longrightarrow$

BV	P	y_1	y_2	y_3	y_4	s_1	s_2	s_3	RHS
y_2	0	1	1	1	0	1	0	0	6.2
s_2	0	−1	0	−1	[1]	−1	1	0	1.2
s_3	0	−1	0	0	1	−1	0	1	2.9
P	1	5	0	3	−5	9	0	0	55.80

$\begin{array}{l} R_3 = -R_2 + r_3 \\ R_4 = 5R_2 + r_4 \end{array} \longrightarrow$

BV	P	y_1	y_2	y_3	y_4	s_1	s_2	s_3	RHS
y_2	0	1	1	1	0	1	0	0	6.2
y_4	0	−1	0	−1	1	−1	1	0	1.2
s_3	0	0	0	[1]	0	0	−1	1	1.7
P	1	0	0	−2	0	4	5	0	61.80

$\begin{array}{l} R_1 = -R_3 + r_1 \\ R_2 = R_3 + r_2 \\ R_4 = 2R_3 + r_4 \end{array} \longrightarrow$

BV	P	y_1	y_2	y_3	y_4	s_1	s_2	s_3	RHS
y_2	0	1	1	0	0	1	1	−1	4.5
y_4	0	−1	0	0	1	−1	0	1	2.9
y_3	0	0	0	1	0	0	−1	1	1.7
P	1	0	0	0	0	4	3	2	65.20

Minimum: $C = 65.20$ when $x_1 = 4$, $x_2 = 3$, and $x_3 = 2$.

c. Mrs. Mintz spends the least amount, $65.20, when she orders four of Lunch #1, three of Lunch #2, and two of Lunch #3.

33. a. Let C be the cost of the supplements; x_1 be the number of pill P needed and x_2 the number of pill Q needed. Mr. Jones wants to minimize costs, $C = 3x_1 + 4x_2$ subject to the nutritional requirement constraints

$$5x_1 + 10x_2 \geq 50$$
$$2x_1 + x_2 \geq 8$$
$$x_1 \geq 0 \qquad x_2 \geq 0 \qquad x_3 \geq 0$$

b.

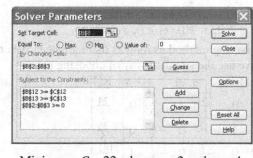

Minimum: $C = 22$ when $x_1 = 2$ and $x_2 = 4$.

	A	B	C
B8	f_x =3*B2+4*B3		
1	**Variables**		
2	number of pill 1, x_1	2	
3	number of pill 2, x_2	4	
4			
5			
6			
7	**Objective**		
8	Minimize cost	22	
9			
10			
11	**Constraints**	Amount used	Minimum
12	amount of calcium	50	50
13	amount of iron	8	8

c. Mr. Jones minimizes his cost at \$0.22 when he adds 2 vitamin P pills and 4 vitamin Q pills to his diet.

35. Let C represent the amount of sodium in the mix, and let x_1, x_2, and x_3 represent the amount (in 100 gram units) of apricots, peaches, and pears respectively in the mix. Katy wants to minimize sodium, $C = 26x_1 + 16x_2 + 7x_3$ subject to the dietary requirement constraints

$$x_1 + x_2 + x_3 \geq 20$$
$$67x_1 + 48x_2 + 35x_3 \geq 400$$
$$12x_1 + 18x_2 + 7x_3 \geq 250$$
$$x_1 + x_3 \geq 5$$
$$x_1 \geq 0 \quad x_2 \geq 0 \quad x_3 \geq 0$$

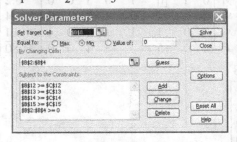

	A	B	C
1	**Variables**		
2	amount of apricots, x_1	0	
3	amount of peaches, x_2	10	
4	amount of pears, x_3	10	
5			
6			
7	**Objective**		
8	Minimize sodium	230	
9			
10			
11	**Constraints**	Amount used	Minimum
12	total amount of fruit	20	20
13	amount of calcium	830	400
14	amount of vitamin C	250	250
15	amount of apricots and pears combined	10	5

B8 ▼ f_x =26*B2+16*B3+7*B4

Minimum: $C = 230$ when $x_1 = 0$, $x_2 = 10$ (100 gram units) and $x_3 = 10$ (100 gram units). Katy should mix 1 kilogram of dried peaches and 1 kilogram of dried pears to meet her requirements and to minimize the sodium content. The minimum sodium is 230 milligrams.

5.4 The Simplex Method for Problems Not in Standard Form

1. False. All variables, including slack variables, must be nonnegative.

3. True

5. Maximize $P = 3x_1 + 4x_2$ subject to the constraints

$$x_1 + x_2 \leq 12$$
$$5x_1 + 2x_2 \geq 36$$
$$7x_1 + 4x_2 \geq 14$$
$$x_1 \geq 0, \; x_2 \geq 0$$

Rewrite the constraints

$$x_1 + x_2 \leq 12$$
$$-5x_1 - 2x_2 \leq -36$$
$$-7x_1 - 4x_2 \leq -14$$
$$x_1 \geq 0, \; x_2 \geq 0$$

Introduce slack variables

$$x_1 + x_2 + s_1 = 12$$
$$-5x_1 - 2x_2 + s_2 = -36$$
$$-7x_1 - 4x_2 + s_3 = -14$$
$$x_1 \geq 0, \; x_2 \geq 0, \; s_1 \geq 0, \; s_2 \geq 0, \; s_3 \geq 0$$

We now set up the initial tableau and use the alternate pivoting method as long as there are negative entries in the RHS. When all the entries in the RHS are positive, we use the standard way of choosing a pivot.

$$
\begin{array}{c|ccccccc|c}
\text{BV} & P & x_1 & x_2 & s_1 & s_2 & s_3 & & \text{RHS} \\
\hline
s_1 & 0 & 1 & 1 & 1 & 0 & 0 & & 12 \\
s_2 & 0 & \boxed{-5} & -2 & 0 & 1 & 0 & & -36 \\
s_3 & 0 & -7 & -4 & 0 & 0 & 1 & & -14 \\
P & 1 & -3 & -4 & 0 & 0 & 0 & & 0 \\
\end{array}
$$

Alternative Pivoting Strategy

$R_2 = -\dfrac{1}{5}r_2$
$R_1 = -R_2 + r_1$
$R_3 = 7R_2 + r_3$
$R_4 = 3R_2 + r_4$

$$
\begin{array}{c|cccccc|c}
\text{BV} & P & x_1 & x_2 & s_1 & s_2 & s_3 & \text{RHS} \\
\hline
s_1 & 0 & 0 & \frac{3}{5} & 1 & \frac{1}{5} & 0 & \frac{24}{5} \\
x_1 & 0 & 1 & \frac{2}{5} & 0 & -\frac{1}{5} & 0 & \frac{36}{5} \\
s_3 & 0 & 0 & -\frac{6}{5} & 0 & -\frac{7}{5} & 1 & \frac{182}{5} \\
P & 1 & 0 & -\frac{14}{5} & 0 & -\frac{3}{5} & 0 & \frac{108}{5} \\
\end{array}
$$

(continued on next page)

(*continued*)

$$
\begin{array}{c|ccccccc|c}
\text{BV} & P & x_1 & x_2 & s_1 & s_2 & s_3 & \text{RHS} \\
\hline
x_2 & 0 & 0 & 1 & \frac{5}{3} & \frac{1}{3} & 0 & 8 \\
x_1 & 0 & 1 & 0 & -\frac{2}{3} & -\frac{1}{3} & 0 & 4 \\
s_3 & 0 & 0 & 0 & 2 & -1 & 1 & 46 \\
\hline
P & 1 & 0 & 0 & \frac{14}{3} & \frac{1}{3} & 0 & 44
\end{array}
$$

Standard Pivoting →
$R_1 = \frac{5}{3} r_1$
$R_2 = -\frac{2}{5} R_1 + r_2$
$R_3 = \frac{6}{5} R_1 + r_3$
$R_4 = \frac{14}{5} R_1 + r_4$

The maximum is $P = 44$, obtained when $x_1 = 4$ and $x_2 = 8$.

7. Maximize $P = 3x_1 + 2x_2 - x_3$ subject to the constraints

$$x_1 + 3x_2 + x_3 \le 9$$
$$2x_1 + 3x_2 - x_3 \ge 2$$
$$3x_1 - 2x_2 + x_3 \ge 5$$
$$x_1 \ge 0, \ x_2 \ge 0, \ x_3 \ge 0$$

Rewrite the constraints

$$x_1 + 3x_2 + x_3 \le 9$$
$$-2x_1 - 3x_2 + x_3 \le -2$$
$$-3x_1 + 2x_2 - x_3 \le -5$$
$$x_1 \ge 0, \ x_2 \ge 0, \ x_3 \ge 0$$

Introduce slack variables

$$x_1 + 3x_2 + x_3 + s_1 \qquad\quad = 9$$
$$-2x_1 - 3x_2 + x_3 \qquad + s_2 \qquad = -2$$
$$-3x_1 + 2x_2 - x_3 \qquad\qquad + s_3 = -5$$
$$x_1 \ge 0, \ x_2 \ge 0, \ x_3 \ge 0, \ s_1 \ge 0, \ s_2 \ge 0, \ s_3 \ge 0$$

We now set up the initial tableau and use the alternate pivoting method as long as there are negative entries in the RHS. When all the entries in the RHS are positive, we use the standard way of choosing a pivot.

$$
\begin{array}{c|ccccccc|c}
\text{BV} & P & x_1 & x_2 & x_3 & s_1 & s_2 & s_3 & \text{RHS} \\
\hline
s_1 & 0 & 1 & 3 & 1 & 1 & 0 & 0 & 9 \\
s_2 & 0 & -2 & -3 & 1 & 0 & 1 & 0 & -2 \\
s_3 & 0 & \boxed{-3} & 2 & -1 & 0 & 0 & 1 & -5 \\
\hline
P & 1 & -3 & -2 & 1 & 0 & 0 & 0 & 0
\end{array}
$$

Alternative Pivoting Strategy
$R_3 = -\frac{1}{3} r_3$
$R_1 = -R_3 + r_1$
$R_2 = 2R_3 + r_2$
$R_4 = 3R_3 + r_4$

$$
\begin{array}{c|ccccccc|c}
\text{BV} & P & x_1 & x_2 & x_3 & s_1 & s_2 & s_3 & \text{RHS} \\
\hline
s_1 & 0 & 0 & \boxed{\frac{11}{3}} & \frac{2}{3} & 1 & 0 & \frac{1}{3} & \frac{22}{3} \\
s_2 & 0 & 0 & -\frac{13}{3} & \frac{5}{3} & 0 & 1 & -\frac{2}{3} & \frac{4}{3} \\
x_1 & 0 & 1 & -\frac{2}{3} & \frac{1}{3} & 0 & 0 & -\frac{1}{3} & \frac{5}{3} \\
\hline
P & 1 & 0 & -4 & 2 & 0 & 0 & -1 & 5
\end{array}
$$

$$
\begin{array}{c|ccccccc|c}
\text{BV} & P & x_1 & x_2 & x_3 & s_1 & s_2 & s_3 & \text{RHS} \\
\hline
x_2 & 0 & 0 & 1 & \frac{2}{11} & \frac{3}{11} & 0 & \boxed{\frac{1}{11}} & 2 \\
s_2 & 0 & 0 & 0 & \frac{27}{11} & \frac{13}{11} & 1 & -\frac{3}{11} & 10 \\
x_1 & 0 & 1 & 0 & \frac{5}{11} & \frac{2}{11} & 0 & -\frac{3}{11} & 3 \\
\hline
P & 1 & 0 & 0 & \frac{30}{11} & \frac{12}{11} & 0 & -\frac{7}{11} & 13
\end{array}
$$

Standard Pivoting →
$R_1 = \frac{3}{11} r_1$
$R_2 = \frac{13}{3} R_1 + r_2$
$R_3 = \frac{2}{3} R_1 + r_3$
$R_4 = 4 R_1 + r_4$

$$
\begin{array}{c|ccccccc|c}
\text{BV} & P & x_1 & x_2 & x_3 & s_1 & s_2 & s_3 & \text{RHS} \\
\hline
s_3 & 0 & 0 & 11 & 2 & 3 & 0 & 1 & 22 \\
s_2 & 0 & 0 & 3 & 3 & 2 & 1 & 0 & 16 \\
x_1 & 0 & 1 & 3 & 1 & 1 & 0 & 0 & 9 \\
\hline
P & 1 & 0 & 7 & 4 & 3 & 0 & 0 & 27
\end{array}
$$

$R_1 = 11 r_1$
$R_2 = \frac{3}{11} R_1 + r_2$
$R_3 = \frac{3}{11} R_1 + r_3$
$R_4 = \frac{7}{11} R_1 + r_4$

The maximum is $P = 27$, obtained when $x_1 = 9$, $x_2 = 0$, and $x_3 = 0$.

9. Maximize $P = 3x_1 + 2x_2$ subject to the constraints

$$2x_1 + x_2 \leq 4$$
$$x_1 + x_2 = 3$$
$$x_1 \geq 0,\ x_2 \geq 0$$

One constraint is an equation, so it gets no slack variable. The equation is the first row to be pivoted.

BV	P	x_1	x_2	s_1	RHS
s_1	0	2	1	1	4
	0	[1]	1	0	3
P	1	−3	−2	0	0

$\xrightarrow[R_3=3R_2+r_3]{R_1=-2R_2+r_1}$

BV	P	x_1	x_2	s_1	RHS
s_1	0	0	[−1]	1	−2
x_1	0	1	1	0	3
P	1	0	1	0	9

We now have a negative value in the RHS, so we use the alternative pivoting strategy.

Alternative Pivoting Strategy
$\xrightarrow[\begin{array}{l}R_1=-r_1\\R_2=-R_1+r_2\\R_3=-R_1+r_3\end{array}]{}$

BV	P	x_1	x_2	s_1	RHS
x_2	0	0	1	−1	2
x_1	0	1	0	1	1
P	1	0	0	1	7

This is the final tableau. The maximum is $P = 7$, obtained when $x_1 = 1$ and $x_2 = 2$.

11. Minimize $z = 6x_1 + 8x_2 + x_3$ subject to the constraints

$$3x_1 + 5x_2 + 3x_3 \geq 20$$
$$x_1 + 3x_2 + 2x_3 \geq 9$$
$$6x_1 + 2x_2 + 5x_3 \geq 30$$
$$x_1 + x_2 + x_3 \leq 10$$
$$x_1 \geq 0,\ x_2 \geq 0,\ x_3 \geq 0$$

We will solve the minimum problem by changing it to a maximum problem. Since it is a maximum problem all the constraints must be written as less than or equal inequalities. We will maximize $P = -z = -6x_1 - 8x_2 - x_3$.

Rewrite the constraints

$$-3x_1 - 5x_2 - 3x_3 \leq -20$$
$$-x_1 - 3x_2 - 2x_3 \leq -9$$
$$-6x_1 - 2x_2 - 5x_3 \leq -30$$
$$x_1 + x_2 + x_3 \leq 10$$
$$x_1 \geq 0,\ x_2 \geq 0,\ x_3 \geq 0$$

Introduce slack variables

$$-3x_1 - 5x_2 - 3x_3 + s_1 = -20$$
$$-x_1 - 3x_2 - 2x_3 + s_2 = -9$$
$$-6x_1 - 2x_2 - 5x_3 + s_3 = -30$$
$$x_1 + x_2 + x_3 + s_4 = 10$$
$$x_1 \geq 0,\ x_2 \geq 0,\ x_3 \geq 0,\ s_1 \geq 0,\ s_2 \geq 0,\ s_3 \geq 0,\ s_4 \geq 0$$

We now set up the initial tableau and use the Alternate Pivoting Strategy as long as there are negative entries in the RHS. When all the entries in the RHS are positive, we use the standard way of choosing a pivot.

BV	P	x_1	x_2	x_3	s_1	s_2	s_3	s_4	RHS
s_1	0	−3	−5	−3	1	0	0	0	−20
s_2	0	−1	−3	−2	0	1	0	0	−9
s_3	0	[−6]	−2	−5	0	0	1	0	−30
s_4	0	1	1	1	0	0	0	1	10
P	1	6	8	1	0	0	0	0	0

Alternative Pivoting Strategy
$\xrightarrow[\begin{array}{l}R_3=-\frac{1}{6}r_3\\R_1=3R_3+r_1\\R_2=R_3+r_2\\R_4=-R_3+r_4\\R_5=-6R_3+r_5\end{array}]{}$

BV	P	x_1	x_2	x_3	s_1	s_2	s_3	s_4	RHS
s_1	0	0	[−4]	$-\frac{1}{2}$	1	0	$-\frac{1}{2}$	0	−5
s_2	0	0	$-\frac{8}{3}$	$-\frac{7}{6}$	0	1	$-\frac{1}{6}$	0	−4
x_1	0	1	$\frac{1}{3}$	$\frac{5}{6}$	0	0	$-\frac{1}{6}$	0	5
s_4	0	0	$\frac{2}{3}$	$\frac{1}{6}$	0	0	$\frac{1}{6}$	1	5
P	1	0	6	−4	0	0	1	0	−30

(continued on next page)

(*continued*)

BV	P	x_1	x_2	x_3	s_1	s_2	s_3	s_4	RHS
x_2	0	0	1	$\frac{1}{8}$	$-\frac{1}{4}$	0	$\frac{1}{8}$	0	$\frac{5}{4}$
s_2	0	0	0	$\boxed{-\frac{5}{6}}$	$-\frac{2}{3}$	1	$\frac{1}{6}$	0	$-\frac{2}{3}$
x_1	0	1	0	$\frac{19}{24}$	$\frac{1}{12}$	0	$-\frac{5}{24}$	0	$\frac{55}{12}$
s_4	0	0	0	$\frac{1}{12}$	$\frac{1}{6}$	0	$\frac{1}{12}$	1	$\frac{25}{6}$
P	1	0	0	$-\frac{19}{4}$	$\frac{3}{2}$	0	$\frac{1}{4}$	0	$-\frac{75}{2}$

Alternative Pivoting Strategy

$R_1 = -\frac{1}{4}r_1$
$R_2 = \frac{8}{3}R_1 + r_2$
$R_3 = -\frac{1}{3}R_1 + r_3$
$R_4 = -\frac{2}{3}R_1 + r_4$
$R_5 = -6R_1 + r_5$

BV	P	x_1	x_2	x_3	s_1	s_2	s_3	s_4	RHS
x_2	0	0	1	0	$-\frac{7}{20}$	$\frac{3}{20}$	$\frac{3}{20}$	0	$\frac{23}{20}$
x_3	0	0	0	1	$\frac{4}{5}$	$-\frac{6}{5}$	$-\frac{1}{5}$	0	$\frac{4}{5}$
x_1	0	1	0	0	$-\frac{11}{20}$	$\boxed{\frac{19}{20}}$	$-\frac{1}{20}$	0	$\frac{79}{20}$
s_4	0	0	0	0	$\frac{1}{10}$	$\frac{1}{10}$	$\frac{1}{10}$	1	$\frac{41}{10}$
P	1	0	0	0	$\frac{53}{10}$	$-\frac{57}{10}$	$-\frac{7}{10}$	0	$\frac{337}{10}$

Alternative Pivoting Strategy

$R_2 = -\frac{6}{5}r_2$
$R_1 = -\frac{1}{8}R_2 + r_1$
$R_3 = -\frac{19}{24}R_2 + r_3$
$R_4 = -\frac{1}{12}R_2 + r_4$
$R_5 = \frac{19}{4}R_2 + r_5$

BV	P	x_1	x_2	x_3	s_1	s_2	s_3	s_4	RHS
x_2	0	$-\frac{3}{19}$	1	0	$-\frac{5}{19}$	0	$\boxed{\frac{3}{19}}$	0	$\frac{10}{19}$
x_3	0	$\frac{24}{19}$	0	1	$\frac{2}{19}$	0	$-\frac{5}{19}$	0	$\frac{110}{19}$
s_2	0	$\frac{20}{19}$	0	0	$-\frac{11}{19}$	1	$-\frac{1}{19}$	0	$\frac{79}{19}$
s_4	0	$-\frac{2}{19}$	0	0	$\frac{3}{19}$	0	$\frac{2}{19}$	1	$\frac{70}{19}$
P	1	6	0	0	2	0	-1	0	-10

Standard Pivoting Strategy

$R_3 = \frac{20}{19}r_3$
$R_1 = -\frac{3}{20}R_3 + r_1$
$R_2 = \frac{6}{5}R_3 + r_2$
$R_4 = -\frac{1}{10}R_3 + r_4$
$R_5 = \frac{57}{10}R_3 + r_5$

BV	P	x_1	x_2	x_3	s_1	s_2	s_3	s_4	RHS
s_3	0	-1	$\frac{19}{3}$	0	$-\frac{5}{3}$	0	1	0	$\frac{10}{3}$
x_3	0	1	$\frac{5}{3}$	1	$-\frac{1}{3}$	0	0	0	$\frac{20}{3}$
s_2	0	1	$\frac{1}{3}$	0	$-\frac{2}{3}$	1	0	0	$\frac{13}{3}$
s_4	0	0	$-\frac{2}{3}$	0	$\frac{1}{3}$	0	0	1	$\frac{10}{3}$
P	1	5	$\frac{19}{3}$	0	$\frac{1}{3}$	0	0	0	$-\frac{20}{3}$

Standard Pivoting Strategy

$R_1 = \frac{19}{3}r_1$
$R_2 = \frac{5}{19}R_1 + r_2$
$R_3 = \frac{1}{19}R_1 + r_3$
$R_4 = -\frac{2}{19}R_1 + r_4$
$R_5 = R_1 + r_5$

The maximum value of $P = -\dfrac{20}{3}$, so the minimum value of $z = \dfrac{20}{3}$. It is obtained when $x_1 = 0$, $x_2 = 0$, and

$x_3 = \dfrac{20}{3}$.

13. a. We define the variables x_1, x_2, x_3, and x_4 so that $x_1 =$ the number of units shipped from M1 to A1, $x_2 =$ the number of units shipped from M1 to A2, $x_3 =$ the number of units shipped from M2 to A1, and $x_4 =$ the number of units shipped from M2 to A2. The objective is to minimize shipping costs, $C = 400x_1 + 100x_2 + 200x_3 + 300x_4$ subject to the constraints

$$
\begin{aligned}
x_1 + x_2 & & & \leq 600 \\
& & x_3 + x_4 & \leq 400 \\
x_1 & + x_3 & & \geq 500 \\
& x_2 & + x_4 & \geq 300
\end{aligned}
$$

$$x_1 \geq 0, \quad x_2 \geq 0, \quad x_3 \geq 0, \quad x_4 \geq 0$$

b. We change the problem to a maximization problem and write the constraints as less than or equal to inequalities. Maximize $P = -C = -400x_1 - 100x_2 - 200x_3 - 300x_4$ subject to the constraints

$$
\begin{aligned}
x_1 + x_2 & & & \leq 600 \\
& & x_3 + x_4 & \leq 400 \\
-x_1 & - x_3 & & \leq -500 \\
& -x_2 & - x_4 & \leq -300
\end{aligned}
$$

$$x_1 \geq 0, \quad x_2 \geq 0, \quad x_3 \geq 0, \quad x_4 \geq 0$$

We introduce nonnegative slack variables and set up the initial tableau.

BV	P	x_1	x_2	x_3	x_4	s_1	s_2	s_3	s_4	RHS
s_1	0	1	1	0	0	1	0	0	0	600
s_2	0	0	0	1	1	0	1	0	0	400
s_3	0	$\boxed{-1}$	0	-1	0	0	0	1	0	-500
s_4	0	0	-1	0	-1	0	0	0	1	-300
P	1	400	100	200	300	0	0	0	0	0

Since there are negative entries in the RHS, we use the Alternative Pivoting Strategy.

Alternative Pivoting Strategy
$R_3 = -r_3$
$R_1 = -R_3 + r_1$
$R_5 = -400R_3 + r_5$

BV	P	x_1	x_2	x_3	x_4	s_1	s_2	s_3	s_4	RHS
s_1	0	0	1	-1	0	1	0	1	0	100
s_2	0	0	0	1	1	0	1	0	0	400
x_1	0	1	0	1	0	0	0	-1	0	500
s_4	0	0	$\boxed{-1}$	0	-1	0	0	0	1	-300
P	1	0	100	-200	300	0	0	400	0	-200,000

Alternative Pivoting Strategy
$R_4 = -r_4$
$R_1 = -R_4 + r_1$
$R_5 = -100R_4 + r_5$

BV	P	x_1	x_2	x_3	x_4	s_1	s_2	s_3	s_4	RHS
s_1	0	0	0	$\boxed{-1}$	-1	1	0	1	1	-200
s_2	0	0	0	1	1	0	1	0	0	400
x_1	0	1	0	1	0	0	0	-1	0	500
x_2	0	0	1	0	1	0	0	0	-1	300
P	1	0	0	-200	200	0	0	400	100	-230,000

Alternative Pivoting Strategy
$R_1 = -r_1$
$R_2 = -R_1 + r_2$
$R_3 = -R_1 + r_3$
$R_5 = 200R_1 + r_5$

BV	P	x_1	x_2	x_3	x_4	s_1	s_2	s_3	s_4	RHS
x_3	0	0	0	1	1	-1	0	-1	-1	200
s_2	0	0	0	0	0	$\boxed{1}$	1	1	1	200
x_1	0	1	0	0	-1	1	0	0	1	300
x_2	0	0	1	0	1	0	0	0	-1	300
P	1	0	0	0	400	-200	0	200	-100	-190,000

(*continued on next page*)

(*continued*)

	BV	P	x_1	x_2	x_3	x_4	s_1	s_2	s_3	s_4	RHS
Standard	x_3	0	0	0	1	1	0	1	0	0	400
Pivoting	s_1	0	0	0	0	0	1	1	1	1	200
Strategy $\rightarrow$	x_1	0	1	0	0	-1	0	-1	-1	0	100
$R_1 = R_2 + r_1$	x_2	0	0	1	0	1	0	0	0	-1	300
$R_3 = -R_2 + r_3$	P	1	0	0	0	400	0	200	400	100	-150,000
$R_4 = 200R_2 + r_4$											

The maximum $P = -150,000$, so the minimum cost $= 150,000$, obtained when $x_1 = 100$, $x_2 = 300$, $x_3 = 400$ and $x_4 = 0$.

c. Private Motors should ship 100 engines from M1 to A1, 300 engines from M1 to A2, 400 engines from M2 to A1, and no engines from M2 to A2 for a minimum shipping charge of $150,000.

15.a. Let C denote the total shipping cost, and x_1 the number of GPS systems shipped from W_1 to D_1; x_2 the number shipped from W_1 to D_2; x_3 the number shipped from W_2 to D_1; and x_4 the number shipped from W_2 to D_2. The manufacturer wants to fill the orders at the lowest possible cost. Minimize $C = 8x_1 + 12x_2 + 13x_3 + 7x_4$ subject to the constraints

$$x_1 + x_3 = 20$$
$$x_2 + x_4 = 30$$
$$x_1 + x_2 \le 40$$
$$x_3 + x_4 \le 15$$

$$x_1 \ge 0, \ x_2 \ge 0, \ x_3 \ge 0, \ x_4 \ge 0$$

b. This is a minimum problem that is not in standard form. We will solve the maximum problem formed by writing $P = -C = -8x_1 - 12x_2 - 13x_3 - 7x_4$. Add nonnegative slack variables and set up the initial tableau. Since there are equality constraints, we first pivot on the nonbasic variables.

BV	P	x_1	x_2	x_3	x_4	s_1	s_2	RHS
	0	[1]	0	1	0	0	0	20
	0	0	1	0	1	0	0	30
s_1	0	1	1	0	0	1	0	40
s_2	0	0	0	1	1	0	1	15
P	1	8	12	13	7	0	0	0

Nonbasic Pivot $\rightarrow$
$R_3 = -R_1 + r_3$
$R_5 = -8R_1 + r_5$

BV	P	x_1	x_2	x_3	x_4	s_1	s_2	RHS
x_1	0	1	0	1	0	0	0	20
	0	0	[1]	0	1	0	0	30
s_1	0	0	1	-1	0	1	0	20
s_2	0	0	0	1	1	0	1	15
P	1	0	12	5	7	0	0	-160

There are two equations, so we again pivot using a nonbasic variable. This time we use x_2.

BV	P	x_1	x_2	x_3	x_4	s_1	s_2	RHS
x_1	0	1	0	1	0	0	0	20
x_2	0	0	1	0	1	0	0	30
s_1	0	0	0	[-1]	-1	1	0	-10
s_2	0	0	0	1	1	0	1	15
P	1	0	0	5	-5	0	0	-520

Nonbasic Pivot $\rightarrow$
$R_3 = -R_2 + r_3$
$R_5 = -12R_2 + r_5$

Since there is a negative entry in the RHS, so we use the alternate pivoting strategy.

BV	P	x_1	x_2	x_3	x_4	s_1	s_2	RHS
x_1	0	1	0	1	0	0	0	20
x_2	0	0	1	-1	0	1	0	20
x_3	0	0	0	1	1	-1	0	10
s_2	0	0	0	0	0	[1]	1	5
P	1	0	0	10	0	-5	0	-470

Alternative Pivoting Strategy $\rightarrow$
$R_3 = -r_3$
$R_1 = R_3 + r_1$
$R_2 = -R_3 + r_2$
$R_4 = -R_3 + r_4$
$R_5 = 10R_3 + r_5$

We now use the standard pivoting strategy since there are no negative entries in the RHS.

(*continued on next page*)

(*continued*)

	BV	P	x_1	x_2	x_3	x_4	s_1	s_2	RHS
Standard	x_1	0	1	0	1	0	0	0	20
Pivoting	x_2	0	0	1	-1	0	0	-1	15
Strategy $\longrightarrow$	x_3	0	0	0	1	1	0	1	15
$R_2 = -R_4 + r_2$	s_2	0	0	0	0	0	1	1	5
$R_3 = R_4 + r_3$									
$R_5 = 5R_4 + r_5$	P	1	0	0	10	0	0	5	-445

Minimum $C = 445$ when $x_1 = 20$, $x_2 = 15$, $x_3 = 0$, $x_4 = 15$.

c. The minimum cost is \$445 when 20 GPS systems are shipped from W_1 to D_1, 15 from W_1 to D_2, 0 from W_2 to D_1, and 15 from W_2 to D_2.

17. a. Let x_1 represent the amount spent on newspaper advertising, and x_2 represent the amount spent on radio advertising. C is the total cost of advertising. We want to minimize $C = x_1 + x_2$ subject to the constraints

$$50x_1 + 70x_2 \geq 100,000$$
$$40x_1 + 20x_2 \geq 120,000$$
$$x_1 \geq 0, \quad x_2 \geq 0$$

b. We change the problem to a maximization problem and write the constraints as less than or equal to inequalities. Maximize $P = -C = -x_1 - x_2$ subject to the constraints

$$-50x_1 - 70x_2 \leq -100,000$$
$$-40x_1 - 20x_2 \leq -120,000$$
$$x_1 \geq 0, \ x_2 \geq 0, \ x_3 \geq 0$$

We introduce nonnegative slack variables and set up the initial tableau. Since there are negative entries in the RHS, we use the Alternative Pivoting Strategy.

BV	P	x_1	x_2	s_1	s_2	RHS
s_1	0	-50	-70	1	0	-100,000
s_2	0	-40	-20	0	1	-120,000
P	1	1	1	0	0	0

Alternative Pivot
$R_2 = -\frac{1}{40}R_2$
$\longrightarrow$
$R_1 = 50R_2 + r_1$
$R_3 = -R_2 + r_3$

BV	P	x_1	x_2	s_1	s_2	RHS
s_1	0	0	-45	1	$-\frac{6}{5}$	50,000
x_1	0	1	$\frac{1}{2}$	0	$-\frac{1}{40}$	3,000
P	1	0	$\frac{1}{2}$	0	$\frac{1}{40}$	-3,000

Maximum $P = -3000$, so the minimum is $C = 3000$ which is obtained when $x_1 = 3000$ and $x_2 = 0$.

c. The appliance store will spend the least on advertising, \$3000, and reach the intended audience, if it spends all \$3000 on newspaper advertising and nothing on radio advertising.

19. Let C represent the total calories in the mixture, and let $x_1, x_2, x_3,$ and x_4 represent the number of cups of corn, peas, green beans, and carrots respectively. We want to minimize $C = 160x_1 + 140x_2 + 40x_3 + 60x_4$ calories per cup subject to the constraints

$$x_1 + x_2 + x_3 + x_4 = 200 \text{ cups}$$
$$4x_1 + 6x_2 + 4x_3 + 4x_4 \geq 4.5(200) \text{ grams of fiber}$$
$$8x_1 + 12x_2 + 4x_3 + 8x_4 \leq 8.5(200) \text{ grams of sugar per cup}$$
$$x_1 - x_3 \geq 0 \quad \text{(at least as much corn as green beans)}$$
$$+ x_2 - 2x_4 \geq 0 \quad \text{(at least half as much carrots as peas)}$$
$$x_1 \geq 0 \quad x_2 \geq 0 \quad x_3 \geq 0 \quad x_4 \geq 0$$

(*continued on next page*)

(continued)

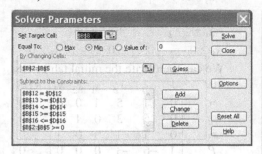

The minimum number of calories per serving is $\dfrac{18000}{400} = 45$. This is obtained by using 25 cups of corn, 50 cups of peas, 25 cups of green beans, and 100 cups of carrots.

21. Let z denote the average price/earning ratio of the stocks purchased, and let x_1, x_2, x_3, and x_4 denote the number of shares of Duke Energy, H.J. Heinz, General Electric, and Ferrellgas Partners respectively purchased. The aim is to minimize $z = 19.79x_1 + 16.33x_2 + 18.23x_3 + 37.78x_4$ subject to the constraints

$$16.26x_1 + 45.99x_2 + 18.71x_3 + 23.14x_4 = 50,000$$
$$45.99x_2 \geq 5,000$$
$$16.26x_1 \geq 10,000$$
$$18.71x_3 \geq 10,000$$
$$23.14x_4 \geq 10,000$$
$$16.26x_1 + 18.71x_3 \leq 25,000$$
$$x_1 \geq 0 \quad x_2 \geq 0 \quad x_3 \geq 0 \quad x_4 \geq 0$$

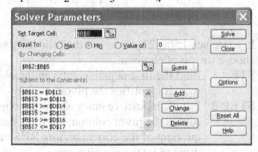

The average price/earning ratio is

$$\frac{615(19.79) + 434.9(16.33) + 534.5(18.23) + 432.2(37.78)}{615 + 434.9 + 534.5 + 432.2} = \frac{45,345.22}{2016.6} \approx 22.49.$$

The annual yield for this optimal investment strategy is

$$0.059(16.26)(615.0) + 0.0365(45.99)(434.9) + 0.0214(18.71)(534.5) + 0.0864(23.14)(432.2) = \$2398.14.$$

Chapter 5 Review Exercises

For exercises 1–7, a maximum problem is in standard form if both of the following conditions are met.

Condition 1   All the variables are nonnegative.
Condition 2   Every other constraint is written as a linear expression that is less than or equal to a positive constant.

1. The problem is in standard form. Both conditions are met.

3. The problem is in standard form. Both conditions are met.

5. This problem is not in standard form. Condition 2 is not met. One of the constraints is written as a greater than or equal to inequality.

7. This problem is not in standard form. Condition 1 is not met. x_2 can be negative.

9. Maximize $P = 2x_1 + x_2 + 3x_3$ subject to the constraints
$$2x_1 + 5x_2 + x_3 \le 100$$
$$x_1 + 3x_2 + x_3 \le 80$$
$$2x_1 + 3x_2 + 3x_3 \le 120$$
$$x_1 \ge 0, \ x_2 \ge 0, \ x_3 \ge 0$$
We add slack variables and rewrite the problem as maximize $P - 2x_1 - x_2 - 3x_3 = 0$ subject to the constraints
$$2x_1 + 5x_2 + x_3 + s_1 \qquad = 100$$
$$x_1 + 3x_2 + x_3 \quad + s_2 \qquad = 80$$
$$2x_1 + 3x_2 + 3x_3 \qquad\quad + s_3 = 120$$
$$x_1 \ge 0, \ x_2 \ge 0, \ x_3 \ge 0, \ s_1 \ge 0, \ s_2 \ge 0, \ s_3 \ge 0$$
We can then write the initial tableau as

BV	P	x_1	x_2	x_3	s_1	s_2	s_3	RHS
s_1	0	2	5	1	1	0	0	100
s_2	0	1	3	1	0	1	0	80
s_3	0	2	3	3	0	0	1	120
P	1	−2	−1	−3	0	0	0	0

11. Maximize $P = 6x_1 + 3x_2$ subject to the constraints
$$x_1 + 5x_2 \le 200$$
$$5x_1 + 3x_2 \le 450$$
$$x_1 + x_2 \le 120$$
$$x_1 \ge 0, \ x_2 \ge 0$$
We add slack variables and rewrite the problem as maximize $P - 6x_1 - 3x_2 = 0$ subject to the constraints

$$x_1 + 5x_2 + s_1 \qquad\quad = 200$$
$$5x_1 + 3x_2 \quad + s_2 \qquad = 450$$
$$x_1 + x_2 \qquad\quad + s_3 = 120$$
$$x_1 \ge 0, \ x_2 \ge 0, \ s_1 \ge 0, \ s_2 \ge 0, \ s_3 \ge 0$$
We can then write the initial tableau as

BV	P	x_1	x_2	s_1	s_2	s_3	RHS
s_1	0	1	5	1	0	0	200
s_2	0	5	3	0	1	0	450
s_3	0	1	1	0	0	1	120
P	1	−6	−3	0	0	0	0

13. Maximize $P = x_1 + 2x_2 + x_3 + 4x_4$ subject to the constraints
$$x_1 + 3x_2 + x_3 + 2x_4 \le 20$$
$$4x_1 + x_2 + x_3 + 6x_4 \le 80$$
$$x_1 \ge 0, \ x_2 \ge 0, \ x_3 \ge 0, \ x_4 \ge 0$$
We add slack variables and rewrite the problem as maximize $P - x_1 - 2x_2 - x_3 - 4x_4 = 0$ subject to the constraints
$$x_1 + 3x_2 + x_3 + 2x_4 + s_1 \qquad = 20$$
$$4x_1 + x_2 + x_3 + 6x_4 \quad + s_2 = 80$$
$$x_1 \ge 0, \ x_2 \ge 0, \ x_3 \ge 0, \ x_4 \ge 0, \ s_1 \ge 0, \ s_2 \ge 0$$
We can then write the initial tableau as

BV	P	x_1	x_2	x_3	x_4	s_1	s_2	RHS
s_1	0	1	3	1	2	1	0	20
s_2	0	4	1	1	6	0	1	80
P	1	−1	−2	−1	−4	0	0	0

In exercises 15(a)-21(a) we use the following steps to choose the pivot element.

STEP 1   Find the first negative entry in the objective row. It identifies the pivot column.

STEP 2   For each positive entry above the objective row in the pivot column, we form the quotient of the corresponding RHS entry divided by the positive entry.

STEP 3   Use the smallest quotient from step 2 to identify the pivot row and the pivot element.

15.a.

BV	P	x_1	x_2	s_1	s_2	RHS
x_1	0	1	5	1	0	40
s_2	0	0	2	2	1	10
P	1	0	−1	0	3	120

We find the pivot element by selecting the first column containing a negative entry in the objective row. Here it is column x_2. We select the pivot row by computing the quotients formed by dividing the entry in the right hand side by the corresponding positive entry of the pivot column. The pivot row has the smallest nonnegative quotient.
$$40 \div 5 = 8 \text{ and } 10 \div 2 = 5$$

(continued on next page)

(*continued*)

The pivot element is 2 in row s_2, column x_2.

$$
\begin{array}{c|cccccc|c}
\text{BV} & P & x_1 & x_2 & s_1 & s_2 & \text{RHS} \\
\hline
x_1 & 0 & 1 & 5 & 1 & 0 & 40 \\
s_2 & 0 & 0 & \boxed{2} & 2 & 1 & 10 \\
\hline
P & 1 & 0 & -1 & 0 & 3 & 120
\end{array}
$$

$$
\xrightarrow[\substack{R_2=\frac{1}{2}r_2 \\ R_1=-5R_2+r_1 \\ R_3=R_2+r_3}]{}
\begin{array}{c|ccccc|c}
\text{BV} & P & x_1 & x_2 & s_1 & s_2 & \text{RHS} \\
\hline
x_1 & 0 & 1 & 0 & -4 & -\frac{5}{2} & 15 \\
x_2 & 0 & 0 & 1 & 1 & \frac{1}{2} & 5 \\
\hline
P & 1 & 0 & 0 & 1 & \frac{7}{2} & 125
\end{array}
$$

b. The resulting system of equations is
$$
\begin{aligned}
x_1 &= 15 + 4s_1 + \frac{5}{2}s_2 \\
x_2 &= 5 - s_1 - \frac{1}{2}s_2 \\
P &= 125 - s_1 - \frac{7}{2}s_2
\end{aligned}
$$

c. The new tableau is the final tableau. The solution is maximum $P = 125$ when $x_1 = 15$ and $x_2 = 5$.

17.a.

$$
\begin{array}{c|cccccc|c}
\text{BV} & P & x_1 & x_2 & x_3 & s_1 & s_2 & \text{RHS} \\
\hline
s_1 & 0 & 1 & 1 & -1 & 1 & 0 & 10 \\
s_2 & 0 & 0 & 1 & 1 & 0 & 1 & 4 \\
\hline
P & 1 & -2 & -1 & -3 & 0 & 0 & 0
\end{array}
$$

We find the pivot element by selecting the first column containing a negative entry in the objective row. Here it is column x_1. In this column there is only one positive entry, so it becomes the pivot. The pivot element is 1 in row s_1, column x_1.

$$
\begin{array}{c|cccccc|c}
\text{BV} & P & x_1 & x_2 & x_3 & s_1 & s_2 & \text{RHS} \\
\hline
s_1 & 0 & \boxed{1} & 1 & -1 & 1 & 0 & 10 \\
s_2 & 0 & 0 & 1 & 1 & 0 & 1 & 4 \\
\hline
P & 1 & -2 & -1 & -3 & 0 & 0 & 0
\end{array}
\xrightarrow{R_3=2R_1+r_3}
\begin{array}{c|cccccc|c}
\text{BV} & P & x_1 & x_2 & x_3 & s_1 & s_2 & \text{RHS} \\
\hline
x_1 & 0 & 1 & 1 & -1 & 1 & 0 & 10 \\
s_2 & 0 & 0 & 1 & 1 & 0 & 1 & 4 \\
\hline
P & 1 & 0 & 1 & -5 & 2 & 0 & 20
\end{array}
$$

b. The resulting system of equations is
$$
\begin{aligned}
x_1 &= 10 - x_2 + x_3 - s_1 \\
s_2 &= 4 - x_2 - x_3 \\
P &= 20 - x_2 + 5x_3 - 2s_1
\end{aligned}
$$

c. This tableau needs further pivoting, since there is still a negative entry in the objective row. The pivot column is x_3, the pivot row is s_2, making 1 the new pivot element.

19. a.

BV	P	x_1	x_2	s_1	s_2	RHS
s_1	0	0.5	0.5	1	0	1
s_2	0	1	1.5	0	1	3
P	1	−2.5	−2	0	0	0

We find the pivot element by selecting the first column containing a negative entry in the objective row. Here it is column x_1. We select the pivot row by computing the quotients formed by dividing the entry in the right hand side by the corresponding positive entry of the pivot column. The pivot row has the smallest nonnegative quotient: $1 \div 0.5 = 2$ and $3 \div 1 = 3$. The pivot element is 0.5 in row s_1, column x_1.

BV	P	x_1	x_2	s_1	s_2	RHS
s_1	0	0.5	0.5	1	0	1
s_2	0	1	1.5	0	1	3
P	1	−2.5	−2	0	0	0

$R_1 = 2r_1$
$R_2 = -R_1 + r_2$
$R_3 = 2.5R_1 + r_3$

BV	P	x_1	x_2	s_1	s_2	RHS
x_1	0	1	1	2	0	2
s_2	0	0	0.5	−2	1	1
P	1	0	0.5	5	0	5

b. The resulting system of equations is
$$x_1 = 2 - x_2 - 2s_1$$
$$s_2 = 1 - 0.5x_2 + 2s_1$$
$$P = 5 - 0.5x_2 - 5s_2$$

c. The new tableau is the final tableau. The solution is maximum $P = 5$ when $x_1 = 2$ and $x_2 = 0$.

21. a.

BV	P	x_1	x_2	x_3	s_1	s_2	s_3	RHS
s_1	0	−1	0	1	1	−1	0	7
x_2	0	−1	1	5	0	1	0	5
s_3	0	1	0	3	0	−5	1	3
P	1	−3	0	4	0	0	0	5

We find the pivot element by selecting the first column containing a negative entry in the objective row. Here it is column x_1. In this column there is only one positive entry, so it becomes the pivot. The pivot element is 1 in row s_3, column x_1.

BV	P	x_1	x_2	x_3	s_1	s_2	s_3	RHS
s_1	0	−1	0	1	1	−1	0	7
x_2	0	−1	1	5	0	1	0	5
s_3	0	1	0	3	0	−5	1	3
P	1	−3	0	4	0	0	0	5

$R_1 = R_3 + r_1$
$R_2 = R_3 + r_2$
$R_4 = 3R_3 + r_4$

BV	P	x_1	x_2	x_3	s_1	s_2	s_3	RHS
s_1	0	0	0	4	1	−6	1	10
x_2	0	0	1	8	0	−4	1	8
x_1	0	1	0	3	0	−5	1	3
P	1	0	0	13	0	−15	3	14

b. The resulting system of equations is
$$s_1 = 10 - 4x_3 + 6s_2 - s_3$$
$$x_2 = 8 - 8x_3 + 4s_2 - s_3$$
$$x_1 = 3 - 3x_3 + 5s_2 - s_3$$
$$P = 14 - 13x_3 + 15s_2 - 3s_3$$

c. This tableau indicates no solution exists for the problem. The pivot column would be column s_2, but every entry in the pivot column is negative.

23. Maximize $P = 100x_1 + 200x_2 + 50x_3$ subject to the constraints

$5x_1 + 5x_2 + 10x_3 \le 1000$

$10x_1 + 8x_2 + 5x_3 \le 2000$

$10x_1 + 5x_2 \quad\quad \le 500$

$x_1 \ge 0,\ x_2 \ge 0,\ x_3 \ge 0$

BV	P	x_1	x_2	x_3	s_1	s_2	s_3	RHS
s_1	0	5	5	10	1	0	0	1000
s_2	0	10	8	5	0	1	0	2000
s_3	0	[10]	5	0	0	0	1	500
P	1	−100	−200	−50	0	0	0	0

$R_3 = \frac{1}{10}r_3$
$R_1 = -5R_3 + r_1$
$R_2 = -10R_3 + r_2$
$R_4 = 100R_3 + r_4$

BV	P	x_1	x_2	x_3	s_1	s_2	s_3	RHS
s_1	0	0	$\frac{5}{2}$	10	1	0	$-\frac{1}{2}$	750
s_2	0	0	3	5	0	1	−1	1500
x_1	0	1	$\frac{1}{2}$	0	0	0	$\frac{1}{10}$	50
P	1	0	−150	−50	0	0	10	5000

$R_3 = 2r_3$
$R_1 = -\frac{5}{2}R_3 + r_1$
$R_2 = -3R_3 + r_2$
$R_4 = 150R_3 + r_4$

BV	P	x_1	x_2	x_3	s_1	s_2	s_3	RHS
s_1	0	−5	0	[10]	1	0	−1	500
s_2	0	−6	0	5	0	1	$-\frac{8}{5}$	1200
x_2	0	2	1	0	0	0	$\frac{1}{5}$	100
P	1	300	0	−50	0	0	40	20000

$R_1 = \frac{1}{10}r_1$
$R_2 = -5R_1 + r_2$
$R_4 = 50R_1 + r_4$

BV	P	x_1	x_2	x_3	s_1	s_2	s_3	RHS
x_3	0	$-\frac{1}{2}$	0	1	$\frac{1}{10}$	0	$-\frac{1}{10}$	50
s_2	0	$-\frac{7}{2}$	0	0	$-\frac{1}{2}$	1	$-\frac{11}{10}$	950
x_2	0	2	1	0	0	0	$\frac{1}{5}$	100
P	1	275	0	0	5	0	35	22250

The maximum value for P is 22,500, obtained when $x_1 = 0$, $x_2 = 100$ and $x_3 = 50$.

25. Maximize $P = 40x_1 + 60x_2 + 50x_3$ subject to the constraints

$2x_1 + 2x_2 + x_3 \le 8$

$x_1 - 4x_2 + 3x_3 \le 12$

$x_1 \ge 0,\ x_2 \ge 0,\ x_3 \ge 0$

BV	P	x_1	x_2	x_3	s_1	s_2	RHS
s_1	0	[2]	2	1	1	0	8
s_2	0	1	−4	3	0	1	12
P	1	−40	−60	−50	0	0	0

$R_1 = \frac{1}{2}r_1$
$R_2 = -R_1 + r_2$
$R_3 = 40R_1 + r_3$

BV	P	x_1	x_2	x_3	s_1	s_2	RHS
x_1	0	1	[1]	$\frac{1}{2}$	$\frac{1}{2}$	0	4
s_2	0	0	−5	$\frac{5}{2}$	$-\frac{1}{2}$	1	8
P	1	0	−20	−30	20	0	160

$R_2 = 5R_1 + r_2$
$R_3 = 20R_1 + r_3$

BV	P	x_1	x_2	x_3	s_1	s_2	RHS
s_1	0	1	1	$\frac{1}{2}$	$\frac{1}{2}$	0	4
s_2	0	5	0	[5]	2	1	28
P	1	20	0	−20	30	0	240

$R_2 = \frac{1}{5}R_2$
$R_1 = -\frac{1}{2}R_2 + r_1$
$R_3 = 20R_2 + r_3$

0	$\frac{1}{2}$	1	0	$\frac{3}{10}$	$-\frac{1}{10}$	$\frac{6}{5}$	
0	1	0	1	$\frac{2}{5}$	$\frac{1}{5}$	$\frac{28}{5}$	
1	40	0	0	38	4	352	

The maximum value for P is 352, obtained when $x_1 = 0$, $x_2 = \dfrac{6}{5}$ and $x_3 = \dfrac{28}{5}$.

For exercises 27–31, a minimum problem is in standard form if the following conditions are met.

Condition 1 All the variables are nonnegative.

Condition 2 Every other constraint is written as a linear expression that is greater than or equal to a positive constant.

Condition 3 The objective function is a linear expression with nonnegative coefficients.

27. The minimum problem is in standard form.

29. The minimum problem is not in standard form. Condition 2 is not met; the constraints are not written as greater than or equal to inequalities.

31. The minimum problem is not in standard form. Condition 2 is not met; one of the constraints is not written as greater than or equal to inequality.

In exercises 33–35 use the following 4 steps to write the Dual Problem.

STEP 1 Write the minimum problem in standard form.

STEP 2 Construct a matrix that represents the constraints and the objective function, placing the objective function in the bottom row.

STEP 3 Interchange the rows and columns to form the matrix of the dual problem.

STEP 4 Translate this matrix into a maximum problem in standard form.

33. **STEP 1**: The problem is in standard form since all conditions are met.

STEP 2: The matrix with the objective function in the bottom row.

$$
\begin{array}{cc}
x_1 & x_2 \\
\end{array}
$$
$$
\left[\begin{array}{cc|c}
2 & 2 & 8 \\
1 & -1 & 2 \\
2 & 1 & 0
\end{array}\right]
$$

STEP 3: The matrix with the rows and columns interchanged.

$$
\left[\begin{array}{cc|c}
2 & 1 & 2 \\
2 & -1 & 1 \\
8 & 2 & 0
\end{array}\right]
$$

STEP 4: The corresponding maximum problem in standard form. Maximize $P = 8y_1 + 2y_2$ subject to the constraints

$$2y_1 + y_2 \le 2$$
$$2y_1 - y_2 \le 1$$
$$y_1 \ge 0 \qquad y_2 \ge 0$$

This maximum problem is the dual of the minimum problem.

35. **STEP 1**: The problem is in standard form since all conditions are met.

STEP 2: The matrix with the objective function in the bottom row.

$$
\begin{array}{ccc}
x_1 & x_2 & x_3 \\
\end{array}
$$
$$
\left[\begin{array}{ccc|c}
1 & 1 & 1 & 100 \\
2 & 1 & 0 & 50 \\
5 & 4 & 2 & 0
\end{array}\right]
$$

STEP 3: The matrix with the rows and columns interchanged.

$$
\left[\begin{array}{cc|c}
1 & 2 & 5 \\
1 & 1 & 4 \\
1 & 0 & 2 \\
100 & 50 & 0
\end{array}\right]
$$

STEP 4: The corresponding maximum problem in standard form. Maximize $P = 100y_1 + 50y_2$ subject to the constraints

$$y_1 + 2y_2 \le 5$$
$$y_1 + y_2 \le 4$$
$$y_1 \qquad \le 2$$
$$y_1 \ge 0 \qquad y_2 \ge 0$$

This maximum problem is the dual of the minimum problem.

37. Using the dual found in Problem 33, we have

BV	P	y_1	y_2	s_1	s_2	RHS
s_1	0	2	1	1	0	2
s_2	0	2	–1	0	1	1
P	1	–8	–2	0	0	0

$$
\begin{array}{l}
R_2 = \frac{1}{2}r_2 \\
R_1 = -2R_2 + r_1 \\
R_3 = 8R_2 + r_3
\end{array}
$$

BV	P	y_1	y_2	s_1	s_2	RHS
s_1	0	0	2	1	–1	1
y_1	0	1	$-\frac{1}{2}$	0	$\frac{1}{2}$	$\frac{1}{2}$
P	1	0	–6	0	4	4

$$
\begin{array}{l}
R_1 = \frac{1}{2}r_1 \\
R_2 = \frac{1}{2}R_1 + r_2 \\
R_3 = 6R_1 + r_3
\end{array}
$$

BV	P	y_1	y_2	s_1	s_2	RHS
y_2	0	0	1	$\frac{1}{2}$	$-\frac{1}{2}$	$\frac{1}{2}$
y_1	0	1	0	$\frac{1}{4}$	$\frac{1}{4}$	$\frac{3}{4}$
P	1	0	0	3	1	7

The minimum value for C is 7, when $x_1 = 3$ and $x_2 = 1$.

39. Using the dual found in Problem 35, we have

BV	P	y_1	y_2	s_1	s_2	s_3	RHS
s_1	0	1	2	1	0	0	5
s_2	0	1	1	0	1	0	4
s_3	0	1	0	0	0	1	2
P	1	−100	−50	0	0	0	0

$$\xrightarrow[\substack{R_1=-R_3+r_1 \\ R_2=-R_3+r_2 \\ R_4=100R_3+r_4}]{}$$

BV	P	y_1	y_2	s_1	s_2	s_3	RHS
s_1	0	0	2	1	0	−1	3
s_2	0	0	1	0	1	−1	2
y_1	0	1	0	0	0	1	2
P	1	0	−50	0	0	100	200

$$\xrightarrow[\substack{R_1=\frac{1}{2}r_1 \\ R_2=-R_1+r_2 \\ R_4=50R_1+r_4}]{}$$

BV	P	y_1	y_2	s_1	s_2	RHS
y_2	0	0	1	$\frac{1}{2}$	0	−$\frac{1}{2}$	$\frac{3}{2}$
s_2	0	0	0	−$\frac{1}{2}$	1	−$\frac{1}{2}$	$\frac{1}{2}$
y_1	0	1	0	0	0	1	2
P	1	0	0	25	0	75	275

The minimum value for C is 275, when $x_1 = 25$, $x_2 = 0$, and $x_3 = 75$.

41. Maximize $P = 3x_1 + 5x_2$ subject to the constraints

$$x_1 + x_2 \geq 2$$
$$2x_1 + 3x_2 \leq 12$$
$$3x_1 + 2x_2 \leq 12$$

$$x_1 \geq 0, \ x_2 \geq 0$$

Rewrite the constraints

$$-x_1 - x_2 \leq -2$$
$$2x_1 + 3x_2 \leq 12$$
$$3x_1 + 2x_2 \leq 12$$
$$x_1 \geq 0, \ x_2 \geq 0$$

Introduce slack variables

$$-x_1 - x_2 + s_1 = -2$$
$$2x_1 + 3x_2 + s_2 = 12$$
$$3x_1 + 2x_2 + s_3 = 12$$
$$x_1 \geq 0, \ x_2 \geq 0, \ s_1 \geq 0, \ s_2 \geq 0, \ s_3 \geq 0$$

We now set up the initial tableau and use the alternate pivoting method as long as there are negative entries in the RHS. When all the entries in the RHS are positive, we use the standard way of choosing a pivot.

BV	P	x_1	x_2	s_1	s_2	s_3	RHS
s_1	0	−1	−1	1	0	0	−2
s_2	0	2	3	0	1	0	12
s_3	0	3	2	0	0	1	12
P	1	−3	−5	0	0	0	0

Alternative Pivoting Strategy
$$\xrightarrow[\substack{R_1=-R_1 \\ R_2=-2R_1+r_2 \\ R_3=-3R_1+r_3 \\ R_4=3R_1+r_4}]{}$$

BV	P	x_1	x_2	s_1	s_2	s_3	RHS
s_1	0	1	1	−1	0	0	2
s_2	0	0	1	2	1	0	8
s_3	0	0	−1	3	0	1	6
P	1	0	−2	−3	0	0	6

The new tableau has only nonnegative entries in the RHS, it represents a maximum problem in standard form. Since the objective row has negative entries, we use the standard pivoting strategy. The pivot column is x_2. We form the quotients: $2 \div 1 = 2$, $8 \div 1 = 8$. The smaller of these is 2, so the pivot row is row x_1.

BV	P	x_1	x_2	s_1	s_2	s_3	RHS
x_1	0	1	1	−1	0	0	2
s_2	0	−1	0	3	1	0	6
s_3	0	1	0	2	0	1	8
P	1	2	0	−5	0	0	10

$$\xrightarrow[\substack{R_2=-R_1+r_2 \\ R_3=R_1+r_3 \\ R_4=2R_1+r_4}]{}$$

$$\xrightarrow[\substack{R_2=\frac{1}{3}r_2 \\ R_1=R_2+r_1 \\ R_3=-2R_2+r_3 \\ R_4=5R_2+r_4}]{}$$

BV	P	x_1	x_2	s_1	s_2	s_3	RHS
x_2	0	$\frac{2}{3}$	1	0	$\frac{1}{3}$	0	4
s_1	0	−$\frac{1}{3}$	0	1	$\frac{1}{3}$	0	2
s_3	0	$\frac{5}{3}$	0	0	−$\frac{2}{3}$	1	4
P	1	$\frac{1}{3}$	0	0	$\frac{5}{3}$	0	20

This is the final tableau. The maximum value of P is 20, obtained when $x_1 = 0$ and $x_2 = 4$.

43. Minimize $C = 2x_1 + 3x_2$ subject to the constraints

$x_1 + x_2 \geq 3$

$x_1 + x_2 \leq 9$

$x_1 \geq 0,\ x_2 \geq 0$

This is a minimum problem not in standard form. We will let $P = -C$ and then maximize P. $P = -2x_1 - 3x_2$

Rewrite the constraints

$-x_1 - x_2 \leq -3$

$x_1 + x_2 \leq 9$

$x_1 \geq 0,\ x_2 \geq 0$

Introduce slack variables

$-x_1 - x_2 + s_1 \quad = -3$

$x_1 + x_2 \quad + s_2 = 9$

$x_1 \geq 0,\ x_2 \geq 0,\ s_1 \geq 0,\ s_2 \geq 0$

We now set up the initial tableau and use the alternate pivoting method as long as there are negative entries in the RHS. The first negative entry is -1 in column x_1. It is the pivot element.

BV	P	x_1	x_2	s_1	s_2	RHS
s_1	0	$\boxed{-1}$	-1	1	0	-3
s_2	0	1	1	0	1	9
P	1	2	3	0	0	0

Alternative Pivoting Strategy

$R_1 = -R_1$
$R_2 = -R_1 + r_2$
$R_3 = -2R_1 + r_3$

BV	P	x_1	x_2	s_1	s_2	RHS
x_1	0	1	1	-1	0	3
s_2	0	0	0	1	1	6
P	1	0	1	2	0	-6

This is the final tableau. The maximum value of P is -6, so the minimum value of C is 6 when $x_1 = 3$ and $x_2 = 0$.

45. Maximize $P = 300x_1 + 200x_2 + 450x_3$ subject to the constraints

$4x_1 + 3x_2 + 5x_3 \leq 140$

$x_1 + x_2 + x_3 = 30$

$x_1 \geq 0,\ x_2 \geq 0,\ x_3 \geq 0$

One constraint is an equation, so it gets no slack variable. The equation is the first row to be pivoted.

$4x_1 + 3x_2 + 5x_3 + s_1 = 140$

$x_1 + x_2 + x_3 \quad = 30$

$x_1 \geq 0,\ x_2 \geq 0,\ x_3 \geq 0,\ s_1 \geq 0$

We now set up the initial tableau and pivot on a nonbasic variable in the row with the equation.

BV	P	x_1	x_2	x_3	s_1	RHS
s_1	0	4	3	5	1	140
s_2	0	$\boxed{1}$	1	1	0	30
P	1	-300	-200	-450	0	0

$R_1 = -4R_2 + r_1$
$R_3 = 300R_2 + r_3$

BV	P	x_1	x_2	x_3	s_1	RHS
s_1	0	0	-1	$\boxed{1}$	1	20
x_1	0	1	1	1	0	30
P	1	0	100	-150	0	9000

Standard Pivoting Strategy

$R_2 = -R_1 + r_2$
$R_3 = 150R_1 + r_3$

BV	P	x_1	x_2	x_3	s_1	RHS
x_3	0	0	-1	1	1	20
x_1	0	1	$\boxed{2}$	0	-1	10
P	1	0	-50	0	150	12,000

Standard Pivoting Strategy

$R_2 = \frac{1}{2} r_2$
$R_1 = R_2 + r_1$
$R_3 = 50R_2 + r_3$

BV	P	x_1	x_2	x_3	s_1	RHS
x_3	0	$\frac{1}{2}$	0	1	$\frac{1}{2}$	25
x_2	0	$\frac{1}{2}$	1	0	$-\frac{1}{2}$	5
P	1	25	0	0	125	12,250

This is the final tableau. The maximum value of P is 12,250, obtained when $x_1 = 0$, $x_2 = 5$, and $x_3 = 25$.

47.a. We let P denote the profit, and let x_1 represent the number of cocktail tables and x_2 the number of end tables made each day. We want to maximize the profit, $P = 20x_1 + 15x_2$, subject to the constraints

$4x_1 + 8x_2 \leq 360$

$4x_1 + 10x_2 \leq 360$

$8x_1 + 4x_2 \leq 360$

$x_1 \geq 0 \quad x_2 \geq 0 \quad x_3 \geq 0$

Note that sanding, staining, and varnishing times are given in minutes, so we must convert 6 hours to minutes.

b. Slack variables are added and the initial tableau is constructed and shown below on the left.

$$
\begin{array}{c|cccccc|c}
BV & P & x_1 & x_2 & s_1 & s_2 & s_3 & RHS \\
\hline
s_1 & 0 & 4 & 8 & 1 & 0 & 0 & 360 \\
s_2 & 0 & 4 & 10 & 0 & 1 & 0 & 360 \\
s_3 & 0 & \boxed{8} & 4 & 0 & 0 & 1 & 360 \\
\hline
P & 1 & -20 & -15 & 0 & 0 & 0 & 0
\end{array}
\qquad
\begin{array}{c}
R_3=\frac{1}{8}r_3 \\
R_1=-4R_3+r_1 \\
R_2=-4R_3+r_2 \\
R_4=20R_3+r_4
\end{array}
\longrightarrow
$$

$$
\begin{array}{c|cccccc|c}
BV & P & x_1 & x_2 & s_1 & s_2 & s_3 & RHS \\
\hline
s_1 & 0 & 0 & 6 & 1 & 0 & -\frac{1}{2} & 180 \\
s_2 & 0 & 0 & \boxed{8} & 0 & 1 & -\frac{1}{2} & 180 \\
x_1 & 0 & 1 & \frac{1}{2} & 0 & 0 & \frac{1}{8} & 45 \\
\hline
P & 1 & 0 & -5 & 0 & 0 & \frac{5}{2} & 900
\end{array}
$$

$$
\begin{array}{c}
R_2=\frac{1}{8}r_2 \\
R_1=-6R_2+r_1 \\
R_3=-\frac{1}{2}R_2+r_3 \\
R_4=5R_2+r_4
\end{array}
\longrightarrow
\begin{array}{c|cccccc|c}
BV & P & x_1 & x_2 & s_1 & s_2 & s_3 & RHS \\
\hline
s_1 & 0 & 0 & 0 & 1 & -\frac{3}{4} & -\frac{1}{8} & 45 \\
x_2 & 0 & 0 & 1 & 0 & \frac{1}{8} & -\frac{1}{16} & \frac{45}{2} \\
x_1 & 0 & 1 & 0 & 0 & -\frac{1}{16} & \frac{5}{32} & \frac{135}{4} \\
\hline
P & 1 & 0 & 0 & 0 & \frac{5}{8} & \frac{35}{16} & \frac{2025}{2}
\end{array}
$$

Since all the entries in the objective row are nonnegative, this is the final tableau. The maximum $P = \dfrac{2025}{2} = 1012.5$, obtained when $x_1 = \dfrac{135}{4} = 33.75$, and $x_2 = \dfrac{45}{2} = 22.5$.

c. The furniture maker can obtain a maximum profit of $1012.50 while maintaining the time constraints if 33.75 cocktail tables and 22.5 end tables are manufactured each day. That is, profit is maximized by producing 135 cocktail tables and 90 end tables over a four-day period. The maximum profit for the four days is $4050.

49. a. Let x_1 and x_2 represent the number of pounds of ground beef and ground pork, respectively, to be combined in a one pound package of meatloaf. The objective is to minimize $C = 1.89x_1 + 1.29x_2$ subject to the constraints

$$
\begin{aligned}
x_1 + x_2 &= 1 \\
0.75x_1 + 0.6x_2 &\geq 0.7 \\
x_1 \geq 0, \quad x_2 &\geq 0
\end{aligned}
$$

b. This is a minimum problem with mixed constraints, we will transform it into a maximization problem. First, rewrite the greater than or equal to constraint as less than or equal to inequality, then introduce a nonnegative slack variable.

$$
\begin{array}{ll}
\begin{aligned}
x_1 + x_2 &= 1 \\
-0.75x_1 - 0.6x_2 &\leq -0.7 \\
x_1 \geq 0, \quad x_2 &\geq 0
\end{aligned}
\qquad
&
\begin{aligned}
x_1 + x_2 &= 1 \\
-0.75x_1 - 0.6x_2 + s_1 &= -0.7 \\
x_1 \geq 0, \quad x_2 \geq 0, \quad s_1 &\geq 0
\end{aligned}
\end{array}
$$

Set up the initial tableau. Since there is an equation in the first row, the first pivot must be in that row. We will use the x_1 column.

$$
\begin{array}{c|cccc|c}
BV & P & x_1 & x_2 & s_1 & RHS \\
\hline
 & 0 & \boxed{1} & 1 & 0 & 1 \\
s_1 & 0 & -0.75 & -0.6 & 1 & -0.7 \\
\hline
P & 1 & 1.89 & 1.29 & 0 & 0
\end{array}
\qquad
\begin{array}{c}
\\
R_2=0.75R_1+r_2 \\
R_3=-1.89R_1+r_3
\end{array}
\longrightarrow
\begin{array}{c|cccc|c}
BV & P & x_1 & x_2 & s_1 & RHS \\
\hline
x_1 & 0 & 1 & 1 & 0 & 1 \\
s_1 & 0 & 0 & \boxed{0.15} & 1 & 0.05 \\
\hline
P & 1 & 0 & -0.6 & 0 & -1.89
\end{array}
$$

The new tableau represents a maximum problem in standard form. Since the objective row has a negative entry, we use the standard pivoting strategy. The pivot column is x_2. The pivot row is row s_1.

$$
\begin{array}{c}
\\
R_2=\frac{1}{0.15}r_2=6.667r_2 \\
R_1=-R_2+r_1 \\
R_3=0.6R_2+r_3
\end{array}
\longrightarrow
\begin{array}{c|cccc|c}
BV & P & x_1 & x_2 & s_1 & RHS \\
\hline
x_1 & 0 & 1 & 0 & -6.667 & 0.667 \\
x_2 & 0 & 0 & 1 & 6.667 & 0.333 \\
\hline
P & 1 & 0 & 0 & 4 & -1.69
\end{array}
$$

This is the final tableau. The maximum $P = -1.69$, so $C = 1.69$ when $x_1 = 0.667$ and $x_2 = 0.333$.

c. The butcher should combine $\dfrac{2}{3}$ of a pound of ground beef with $\dfrac{1}{3}$ pound ground pork for a meatloaf mixture of minimum cost of $1.69 per pound.

Chapter 5 Project

1. The objective is to maximize the carbohydrates in the trail mix. The objective function is
$C = 31.4x_1 + 114.74x_2 + 148.12x_3 + 33.68x_4$.

3. This is a maximum problem with mixed constraints. Since the problem is so large we will use Excel to find the solution.

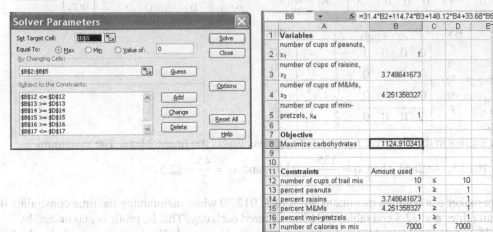

We find that the trail mix will maximize carbohydrates while meeting the constraints on the mix if 1 cup of peanuts, 3.7 cups of raisins, 4.3 cups of M&M's, and 1 cup of pretzels are used in the mix. Then the maximum carbohydrates will be 1124.9 grams.

5. To find the mix that will maximize the protein, we change the objective function. We maximize
$P = 34.57x_1 + 4.67x_2 + 9.01x_3 + 3.87x_4$.

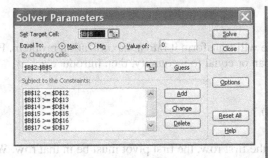

The trail mix which maximizes protein while meeting the constraints will contain about 6.4 cups of peanuts, 0.9 cups of raisins, 0.9 cups of M&M's, and 0.9 cups of pretzels. The maximum protein in the mix will be 268.6 grams. This trail mix contains 9.2 cups of mixture. So there are $\dfrac{239.1}{9.2} \approx 26.0$ grams of protein per cup.

There are $\dfrac{(854.10)(6.4) + (435)(0.9) + (1023.96)(0.9) + (162.02)(0.9)}{9.2} \approx 752.7$ calories per cup of mix.

7. The mix that minimizes the fat contains

$$\frac{(72.5)(1)+(0.67)(6.9)+(43.95)(1)+(1.49)(1)}{9.8} \approx 12.5 \text{ grams of fat per cup of mix,}$$

$$\frac{(34.57)(1)+(4.67)(6.9)+(9.01)(1)+(3.87)(1)}{9.8} \approx 8.1 \text{ grams of protein per cup of mix, and}$$

$$\frac{(31.40)(1)+(114.74)(6.9)+(148.12)(1)+(33.68)(1)}{9.8} \approx 102.5 \text{ grams of carbohydrates per cup of mix.}$$

Mathematical Questions from Professional Exams

1. C	**3.** C	**5.** B
7. C	**9.** D	**11.** D

Chapter 6 Finance

6.1 Interest

1. If a principal P is borrowed at a simple interest rate r (expressed as a decimal) for a period of t years, the interest charge I is <u>Prt</u>.

3. False. For a simple interest loan with interest rate r, the amount A due at the end of t years on a principal P borrowed is $A = P(1 + rt)$.

5. $0.60 = \dfrac{60}{100} = 60\%$

7. $1.1 = \dfrac{110}{100} = 110\%$

9. $0.06 = \dfrac{6}{100} = 6\%$

11. $0.0025 = \dfrac{25}{10000} = \dfrac{0.25}{100} = 0.25\%$

13. $25\% = \dfrac{25}{100} = 0.25$

15. $100\% = \dfrac{100}{100} = 1.00$

17. $6.5\% = \dfrac{6.5}{100} = 0.065$

19. $0.05\% = \dfrac{0.05}{100} = 0.0005$

21. 15% of $1000 = (0.15) \cdot (1000) = 150$

23. 18% of $100 = (0.18) \cdot (100) = 18$

25. 210% of $50 = (2.10) \cdot (50) = 105$

27. $x\%$ of $80 = 4$

$\dfrac{x}{100} \cdot 80 = 4 \Rightarrow 80x = 400 \Rightarrow x = 5$

4 is 5% of 80.

29. $x\%$ of $5 = 8$

$\dfrac{x}{100} \cdot 5 = 8 \Rightarrow 5x = 800 \Rightarrow x = 160$

8 is 160% of 5.

31. $20 = 8\%$ of x

$20 = 0.08x \Rightarrow x = \dfrac{20}{0.08} = 250$

20 is 8% of 250

33. $50 = 15\%$ of x

$50 = 0.15x \Rightarrow x = \dfrac{50}{0.15} \approx 333.333$

50 is 15% of 333.33.

35. $I = Prt = (\$1000)(0.04)\left(\dfrac{3}{12}\right) = \$10$

37. $I = Prt = (\$500)(0.12)\left(\dfrac{9}{12}\right) = \$45$

39. $I = Prt = (\$1000)(0.10)\left(\dfrac{18}{12}\right) = \$150$

41. $\quad A = P + Prt$

$1050 = 1000 + 1000r\left(\dfrac{6}{12}\right)$

$\quad 50 = 500r$

$\quad \dfrac{50}{500} = 0.1 = r$

The per annum rate of interest is 10%.

43. $\quad A = P + Prt$

$400 = 300 + 300r\left(\dfrac{12}{12}\right)$

$100 = 300r$

$\dfrac{100}{300} \approx 0.333 = r$

The per annum rate of interest is 33.3%.

45. $\quad A = P + Prt$

$1000 = 900 + 900r\left(\dfrac{10}{12}\right)$

$100 = 750r$

$\dfrac{100}{750} \approx 0.133 = r$

The per annum rate of interest is 13.3%.

47. $R = L - Lrt = 1200 - 1200(0.10)\left(\dfrac{6}{12}\right) = 1140$

The proceeds of the loan is \$1140.

49. $R = L - Lrt = 2000 - 2000(0.08)\left(\dfrac{24}{12}\right) = 1680$

The proceeds of the loan is \$1680.

51. $R = L(1 - rt)$

$$1200 = L\left[1 - (0.10)\left(\frac{1}{2}\right)\right] = L(0.95)$$

$$\frac{1200}{0.95} \approx 1263.16 = L$$

You must repay $1263.16 for the discounted loan.

The equivalent simple interest on the loan:

$$A = P + Prt$$

$$1263.16 = 1200 + 1200r\left(\frac{1}{2}\right)$$

$$63.16 = 600r$$

$$\frac{63.16}{600} \approx 0.1053 = r$$

The simple rate of interest for the loan is 10.53%.

53. $R = L(1 - rt)$

$$2000 = L\left[1 - (0.08)\left(\frac{24}{12}\right)\right] = L(0.84)$$

$$\frac{2000}{0.84} \approx 2380.95 = L$$

You must repay $2380.95 for the discounted loan.

The equivalent simple interest on the loan:

$$A = P + Prt$$

$$2380.95 = 2000 + 2000r\left(\frac{24}{12}\right)$$

$$380.95 = 4000r$$

$$\frac{380.95}{4000} \approx 0.0952 = r$$

The simple rate of interest for the loan is 9.52%.

55. $A = \$500$, $r = 0.03$, $t = \dfrac{9}{12}$

We want to know P, the principal.

$$A = P + Prt = P(1 + rt)$$

$$500 = P\left(1 + 0.03 \cdot \frac{9}{12}\right)$$

$$500 = 1.0225P$$

$$\frac{500}{1.0225} \approx 489$$

Madalyn should invest $489 if she wants to buy the stereo.

57.   The sheets that list for $130 are selling for $84.50. The difference between the list price and the sale price is the discount. Let x denote the discount written as a decimal.

$$130 - 130x = 84.50$$

$$130(1 - x) = 84.50$$

$$1 - x = \frac{84.50}{130} \Rightarrow x = 1 - \frac{84.50}{130} = \frac{7}{20} = 0.35$$

The list price of the sheets was reduced 35%.

59.   Let L denote the amount that Sarah Jane borrows. We know $R = \$5000$, $r = 0.0899$, and

$$t = \frac{18}{12} = 1.5.$$

$$R = L(1 - rt)$$

$$5000 = L[1 - 0.0899(1.5)] = L(0.86515)$$

$$\frac{5000}{0.86515} \approx 5779.3446 = L$$

Sarah Jane should get a loan of $5779.34.

61.   We are told that this is a discounted loan with

$$R = 250, L = 296.55, \text{ and } t = \frac{2}{52} = \frac{1}{26} \text{ of a year.}$$

We seek r.

$$R = L(1 - rt)$$

$$250 = 296.55\left[1 - r\left(\frac{1}{26}\right)\right]$$

$$\frac{250}{296.55} = 1 - r\left(\frac{1}{26}\right)$$

$$\frac{1}{26}r = 1 - \frac{250}{296.55}$$

$$r = 26\left(1 - \frac{250}{296.55}\right) \approx 4.0813$$

The per annum interest rate of this loan is 408.13%.

63.   We need to compare both loans and choose the one with the least interest due.

Discounted loan: $R = 1000$, $r = 0.09$, $t = \dfrac{1}{2}$.

$$R = L(1 - rt)$$

$$1000 = L\left[1 - 0.09\left(\frac{1}{2}\right)\right] = 0.955L$$

$$\frac{1000}{0.955} \approx 1047.12 = L$$

The interest on the discounted loan is $1047.12 - $1000 = $47.12.

Simple interest loan: $P = 1000$, $r = 0.1$, $t = \dfrac{1}{2}$.

$$I = Prt = (\$1000)(0.1)\left(\frac{1}{2}\right) = 50$$

The interest on the simple interest loan is $50. You should choose the discounted loan at 9% per annum. You will save $50 - $47.12 = $2.88.

65. a. $P = 8000$, $r = 0.0574$, $t = 3$
$I = Prt = (8000)(0.0574)(3) = 1377.60$
The interest is $1377.60.

b. The total amount of the loan is
$8000 + $1377.60 = $9377.60.

c. Assuming that the loan is repaid in 36 equal monthly payments, each payment will be
$\dfrac{\$9377.60}{36} = \$260.49$.

67. This is an example of a discounted loan. The bank's $4,000,000 is L; the amount the bank pays (its bid) $3,996,800 is R. In this problem $t = 3$ months $= \dfrac{3}{12}$ year. We are seeking r.

$$R = L(1 - rt)$$

$$3,996,800 = 4,000,000\left[1 - r\left(\frac{3}{12}\right)\right]$$

$$\frac{3,996,800}{4,000,000} = 1 - \frac{r}{4}$$

$$\frac{r}{4} = 1 - \frac{3,996,800}{4,000,000}$$

$$r = 4\left(1 - \frac{3,840,000}{4,000,000}\right) = 0.0032$$

The interest rate is 0.0032 or 0.32%.

69. Since the price of the T-bills were $993.25 per $1000, and the investor purchased $100,000 worth of bills, the investor paid ($993.25)(100) = $99,325 for the treasuries. The interest earned on the maturity date was $100,000 − $99,325 = $675.
The per annum simple interest rate on the investment is $4\left(\dfrac{675}{99,325}\right) \approx 0.027 = 2.7\%$.

71. No. Let x denote the original price of the item. Then $1.1x$ is the marked up price. When the store later discounts it by 10%, the resulting price is 90% of the marked up price. $0.90(1.1x) = 0.99x \neq x$

73. No. Let x denote the original selling price of the item. When the store discounts the price by 40%, the resulting price is $0.60x$. If the store then reduces the price another 10%, the final price is given by $0.90(0.60x) = 0.54x \neq 0.50x$.

6.2 Compound Interest

1. a. $(1 + 0.05)^8 \approx 1.477$

b. $\left(1 + \dfrac{0.08}{12}\right)^{-24} \approx 0.853$

3. $2 = 1.05^t \Rightarrow t = \log_{1.05} 2$

5. True

7. When interest at r% is compounded on a principal P so that the amount after one year is Pe^r, the interest is said to be <u>compounded continuously</u>.

9. $P = \$100$; $r = 0.04$; $n = 4$; $t = 2$
$$A = 100\left(1 + \frac{0.04}{4}\right)^{4 \cdot 2} \approx 108.2857$$
The amount accumulated in the investment is $A = \$108.29$.

11. $P = \$500$; $r = 0.08$; $n = 4$; $t = 2.5$
$$A = 500\left(1 + \frac{0.08}{4}\right)^{4 \cdot 2.5} \approx 609.4972$$
The amount accumulated in the investment is $A = \$609.50$.

13. $P = \$600$; $r = 0.05$; $n = 365$; $t = 3$
$$A = 600\left(1 + \frac{0.05}{365}\right)^{365 \cdot 3} \approx 697.0934$$
The amount accumulated in the investment is $A = \$697.09$.

15. $P = \$10$; $r = 0.02$; compounded continuously; $t = 2$
$$A = 10e^{0.02 \cdot 2} \approx 10.4081$$
The amount accumulated in the investment is $A = \$10.41$.

17. $P = \$100$; $r = 0.03$; compounded continuously; $t = 2.25$
$$A = 100e^{0.03 \cdot 2.25} \approx 106.9830$$
The amount accumulated in the investment is $A = \$106.98$.

In Problems 19–27, use the formula
$$P = A \cdot \left(1 + \frac{r}{n}\right)^{-nt}$$
if the interest is compounded n times per annum and the formula $P = Ae^{-rt}$ if the interest is compounded continuously.

19. $A = \$100$; $r = 0.02$; $n = 12$; $t = 2$
$$P = 100\left(1 + \frac{0.02}{12}\right)^{-12 \cdot 2} \approx 96.0821$$
The present value of the investment is $96.08.

21. $A = \$1000$; $r = 0.06$; $n = 365$; $t = 2.5$

$$P = 1000\left(1 + \frac{0.06}{365}\right)^{-365 \cdot 2.5} \approx 860.7186$$

The present value of the investment is \$860.72.

23. $A = \$600$; $r = 0.04$; $n = 4$; $t = 2$

$$P = 600\left(1 + \frac{0.04}{4}\right)^{-4 \cdot 2} \approx 554.0899$$

The present value of the investment is \$554.09.

25. $P = \$80$; $r = 0.02$; compounded continuously; $t = 3.25$

$$A = 80e^{-0.02 \cdot 3.25} \approx 74.9654$$

The present value of the investment is $A = \$74.97$.

27. $P = \$400$; $r = 0.04$; compounded continuously; $t = 1$

$$A = 400e^{-0.04 \cdot 1} \approx 384.3158$$

The present value of the investment is $A = \$384.32$.

For exercises 29–31, to decide which rate yields the larger amount in 1 year, we use $A = P\left(1 + \dfrac{r}{n}\right)^{nt}$ when $t = 1$ for each rate and compare the results. We use $P = \$10,000$ as suggested in the text.

29. $r = 6\%$; $n = 4$:

$$A = 10,000\left(1 + \frac{0.06}{4}\right)^{4 \cdot 1} \approx \$10,613.64$$

$r = 6\frac{1}{4}\%$, $n = 1$:

$$A = 10,000\left(1 + \frac{0.0625}{1}\right)^{1 \cdot 1} = \$10,625$$

$6\frac{1}{4}\%$ compounded annually has the greater yield.

31. $r = 9\%$; $n = 12$:

$$A = 10,000\left(1 + \frac{0.09}{12}\right)^{12 \cdot 1} \approx \$10,938.07$$

$r = 8.8\%$; $n = 365$:

$$A = 10,000\left(1 + \frac{0.088}{365}\right)^{365 \cdot 1} \approx \$10,919.77$$

9% compounded monthly has the greater yield.

For exercises 33–35, let $P = 1$ and $t = 1$. Then the interest earned is $\left(1 + \dfrac{r}{n}\right)^n - 1$ or $e^n - 1$ if the compounding is continuous. Then, using the simple interest formula, $I = PRt$, we have

$$\left(1 + \frac{r}{n}\right)^n - 1 = (1)R(1) = R \text{ or } e^n - 1 = (1)R(1) = R.$$

Thus, the effective rate of interest is $R = \left(1 + \dfrac{r}{n}\right)^n - 1$ or $e^n - 1 = R$ for continuous compounding.

33. $r = 0.05$, $n = 4$

$$\text{Effective rate of interest} = \left(1 + \frac{0.05}{4}\right)^4 - 1$$
$$\approx 0.05095 \text{ or } 5.095\%$$

35. $r = 0.05$, compounded continuously

$$\text{Effective rate of interest} = e^{0.05} - 1$$
$$\approx 0.05127 \text{ or } 5.127\%$$

37. To double an investment in 3 years, let $P = 1$, $A = 2P = 2$, $n = 1$, and $t = 3$. Find r.

$$A = P(1 + r)^t$$
$$2 = (1 + r)^3$$
$$1 + r = \sqrt[3]{2} \Rightarrow r = \sqrt[3]{2} - 1 \approx 0.25992$$

The annual rate of interest needed to double an investment in 3 years is 25.99%.

39. To triple an investment in 5 years, let $P = 1$, $A = 3P = 3$, $n = 1$, and $t = 5$. Find r.

$$A = P(1 + r)^t \Rightarrow 3 = (1 + r)^5 \Rightarrow$$
$$1 + r = \sqrt[5]{3} \Rightarrow r = \sqrt[5]{3} - 1 \approx 0.24573$$

The annual rate of interest needed to triple an investment in 5 years is 24.573%.

41. a. To double an investment when $r = 0.08$ and $n = 12$, let $P = 1$ and $A = 2P = 2$. Find t.

$$A = P\left(1 + \frac{r}{n}\right)^{nt}$$
$$2 = \left(1 + \frac{0.08}{12}\right)^{12t} = (1.00667)^{12t}$$
$$12t = \log_{1.00667} 2$$
$$t = \frac{\log 2}{12 \cdot \log 1.00667} \approx 8.688$$

It will take approximately 8.69 years to double money invested at 8% compounded monthly.

b. To double an investment when $r = 0.08$ compounded continuously, let $P = 1$ and $A = 2P = 2$. Find t.

$$A = Pe^{rt}$$
$$2 = e^{0.08t} \Rightarrow 0.08t = \ln 2 \Rightarrow t = \frac{\ln 2}{0.08} \approx 8.664$$

It will take approximately 8.66 years to double money invested at 8% compounded continuously.

For exercises 43–44, let $P = 1$ and $t = 1$. Then the interest earned is $\left(1 + \dfrac{r}{n}\right)^n - 1$ or $e^n - 1$ if the compounding is continuous. Then, using the simple interest formula, $I = PRt$, we have

$$\left(1 + \frac{r}{n}\right)^n - 1 = (1)R(1) = R \text{ or } e^n - 1 = (1)R(1) = R.$$

Thus, the effective rate of interest is $R = \left(1 + \dfrac{r}{n}\right)^n - 1$ or $e^n - 1 = R$ for continuous compounding.

43. $0.07 = \left(1 + \dfrac{r}{4}\right)^4 - 1$

$1.07 = \left(1 + \dfrac{r}{4}\right)^4$

$\sqrt[4]{1.07} = 1 + \dfrac{r}{4}$

$r = 4\left(\sqrt[4]{1.07} - 1\right) \approx 0.06823$

6.823% compounded quarterly has an effective rate of 7%.

45.   The principal $P = \$1000$, and the interest rate $r = 0.04$. The investment earns interest for $t = 3$ years.

a.   For annual compounding, $n = 1$.

$$A = P\left(1 + \frac{r}{n}\right)^{nt} = 1000(1.04)^3 \approx 1124.86$$

The interest earned is $\$1124.86 - \$1000 = \$124.86$.

b.   For monthly compounding, $n = 12$.

$$A = P\left(1 + \frac{r}{n}\right)^{nt} = 1000\left(1 + \frac{0.04}{12}\right)^{12 \cdot 3}$$
$$\approx 1127.27$$

The interest earned is $\$1127.27 - \$1000 = \$127.27$.

47.   The principal $P = \$1000$, and the interest rate $r = 0.02$ is compounded quarterly, so $n = 4$.

a.   The amount A after $t = 2$ years is

$$A = P\left(1 + \frac{r}{n}\right)^{nt} = 1000\left(1 + \frac{0.02}{4}\right)^{4 \cdot 2}$$
$$\approx \$1040.71$$

b.   The amount A after $t = 3$ years is

$$A = P\left(1 + \frac{r}{n}\right)^{nt} = 1000\left(1 + \frac{0.02}{4}\right)^{4 \cdot 3}$$
$$\approx \$1061.68$$

c.   The amount A after $t = 4$ years is

$$A = P\left(1 + \frac{r}{n}\right)^{nt} = 1000\left(1 + \frac{0.02}{4}\right)^{4 \cdot 4}$$
$$\approx \$1083.07$$

49.   When $r = 0.03$ and $n = 2$, the deposit necessary to have $\$5000$ (the present value)

a.   in 4 years is

$$P = A\left(1 + \frac{r}{n}\right)^{-nt} = 5000\left(1 + \frac{0.03}{2}\right)^{-2 \cdot 4}$$
$$\approx \$4438.56$$

b.   in 8 years is

$$P = A\left(1 + \frac{r}{n}\right)^{-nt} = 5000\left(1 + \frac{0.03}{2}\right)^{-2 \cdot 8}$$
$$\approx \$3940.16$$

51.   We find the interest due on each loan and compare the amounts. We have $P = \$1000$, $t = 2$ years.
With a simple interest rate $r = 12\% = 0.12$, the interest due after two years is
$I = Prt = (\$1000)(0.12)(2) = \$240$.
With an $r = 10\% = 0.10$ compounded monthly, the interest due after 2 years is

$$I = A - P = 1000\left(1 + \frac{0.10}{12}\right)^{24} - 1000$$
$$= \$220.39$$

Mr. Nielsen will pay less interest with a loan charging 10% interest compounded monthly.

53. $P = 100$, $r = 0.03$, $A = 150$
Find t when the interest is compounded monthly ($n = 12$).

$$A = P\left(1 + \frac{r}{n}\right)^{nt}$$
$$150 = 100\left(1 + \frac{0.03}{12}\right)^{12t}$$
$$1.5 = (1.0025)^{12t}$$
$$12t = \log_{1.0025} 1.5$$
$$t = \frac{\log 1.5}{12\log 1.0025} \approx 13.53 \text{ years}$$

Find t when the interest is compounded continuously.

$$A = Pe^{rt}$$
$$150 = 100e^{0.03t}$$
$$1.5 = e^{0.03t}$$
$$0.03t = \ln 1.5$$
$$t = \frac{\ln 1.5}{0.03} \approx 13.52 \text{ years}$$

55. $P = \$10,000$, $A = \$25,000$, $r = 0.06$ compounded continuously

$$A = Pe^{rt}$$
$$25,000 = 10,000e^{0.06t}$$
$$2.5 = e^{0.06t}$$
$$0.06t = \ln 2.5$$
$$t = \frac{\ln 2.5}{0.06} \approx 15.27 \text{ years}$$

It takes 15.27 years for $10,000 to grow to $25,000 at 6% compounded continuously.

57. In this problem we are looking for the amount of a $90,000 investment after 5 years if interest is 3% compounded annually. So, $P = 90,000$, $r = 0.03$, $n = 1$, $t = 5$.

$$A = P\left(1+\frac{r}{n}\right)^{nt} = 90,000(1.03)^5 \approx 104,334.67$$

The house will appreciate to about $104,335 after 5 years.

59. We are looking for the present value of $15,000, if $r = 0.05$ compounded continuously and $t = 3$.

$$P = Ae^{-rt} = 15,000e^{-0.05 \cdot 3} \approx 12,910.62$$

Jerome should ask his parents for $12,910.62.

61. Assuming that the stock continues to grow at 15%, we are looking for A when $P = 100 \cdot \$15 = \$1500$, $r = 0.15$, $n = 1$, and $t = 5$.

$$A = P\left(1+\frac{r}{n}\right)^{nt} = 1500(1.15)^5 \approx 3017.04$$

The 100 shares of the stock should be worth $3017.04 in 5 years.

63. Find A if $P = \$1000$, $r = 0.056$ compounded continuously, and $t = 1$ year.

$$A = Pe^{rt}$$
$$A = 1000e^{0.056} \approx 1057.5977$$

Jim's investment will accumulate to $1057.60. He will not have enough to purchase the computer; he will be short $2.40.
If $r = 0.059$ and $n = 12$, then A will become

$$A = P\left(1+\frac{r}{n}\right)^{nt}$$
$$A = 1000\left(1+\frac{0.059}{12}\right)^{12} = 1060.6219$$

The new interest rate is a better deal. Jim's investment will accumulate to $1060.62, and he will be able to purchase his computer.

65. Will: $P = \$2000$, $r = 0.05$, $n = 2$, and $t = 20$

$$A = P\left(1+\frac{r}{n}\right)^{nt}$$
$$A = 2000\left(1+\frac{0.05}{2}\right)^{2 \cdot 20} \approx 5370.1277$$

After 20 years, Will will have $5370.13 in his IRA.
Henry: $P = \$2000$, $r = 0.045$ compounded continuously, $t = 20$.

$$A = Pe^{rt}$$
$$A = 2000e^{0.045 \cdot 20} \approx 4919.2062$$

After 20 years, Henry will have $4919.21 in his IRA.
After 20 years, Will's IRA will be worth $450.92 more than Henry's IRA.

67. We need the present value of $40,000 after 4 years. We are told $A = \$40,000$, $r = 0.03$, $n = 4$, and $t = 4$.

$$P = A\left(1+\frac{r}{n}\right)^{-nt} = 40,000\left(1+\frac{0.03}{4}\right)^{-4 \cdot 4}$$
$$\approx \$35,492.7062$$

Tami and Todd should deposit $35,492.71 now if they want the $40,000 down payment for the house.

69. We want the accumulated value of the $6000 investment after 25 years. $P = \$6000$, $r = 0.03$, $n = 2$, and $t = 25$.

$$A = P\left(1+\frac{r}{n}\right)^{nt}$$
$$A = 6000\left(1+\frac{0.03}{2}\right)^{2 \cdot 25} = 12,631.4545$$

The account will be worth $12,631.45 when the child is 25 years old.

71. $P = \$10,000$, $A = \$25,000$, $r = 0.03$, $n = 365$. Find t.

$$A = P\left(1+\frac{r}{n}\right)^{nt}$$
$$25,000 = 10,000\left(1+\frac{0.03}{365}\right)^{365t}$$
$$2.5 = 1.000082192^{365t}$$
$$365t = \log_{1.000082192} 2.5$$
$$t = \frac{\log 2.5}{365\log 1.000082192} \approx 30.5442$$

It will take about 30.54 years for $10,000 to grow to $25,000 at 3% compounded daily.

73. $P = \$4000$, $r = 0.04$, $n = 1$, $t = 30$. Find A.

$$A = P\left(1 + \frac{r}{n}\right)^{nt}$$

$$A = 4000(1.04)^{30} \approx 12{,}973.5900$$

After 30 years, the investment will be worth $12,973.59.

75. We want the present value P of $A = \$43{,}423$ when $r = 0.08$ compounded continuously and $t = 18$ years.

$$P = Ae^{-rt}$$

$$P = 43{,}423\,e^{-0.08 \cdot 18} \approx 10{,}288.1141$$

Tom and Anita should invest $10,288.11 now to have the money needed in 18 years.

77. a. We are looking for r the annual growth rate of the debt, given $A = \$12.31$ trillion, $P = \$5.77$ trillion, $t = 10$.

$$A = P(1 + r)^t$$

$$12.31 = 5.77(1 + r)^{10}$$

$$\frac{12.31}{5.77} = (1 + r)^{10}$$

$$r = \sqrt[10]{\frac{12.31}{5.77}} - 1 \approx 0.078719$$

The national debt grew at an annualized rate of 7.87%.

b. Assuming the growth rate continues, then in 2020 ($t = 20$), the national debt will be

$$A = 5.77(1.078719)^{20} \approx 26.2629.$$

The national debt will be about $26.26 trillion on January 1, 2020.

79. We will find the accumulated amount to which $100,000 grows with all three options and choose the smallest.

a. 6% simple interest: $P = \$100{,}000$, $r = 0.06$, $t = 5$

$$A = P(1 + rt) = 100{,}000(1 + 0.06 \cdot 5)$$
$$= \$130{,}000$$

b. $P = \$100{,}000$, $r = 0.0575$, $n = 12$, $t = 5$

$$A = P\left(1 + \frac{r}{n}\right)^{nt} = 100{,}000\left(1 + \frac{0.0575}{12}\right)^{12 \cdot 5}$$
$$\approx 133{,}217.56$$

c. $P = \$100{,}000$, $r = 0.055$ compounded continuously, $t = 5$

$$A = Pe^{rt} = 100{,}000e^{0.055 \cdot 5}$$
$$\approx 131{,}653.07$$

Option (a), paying 6% simple interest, is best.

81. We are given $A = \$950$, $P = \$1000$, and $n = 2$. We solve for r.

$$A = P(1 - r)^n$$
$$950 = 1000(1 - r)^2$$
$$0.95 = (1 - r)^2$$
$$r = 1 - \sqrt{0.95} \approx 0.02532$$

The inflation rate is about 2.5%.

83. We are given $r = 0.02$; we seek n. Let $P = 2$ and $A = \frac{1}{2}P = 1$.

$$A = P(1 - r)^n$$
$$1 = 2(1 - 0.02)^n$$
$$0.5 = 0.98^n$$
$$n = \log_{0.98} 0.5 = \frac{\log 0.5}{\log 0.98} \approx 34.3096$$

At a 2% annual inflation rate, money's value is halved in 34.3 years.

85. $A = \$10{,}000$, $t = 20$, we need to find P, the present value of the bond.

a. When the return is 5% compounded monthly, $r = 0.05$ and $n = 12$.

$$P = A\left(1 + \frac{r}{n}\right)^{-nt} = 10{,}000\left(1 + \frac{0.05}{12}\right)^{-12 \cdot 20}$$
$$\approx 3686.45$$

You should pay $3686.45 for the bond if the interest is 5% compounded monthly.

b. When the return is 5% compounded continuously,

$$P = Ae^{-rt} = 10{,}000e^{-0.05 \cdot 20} \approx 3678.79$$

You should pay $3678.79 for the bond if the interest is 5% compounded continuously.

87. $A = \$10{,}000$, $t = 10$, $r = 0.04$, and $n = 1$, we need to find P, the present value of the bond.

$$P = A\left(1 + \frac{r}{n}\right)^{-nt} = 10{,}000(1.04)^{-10} \approx 6755.64$$

The $10,000 bond should be sold for $6755.64.

89. a. $m = 2$, $r = 0.04$, $n = 1$

$$t = \frac{\ln m}{n \ln\left(1 + \frac{r}{n}\right)} = \frac{\ln 2}{1 \cdot \ln\left(1 + \frac{0.04}{1}\right)} = \frac{\ln 2}{\ln 1.04}$$
$$\approx 17.6730$$

It takes about 17.7 years to double an investment that earns 4% per annum.

b. $m = 3$, $r = 0.03$, $n = 4$

$$t = \frac{\ln m}{n \ln\left(1 + \dfrac{r}{n}\right)} = \frac{\ln 3}{4 \cdot \ln\left(1 + \dfrac{0.03}{4}\right)}$$

$$= \frac{\ln 3}{4 \ln 1.0075} \approx 36.7576$$

It takes about 36.8 years to triple an investment that earns 3% per annum.

c. Given a principal P and interest rate r compounded n times per annum, the time t it takes the investment to multiply the investment m times is given by

$$A = mP = P\left(1 + \frac{r}{n}\right)^{nt}. \text{ Then we have}$$

$$m = \left(1 + \frac{r}{n}\right)^{nt} \Rightarrow \ln m = nt \ln\left(1 + \frac{r}{n}\right) \Rightarrow$$

$$t = \frac{\ln m}{n \ln\left(1 + \dfrac{r}{n}\right)}.$$

91. a. There are 10 years between 2000 and 2010. To find the inflation rate we solve for r.

$$CPI = CPI_0\left(1 + \frac{r}{100}\right)^n$$

$$217.6 = 169.3\left(1 + \frac{r}{100}\right)^{10}$$

$$1 + \frac{r}{100} = \sqrt[10]{\frac{217.6}{169.3}}$$

$$r = 100\left(\sqrt[10]{\frac{217.6}{169.3}} - 1\right) \approx 2.5416$$

The average rate of inflation from 2000 to 2010 was 2.54%.

b. Using $r = 2.54\%$ and CPI = 300, we solve for n.

$$CPI = CPI_0\left(1 + \frac{r}{100}\right)^n$$

$$300 = 169.3\left(1 + \frac{2.54}{100}\right)^n$$

$$\frac{300}{169.3} = (1.0254)^n$$

$$n = \log_{1.0254}\left(\frac{300}{169.3}\right) = \frac{\log\dfrac{300}{169.3}}{\log 1.0254}$$

$$\approx 22.8089$$

The CPI will reach 300 22 years after 2000, which is during the year 2022.

93. With $r = 2.3\%$, we solve for n.

$$CPI = CPI_0\left(1 + \frac{r}{100}\right)^n$$

$$2 = 1\left(1 + \frac{2.3}{100}\right)^n$$

$$2 = (1.023)^n$$

$$n = \log_{1.023} 2 = \frac{\log 2}{\log 1.023} \approx 30.4821$$

It will take about 30.5 years for the CPI to double if the inflation rate is 2.3%

95. Answers will vary. Sample answer:
Compound interest is interest paid on the initial principal and previously earned interest. Continuous compounding is the process of calculating interest and adding it to existing principal and interest at infinitely short time intervals.

97. Answers will vary. Sample answer:
You may want to consider the total amount to be repaid.
Bank 1: $I = 541.60(12)(30) = 194,976$
Total to be repaid = \$194,976 + \$1750
$\qquad\qquad\qquad = \$196,726.$
Bank 2:
$I = 755.32(12)(15) = 135,957.60$
Total to be repaid = \$135,957.60 + \$1500
$\qquad\qquad\qquad = \$137,457.60.$
Bank 3: $I = 549.05(12)(30) = 197,685$
Total to be repaid = \$197,685.
Bank 4: $I = 782.48(12)(15) = 140,846.40$
Total to be repaid = \$140,846.40.
The loan from Bank 2 will cost the least overall.

6.3 Annuities; Sinking Funds

1. $(1 + 0.04)^4 \approx 1.16986$

3. An <u>annuity</u> is a sequence of equal periodic deposits.

5. The deposit is $P = \$100$. The number of deposits is $n = 10$ and the interest per payment period is $i = 0.10$.

$$A = P\frac{(1 + i)^n - 1}{i} = 100\left[\frac{(1 + 0.10)^{10} - 1}{0.10}\right]$$

$$\approx 1593.74$$

There is \$1593.74 in the account after 10 years.

7. The deposit is $P = \$400$. The number of deposits is $n = 12$ and the interest per payment period is $i = \frac{0.02}{12}$.

$$A = P\frac{(1+i)^n - 1}{i} = 400\left[\frac{\left(1+\frac{0.02}{12}\right)^{12} - 1}{\frac{0.02}{12}}\right]$$

$$\approx 4844.25$$

There is \$4844.25 in the account after 12 months (1 year).

9. The deposit is $P = \$200$. The number of deposits is $n = 36$ and the interest per payment period is $i = \frac{0.03}{12}$.

$$A = P\frac{(1+i)^n - 1}{i} = 200\left[\frac{\left(1+\frac{0.03}{12}\right)^{36} - 1}{\frac{0.03}{12}}\right]$$

$$\approx 7524.11$$

There is \$7524.11 in the account after 36 months (3 years).

11. The deposit is $P = \$100$. The number of deposits is $n = 60$ and the interest per payment period is $i = \frac{0.04}{12}$.

$$A = P\frac{(1+i)^n - 1}{i} = 100\left[\frac{\left(1+\frac{0.04}{12}\right)^{60} - 1}{\frac{0.04}{12}}\right]$$

$$\approx 6629.90$$

There is \$6629.90 in the account after 60 deposits (5 years).

13. The deposit is $P = \$9000$. The number of deposits is $n = 10$ and the interest per payment period is $i = 0.05$.

$$A = P\frac{(1+i)^n - 1}{i} = 9000\left[\frac{(1.05)^{10} - 1}{0.05}\right]$$

$$\approx 113,201.03$$

There is \$113,201.03 in the account after 10 deposits (10 years).

15. The amount required is $A = \$10,000$ after $n = 12 \cdot 5 = 60$ monthly payments. The interest rate per payment period is $i = \frac{0.03}{12} = 0.0025$. To find the monthly payment use $A = P\frac{(1+i)^n - 1}{i}$ and solve for P.

$$10,000 = P\left[\frac{\left(1+\frac{0.03}{12}\right)^{60} - 1}{\frac{0.03}{12}}\right]$$

$$P = 10,000\left[\frac{0.0025}{(1.0025)^{60} - 1}\right] \approx 154.69$$

Sixty monthly payments of \$154.69 are needed to accumulate \$10,000.

17. The amount required is $A = \$20,000$ after $n = 4 \cdot 2.5 = 10$ quarterly payments. The interest rate per payment period is $i = \frac{0.02}{4} = 0.005$. To find the monthly payment use $A = P\frac{(1+i)^n - 1}{i}$ and solve for P.

$$20,000 = P\left[\frac{(1.005)^{10} - 1}{0.005}\right]$$

$$P = 20,000\left[\frac{0.005}{(1.005)^{10} - 1}\right] \approx 1955.41$$

Ten quarterly payments of \$1955.41 are needed to accumulate \$20,000.

19. The amount required is $A = \$25,000$ after $n = 6$ monthly payments. The interest rate per payment period is $i = \frac{0.055}{12}$. To find the monthly payment we use $A = P\frac{(1+i)^n - 1}{i}$ and solve for P.

$$25,000 = P\left[\frac{\left(1+\frac{0.055}{12}\right)^{6} - 1}{\frac{0.055}{12}}\right]$$

$$P = 25,000\left[\frac{\frac{0.055}{12}}{\left(1+\frac{0.055}{12}\right)^{6} - 1}\right] \approx 4119.18$$

Six monthly payments of \$4119.18 are needed to accumulate \$25,000.

21. The amount required is $A = \$5000$ after $n = 12 \cdot 2 = 24$ monthly payments. The interest rate per payment period is $i = \frac{0.04}{12}$. To find the monthly payment we use $A = P\frac{(1+i)^n - 1}{i}$ and solve for P.

$$5000 = P\left[\frac{\left(1+\frac{0.04}{12}\right)^{24} - 1}{\frac{0.04}{12}}\right]$$

$$P = 5000\left[\frac{\frac{0.04}{12}}{\left(1+\frac{0.04}{12}\right)^{24} - 1}\right] \approx 200.46$$

Twenty-four monthly payments of \$200.46 are needed to accumulate \$5000.

23. The amount required is $A = \$9000$ after $n = 4$ annual payments. The interest rate per payment period is $i = 0.05$ To find the monthly payment we use $A = P\dfrac{(1+i)^n - 1}{i}$ and solve for P.

$$9000 = P\left[\frac{(1.05)^4 - 1}{0.05}\right]$$

$$P = 9000\left[\frac{0.05}{(1.05)^4 - 1}\right] \approx 2088.11$$

Four annual payments of $2088.11 are needed to accumulate $9000.

25. Al's investment is an example of an annuity. The deposit is $P = \$2500$, the number of deposits is $n = 15$ and the interest per payment period is $i = 0.05$.

$$A = P\frac{(1+i)^n - 1}{i} = 2500\left[\frac{(1+0.05)^{15} - 1}{0.05}\right]$$
$$\approx 53,946.41$$

The mutual fund will be worth $53,946.41.

27. Sheila is setting up a sinking fund. The amount required is $A = \$12,000$ after $n = 4 \cdot 3 = 12$ quarterly payments. The interest rate per payment period is $i = \frac{0.02}{4} = 0.005$. To find the quarterly payment we use $A = P\dfrac{(1+i)^n - 1}{i}$ and solve for P.

$$12,000 = P\left[\frac{(1.005)^{12} - 1}{0.005}\right]$$

$$P = 12,000\left[\frac{0.005}{(1.005)^{12} - 1}\right] \approx 972.80$$

Sheila needs to save $972.80 per quarter to accumulate $12,000.

29. Dan's pension fund can be thought of as a sinking fund. The amount required is $A = \$350,000$ after $n = 12 \cdot 20 = 240$ monthly payments. The interest rate per payment period is $i = \frac{0.05}{12}$. To find the monthly payment we use $A = P\dfrac{(1+i)^n - 1}{i}$ and solve for P.

$$350,000 = P\left[\frac{\left(1 + \frac{0.05}{12}\right)^{240} - 1}{\frac{0.05}{12}}\right]$$

$$P = 350,000\left[\frac{\frac{0.05}{12}}{\left(1 + \frac{0.05}{12}\right)^{240} - 1}\right] \approx 851.51$$

Dan needs to save $851.51 per month to have $350,000 in 20 years.

31. In this sinking fund the amount required is $A = \$100,000$ after $n = 4$ annual payments. The interest rate per payment period is $i = 0.03$. To find the annual payment we use

$$A = P\frac{(1+i)^n - 1}{i}$$ and solve for P.

$$100,000 = P\left[\frac{(1.03)^4 - 1}{0.03}\right]$$

$$P = 100,000\left[\frac{0.03}{(1.03)^4 - 1}\right] \approx 23,902.70$$

An annual payment of $23,902.70 is necessary to accumulate the $100,000 needed.
The table shows the growth of the sinking fund over time.

Payment Number	Deposit $	Cumulative Deposits	Accumulated Interest	Total $
1	23,902.70	23,902.70	0	23,902.70
2	23,902.70	47,805.40	717.081	48,522.48
3	23,902.70	71,708.10	2172.76	73,880.86
4	23,902.70	95,610.80	4389.18	99,999.98

33. Let x denote the price to be paid for the oil well. Then $0.14x$ represents a 14% annual Return on Investment (ROI). The annual Sinking Fund Contribution (SFC) needed to recover the purchase price x in 30 years can be calculated letting $A = x$, $n = 30$, $i = 0.15$, and $P = $ SFC.

$$A = P\frac{(1+i)^n - 1}{i}$$

$$x = \text{SFC}\frac{1.05^{30} - 1}{0.05}$$

$$\text{SFC} = \frac{0.05x}{1.05^{30} - 1} \approx 0.0150514351x$$

The required annual ROI plus the annual SFC equals the annual net income of $30,000.
$$0.14x + 0.0150514351x = 30,000$$
$$0.1550514351x = 30,000$$
$$x \approx 193,484.18$$

The investor should pay $193,484 for the oil well.

35. Since the fund will be started after 7 years and the bonds are payable in 10 years, $n = 3 \cdot 4 = 12$ quarterly payments. The amount required is $A = 1,000,000$, and the interest rate per payment period is $i = \dfrac{0.04}{4} = 0.01$.

$$A = P\frac{(1+i)^n - 1}{i}$$

$$1,000,000 = P\left[\frac{(1.01)^{12} - 1}{0.01}\right]$$

$$P = 1,000,000\left(\frac{0.01}{1.01^{12} - 1}\right)$$

$$\approx 78,848.79$$

Twelve quarterly payments of $78,848.79 are necessary to accumulate $1,000,000.

37. a. The projected cost of the roof in 20 years is given by $A = P(1 + i)^n$. We have $P = 100,000$, $i = 0.03$, and $n = 20$. The roof will cost $A = \$100,000(1.03)^{20} = \$180,611.12$.

b. The Condo Association makes $n = 2 \cdot 20 = 40$ payments earning interest at $i = \dfrac{0.04}{2} = 0.02$ per payment period.

$$A = P\frac{(1+i)^n - 1}{i}$$

$$180,611.12 = P\left[\frac{(1.02)^{40} - 1}{0.02}\right]$$

$$P = 180,611.12\left[\frac{0.02}{(1.02)^{40} - 1}\right]$$

$$\approx 2990.15$$

Each semiannual payment should be $2990.15.

39. This is an annuity with monthly payments.

$A = \$1,000,000$, $P = \$600$, $i = \dfrac{0.07}{12}$.

$$A = P\frac{(1+i)^n - 1}{i}$$

$$1,000,000 = 600\left[\frac{\left(1 + \frac{0.07}{12}\right)^n - 1}{\frac{0.07}{12}}\right]$$

$$\frac{1,000,000}{600}\left(\frac{0.07}{12}\right) + 1 = \left(1 + \frac{0.07}{12}\right)^n$$

$$10.722222 = 1.0058333^n$$

$$n = \log_{1.0058333} 10.722222$$

$$= \frac{\log 10.722222}{\log 1.058333}$$

$$\approx 407.868$$

It will take approximately 407.868 months or almost 34 years to accumulate $1,000,000.

41. Angie is setting up a sinking fund where $A = \$9500$, $n = 2 \cdot 12 = 24$ monthly payments, and $i = \dfrac{0.0275}{12}$. We need to find P.

$$A = P\frac{(1+i)^n - 1}{i}$$

$$9500 = P\left[\frac{\left(1 + \frac{0.0275}{12}\right)^{24} - 1}{\frac{0.0275}{12}}\right]$$

$$P = 9500\left[\frac{\frac{0.0275}{12}}{\left(1 + \frac{0.0275}{12}\right)^{24} - 1}\right] \approx 385.50$$

Angie should save $385.50 each month to pay for the trip.

43. In this sinking fund $A = \$1,000,000$, $n = 4 \cdot 5 = 20$ payments, and $i = \dfrac{0.05}{4} = 0.0125$. We need P.

$$A = P\frac{(1+i)^n - 1}{i}$$

$$1,000,000 = P\left[\frac{(1.0125)^{20} - 1}{0.0125}\right]$$

$$P = 1,000,000\left[\frac{0.0125}{(1.0125)^{20} - 1}\right]$$

$$\approx 44,320.39$$

The school district should make quarterly deposits of $44,320.39 into the fund.

45. The value of the plan after 180 deposits is the amount accumulated by the initial investment of $400 invested for 181 interest periods plus the accumulated amount of an annuity with 180 monthly deposits of $100.

$$A = 400\left(1 + \frac{0.06}{12}\right)^{181} + 100\frac{\left(1 + \frac{0.06}{12}\right)^{180} - 1}{\frac{0.06}{12}}$$

$$\approx 986.55 + 29,081.87 = 30,068.42$$

The account will be worth $30,068.32.

47. Option 1: $P = \$1,326,000$ invested for $t = 25$ years at $r = 0.085$ compounded annually.

$$A = \$1,326,000(1.085)^{25} = \$10,192,646.88$$

Option 2: An annuity with $P = \$100,000$, $i = 0.085$, $n = 25$

$$A = \$100,000\frac{(1.085)^{26} - 1}{0.085} = \$8,635,455.48$$

Dan should select the lump sum payment.

49. a. $P = \$23,830$, $r = 0.0323$, $n = 2014 - 2010 = 4$. We want A.

$$A = \$23,830(1.0323)^4 = \$27,061.24$$

The projected price of a Honda Accord in 2014 is $27,061.24.

b. The final projected price, including sales tax is $27,061.24\ (1.0925) = \$29,564.40$.

c. The monthly payment into a sinking fund needed to have the projected purchase price when $i = \dfrac{0.0275}{12}$, $n = 12 \cdot 4 = 48$, is given by

$$29,564.40 = P\frac{\left(1 + \frac{0.0275}{12}\right)^{48} - 1}{\frac{0.0275}{12}}$$

$$P = 29,564.40\left[\frac{\frac{0.0275}{12}}{\left(1 + \frac{0.0275}{12}\right)^{48} - 1}\right]$$

$$\approx \$583.37$$

The monthly payment should be $583.37.

51. a. If $P = \$7020$ and $r = 0.065$, then the projected annual undergraduate college tuition and fees for public 4-year institutions are

Year	n	$A = P(1 + r)^n = 7020(1.065)^n$
2023–2024	14	16,952.42
2024–2025	15	18,054.32
2025–2026	16	19,227.85
2026–2027	17	20,477.67

b. The total projected cost of the four years tuition is $74,712.26. $i = \dfrac{0.042}{4} = 0.0105$.

$$A = P\frac{(1 + i)^n - 1}{i}$$

$$74,712.26 = P\left(\frac{1.0105^{45} - 1}{0.0105}\right)$$

$$P = 74,712.26\left(\frac{0.0105}{1.0105^{45} - 1}\right)$$

$$\approx 1307.35$$

The quarterly payments should be $1307.35.

6.4 Present Value of an Annuity; Amortization

1. $P = \$500$, $n = 36$, $i = \dfrac{0.04}{12}$

$$V = P \cdot \frac{1 - (1 + i)^{-n}}{i}$$

$$V = \$500\left[\frac{1 - \left(1 + \frac{0.04}{12}\right)^{-36}}{\frac{0.04}{12}}\right] \approx \$16,935.38$$

3. $P = \$100$, $n = 9$, $i = \dfrac{0.02}{12}$

$$V = P \cdot \frac{1 - (1 + i)^{-n}}{i}$$

$$V = \$100\left[\frac{1 - \left(1 + \frac{0.02}{12}\right)^{-9}}{\frac{0.02}{12}}\right] \approx \$892.55$$

5. $P = \$10,000$, $n = 20$, $i = 0.05$

$$V = P \cdot \frac{1 - (1 + i)^{-n}}{i}$$

$$V = \$10,000\left[\frac{1 - (1.05)^{-20}}{0.05}\right] \approx \$124,622.10$$

7. $V = \$10,000$, $n = 48$, $i = \dfrac{0.08}{12}$

$$P = V\left[\frac{i}{1 - (1 + i)^{-n}}\right]$$

$$P = \$10,000\left[\frac{\frac{0.08}{12}}{1 - \left(1 + \frac{0.08}{12}\right)^{-48}}\right] \approx \$244.13$$

9. $V = \$500,000$, $n = 360$, $i = \dfrac{0.10}{12}$

$$P = V\left[\frac{i}{1 - (1 + i)^{-n}}\right]$$

$$P = \$500,000\left[\frac{\frac{0.10}{12}}{1 - \left(1 + \frac{0.10}{12}\right)^{-360}}\right] \approx \$4387.86$$

11. $V = \$1,000,000$, $n = 360$, $i = \dfrac{0.08}{12}$

$$P = V\left[\frac{i}{1 - (1 + i)^{-n}}\right]$$

$$P = \$1,000,000\left[\frac{\frac{0.08}{12}}{1 - \left(1 + \frac{0.08}{12}\right)^{-360}}\right] \approx \$7337.65$$

13. For the two-year loan, $V = \$10,000$,

$$n = 2 \cdot 12 = 24, \quad i = \frac{0.12}{12} = 0.01$$

$$P = V\left[\frac{i}{1-(1+i)^{-n}}\right]$$

$$P = \$10000\left[\frac{0.01}{1-(1.01)^{-24}}\right] \approx \$470.73$$

The monthly payment is $470.73.

15. Mr. Doody is interested in finding the present value of an annuity that has an interest rate of $i = \frac{0.05}{12}$ and from which he withdraws $P = \$250$ each month for $n = 20 \cdot 12 = 240$ months.

$$V = P \cdot \frac{1-(1+i)^{-n}}{i}$$

$$V = \$250\left[\frac{1-\left(1+\frac{0.05}{12}\right)^{-240}}{\frac{0.05}{12}}\right] \approx \$37,881.33$$

Mr. Doody needs $37,881.33 now to guarantee his future income.

17. $V = \$8000$, $i = \frac{0.074}{12}$, $n = 36$ months

a. $P = V\left[\dfrac{i}{1-(1+i)^{-n}}\right]$

$$P = \$8000\left[\frac{\frac{0.074}{12}}{1-\left(1+\frac{0.074}{12}\right)^{-36}}\right] \approx \$248.48$$

The monthly payment is $248.48.

b. The total amount you pay is
$36 \cdot 248.48 = \$8945.28$.

c. The interest, I, you pay is the difference between the total amount paid and the amount of the loan V.
$I = \$8945.28 - \$8000 = \$945.28$
You pay $945.28 in interest on the loan.

19.a. **Option 1:** 20 year mortgage at 8% interest, $i = \frac{0.08}{12}$, and $n = 12 \cdot 20 = 240$ monthly payments, Their monthly payments will be

$$P = V\left[\frac{i}{1-(1+i)^{-n}}\right]$$

$$P = \$160,000\left[\frac{\frac{0.08}{12}}{1-\left(1+\frac{0.08}{12}\right)^{-240}}\right] \approx \$1338.30$$

Option 2: 25 year mortgage at 9% interest, $i = \frac{0.09}{12} = 0.0075$, and $n = 12 \cdot 25 = 300$ monthly payments. Their monthly payments will be

$$P = V\left[\frac{i}{1-(1+i)^{-n}}\right]$$

$$P = \$160,000\left[\frac{0.0075}{1-(1.0075)^{-300}}\right] \approx \$1342.71$$

The 25 year loan at 9% interest has a larger monthly payment.

b. **Option 1:** The total interest paid on this loan is ($1338.30)(240) − $160,000 = $161,192.
Option 2: The total interest paid on this loan is ($1342.71)(300) − $160,000 = $242,813.
The total interest paid on the 25 year loan is $81,621 more than that on the 20 year loan.

c. The equity in the house is the sum of the down payment and the amount paid on the loan.
Option 1: After 10 years the couple still owes 120 payments. The amount still owed on the loan is the present value of these payments.

$$V = P \cdot \frac{1-(1+i)^{-n}}{i}$$

$$V = \$1338.30\left[\frac{1-\left(1+\frac{0.08}{12}\right)^{-120}}{\frac{0.08}{12}}\right]$$

$$\approx \$110,304.67$$

This means that the amount paid on the loan $160,000 − $110,304.67 = $49,695.33.
The couple's equity after 10 years is $40,000 + $49,695.33 = $89,695.33.
Option 2: After 10 years the couple still owes 180 payments. The amount still owed on the loan is the present value of these payments.

$$V = P \cdot \frac{1-(1+i)^{-n}}{i}$$

$$V = \$1342.71\left[\frac{1-(1.0075)^{-180}}{0.0075}\right]$$

$$\approx \$132,382.36$$

This means that the amount paid on the loan $160,000 − $132,382.36 = $27,617.64.
The couple's equity after 10 years is $40,000 + $27,617.64 = $67,617.64.
After 10 years the couple would have more equity in their purchase if they had the 8% loan.

21. We first find how much John needs to accumulate prior to retirement. This is the present value V of $n = 12 \cdot 30 = 360$ payments of $300 at $i = \dfrac{0.04}{12}$ per month.

$$V = P \cdot \frac{1-(1+i)^{-n}}{i}$$

$$V = \$300 \left[\frac{1-\left(1+\frac{0.04}{12}\right)^{-360}}{\frac{0.04}{12}} \right] \approx \$62,838.37$$

We can now find the $n = 12 \cdot 20 = 240$ monthly payments necessary to meet this goal.

$$A = P \cdot \frac{(1+i)^n - 1}{i}$$

$$62,838.37 = P \left[\frac{\left(1+\frac{0.04}{12}\right)^{240} - 1}{\frac{0.04}{12}} \right]$$

$$P = 62,838.37 \left[\frac{\frac{0.04}{12}}{\left(1+\frac{0.04}{12}\right)^{240} - 1} \right]$$

$$\approx 171.33$$

John should save $171.33 each month to achieve his retirement goal.

23. a. Dan's accumulated savings is actually an annuity with $P = \$100$, $n = 12$, and $i = \dfrac{0.01}{52}$ per week.

$$A = P\frac{(1+i)^n - 1}{i} = 100 \left[\frac{\left(1+\frac{0.01}{52}\right)^{12} - 1}{\frac{0.01}{52}} \right]$$

$$\approx 1201.27$$

After 12 weeks Dan has $1201.27.

b. Dan now amortizes his savings, he needs to find P if $V = \$1201.27$, $n = 34$, and $i = \dfrac{0.01}{52}$.

$$P = V \left[\frac{i}{1-(1+i)^{-n}} \right]$$

$$P = \$1201.27 \left[\frac{\frac{0.01}{52}}{1-\left(1+\frac{0.01}{52}\right)^{-34}} \right] \approx \$35.45$$

Dan can withdraw $35.45 each week for 34 weeks.

25. a. Mike and Yola's down payment is 20% of $200,000.
(0.20)($200,000) = $40,000

b. Their loan amount is the purchase price less the down payment.
$200,000 − $40,000 = $160,000

c. The 30 year loan will have $n = 12 \cdot 30 = 360$ monthly payments with $i = \dfrac{0.09}{12} = 0.0075$ per month.

$$P = V \left[\frac{i}{1-(1+i)^{-n}} \right]$$

$$P = \$160,000 \left[\frac{0.0075}{1-(1.0075)^{-360}} \right] \approx \$1287.40$$

d. The total interest paid is the difference between the total payments and the amount borrowed. Mike and Yola will pay ($1287.40)(360) − $160,000 = $303,464 in interest.

e. If they pay an additional $100 each month, the payment $P = \$1387.40$. The present value V of the loan remains $160,000 and the interest rate $i = 0.0075$ per month. We solve for n.

$$V = P \cdot \frac{1-(1+i)^{-n}}{i}$$

$$160,000 = 1387.40 \left[\frac{1-(1.0075)^{-n}}{0.0075} \right]$$

$$1.0075^{-n} = 1 - \frac{160,000}{1387.40} \cdot (0.0075)$$

$$1.0075^{-n} = 0.1350728$$

$$-n = \log_{1.0075} 0.1350728$$

$$n = -\frac{\log 0.1350728}{\log 1.0075} \approx 267.93$$

If Mike and Yola increase their monthly payments by $100, they will pay off the loan in 268 months or 22 years and 4 months.

f. The total interest paid on the loan with the revised monthly payment of $1387.40 is ($1387.40)(268) − $160,000 = $211,823.20.

27. The down payment on the car is 20% of $12,000 or (0.20)($12,000) = $2400.
The amount of money borrowed is $12,000 − $2400 = $9600 = V$.

$$n = 12 \cdot 3 = 36 \text{ and } i = \frac{0.15}{12} = 0.0125.$$

$$P = V \left[\frac{i}{1-(1+i)^{-n}} \right]$$

$$P = \$9600 \left[\frac{0.0125}{1-(1.0125)^{-36}} \right] \approx \$332.79$$

The monthly payment is $332.79.

29. a. The down payment on the equipment is 10% of \$20,000 or $(0.10)(\$20,000) = \$2000$. The amount of money the restaurant owner borrowed is $\$20,000 - \$2000 = \$18,000 = V$.

$n = 12 \cdot 4 = 48$ and $i = \dfrac{0.12}{12} = 0.01$.

$$P = V\left[\frac{i}{1-(1+i)^{-n}}\right]$$

$$P = \$18,000\left[\frac{0.01}{1-(1.01)^{-48}}\right] \approx \$474.01$$

The monthly payment will be \$474.01.

b. The restaurant owner will pay a total of $(\$474.01)(48) - \$18,000 = \$4752.48$ in interest on the loan.

31. The home buyer made a down payment of 20% of \$140,000 or $(0.20)(\$140,000) = \$28,000$ on the house, and financed the remainder $(\$140,000 - \$28,000 = \$112,000)$. The terms of the mortgage were $n = 12 \cdot 30 = 360$ payments and $i = \dfrac{0.06}{12} = 0.005$ per month.

$$P = V\left[\frac{i}{1-(1+i)^{-n}}\right]$$

$$P = \$112,000\left[\frac{0.005}{1-(1.005)^{-360}}\right] \approx \$671.50$$

The total interest paid is
$I = 360(\$671.50) - \$112,000 = \$129,740.$

If the term of the loan is reduced to 15 years, then $n = 12 \cdot 15 = 180$ and the monthly payment is

$$P = V\left[\frac{i}{1-(1+i)^{-n}}\right]$$

$$P = \$112,000\left[\frac{0.005}{1-(1.005)^{-180}}\right] \approx \$945.12$$

The total interest paid will be
$I = 180(\$945.12) - \$112,000 = \$58,121.60.$

Decreasing the term of the loan from 30 years to 15 years increases the monthly payment by \$273.62, but saves \$71,618.40 in interest.

33. We have $V = \$100,000$, $P = \$2000$ per month and $i = \dfrac{0.05}{12}$ per month. To determine how long the IRA payments will last need to find n.

$$V = P \cdot \frac{1-(1+i)^{-n}}{i}$$

$$100,000 = 2000\left[\frac{1-\left(1+\frac{0.05}{12}\right)^{-n}}{\frac{0.05}{12}}\right]$$

$$1 - \frac{0.05}{12}(50) = \left(1+\frac{0.05}{12}\right)^{-n}$$

$$0.7916667 = 1.00416667^{-n}$$

$$-n = \log_{1.00416667} 0.7916667$$

$$n = -\frac{\log 0.7916667}{\log 1.00416667} \approx 56.18$$

The payments will last 56.18 months, or about 4 years and 8 months.

35. We need the present value V of an annuity, if $P = \$1690$, $n = 9$ months, and

$i = \dfrac{0.048}{12} = 0.004$.

$$V = P\left[\frac{1-(1+i)^{-n}}{i}\right] = \$1690\left[\frac{1-1.004^{-9}}{0.004}\right]$$

$$\approx \$14,910.21$$

There must be \$14,910.21 in the account on August 1.

37. Each withdrawal is $\dfrac{\$1,000,000}{20} = \$50,000$.

We need the present value of the annuity when

$n = 5 \cdot 4 = 20$ and $i = \dfrac{0.078}{4} = 0.0195$.

$$V = P\left[\frac{1-(1+i)^{-n}}{i}\right] = \$50,000\left[\frac{1-1.0195^{-20}}{0.0195}\right]$$

$$= \$821,530.33$$

The foundation must invest \$821,530.33 to meet the project's goals.

39. The original loan had 360 payments at an interest rate $i = \dfrac{0.06825}{12} = 0.0056875$ per month. The loan payments were

$$P = V\left[\frac{i}{1-(1+i)^{-n}}\right]$$

$$P = 312,000\left[\frac{0.0056875}{1-(1.0056875)^{-360}}\right] \approx \$2039.20$$

After 60 months there are 300 payments still to be made, and the present value of the loan is

$$V = P\left[\frac{1-(1+i)^{-n}}{i}\right]$$

$$= \$2039.20\left[\frac{1-1.0056875^{-300}}{0.0056875}\right]$$

$$= \$293,133.88$$

(continued on next page)

(*continued*)

Amount of the debt paid:

$312,000 - $293,133.88 = $18,866.12

Interest paid:

($2039.20)(60) - $18,866.12 = $103,485.88

The new loan has a present value

$V = $293,133.88, $n = 300$ monthly payments at

an interest rate $i = \dfrac{0.06125}{12}$ per month. The new

monthly payment is

$$P = V\left[\dfrac{i}{1-(1+i)^{-n}}\right]$$

$$P = 293,133.88\left[\dfrac{\frac{0.06125}{12}}{1-\left(1+\frac{0.06125}{12}\right)^{-300}}\right]$$

$\approx \$1911.13$

With the refinancing the monthly payments are reduced by $2039.20 - $1911.13 = $128.07 per month.

The interest that will be saved by refinancing at a lower interest rate is the difference between the interest that would have been paid without refinancing and the interest that will be paid with refinancing.

Interest without refinancing:

($2039.20)(360) - $312,000 = $422,112

Interest with refinancing: interest paid on the old loan + interest on new loan

$$I = 103,485.88 + \left[(1911.33)(300) - 293,133.88\right]$$

$\quad = \$383,751$

The interest saved by refinancing:

$422,112 - $383,751 = $38,361

41. The $V = $235,000 loan had $n = 12 \cdot 30 = 360$

payment at an interest rate $i = \dfrac{0.06125}{12}$ per

month. The minimum monthly payment on the loan was

$$P = V\left[\dfrac{i}{1-(1+i)^{-n}}\right]$$

$$P = 235,000\left[\dfrac{\frac{0.06125}{12}}{1-\left(1+\frac{0.06125}{12}\right)^{-360}}\right] \approx \$1427.88$$

After paying the loan for 4 years (48 payments) there is

$$V = P\left[\dfrac{1-(1+i)^{-n}}{i}\right]$$

$$= 1427.88\left[\dfrac{1-\left(1+\frac{0.06125}{12}\right)^{-312}}{\frac{0.06125}{12}}\right]$$

$\approx \$222,611.64$

remaining on the loan.

Amount of debt paid:

$235,000 - $222,611.64 = $12,388.36

Interest paid:

($1427.88)(48) - $12,388.36 = $56,149.88

To find the number of months that the term of the mortgage will be reduced, we find n with the new payment $P = $1427.88 + $150 = $1577.88.

$$V = P\left[\dfrac{1-(1+i)^{-n}}{i}\right]$$

$$222,611.64 = 1577.88\left[\dfrac{1-\left(1+\frac{0.06125}{12}\right)^{-n}}{\frac{0.06125}{12}}\right]$$

$$\left(1+\frac{0.06125}{12}\right)^{-n} = 1 - \frac{222,611.64}{1577.88}\left(\frac{0.06125}{12}\right)$$

$1.005104167^{-n} = 0.2798901612$

$-n = \log_{1.005104167} 0.2798901612$

$$n = -\dfrac{\log 0.2798901612}{\log 1.005104167}$$

≈ 250.1

The term of the loan will be reduced by

$312 - 250.1 = 61.9$ months.

The interest that will be saved by increasing the monthly payment is the difference between the interest that would have been paid if the couple continued to make minimum payments and the interest that will be paid with their revised payment.

Interest on the existing loan:

($1427.88)(360) - $235,000 = $279,036.80

Interest with increased payments:
interest paid before + interest paid with new payments

$$I = 56,149.88 + \left[(1577.88)(250.1) - 222,611.64\right]$$

$\approx \$228,166.03$

The interest saved by increasing monthly payments:

$279,036.80 - $228,166.03 = $50,870.77

43. We compare the present value of leasing using the money which is invested at $r = 5.4\%$ compounded monthly with the cost of buying the copier outright.

$$V = P\left[\dfrac{1-(1+i)^{-n}}{i}\right]$$

$$= \$196.88\left[\dfrac{1-\left(1+\frac{0.054}{12}\right)^{-36}}{\frac{0.054}{12}}\right] \approx \$6529.84$$

The state should purchase the equipment because the purchase price is less than the total cost of the lease, V.

45. The present value of a 4 year annuity invested at 5% per annum with annual payments of $50,000 is

$$V = P\left[\frac{1-(1+i)^{-n}}{i}\right]$$

$$= 50,000\left[\frac{1-(1.05)^{-4}}{0.05}\right] \approx \$177,297.53$$

Since the present value of the annuity exceeds the purchase price of the trucks, purchasing the trucks is preferable.

47.a. We first find the equivalent annual cost of each machine.

Machine A: Machine A has an expected life of 8 years. We determine that at 10% interest the average annual cost of Machine A is

$$P = \$10,000\left[\frac{0.10}{1-1.10^{-8}}\right] \approx \$1874.44.$$

Machine B: Machine B has an expected life of 6 years. The average annual cost of B is

$$P = \$8000\left[\frac{0.10}{1-1.10^{-6}}\right] = \$1836.86.$$

The net annual savings of each machine is

Machine	A	B
Labor Savings	$2000.00	$1800.00
Equivalent Annual Cost	$1874.44	$1836.86
Net Savings	$ 125.56	($36.86)

Machine A has a greater net savings, so Machine A is preferable. Note that Machine B results in a net loss.

b. If the time value of money is 14% per annum, then Machine A has an average annual cost of

$$P = \$10,000\left[\frac{0.14}{1-1.14^{-8}}\right] = \$2155.70, \text{ and}$$

Machine B has an average annual cost of

$$P = \$8000\left[\frac{0.14}{1-1.14^{-6}}\right] = \$2057.56.$$

The net annual savings of each machine is

Machine	A	B
Labor Savings	$2000.00	$1800.00
Equivalent Annual Cost	$2155.17	$2057.26
Net Savings	($155.17)	($257.26)

Neither machine results in a savings. Of the two, Machine A has a smaller increased cost and so is the preferable option.

49. We have a $10,000 bond with a nominal interest rate of 6.25% that matures in 8 years. We want to find the price that will yield a true rate of 6.5%. To find the price we follow the steps outlined in the text.

STEP 1. Calculate the semi-annual interest payments,

$$I = Prt = (\$10,000)(0.0625)\left(\tfrac{1}{2}\right) = \$312.50.$$

STEP 2. Calculate the present value of the semi-annual payments. $P = \$312.50$,

$n = 2 \cdot 8 = 16$, $i = \frac{0.065}{2} = 0.0325$.

$$PV_1 = P\left[\frac{1-(1+i)^{-n}}{i}\right]$$

$$= 312.50\left[\frac{1-(1.0325)^{-16}}{0.0325}\right] \approx \$3851.36$$

STEP 3. Calculate the present value of the bond at maturity.

$$PV_2 = A(1+i)^{-n} = 10,000(1.0325)^{-16} \approx 5994.58$$

STEP 4. Calculate the price of the bond by summing the results of steps 2 and 3.
$3851.36 + $5994.58 = $9845.94.

51. The face value: $1000, 5-year bond, nominal interest rate: $r = 2.125\%$, $t = \frac{1}{2}$; true interest rate: 2.130%, $n = 2$, $i = \frac{0.02130}{2} = 0.01065$

STEP 1. Calculate the semi-annual interest payments,

$$I = Prt = (\$1000)(0.02125)\left(\tfrac{1}{2}\right) = \$10.625.$$

STEP 2. Calculate the present value of the semi-annual payments. $P = \$10.625$, $n = 2 \cdot 5 = 10$

$$PV_1 = P\left[\frac{1-(1+i)^{-n}}{i}\right]$$

$$= 10.625\left[\frac{1-(1.01065)^{-10}}{0.01065}\right] \approx \$100.28$$

STEP 3. Calculate the present value of the bond at maturity.

$$PV_2 = A(1+i)^{-n} = 1000(1.01065)^{-10} \approx 899.48$$

STEP 4. Calculate the price of the bond by summing the results of steps 2 and 3.
$100.28 + 899.48 = $999.76.

53. The face value: $1000, 5-year bond, nominal interest rate: $r = 2.625\%$, $t = \frac{1}{2}$; true interest rate: 2.665%, $n = 2$, $i = \frac{0.02665}{2} = 0.013325$

STEP 1. Calculate the semi-annual interest payments,

$$I = Prt = (\$1000)(0.02625)\left(\tfrac{1}{2}\right) = \$13.125.$$

(*continued on next page*)

(continued)

STEP 2. Calculate the present value of the semi-annual payments. $P = \$13.125$, $n = 2 \cdot 5 = 10$

$$PV_1 = P\left[\frac{1-(1+i)^{-n}}{i}\right]$$

$$= 13.125\left[\frac{1-(1.013325)^{-10}}{0.013325}\right] \approx \$122.12$$

STEP 3. Calculate the present value of the bond at maturity.

$$PV_2 = A(1+i)^{-n} = 1000(1.013325)^{-10} \approx 876.02$$

STEP 4. Calculate the price of the bond by summing the results of steps 2 and 3.
$\$122.12 + 876.02 = \$998.14$.

6.5 Annuities and Amortization Using Recursive Sequences

1. $s_1 = 2$, $s_{n+1} = 0.5s_n + 3$
$s_2 = 0.5(2) + 3 = 4$
$s_3 = 0.5(4) + 3 = 5$

3.a. If $B_0 = \$3000$ and $B_n = 1.01B_{n-1} - 100$, then John's balance after 1 payment is
$B_1 = 1.01(\$3000) - \$100 = \$2930$.

b. Use the sequence mode, with n Min $= 0$,
$u(n) = 1.01u(n-1) - 100$, and
$u(n \text{ Min}) = 3000$.

n	u(n)
10	2267.6
11	2190.3
12	2112.2
13	2033.3
14	1953.7
15	1873.2
16	1791.9

u(n)=1953.680507

The balance is below \$2000 after 14 payments. The balance is \$1953.70.

c.

n	u(n)
31	470.71
32	375.42
33	279.17
34	181.96
35	83.781
36	-15.38
37	-115.5

n=36

John will pay off the debt on the 36th payment of \$83.78. The total of all payments is $(35)(\$100) + \$83.78 = \$3583.78$.

d. John's interest expense was
$\$3584.62 - \$3000 = \$584.62$

5.a. If $p_0 = 2000$, and $p_n = 1.03p_{n-1} + 20$, then after 2 months there are $p_2 = 1.03p_1 + 20$ trout, where $p_1 = 1.03p_0 + 20 = 1.03(2000) + 20$. So, $p_1 = 2080$, and $p_2 = 1.03(2080) + 20 = 2162$ trout in the pond.

b. Use the sequence mode, with n Min $= 0$,
$u(n) = 1.03u(n-1) + 20$, and
$u(n \text{ Min}) = 2000$.

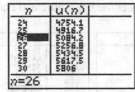

n	u(n)
24	4754.1
25	4916.7
26	5084.2
27	5256.8
28	5434.5
29	5617.5
30	5806

n=26

The trout population reaches 5000 after 26 months (population $p_{26} = 5084.2$).

7.a. If the payments are $P = \$500$ per quarter and the interest rate per payment period is $i = \frac{0.08}{4} = 0.02$, then a recursive formula that represents the balance at the end of each quarter is $A_0 = \$500$,
$A_n = (1 + 0.02)A_{n-1} + P = 1.02A_{n-1} + \$500$.

b. Use the sequence mode, with n Min $= 0$,
$u(n) = 1.02u(n-1) + 500$, and
$u(n \text{ Min}) = 500$. The table below shows that the value of the account will exceed \$100,000 after the 81th payment.

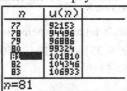

n	u(n)
77	92153
78	94496
79	96886
80	99324
81	101810
82	104346
83	106933

n=81

c. When Bob retires in 25 years, $n = 25 \cdot 4 = 100$. The value of the account will be $A = \$159,738.48$.

n	u(n)
98	152565
99	156116
100	159738
101	163433
102	167202
103	171046
104	174967

u(n)=159738.476

9.a. Since Bill and Laura borrowed \$150,000 at an interest rate $i = \frac{0.06}{12} = 0.005$ per month and are repaying the loan with $n = 360$ monthly payments of \$899.33, a recursive formula which represents their balance after each payment is $A_0 = \$150,000$,
$A_n = (1 + 0.005)A_{n-1} - P$
$= 1.005A_{n-1} - \$899.33$.

b. After 1 payment, Bill and Laura's balance is
$A_1 = (1.005)(150,000) - \$899.33$
$= \$149,850.67$.

c. Use the sequence mode, with n Min = 0, $u(n) = 1.005u(n-1) - 899.33$, and $u(n$ Min$) = 150{,}000$. The tables below show the balances at the beginning and end of the loan.

n	$u(n)$
0	150000
1	149851
2	149701
3	149550
4	149398
5	149246
6	149093

$u(n)=150000$

n	$u(n)$
354	5298.7
355	4425.9
356	3548.7
357	2667.1
358	1781.1
359	890.65
360	-4.231

$u(n)=890.645483$

d. Using the recursive formula from part (c) and the TABLE, we find that after 4 years and 10 months (58 payments) the balance on the loan is $139{,}981.09.

n	$u(n)$
56	140377
57	140180
58	139981
59	139782
60	139581
61	139380
62	139177

$u(n)=139981.0876$

e. Bill and Laura pay off the balance with the 360th payment. It is a reduced payment of $890.65(1.005) = 895.10.

f. Bill and Laura's interest expense is
$$I = \left[(359)(899.33) + 895.10\right] - 150{,}000$$
$$= \$173{,}754.57$$

ga. If Bill and Laura make payments of $999.33 instead of $899.33, the recursive formula representing their balance after each payment is $A_0 = \$150{,}000$,
$$A_n = (1 + 0.005)A_{n-1} - P$$
$$= 1.005A_{n-1} - \$999.33.$$

gb. After 1 payment, Bill and Laura's balance is
$$A_1 = (1.005)(150{,}000) - \$999.33$$
$$= \$149{,}750.67.$$

gc. Use the sequence mode, with n Min = 0, $u(n) = 1.005u(n-1) - 999.33$, and $u(n$ Min$) = 150{,}000$.

n	$u(n)$
0	150000
1	149751
2	149500
3	149248
4	148995
5	148741
6	148485

$u(n)=149750.67$

n	$u(n)$
35	140489
36	140192
37	139894
38	139594
39	139293
40	138990
41	138685

$u(n)=139893.9606$

gd. Using the recursive formula from part (c) and the TABLE, we find that after 3 years and one month (37 payments) the balance on the loan is $139{,}894.

ge. Bill and Laura pay off the balance with the 279th payment. It is a reduced payment of $353.69(1.005) = \$355.37.

n	$u(n)$
275	3316.7
276	2333.9
277	1346.3
278	353.69
279	-643.9
280	-1646
281	-2654

$u(n)=353.6881682$

gf. Bill and Laura's interest expense is
$$I = \left[(278)(999.33) + 355.37\right] - 150{,}000$$
$$= \$128{,}169.11$$

h. Answers will vary. Sample answer: If Bill and Laura pay the additional $100 per month, the loan is paid off 81 months sooner and they save $45,585.46 in interest.

Chapter 6 Review Exercises

1. 3% of 500 = (0.03) $\cdot$ (500) = 15

3. 140% of 250 = (1.40) $\cdot$ (250) = 350

5. $x\%$ of 350 = 75
$$\frac{x}{100} \cdot 350 = 75$$
$$\frac{7}{2}x = 75 \Rightarrow x = \frac{75 \cdot 2}{7} = \frac{150}{7} = 21\frac{3}{7}$$
75 is $21\frac{3}{7}$ % of 350.

7. 12 is 15% of x.
$$12 = 0.15x \Rightarrow x = \frac{12}{0.15} = 80$$
12 is 15% of 80.

9. 11 is 0.5% of x.
$$11 = 0.005x \Rightarrow x = \frac{11}{0.005} = 2200$$
11 is 0.5% of 2200.

11. Dan will have to pay 7% of $330.00 in sales tax. (0.07)($330.00) = $23.10. Dan will pay $23.10 in sales tax.

13. If Dan borrows $500 for 1 year and 2 months (14 months) at 9% simple interest, the interest charged is
$$I = Prt = (\$500)(0.09)\left(\frac{14}{12}\right) = \$52.50 \text{ and the}$$
amount due is
$$A = P + I = \$500 + \$52.50 = \$552.50.$$

15. The proceeds of Warren's loan is $15,000. If the loan has an interest rate of 12%, Warren must repay L at the end of 2 years where $R = L(1 - rt)$.

$$15,000 = L\left[1 - (0.12)(2)\right] = 0.76L$$

$$L = \frac{15,000}{0.76} \approx 19,736.84$$

Warren must pay $19,736.84 to settle his debt.

17. If $P = \$100$ is invested for 2 years and 3 months ($n = 27$ months) at an interest rate

$i = \dfrac{0.03}{12} = 0.0025$ per month, it will accumulate

to $A = P(1+i)^n = 100(1.0025)^{27} \approx \$106.97$.

19. Mike is choosing between two loans.
Loan (a): $3000 for $t = 3$ years at $r = 12\%$ simple interest. This loan will cost Mike

$$I = P(1 + rt) = \$3000\left[1 + (0.12)(3)\right] = \$4080$$

Loan (b): $3000 for $n = 12 \cdot 3 = 36$ months at

interest rate $i = \dfrac{0.10}{12}$ per month. This loan will

cost

$$A = P(1+i)^n = 3000\left(1 + \frac{0.10}{12}\right)^{36} \approx \$4044.55.$$

Loan (b) costs Mike less.

21. Katy wants the amount $A = \$75$ in $n = 6$ months.

She can earn $i = \dfrac{0.03}{12} = 0.0025$ per month.

$$P = A(1+i)^{-n} = 75(1.0025)^{-6} \approx 73.88$$

Katy should deposit $73.88 in order to reach her goal.

23. To double an investment in 12 years, let $P = 1$, $A = 2P = 2$, and $n = 12$. Find r.

$$A = P(1+r)^n \Rightarrow 2 = (1+r)^{12} \Rightarrow$$
$$1 + r = \sqrt[12]{2} \Rightarrow r = \sqrt[12]{2} - 1 \approx 0.0595$$

Money will double in 12 years when invested at 5.95% per annum.

25. The effective rate of interest is the simple interest rate that yields the same amount as the actual compounded rate. We want the interest rate compounded quarterly that is equivalent to the simple interest rate $r = 0.06$.

$$r_{\text{eff}} = (1+i)^n - 1 \Rightarrow 0.06 = (1+i)^4 - 1 \Rightarrow$$
$$1.06 = (1+i)^4$$
$$\sqrt[4]{1.06} - 1 \approx 0.0146738 \text{ per quarter}$$

The annual interest rate compounded quarterly equivalent to 6% simple interest is
$4i = 4(0.0146738) = 0.05870 = 5.87\%$.

27. Mr. and Mrs. Corey are setting up a sinking fund to save the down payment for the house, where $A = \$40,000$, $n = 12 \cdot 2 = 24$ months, and the interest rate per payment period is

$i = \dfrac{0.03}{12} = 0.0025$. We are looking for P, the

Corey's monthly payment.

$$A = P\frac{(1+i)^n - 1}{i}$$

$$40,000 = P\left[\frac{(1.0025)^{24} - 1}{0.0025}\right]$$

$$P = 40,000\left[\frac{0.0025}{(1.0025)^{24} - 1}\right] \approx 1619.25$$

Mr. and Mrs. Corey must save $1619.25 per month in order to have the $40,000 they need for the down payment.

29. Mr. and Mrs. Ostedts' new house cost $400,000. They made a $100,000 down payment, and are financing $300,000 at 10% for 25 years.

a. The monthly payments are found by

evaluating $P = V\dfrac{i}{1 - (1+i)^{-n}}$.

$V = 300,000$, $n = 12 \cdot 25 = 300$ months,

$i = \dfrac{0.10}{12}$.

$$P = 300,000\left[\frac{\frac{0.10}{12}}{1 - \left(1 + \frac{0.10}{12}\right)^{-300}}\right] \approx 2726.10$$

The Ostedts' monthly payments are $2726.10.

b. The Ostedts' will pay
$(300)(\$2726.10) - \$300,000 = \$517,830$ in interest.

c. The equity in a house is the sum of the down payment and the amount paid on the loan. After 5 years, the Ostedt's have made 60 payments. The present value of the remaining 240 payments is

$$V = P \cdot \frac{1 - (1+i)^{-n}}{i}$$

$$= 2726.10\left[\frac{1 - \left(1 + \frac{0.10}{12}\right)^{-240}}{\frac{0.10}{12}}\right] \approx 282,491.07$$

The Ostedt's have paid
$300,000 - \$282,491.07 = \$17,508.93$ of the principal. Their equity is
$100,000 + \$17,508.93 = \$117,508.93$.

31. If $125,000 is to be amortized at a periodic interest rate $i = \dfrac{0.09}{12} = 0.0075$ per month over $n = 12 \cdot 25 = 300$ months, the monthly payment will be

$$P = V \frac{i}{1 - (1+i)^{-n}}$$

$$= 125,000 \left[\frac{0.0075}{1 - (1.0075)^{-300}} \right] \approx \$1049.00$$

$$V = P \cdot \frac{1 - (1+i)^{-n}}{i}$$

$$= 1049 \left[\frac{1 - (1.0075)^{-180}}{0.0075} \right] \approx 103,424.49$$

Thus, $125,000 - $103,424.49 = $21,575.51 of the loan has been paid. The equity in the property after 10 years is about $21,576 plus any down payment.

33. Let x denote the price Mr. Graff should pay for the gold mine. Then $0.15x$ represents a 15% annual Return on Investment (ROI). The annual Sinking Fund Contribution (SFC) needed to recover the purchase price x in 20 years can be calculated letting $A = x$, $n = 20$, $i = 0.10$, and $P = $ SFC.

$$A = P \frac{(1+i)^n - 1}{i} \Rightarrow x = \text{SFC} \left[\frac{(1.10)^{20} - 1}{0.10} \right]$$

$$\text{SFC} = x \left[\frac{0.10}{(1.10)^{20} - 1} \right] \approx 0.0174596248x$$

The required annual ROI plus the annual SFC equals the annual net income of $20,000.
$$0.15x + 0.0174596248x = 20,000$$
$$0.1674596248x = 20,000$$
$$x \approx 119,431.78$$
Mr. Graff should pay $119,432 for the gold mine.

35. Mr. Doody wants to receive $n = 12 \cdot 15 = 180$ monthly payments of $300. He can invest the money now at an interest rate $i = \dfrac{0.04}{12}$ per month.

$$V = P \cdot \frac{1 - (1+i)^{-n}}{i}$$

$$= 300 \left[\frac{1 - \left(1 + \frac{0.04}{12}\right)^{-180}}{\frac{0.04}{12}} \right] \approx 40,557.64$$

He needs $40,557.64 to insure his retirement goals are met.

37. Mr. Jones is saving $500 every six months at an interest rate $i = \dfrac{0.03}{2} = 0.015$ per payment period. $n = 2 \cdot 8 = 16$

$$A = P \frac{(1+i)^n - 1}{i} = 500 \left[\frac{(1.015)^{16} - 1}{0.015} \right]$$

$$\approx 8966.18$$

After 8 years, Mr. Jones will have $8966.18.

39. A loan of $3000 at an interest rate per payment period of $i = \dfrac{0.12}{12} = 0.01$ is to be amortized in 2 years with $n = 2 \cdot 12 = 24$ equal monthly payments.

$$P = V \left[\frac{i}{1 - (1+i)^{-n}} \right]$$

$$P = 3000 \left[\frac{0.01}{1 - (1.01)^{-24}} \right] \approx \$141.22$$

Each payment will be $141.22.

41. The effective rate of interest is the simple interest rate that yields the same amount as the actual compounded rate. The annual interest rate of 9% compounded monthly gives an interest rate of $i = \dfrac{0.09}{12} = 0.0075$ per month.
$$I_{\text{effective}} = (1.0075)^{12} - 1 \approx 0.0938$$
The effective rate of interest is 9.38%.

43. If John's trust fund is earning an interest rate $i = \dfrac{0.08}{2} = 0.04$ very 6 months it will be amortized in 15 years with $n = 2 \cdot 15 = 30$ equal payments.

$$P = V \left[\frac{i}{1 - (1+i)^{-n}} \right]$$

$$P = 20,000 \left[\frac{0.04}{1 - (1.04)^{-30}} \right] \approx \$1156.60$$

Each semiannual payment will be $1156.60.

45. $P = 60, $i = 0.01$ per month, $n = 30$ months

$$A = P \frac{(1+i)^n - 1}{i} = 60 \left[\frac{(1.01)^{30} - 1}{0.01} \right] \approx 2087.09$$

There will be $2087.09 in the employee's retirement fund at the end of 30 months.

47. The student is to amortize the $4000 loan with $n = 4 \cdot 4 = 16$ equal quarterly payments. The interest rate per payment period of the loan is $i = \dfrac{0.14}{4} = 0.035$ per quarter.

$$P = V\left[\frac{i}{1-(1+i)^{-n}}\right]$$

$$P = 4000\left[\frac{0.035}{1-(1.035)^{-16}}\right] \approx \$330.74$$

The student's quarterly payment is $330.74.

Chapter 6 Project

1. The purchase price of the car is $23,700 and the down payment is $3000. Then, the amount to be financed is $V = \$20,700$. We must find the monthly payment for $n = 12 \cdot 5 = 60$ months at 5% per year.

$$P = V\left[\frac{i}{1-(1+i)^{-n}}\right]$$

$$P = 20,700\left[\frac{\frac{0.05}{12}}{1-\left(1+\frac{0.05}{12}\right)^{-60}}\right] \approx 390.63$$

The monthly payment is $390.63. The total amount of interest paid over the term of the loan is $390.63(60) - \$20,700 = \$2737.80$.

3. 20% of $23,700 is $4740, so the down payment is $4740 and the principal amount for the loan is $18,960. $i = \dfrac{0.055}{12}$ and $n = 12 \cdot 7 = 84$ months.

$$P = V\left[\frac{i}{1-(1+i)^{-n}}\right]$$

$$= 18,960\left[\frac{\frac{0.055}{12}}{1-\left(1+\frac{0.055}{12}\right)^{-84}}\right] \approx 272.46$$

The monthly payment is $272.46. The total amount of interest paid is $84(272.46) - 18,960 = \$3926.64$.

5. $i = \dfrac{0.08}{12}$ per month

For Option A: $n = 60$ months, $P = \$390.63$

$$V = P \cdot \frac{1-(1+i)^{-n}}{i} = 390.63\left[\frac{1-\left(1+\frac{0.08}{12}\right)^{-60}}{\frac{0.08}{12}}\right]$$

$$\approx \$19,265.26$$

The present value for Option A is $19,265.26 + 3000 = \$22,265.26$.

For Option B: $n = 72$ months, $P = \$283.42$

$$V = P \cdot \frac{1-(1+i)^{-n}}{i} = 283.42\left[\frac{1-\left(1+\frac{0.08}{12}\right)^{-72}}{\frac{0.08}{12}}\right]$$

$$\approx \$16,164.72$$

The present value for Option B is $16,164.72 + \$6000 = \$22,164.72$

For Option C: $n = 84$ months, $P = 272.46$

$$V = P \cdot \frac{1-(1+i)^{-n}}{i} = 272.46\left[\frac{1-\left(1+\frac{0.08}{12}\right)^{-84}}{\frac{0.08}{12}}\right]$$

$$\approx \$17,480.83$$

The present value for Option C is $17,480.83 + \$4740 = \$22,220.83$. Option B minimizes total present value.

7. The depreciated price after three years is $0.9^3(\$23,700) \approx \$17,277.30$.

The depreciated price after four years is $0.9^4(\$23,700) \approx \$15,549.57$.

9. Option B has the lowest total present value.

11. The depreciated price after three years is $17,277.30. We want to determine the monthly payment P for a loan of $17,277.30 with $i = \dfrac{0.08}{12}$ per month for $n = 36$ months.

$$A = P\left[\frac{(1+i)^n - 1}{i}\right] \Rightarrow$$

$$\$17,277.30 = P\left[\frac{\left(1+\frac{0.08}{12}\right)^{36} - 1}{\frac{0.08}{12}}\right] \Rightarrow$$

$$P = \$17,277.30\left[\frac{\frac{0.08}{12}}{\left(1+\frac{0.08}{12}\right)^{36} - 1}\right] \approx \$426.23$$

The total monthly payment = $225 + \$426.23 = \$651.23$.

Mathematical Questions from Professional Exams

1. B   **3.** B   **5.** A

Chapter 7 Probability

7.1 Sets

1. $\{1, 2, 3, 4\} \subseteq \{1, 2, 3, 4\}$ or
$\{1, 2, 3, 4\} = \{1, 2, 3, 4\}$

3. $\{1, 2, 3\} \cap \{2, 3, 4, 5\} = \{2, 3\}$

5. True   **7.** False

9. False   **11.** True

13. True   **15.** True

17. $\{1, 2, 3\} \cap \{2, 3, 4, 5\} = \{2, 3\}$

19. $\{1, 2, 3\} \cup \{2, 3, 4, 5\} = \{1, 2, 3, 4, 5\}$

21. $\{2, 4, 6, 8\} \cap \{1, 3, 5, 7\} = \varnothing$

23. $\{a, b, e\} \cup \{d, e, f, g\} = \{a, b, d, e, f, g\}$

25. a. $A \cup B = \{0, 1, 2, 3, 5, 7, 8\}$

    b. $B \cap C = \{5\}$

    c. $A \cap B = \{5\}$

    d. $\overline{A \cap B} = \{0, 1, 2, 3, 4, 6, 7, 8, 9\}$

    e. $\overline{A} \cap \overline{B} = \{2, 3, 4, 6, 8, 9\} \cap \{0, 1, 4, 6, 7, 9\}$
$= \{4, 6, 9\}$

    f. $A \cup (B \cap A) = \{0, 1, 5, 7\} \cup \{5\}$
$= \{0, 1, 5, 7\} = A$

    g. $(C \cap A) \cap (\overline{A}) = \{5\} \cap \{2, 3, 4, 6, 8, 9\} = \varnothing$

    h. $(A \cap B) \cup (B \cap C) = \{5\} \cup \{5\} = \{5\}$

27. a. $A \cup B = \{b, c, d, e, f, g\}$

    b. $A \cap B = \{c\}$

    c. $\overline{A} \cap \overline{B} = \overline{A \cup B}$
$= \{a, h, i, j, k, l, m, n, o, p, q, r,$
$s, t, u, v, w, x, y, z\}$

    d. $\overline{A} \cup \overline{B} = \overline{A \cap B}$
$= \{a, b, d, e, f, g, h, i, j, k, l, m, n,$
$o, p, q, r, s, t, u, v, w, x, y, z\}$

29.

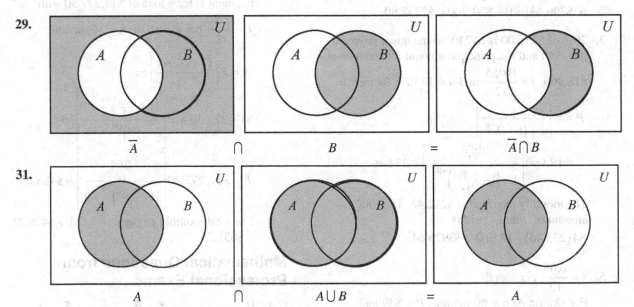

31.

168

33.

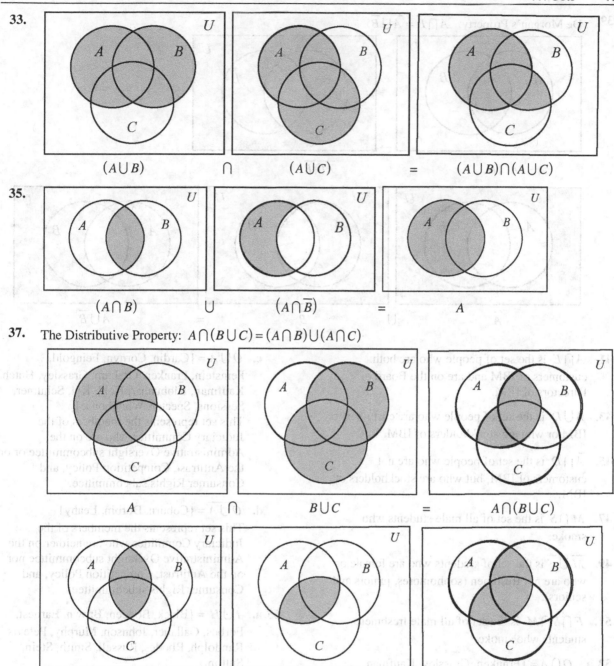

$(A \cup B)$ $\cap$ $(A \cup C)$ = $(A \cup B) \cap (A \cup C)$

35.

$(A \cap B)$ $\cup$ $(A \cap \overline{B})$ = A

37. The Distributive Property: $A \cap (B \cup C) = (A \cap B) \cup (A \cap C)$

A $\cap$ $B \cup C$ = $A \cap (B \cup C)$

$A \cap B$ $\cup$ $A \cap C$ = $(A \cap B) \cup (A \cap C)$

39. De Morgan's Property: $\overline{A \cap B} = \overline{A} \cup \overline{B}$

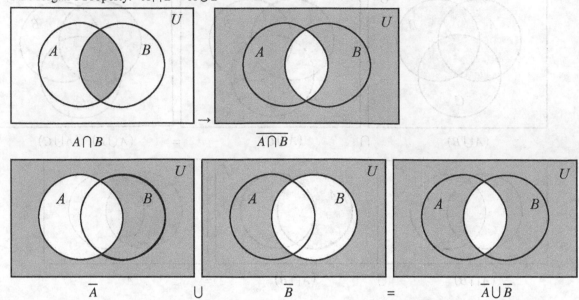

$$A \cap B \qquad \rightarrow \qquad \overline{A \cap B}$$

$$\overline{A} \qquad \cup \qquad \overline{B} \qquad = \qquad \overline{A} \cup \overline{B}$$

41. $A \cap E$ is the set of people who are both customers of IBM and are on the Board of Directors of IBM.

43. $A \cup D$ is the set of people who are customers of IBM or who are stockholders of IBM.

45. $\overline{A} \cap D$ is the set of people who are not customers of IBM, but who are stockholders of IBM.

47. $M \cap S$ is the set of all male students who smoke.

49. $\overline{M} \cup F$ is the set of students who are female or who are not freshmen (sophomores, juniors and seniors).

51. $F \cap S \cap M$ is the set of all male freshmen students who smoke.

53. a. $O \cap A$ = {Franken, Grassley, Kaufman, Schumer, Whitehouse}
This set represents the members of the Judiciary Committee who are on both the Administrative Oversight and the Antitrust, Competition Policy and Consumer Rights subcommittees.

b. $\overline{O}$ = {Coburn, Cornyn, Durbin, Hatch, Klobuchar, Kohl, Leahy, Specter}
This set represents the members of the Judiciary Committee who are not on the Administrative Oversight subcommittee.

c. $O \cup A$ = {Cardin, Cornyn, Feingold, Feinstein, Franken, Graham, Grassley, Hatch, Kaufman, Klobuchar, Kohl, Kyl, Schumer, Sessions, Specter, Whitehouse}
This set represents the members of the Judiciary Committee who are on the Administrative Oversight subcommittee or on the Antitrust, Competition Policy, and Consumer Rights subcommittee.

d. $\overline{O \cup A}$ = {Coburn, Durbin, Leahy}
This set represents the members of the Judiciary Committee who are neither on the Administrative Oversight subcommittee nor on the Antitrust, Competition Policy, and Consumer Rights subcommittee.

55. a. $I \cup H$ = {Black, Bodden, Brown, Earnest, Forbes, Gallaher, Johnson, Murphy, Petevis Randolph, Rhodes, Russell, Smith, Stein, Sutton,}
This set represents the clients who own either Intel Corp. stock or Hewlett-Packard stock.

b. $I \cap H$ = {Bodden, Gallaher, Sutton}
This set is the persons in the advisors data base who own both Intel Corp. and Hewlett-Packard stock.

57. The subsets of $\{a, b, c\}$ are $\emptyset$, $\{a\}$, $\{b\}$, $\{c\}$, $\{a, b\}$, $\{a, c\}$, $\{b, c\}$, and $\{a, b, c\}$.

59. Answers will vary.

7.2 The Number of Elements in a Set

1. False. If A and B are sets, then
$n(A\cup B)=n(A)+n(B)-n(A\cap B)$.

3. $n(A)=6$

5. $A\cap B=\{2,\ 4,\ 6\}$
$n(A\cap B)=3$

7. $(A\cap B)\cup A=\{2,\ 4,\ 6\}\cup\{1,\ 2,\ 3,\ 4,\ 5,\ 6\}$
$=\{1,\ 2,\ 3,\ 4,\ 5,\ 6\}$
$=A$
$n[(A\cap B)\cup A]=n(A)=6$

9. $n(A\cup B)=n(A)+n(B)-n(A\cap B)$
$=4+3-2=5$

11. $n(A\cup B)=n(A)+n(B)-n(A\cap B)$
$n(A\cap B)=n(A)+n(B)-n(A\cup B)$
$=5+4-7=2$

13. $n(A\cup B)=n(A)+n(B)-n(A\cap B)$
$n(A)=n(A\cup B)-n(B)+n(A\cap B)$
$=14-8+4=10$

15. Here we are looking for the number of cars that have either GPS system or satellite radio. It is the same as finding $n(A\cup B)$. So the number of cars we are looking for is

$\left(\begin{array}{c}\text{the number of cars}\\ \text{with GPS systems}\end{array}\right)+\left(\begin{array}{c}\text{the number of cars}\\ \text{with satellite radio}\end{array}\right)$
$-\left(\begin{array}{c}\text{the number of cars}\\ \text{with both options}\end{array}\right)$
$=325+216-89=452$

The company manufactured 452 cars.

17. $n(A)=10+6+3+5=24$

19. Before determining the number of elements in A or B, we need to find the number of elements in $A\cap B$.
$n(A\cap B)=6+3=9$
Thus,
$n(A\cup B)=n(A)+n(B)-n(A\cap B)$
$=24+19-9=34$

21. To find the number of elements in A but not in B, subtract $n(A\cap B)$ from $n(A)$.
$n(A)-n(A\cap B)=24-9=15$

23. The number of elements in A or B or C is found by expanding the formula for finding the number of elements in A or C.

$n(A\cup B\cup C)=n(A)+n(B)+n(C)-n(A\cap B)$
$-n(A\cap C)-n(B\cap C)$
$+n(A\cap B\cap C)$
$=24+19+30-9-8-5+3$
$=54$

25. The number of elements in A and B and C is the number of elements common to all three sets.
$n(A\cap B\cap C)=3$.

27.a. The number of households in the South that are maintained by the mother is given by $1782+1279+833+156=4050$. Since the numbers are given in thousands, there are 4,050,000 households in the South that are maintained by the mother.

b. The number of households in the Midwest that are maintained by a divorced parent is given by $202+72=274$. Since the numbers are given in thousands, there are 274,000 households in the Midwest that are maintained by a divorced parent.

c. The number of households that are maintained by a father who was never married or widowed is given by
$99+132+174+105+19+18+33+20$
$=600$.
Since the number are given in thousands, there are 600,000 households that are maintained by a father who is never married or widowed.

29.a. Number who are unemployed:
$n(\text{unemployed males})+n(\text{unemployed females})=7882+5532=13,414$
Since the numbers are given in thousands, there were 13,414,000 U.S. citizens 20 years or older who were unemployed.

b. Number who are unemployed or not in the labor force:
$n(\text{unemployed})+n(\text{males not in labor force})$
$+n(\text{females not in labor force})$
$=13,414+27,403+44,947=85,764$
Since the numbers are given in thousands, 85,764,000 U.S. citizens 20 years or older were unemployed or not in the labor force.

c. Number who are female or employed:
$n(\text{females})+n(\text{employed})-n(\text{females and employed})$
$=(63,495+5532+44,947)$
$+(70,913+63,495)-(63,495)$
$=184,887$
Since the numbers are given in thousands, 184,887,000 U.S. citizens 20 years or older were female or employed.

d. There are 27,403,000 males not in the labor force.

31. Before doing this problem, it is suggested that you add the numbers in the rows and in the columns of the table.

a. Layoffs who are executive managerial or faculty or professional nonfaculty is the sum of the first three rows of the table.
$11 + 84 + 161 = 256$
256 layoffs were executive managerial or faculty or professional nonfaculty.

b. Layoffs who are female is the sum of the female columns of the table.
$134 + 12 + 8 + 14 = 168$
168 females were laid off.

c. n(black female column) + n(faculty row) $- n$(black, female, faculty) $= 12 + 84 - 3 = 93$
93 black female or faculty were laid off.

d. n(white male column) + n(executive managerial row) $- n$(white male managerial hires) $= 119 + 11 - 5 = 125$
125 white males or executive managerial were laid off.

33.a.

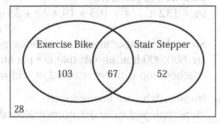

For parts (b) through (e) let E denote exercise bike and let S denote stair stepper.

b. $n(E \cup S) = 103 + 67 + 52 = 222$ members use an exercise bike or a stair stepper regularly.

c. $n(E \cap \overline{S}) = 103$ members use an exercise bike regularly, but not a stair stepper.

d. $n(S \cap \overline{E}) = 52$ members use a stair stepper regularly, but not an exercise bike.

e. $250 - n(E \cup S) = 250 - 222 = 28$ members use neither an exercise bike nor a stair stepper regularly.

35.a.

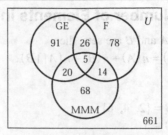

b. 91 clients own only GE.

c. 78 clients own only F.

d. 68 clients own only MMM.

e. $n\left[(\text{GE} \cup \text{MMM}) \cap \overline{\text{F}}\right] = 91 + 20 + 68 = 179$ clients own GE or MMM but not F.

f. $n\left[(\text{GE} \cap \text{MMM}) \cap \overline{\text{F}}\right] = 20$ clients own GE and MMM but not F.

g. $n(\text{GE}) + n(\text{MMM}) + n(\text{F}) - n(\text{GE} \cap \text{MMM})$
$- n(\text{GE} \cap \text{F}) - n(\text{MMM} \cap \text{F})$
$+ n(\text{GE} \cap \text{MMM} \cap \text{F})$
$= 142 + 107 + 123 - 25 - 31 - 19 + 5 = 302$
302 clients own at least one of the three stocks. Thus,
$n\left(\overline{\text{GM} \cup \text{DCX} \cup \text{F}}\right) = 963 - 302 = 661$ of the clients own none of the stocks.

37. We denote the sets of females, seniors, and students on the dean's list by F, S, and D respectively. From the data given we determine that
$n(F \cap S \cap D) = 31 \qquad n(F \cap \overline{S} \cap D) = 62$
$n(\overline{F} \cap S \cap D) = 45 \qquad n(F \cap \overline{S} \cap \overline{D}) = 275$
$n(F \cap S \cap \overline{D}) = 87 \qquad n(\overline{F} \cap S \cap \overline{D}) = 96$
$n(\overline{F} \cap \overline{S} \cap D) = 89 \qquad n(\overline{F} \cap \overline{S} \cap \overline{D}) = 227$
We next place the values on a Venn diagram, so that we can answer the questions.

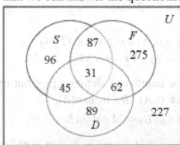

a. There were $96 + 87 + 31 + 45 = 259$ seniors at the college.

b. There were $275 + 87 + 31 + 62 = 455$ women at the college.

c. There were $89 + 45 + 31 + 62 = 227$ students on the dean's list.

d. There were $45 + 31 = 76$ seniors on the dean's list.

e. There were $87 + 31 = 118$ female seniors.

f. There were $31 + 62 = 93$ women on the dean's list.

g. There were $96 + 87 + 275 + 45 + 31 + 62 + 89 + 227 = 912$ students at the college.

39. We denote the sets of cars with heated seats, GPS, and satellite radio by H, G, and R respectively. From the data given we determine that

$$n(H) = 90 \qquad\qquad n(G) = 100$$
$$n(R) = 75 \quad n(H \cap G \cap R) = 5$$
$$n(\overline{H \cup G \cup R}) = 20 \quad n(H \cap \overline{G} \cap \overline{R}) = 20$$
$$n(\overline{H} \cap G \cap \overline{R}) = 60 \quad n(\overline{H} \cap \overline{G} \cap R) = 30$$
$$n(G \cap R) = 10$$

We next place the values on a Venn diagram, so that we can answer the questions.

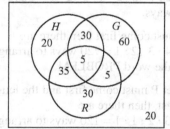

a. $35 + 5 = 40$ cars had satellite radio and heated seats.

b. $30 + 5 = 35$ cars had GPS and heated seats.

c. $20 + 20 = 40$ cars had neither satellite radio nor GPS.

d. There were
$20 + 20 + 30 + 5 + 35 + 30 + 5 + 60 = 205$
cars sold in July.

e. $100 + 90 - 35 = 155$ cars were sold with GPS or heated seats or both.

41. There are 8 different possible blood types, as shown in the Venn diagram below. They can be listed as A^+, A^-, B^+, B^-, AB^+, AB^-, O^+, and O^-.

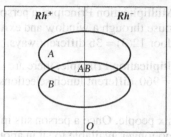

43. We denote the sets of users of HP, Mac, and Dell by H, M, and D respectively. From the information given, we determine that
$n(H) = 27$, $n(M) = 35$, $n(D) = 35$,
$n(H \cap M) = 10$, $n(H \cap D) = 10$,
$n(M \cap D) = 10$, $n(H \cap M \cap D) = 3$, and
$n(\overline{H \cup M \cup D}) = 30$.

Putting this information onto a Venn diagram will help answer the question.

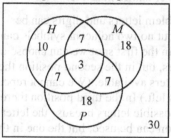

There are $10 + 18 + 18 = 46$ users who exclusively use only one brand of computer.

45. The subsets of $\{a, b, c, d\}$ are $\emptyset$, $\{a\}$, $\{b\}$, $\{c\}$, $\{d\}$, $\{a, b\}$, $\{a, c\}$, $\{a, d\}$, $\{b, c\}$, $\{b, d\}$, $\{c, d\}$, $\{a, b, c\}$, $\{a, b, d\}$, $\{a, c, d\}$, $\{b, c, d\}$, and $\{a, b, c, d\}$. There are $2^4 = 16$ subsets, where 4 is the number of elements in the original set.

7.3 The Multiplication Principle

1. By the Multiplication Principle, the man can wear $5(3) = 15$ different shirt and tie combinations.

3. There are nine choices for the first digit and ten choices for each of the remaining three digits. By the Multiplication Principle, $9 \cdot 10 \cdot 10 \cdot 10 = 9000$ different four-digit numbers can be formed.

5. By the Multiplication Principle, one can travel $2(4) = 8$ different routes from town A to town C through town B.

7. If there are 3 different car models and 8 different color schemes, then by the Multiplication Principle, you must display $3(8) = 24$ different cars to show all possibilities.

9. Using the Multiplication Principle, a person can enter the house through a window and exit through a door $12(3) = 36$ different ways.

11. By the Multiplication Principle, there are $3 \cdot 8 \cdot 10 \cdot 4 = 960$ different lunch selections available.

13. There are six people. Once a person sits in a seat, he is no longer available to sit in another seat. So by the Multiplication Principle, there are $6 \cdot 5 \cdot 4 \cdot 3 \cdot 2 \cdot 1 = 720$ different ways six people can sit in a row of six seats.

15. Since there are 26 different letters and 10 different digits, by the Multiplication Principle, a user name formed by choosing 4 letters followed by four digits can be chosen in $26 \cdot 26 \cdot 26 \cdot 26 \cdot 10 \cdot 10 \cdot 10 \cdot 10 = 4,569,760,000$ ways. So, theoretically, there are 4,569,760,000 different user names possible.

17. In this problem letters and digits can be repeated, but no two adjacent symbols can be the same. In the first spot we can choose any one of 26 letters, but in the second position there are only 25 letters available. (We cannot repeat the letter to its left.) In the third position there are again 25 possible letters because the letter in the first position can be used, but the one in the second position cannot. The fourth position also has 25 available letters. Similar reasoning is used for choosing the digits. Any one of ten digits can be chosen for the fifth position, but only nine that can be used in the sixth position. The seventh and eighth positions each have nine possible digits that can be used. Using this reasoning and the Multiplication Principle, we find that there are $26 \cdot 25 \cdot 25 \cdot 25 \cdot 10 \cdot 9 \cdot 9 \cdot 9 = 2,961,562,500$ user names for the system.

19. Since in the World Series one team from the National League plays one team from the American League, there are $16 \cdot 14 = 224$ different match-ups possible.

21. By the Multiplication Principle, we see that Adam has $9 \cdot 9 \cdot 3 \cdot 2 = 486$ different insurance options from which to choose.

23. The pharmaceutical sales rep can have $5 \cdot 2 \cdot 3 = 30$ different sales portfolios.

25. David must choose among $4 \cdot 2 \cdot 25 = 200$ different one-topping pizzas.

27. Since a network doesn't show the same show twice in one week, there were $19 \cdot 18 \cdot 17 = 5814$ different new-comedy lineups possible.

29. According to the Multiplication Principle, the audits can be scheduled in $10 \cdot 9 \cdot 8 \cdot 7 \cdot 6 \cdot 5 \cdot 4 \cdot 3 \cdot 2 \cdot 1 = 3,628,800$ different ways.

31. a. According to the Multiplication Principle there can be $10 \cdot 9 \cdot 8 \cdot 7 \cdot 6 \cdot 5 \cdot 4 = 604,800$ different 7-digit telephone numbers if no digit is allowed to repeat.

    b. If the first number cannot be 0 and digits cannot be repeated, there can be only $9 \cdot 9 \cdot 8 \cdot 7 \cdot 6 \cdot 5 \cdot 4 = 544,320$ different 7-digit telephone numbers.

    c. The number of 7-digit telephone numbers possible if digits can be repeated and if 0 is allowed in the first position, is $10 \cdot 10 \cdot 10 \cdot 10 \cdot 10 \cdot 10 \cdot 10 = 10,000,000$.

33. a. The seven letters of the word PROBLEM can be arranged $7 \cdot 6 \cdot 5 \cdot 4 \cdot 3 \cdot 2 \cdot 1 = 5040$ different ways.

    b. If the P must come first then there are $1 \cdot 6 \cdot 5 \cdot 4 \cdot 3 \cdot 2 \cdot 1 = 720$ ways to arrange the letters of the word PROBLEM.

    c. If the letter P must come first and the letter M must be last, then there are $1 \cdot 5 \cdot 4 \cdot 3 \cdot 2 \cdot 1 \cdot 1 = 120$ ways to arrange the letters of the word PROBLEM.

35. a. If the letters and digits can be repeated, then using the Multiplication Principle, we find that $26 \cdot 26 \cdot 10 \cdot 10 \cdot 10 \cdot 10 = 6,760,000$ different license plates can be made.

    b. If the letters can be repeated, but the digits cannot be repeated, then using the Multiplication Principle we find that $26 \cdot 26 \cdot 10 \cdot 9 \cdot 8 \cdot 7 = 3,407,040$ different license plates can be made.

    c. If neither the letters nor the digits can be repeated, then using the Multiplication Principle we find that $26 \cdot 25 \cdot 10 \cdot 9 \cdot 8 \cdot 7 = 3,276,000$ different license plates can be made.

37. This problem involves two steps since two models have 3 body styles and one model has only two body styles. It is easiest to see using a tree diagram.

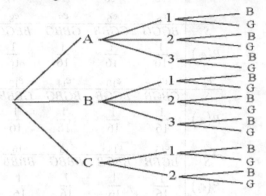

From the tree we see that there are 16 distinguishable car types.

39. By the Multiplication Principle, the contractor can build $5 \cdot 3 \cdot 4 = 60$ different types of homes.

41. By the Multiplication Principle, there can be $2^8 = 256$ different numbers (called bytes by computer scientists) possible.

43. Since there are 50 numbers on the lock and repetition is allowed, there are $50^3 = 125,000$ different lock combinations possible.

45. Think of the maze as if it were a tree diagram with each door of the maze as a branch on the tree. Then using the Multiplication Principle, we find that there are $4 \cdot 2 \cdot 1 = 8$ different paths from start to finish.

7.4 Sample Spaces and the Assignment of Probabilities

1. False. $P(e) \geq 0$ and the sum of all the probabilities assigned to the outcomes equals 1.

3. A <u>sample space</u> of an experiment is the set of all possibilities that can result from performing the experiment.

5. A(n) <u>event</u> is a subset of a sample space.

7. Let G stand for green and R stand for red.
$S = \{GA, GB, GC, RA, RB, RC\}$

9. $S = \{AA, AB, AC, BA, BB, BC, CA, CB, CC\}$

11. $S = \{$AA1, AA2, AA3, AA4, AB1, AB2, AB3, AB4, AC1, AC2, AC3, AC4, BA1, BA2, BA3, BA4, BB1, BB2, BB3, BB4, BC1, BC2, BC3, BC4, CA1, CA2, CA3, CA4, CB1, CB2, CB3, CB4, CC1, CC2, CC3, CC4$\}$

13. Let G stand for green and R stand for red.
$S = \{$GA1, GA2, GA3, GA4, GB1, GB2, GB3, GB4, GC1, GC2, GC3, GC4, RA1, RA2, RA3, RA4, RB1, RB2, RB3, RB4, RC1, RC2, RC3, RC4$\}$

15. Each time a coin is tossed there are 2 possible outcomes. So using the Multiplication Principle, when a coin is tossed 4 times there are $2 \cdot 2 \cdot 2 \cdot 2 = 2^4 = 16$ outcomes in the sample space.

17. Each time a die is tossed there are 6 possible outcomes. Using the Multiplication Principle there are $6 \cdot 6 \cdot 6 = 6^3 = 216$ outcomes in the sample space when 3 dice are tossed.

19. A regular deck contains 52 cards. There are 52 possibilities for the first outcome and 51 for the second outcome. So using the Multiplication Principle there are $52 \cdot 51 = 2652$ outcomes in the sample space.

21. There are 26 possible outcomes for the first letter chosen and 26 possible outcomes for the second letter chosen. Using the Multiplication Principle, there are $26 \cdot 26 = 26^2 = 676$ outcomes in the sample space.

23. Valid assignments must be nonnegative, and the sum of all probabilities in the sample space must equal one. Assignments A, B, C, and F are valid.

25. If the coin always comes up tails HH, HT, and TH are impossible events and have probabilities of 0. Assignment B should be used.

For exercises 35–43, S is the sample space and $n(S) = 23$.

27. Define event E: A white ball is picked. There are 3 ways E can occur, so $n(E) = 3$, and
$$P(E) = \frac{n(E)}{n(S)} = \frac{3}{23}.$$

29. Define event E: A green ball is picked. There are 7 ways E can occur, so $n(E) = 7$, and
$$P(E) = \frac{n(E)}{n(S)} = \frac{7}{23}.$$

31. Define event E: A white or a red ball is picked. There are 8 ways E can occur, so $n(E) = 8$, and
$$P(E) = \frac{n(E)}{n(S)} = \frac{8}{23}.$$

33. Define event E: A white or a blue ball is picked. There are 11 ways E can occur, so $n(E) = 11$, and $P(E) = \frac{n(E)}{n(S)} = \frac{11}{23}.$

For exercises 35–43, a regular deck of cards has 52 cards, so $n(S) = 52$.

35. Define event E: The ace of hearts is drawn. There is one ace of hearts, so $n(E) = 1$, and
$$P(E) = \frac{n(E)}{n(S)} = \frac{1}{52}.$$

37. Define event E: A spade is drawn. There are 13 spades, so $n(E) = 13$, and
$$P(E) = \frac{n(E)}{n(S)} = \frac{13}{52} = \frac{1}{4}.$$

39. Define event E: A picture card is drawn. There are 12 picture cards, so $n(E) = 12$, and
$$P(E) = \frac{n(E)}{n(S)} = \frac{12}{52} = \frac{3}{13}.$$

41. Define event E: A card with a number less than 6 is drawn. There are 5 numbers less than 6 and 4 of each number, so $n(E) = 20$, and
$$P(E) = \frac{n(E)}{n(S)} = \frac{20}{52} = \frac{5}{13}.$$

43. Define event E: A card that is not an ace is drawn. There are 48 cards that are not aces, so $n(E) = 48$, and $P(E) = \frac{n(E)}{n(S)} = \frac{48}{52} = \frac{12}{13}.$

45. Let F denote female and M denote male, and let D denote dry, O denote oil, and R denote regular. A sample space describing this experiment is $S = \{FD, FO, FR, MD, MO, MR\}$. There are six outcomes in the sample space. That is $n(S) = 6$.

47. a. Let G represent girl and B represent boy. We form sample space below. There are $n(S) = 16$ outcomes in the sample space. We assume the probability of a boy child equals the probability of a girl child. So we assign a probability of $\frac{1}{16}$ to each outcome.

e_i	e_1	e_2	e_3	e_4
S	GGGG	GGGB	GGBG	GBGG
$P(e_i)$	$\frac{1}{16}$	$\frac{1}{16}$	$\frac{1}{16}$	$\frac{1}{16}$

e_i	e_5	e_6	e_7	e_8
S	BGGG	GGBB	GBBG	BBGG
$P(e_i)$	$\frac{1}{16}$	$\frac{1}{16}$	$\frac{1}{16}$	$\frac{1}{16}$

e_i	e_9	e_{10}	e_{11}	e_{12}
S	GBGB	BGGB	BGBG	GBBB
$P(e_i)$	$\frac{1}{16}$	$\frac{1}{16}$	$\frac{1}{16}$	$\frac{1}{16}$

e_i	e_{13}	e_{14}	e_{15}	e_{16}
S	BGBB	BBGB	BBBG	BBBB
$P(e_i)$	$\frac{1}{16}$	$\frac{1}{16}$	$\frac{1}{16}$	$\frac{1}{16}$

b. i. If E is the event "the first two children are girls, then $E = \{GGGG, GGGB, GGBG, GGBB\}$, and $n(E) = 4$. So, the probability the first two children are girls is
$$P(E) = \frac{n(E)}{n(S)} = \frac{4}{16} = \frac{1}{4}.$$

ii. If F is the event all children are boys, then $F = \{BBBB\}$ and $n(F) = 1$. So the probability that all the children are boys is
$$P(F) = \frac{n(F)}{n(S)} = \frac{1}{16}.$$

iii. If G is the event at least one child is a girl then $n(G) = 15$ and the probability that at least one child is a girl is
$$P(G) = \frac{n(G)}{n(S)} = \frac{15}{16}.$$

iv. If H is the event the first child and the last child are girls, then $H = \{GGGG, GGBG, GBGG, GBBG\}$ and $n(G) = 4$. So the probability the first and last children are girls is $P(H) = \frac{n(H)}{n(S)} = \frac{4}{16} = \frac{1}{4}.$

49. To form the probability model, we assign a probability to each income level that is proportional to the number of families in the category. First we add the right hand column to find $n(S) = 117,063.8$.

$$P(<\$25,000) = \frac{28943.7}{117063.8} \approx 0.247$$

$$P(\$25,000-\$49,999) = \frac{29178.1}{117063.8} \approx 0.249$$

$$P(\$50,000-\$74,999) = \frac{20975.4}{117063.8} \approx 0.179$$

$$P(\$75,000-\$99,999) = \frac{13944.5}{117063.8} \approx 0.119$$

$$P(\geq\$100,000) = \frac{24022.1}{117063.8} \approx 0.205$$

51. $P(\$50,000-\$99,999)$

$$= \frac{20975.4 + 13944.5}{117063.8} \approx 0.298 \text{ or}$$

$P(\$50,000-\$99,999) = 0.179 + 0.119 = 0.298$

53. $P(<\$50,000) = 0.247 + 0.249 = 0.496$

For exercises 55–57,
$S = \{$Visa, MasterCard, American Express, Discover$\}$
and $n(S) = 270.1 + 203 + 48.9 + 54.4 = 576.4$.

55. $P(\text{MasterCard}) = \dfrac{203}{576.4} \approx 0.352$

57. $P(\text{American Express or Discover})$

$$= \frac{48.9}{576.4} + \frac{54.4}{576.4} \approx 0.179$$

59. There are 14 stocks have prices that are greater than \$50 per share. So, $P(>\$50) = \dfrac{14}{30} = \dfrac{7}{15}$.

61. There are two stocks with prices between \$40 and \$40 per share. So,

$$P(\text{price between \$40 and \$50}) = \frac{2}{30} = \frac{1}{15}.$$

For exercises 63–69, the sample space is the set of all homes that were sold.
$n(S) = 47 + 82 + 114 + 102 + 67 + 94 + 607 + 179$
$\qquad + 174 + 91 + 402 + 607 + 623 + 578 + 299$
$\qquad + 143 + 278 + 302 + 253 + 170 = 5212$

63. The event E = a home that was sold from Feb–Mar 2009.
$n(E) = 47 + 94 + 402 + 143 = 686$.

$$P(E) = \frac{686}{5212} \approx 0.132$$

65. The event E = a home that was sold from Oct–Dec 2009.
$n(E) = 102 + 174 + 578 + 253 = 1107$.

$$P(E) = \frac{1107}{5212} \approx 0.212$$

67. The event E = a home that was sold in the West from July–Sept 2009.
$n(E) = 302$.

$$P(E) = \frac{302}{5212} \approx 0.058$$

69. The event E = a home that was sold in the Northeast from Jan–Feb 2010.
$n(E) = 67$.

$$P(E) = \frac{67}{5212} \approx 0.013$$

71.a. $P(\text{insurance}) = \dfrac{255,143,000}{255,143,000 + 46,340,000}$
$\qquad\qquad\qquad \approx 0.846$

   b. $P(\text{no insurance}) = 1 - P(\text{insurance})$
$\qquad\qquad\qquad\qquad = 1 - 0.846 = 0.154$

73. Simulation: Answers will vary but should be close to the actual probabilities.
Actual Probabilities: $n(S) = 30$, $n(R) = 18$, and $n(W) = 12$.

$$P(R) = \frac{n(R)}{n(S)} = \frac{18}{30} = \frac{3}{5} = 0.6 \text{ and}$$

$$P(W) = \frac{n(W)}{n(S)} = \frac{12}{30} = \frac{2}{5} = 0.4.$$

75. Simulation: Answers will vary but should be close to the actual probabilities.
Actual Probabilities: $n(S) = 50$, $n(R) = 15$, and $n(W) = 35$.

$$P(R) = \frac{n(R)}{n(S)} = \frac{15}{50} = \frac{3}{10} = 0.3 \text{ and}$$

$$P(W) = \frac{n(W)}{n(S)} = \frac{35}{50} = \frac{7}{10} = 0.7.$$

77. Simulation: Answers will vary but should be close to the actual probabilities.
Actual Probabilities: $P(\text{Adam wins}) = 0.22$
$P(\text{ Beatrice wins}) = 0.60$
$P(\text{Cathy wins}) = 0.18$.

7.5 Properties of the Probability of an Event

1. A(n) <u>event</u> is a subset of a sample space.

3. False. If E is an event, then $P(E) = 1 - P(\overline{E})$.

5. $P(E \cup F) = P(E) + P(F) - P(E \cap F)$
$= 0.4 + 0.5 - 0.2 = 0.7$

7. $P(E \cup F) = P(E) + P(F) - P(E \cap F)$
$P(E \cap F) = P(E) + P(F) - P(E \cup F)$
$= 0.7 + 0.5 - 0.8 = 0.4$

9. $P(E \cup F) = P(E) + P(F) - P(E \cap F)$
$P(F) = P(E \cup F) + P(E \cap F) - P(E)$
$= 0.6 + 0.1 - 0.4 = 0.3$

11. $P(\overline{E}) = 1 - P(E) = 1 - 0.4 = 0.6$

13. $P(A) = 0.5$, $P(B) = 0.4$ and $P(A \cap B) = 0.2$

   a. $P(A \text{ or } B) = P(A) + P(B) - P(A \cap B)$
$= 0.5 + 0.4 - 0.2 = 0.7$

   b. $P(A \text{ but not } B) = P(A) - P(A \cap B)$
$= 0.5 - 0.2 = 0.3$

   c. $P(B \text{ but not } A) = P(B) - P(A \cap B)$
$= 0.4 - 0.2 = 0.2$

   d. $P(\text{neither } A \text{ nor } B) = P(\overline{A \cup B})$
$= 1 - P(A \cup B)$
$= 1 - 0.7 = 0.3$

15. $A \cap B = \varnothing$, $P(A) = 0.6$, and $P(B) = 0.2$

   a. $P(A \cap B) = 0$

   b. $P(A \cup B) = P(A) + P(B) = 0.6 + 0.2 = 0.8$

   c. $P(\overline{A \cup B}) = 1 - P(A \cup B) = 1 - 0.8 = 0.2$

   d. $P(\overline{B}) = 1 - P(B) = 1 - 0.2 = 0.8$

   e. $P(\overline{A}) = 1 - P(A) = 1 - 0.6 = 0.4$

   f. $P(\overline{A \cap B}) = 1 - P(A \cap B) = 1 - 0 = 1$

17. $P(E) = \dfrac{3}{3+1} = \dfrac{3}{4}$

19. $P(E) = \dfrac{5}{5+7} = \dfrac{5}{12}$

21. $P(E) = \dfrac{1}{1+1} = \dfrac{1}{2}$

23. $P(E) = 0.6$
$P(\overline{E}) = 1 - P(E) = 1 - 0.6 = 0.4$

Odds for E: $\dfrac{P(E)}{P(\overline{E})} = \dfrac{6}{4}$ or 3 to 2

Odds against E: $\dfrac{P(\overline{E})}{P(E)} = \dfrac{4}{6}$ or 2 to 3

25. $P(F) = \dfrac{3}{4}$

$P(\overline{F}) = 1 - P(F) = 1 - \dfrac{3}{4} = \dfrac{1}{4}$

Odds for F: $\dfrac{P(F)}{P(\overline{F})} = \dfrac{\frac{3}{4}}{\frac{1}{4}} = \dfrac{3}{1}$ or 3 to 1

Odds against F: $\dfrac{P(\overline{F})}{P(F)} = \dfrac{\frac{1}{4}}{\frac{3}{4}} = \dfrac{1}{3}$ or 1 to 3

27. Let E and F be the events.
E: The Bears win; F: The Bears tie.
E and F are mutually exclusive, so the probability that the Bears either win or tie is
$P(E \cup F) = P(E) + P(F) = 0.65 + 0.05 = 0.70$.
The probability that the Bears lose is the complement of their winning or tying.
$P(\overline{E \cup F}) = 1 - P(E \cup F) = 1 - 0.70 = 0.30$.

29. Let M and E be the events Anne passes mathematics and Anne passes English respectively. We are given $P(M) = 0.4$, $P(E) = 0.6$, and $P(M \cup E) = 0.8$. The probability that Anne passes both courses is
$P(M \cap E) = P(M) + P(E) - P(M \cup E)$
$= 0.4 + 0.6 - 0.8 = 0.2$

31. Let T be the event that the car needs a tune-up and let B be the event that the car needs a brake job. $P(T) = 0.6$, $P(B) = 0.1$, and $P(T \cap B) = 0.02$.

   a. The probability the car needs either a tune-up or a brake job is
$P(T \cup B) = P(T) + P(B) - P(T \cap B)$
$= 0.6 + 0.1 - 0.02 = 0.68$

   b. The probability that a car requires a tune-up but not a brake job is
$P(T \cap \overline{B}) = P(T) - P(T \cap B)$
$= 0.6 - 0.02 = 0.58$

c. The probability that a car requires neither type of repair is
$$P(\overline{T \cup B}) = 1 - P(T \cup B) = 1 - 0.68 = 0.32.$$

33. a. $P(1 \text{ or } 2 \text{ TVs}) = P(1 \text{ TV}) + P(2 \text{ TVs})$
$$= 0.24 + 0.33 = 0.57$$

b. $P(1 \text{ or more TVs}) = 1 - P(0 \text{ TVs})$
$$= 1 - 0.05 = 0.95$$

c. $P(3 \text{ or fewer TVs}) = 1 - P(4 \text{ TVs})$
$$= 1 - 0.17 = 0.83$$

d. $P(3 \text{ or more TVs}) = P(3 \text{ TVs}) + P(4 \text{ TVs})$
$$= 0.21 + 0.17 = 0.38$$

e. $P(\text{fewer than 2 TVs}) = P(0 \text{ TVs}) + P(1 \text{ TV})$
$$= 0.05 - 0.24 = 0.29$$

f. $P(\text{not even 1 TV}) = P(0 \text{ TVs}) = 0.05$

g. $P(1, 2, \text{ or } 3 \text{ TVs})$
$$= P(1 \text{ TV}) + P(2 \text{ TVs}) + P(3 \text{ TVs})$$
$$= 0.24 + 0.33 + 0.21 = 0.78$$

h. $P(2 \text{ or more TVs})$
$$= P(2 \text{ TVs}) + P(3 \text{ TVs}) + P(4 \text{ TVs})$$
$$= 0.33 + 0.21 + 0.17 = 0.71$$

35. $P(\text{RH}^+) = P(\text{O}^+) + P(\text{A}^+) + P(\text{B}^+) + P(\text{AB}^+)$
$$= 0.39 + 0.31 + 0.09 + 0.03 = 0.82$$

37. $P(\text{A}) = P(\text{A}^+) + P(\text{A}^-) + P(\text{AB}^+) + P(\text{AB}^-)$
$$= 0.31 + 0.06 + 0.03 + 0.01 = 0.41$$

39. From exercise 35, we have $P(\text{RH}^+) = 0.82$.

$$P(\text{O or Rh}^+) = P(\text{O}) + P(\text{Rh}^+) - P(\text{O} \cap \text{Rh}^+)$$
$$= 0.39 + 0.09 + 0.82 - 0.39$$
$$= 0.91$$

41. Define E to be the event, a woman giving birth in 2006 gave birth to her first child.
$$P(E) = 0.457$$
$$P(\overline{E}) = 1 - P(E) = 1 - 0.457 = 0.543$$

43. If 150,000 were rated deficient or obsolete, then $603,000 - 150,000 = 453,000$ were rated as neither deficient nor obsolete. The odds that a randomly selected bridge is either deficient or obsolete is $\dfrac{150,000}{453,000} = \dfrac{150}{453}$ or 150 to 453.

45. Let E define the event, a smartphone owner has an iPhone. If $P(E) = 0.144$, then the probability a smartphone owner does not have an iPhone is
$$P(\overline{E}) = 1 - P(E) = 1 - 0.144 = 0.856 \text{ or } 85.6\%.$$

47. Assuming every type of patent is mutually exclusive with each of the other types, the probability that a randomly selected patent was either for a design or a botanical plant is

$P(\text{design or botanical plant})$
$$= P(\text{design}) + P(\text{botanical plant}).$$
$P(\text{design} \cup \text{botanical plant})$
$$= \frac{n(\text{design})}{n(\text{patents})} + \frac{n(\text{botanical plants})}{n(\text{patents})}$$
$$= \frac{23,915}{190,621} + \frac{1096}{190,621} = \frac{25,011}{190,621} \approx 0.1312$$

49. The probability the fund will outperform the market in at least one of the next two years is the probability it outperforms the market in the first year plus the probability it outperforms the market in the second year minus the probability it outperforms the market in both years.
$P(\text{fund outperforms in at least one of the next 2 years}) = 0.12 + 0.12 - 0.05 = 0.19.$

51. If the odds A wins are 1 to 2, the probability A wins is $P(A) = \dfrac{1}{3}$. If the odds B wins are 2 to 3, then the probability B wins is $P(B) = \dfrac{2}{5}$. Since only one person can win the race the events winning are mutually exclusive and
$$P(A \text{ or } B \text{ wins}) = P(A \cup B) = P(A) + P(B)$$
$$= \frac{1}{3} + \frac{2}{5} = \frac{11}{15}$$

53. The number of elements in the sample space, S, is found by using the Multiplication Principle. Since there are 7 digits in the phone number, and there are no restrictions, $n(S) = 10^7$. Define event E as a phone number has one or more repeated digits.. It easier to do this problem by looking at the complement of E. $\overline{E}$ is a phone number has no repeated digits.

$$n(\overline{E}) = P(10, 7).$$

$$P(E) = 1 - P(\overline{E}) = 1 - \frac{P(10, \ 7)}{10^7}$$
$$= 1 - \frac{604,800}{10,000,000} \approx 0.940$$

55. The number of elements in the sample space, S, is found by using the Multiplication Principle. Since there are 26 letters in the alphabet and we are going to select 5, allowing repetitions, $n(S) = 26^5$. Define event E as no letters are repeated. $n(E) = C(26, 5) = 65,780$.

$$P(E) = \frac{n(E)}{n(S)} = \frac{C(26, 5)}{26^5} \approx 0.0055$$

57. The sample space consists of the 60 numbers. $n(S) = 60$. Define the event E as at least two students choose the same number. The easiest way to approach this problem is to consider its complement: no students choose the same number.

$$P(\overline{E}) = \frac{P(60, 30)}{60^{30}} \approx 0, \text{ so } P(E) \approx 1.$$

59. The number of elements in the sample space, S, is found by using the Multiplication Principle. Since there are 12 months in the year, and we are going to select 3, $n(S) = 12^3 = 1728$. Define event E: At least 2 were born in the same month. It is easier to do this problem by looking at the complement of E, $\overline{E}$: no two people were born in the same month.

$n(\overline{E}) = P(12, 3) = 12 \cdot 11 \cdot 10 = 1320$
The probability that at least 2 were born in the same month is

$$P(E) = 1 - P(\overline{E}) = 1 - \frac{n(\overline{E})}{n(S)} = 1 - \frac{1320}{1728} = 0.236.$$

61. The number of elements in the sample space, S, is found by using the Multiplication Principle. Since there are 365 days in a year, and 100 senators, $n(S) = 365^{100}$. Define event E as at least two senators have the same birthday. It is easier to do this problem by looking at the complement of E, $\overline{E}$: no two senators have the same birthday.

$n(\overline{E}) = P(365, 100) = 365 \cdot 364 \cdot 363 \cdot \ldots \cdot 266$
The probability that at least 2 senators have the same birthday is

$$P(E) = 1 - P(\overline{E}) = 1 - \frac{n(\overline{E})}{n(S)}$$

$$= 1 - \frac{P(365, 100)}{365^{100}} = 0.99999 \approx 1$$

63. Prove $P(E \cup F) = P(E) + P(F) - P(E \cap F)$.
From example 9 in section 7.1, we have
$E \cup F = \left(E \cap \overline{F}\right) \cup (E \cap F) \cup \left(\overline{E} \cap F\right)$. Then,

$$P(E \cup F) = P\left[\left(E \cap \overline{F}\right) \cup (E \cap F) \cup \left(\overline{E} \cap F\right)\right].$$

Since the sets are disjoint, we have
$P(E \cup F)$

$$= P\left(E \cap \overline{F}\right) + P(E \cap F) + P\left(\overline{E} \cap F\right). \quad (1)$$

Now write E and F as the union of disjoint sets using the hint given in the exercise.

$E = (E \cap F) \cup \left(E \cap \overline{F}\right)$ and
$F = (E \cap F) \cup \left(\overline{E} \cap F\right)$.

Since $E \cap F$ and $E \cap \overline{F}$ are disjoint, and $E \cap F$ and $\overline{E} \cap F$ are disjoint, we have

$P(E) = P(E \cap F) + P\left(E \cap \overline{F}\right) \Rightarrow$
$P\left(E \cap \overline{F}\right) = P(E) - P(E \cap F)$ and
$P(F) = P(E \cap F) + P\left(\overline{E} \cap F\right) \Rightarrow$
$P\left(\overline{E} \cap F\right) = P(F) - P(E \cap F)$.

Substituting these equations into (1), we have
$P(E \cup F)$

$$= \left[P(E) - P(E \cap F)\right] + P(E \cap F)$$
$$+ \left[P(F) - P(E \cap F)\right]$$
$$= P(E) + P(F) - P(E \cap F)$$

65. Prove
$P(A \cup B \cup C)$
$= P(A) + P(B) + P(C) - P(A \cap B) - P(A \cap C)$
$- P(B \cap C) + P(A \cap B \cap C)$.

Let $E = A \cup B$. The addition rule states
$P(E \cup C) = P(E) + P(C) - P(E \cap C)$.
Substitute $A \cup B$ for E to obtain
$P\left[(A \cup B) \cup C\right]$
$= P(A \cup B) + P(C) - P\left[(A \cup B) \cap C\right]$.

Apply the addition rule to $P(A \cup B)$.
$P\left[(A \cup B) \cup C\right]$
$= \left[P(A) + P(B) - P(A \cap B)\right]$
$+ P(C) - P\left[(A \cup B) \cap C\right]$.

Now apply the distributive property to the sets in the last bracket.
$P\left[(A \cup B) \cup C\right]$
$= \left[P(A) + P(B) - P(A \cap B)\right] + P(C)$
$- P\left[(A \cap C) \cup (B \cap C)\right]$.

Apply the addition rule to the union of sets in the last bracket.
$P\left[(A \cup B) \cup C\right]$
$= \left[P(A) + P(B) - P(A \cap B)\right] + P(C)$
$- \left[P(A \cap C) + P(B \cap C) - P(A \cap B \cap C)\right]$.

(continued on next page)

(continued)

Remove the brackets and rearrange terms.

$$P(A \cup B \cup C)$$
$$= P(A) + P(B) + P(C) - P(A \cap B) - P(A \cap C)$$
$$- P(B \cap C) + P(A \cap B \cap C).$$

67. The number of elements in the sample space, S, is found by using the Multiplication Principle. Since there are 60 numbers, and we are going to select 2, $n(S) = 60^2 = 3600$. Define event E as at least 2 will choose the same number. It is easier to do this problem by looking at the complement of E, $\overline{E}$: no two people will choose the same number.

$$n(\overline{E}) = P(60, 2) = 60 \cdot 59 = 3540$$

The probability that at least two will choose the same number is

$$P(E) = 1 - P(\overline{E}) = 1 - \frac{n(\overline{E})}{n(S)} = 1 - \frac{3540}{3600} \approx 0.017.$$

7.6 Expected Value

1. True

3. $E = 2(0.4) + 3(0.2) + (-2)(0.1) + 0(0.3) = 1.2$

5. $E = 30,000(0.08) + 40,000(0.42)$
$$+ 60,000(0.42) + 80,000(0.08)$$
$$= 50,800$$

50,800 fans are expected to attend the game.

7. $E = 8(0.1) + 0(0.90) = 0.8$

Mary should pay \$0.80 for one draw.

9. $P(\text{double when throwing 2 dice}) = \dfrac{1}{6}$

$$E = (12)\left(\frac{1}{6}\right) + (0)\left(\frac{5}{6}\right) = 2$$

David should pay \$2 for a throw.

11. $E = 100(0.001) + 50(0.003) + 0(0.996) = 0.25$
The price of a ticket exceeds the expected value by \$0.75.

13. $P(3 \text{ tails}) = \dfrac{1}{8}$, $P(2 \text{ tails}) = \dfrac{3}{8}$, $P(1 \text{ tail}) = \dfrac{3}{8}$,

$$P(0 \text{ tails}) = \frac{1}{8}$$

a. $E = (3)\left(\dfrac{1}{8}\right) + (2)\left(\dfrac{3}{8}\right) + (0)\left(\dfrac{3}{8}\right) + (-3)\left(\dfrac{1}{8}\right)$
$$= 0.75$$
The expected value of the game is \$0.75.

b. The game is not fair. Fair games have an expected value of 0.

c. Let x represent the payoff for tossing 1 tail.

$$x\left(\frac{3}{8}\right) = -\frac{6}{8} \Rightarrow x = -2$$

To make the game fair the player should lose \$2.00 if one tail is thrown.

15. $P(\text{team A wins}) = \dfrac{9}{14}$, $P(\text{team B wins}) = \dfrac{5}{14}$

If team A wins you lose \$4; if team B wins you win \$6. The expected value of the game is

$$E = (-4)\left(\frac{9}{14}\right) + (6)\left(\frac{5}{14}\right) \approx -0.4286.$$

The bet is not fair to you. You should expect to lose 43 cents.

17. $P(\text{selecting a heart other than the ace}) = \dfrac{12}{52}$,

$P(\text{selecting an ace other than the heart}) = \dfrac{3}{52}$,

$P(\text{selecting the ace of hearts}) = \dfrac{1}{52}$

$$E = (40)\left(\frac{12}{52}\right) + (50)\left(\frac{3}{52}\right) + (90)\left(\frac{1}{52}\right)$$
$$\approx 13.8462$$

Sarah's expected winnings are $13.8 - 15 = -1.2$ cents, so she should not play the game.

19. First we find the distribution of the corporations by size.
$$954,123 + 650,138 + 707,229 + 545,457$$
$$+ 443,573 = 3,300,520$$

Size	Number	Proportion
<\$25,000	954,123	$\dfrac{954,123}{3,000,520}$
\$25,000-\$99,999	650,138	$\dfrac{650,138}{3,000,520}$
\$100,000-\$249,000	707,229	$\dfrac{707,229}{3,000,520}$
\$250,000-\$499,000	545,457	$\dfrac{545,457}{3,000,520}$
>\$500,000	443,573	$\dfrac{443,573}{3,000,520}$
Total	3,000,520	

$$E = \frac{954,123}{3,000,520}(12,500) + \frac{650,138}{3,000,520}(62,500)$$
$$+ \frac{707,229}{3,000,520}(175,000) + \frac{545,457}{3,000,520}(375,000)$$
$$+ \frac{443,573}{3,000,520}(750,000) \approx 216,194$$

The expected receipts of the chosen corporation are \$216,194.

21. First we find the distribution of the number of houses sold.

$$66 + 96 + 114 + 46 + 23 + 20 = 365$$

Size	Number	Proportion
\$0-\$149,999	66	$\frac{66}{365}$
\$150,000-\$199,999	96	$\frac{96}{365}$
\$200,000-\$299,999	114	$\frac{114}{365}$
\$300,000-\$399,999	46	$\frac{46}{365}$
\$400,000-\$499,999	23	$\frac{24}{365}$
\$500,000-\$749,999	20	$\frac{20}{365}$
Total	365	

$$E = \frac{66}{365}(75,000) + \frac{96}{365}(175,000)$$
$$+ \frac{114}{365}(250,000) + \frac{46}{365}(350,000)$$
$$+ \frac{23}{365}(450,000) + \frac{20}{365}(625,000)$$
$$\approx 244,384$$

The expected sale price of a new house is about \$244,384.

23.a. $E = 7(0.10) + 8(0.20) + 9(0.40)$
$$+ 10(0.20) + 11(0.10) = 9$$

The expected number of customers is 9.

b. To decide the optimal number of trucks to have on hand, we need to find the expected profit for each possible number of trucks.

Profit = Revenue − Cost

Expected profit is the product of the profit and the probability of obtaining it.

If 7 trucks are on hand, all will be rented, and the expected profit will be
$E(\text{profit}) = (7)(90 - 20) = \$490$

If 8 trucks are on hand, the expected profit is given as follows:
For 7 customers, the revenue generated is $7(90) = 630$ and the cost is $8(20) = 160$ (since one truck is not rented), for a profit of \$470.
For 8 customers, the revenue generated is $8(90) = 720$ and the cost is $8(20) = 160$, for a profit of \$560.
For 9, 10, or 11 customers, the profit is also \$560 since all the trucks are rented.
Thus, the expected profit is
$E = 0.10(490) + 0.20(560) + 0.40(560)$
$$+ 0.20(560) + 0.10(560) = \$553$$

If 9 trucks are on hand, the expected profit is given as follows:
For 7 customers, the revenue generated is $7(90) = 630$ and the cost is $9(20) = 180$ (since two trucks are not rented), for a profit of \$450.
For 8 customers, the revenue generated is $8(90) = 720$ and the cost is $9(20) = 180$ (since one truck is not rented), for a profit of \$540.
For 9, 10, or 11 customers, the profit is $9(90 - 20) = \$630$. Thus, the expected profit is
$E = 0.10(450) + 0.20(540) + 0.40(630)$
$$+ 0.20(630) + 0.10(630) = \$594$$

If 10 trucks are on hand, the expected profit is given as follows:
For 7 customers, the revenue generated is $7(90) = 630$ and the cost is $10(20) = 200$ (since three trucks are not rented), for a profit of \$430.
For 8 customers, the revenue generated is $8(90) = 720$ and the cost is $10(20) = 200$ (since two trucks are not rented), for a profit of \$520.
For 9 customers, the revenue generated is $9(90) = 810$ and the cost is $10(20) = 200$ (since one truck is not rented), for a profit of \$610. For 10 or 11 customers, the profit is $10(90 - 70) = 700$ Thus, the expected profit is
$E = 0.10(430) + 0.20(520) + 0.40(610)$
$$+ 0.20(700) + 0.10(700) = \$601$$

If 11 trucks are on hand, the expected profit is given as follows:
For 7 customers, the revenue generated is $7(90) = 630$ and the cost is $11(20) = 220$, for a profit of \$410.
For 8 customers, the revenue generated is $8(90) = 720$ and the cost is $11(20) = 220$, for a profit of \$500.
For 9 customers, the revenue generated is $9(90) = 810$ and the cost is $11(20) = 220$, for a profit of \$590.
For 10 customers, the revenue generated is $10(90) = 900$ and the cost is $11(20) = 220$, for a profit of \$680.
For 11 customers, the revenue generated is $11(90) = 990$ and the cost is $11(20) = 220$, for a profit of \$770. Thus, the expected profit is
$E = 0.10(410) + 0.20(500) + 0.40(590)$
$$+ 0.20(680) + 0.10(770) = \$590$$

Thus, to maximize the profit, the company should have 10 trucks on hand each day.

c. The largest expected profit is \$601.

25. a. If the person dies the insurance company has a loss of $250,000 - $450 = $249,550. If the person survives the company has a gain of $450.
$$E = 450(0.9986) - 249,550(1 - 0.9986) = 100$$
The insurance company can expect a profit of $100.

b. To have a profit of $250 per policy, the company's expected value must be 250. Let x be the premium necessary to obtain an expected value of 250.
$$x(0.9986) - (250,000 - x)(1 - 0.9986) = 250$$
$$0.9986x - 250,000$$
$$+249,605 + x - 0.9986x = 250$$
$$-350 + x = 250$$
$$x = 600$$
The company should set the premium at $600.

27. a. We compare the two expected profits.
$$P(\text{success}) = \frac{1}{2}$$
Location 1:
$$E = (15,000)\left(\frac{1}{2}\right) + (-3000)\left(\frac{1}{2}\right) = 6000$$
The first location will provide an expected profit of $6000.
Location 2:
$$E = (20,000)\left(\frac{1}{2}\right) + (-6000)\left(\frac{1}{2}\right) = 7000$$
The second location will provide an expected profit of $7000.
The management should choose location 2 since it has a higher expected profit.

b. We compare the two expected profits.
Location 1:
$$P(\text{success}) = \frac{2}{3}$$
$$E = (15,000)\left(\frac{2}{3}\right) + (-3000)\left(\frac{1}{3}\right) = 9000$$
The first location will provide an expected profit of $9000.
Location 2:
$$P(\text{success}) = \frac{1}{3}$$
$$E = (20,000)\left(\frac{1}{3}\right) + (-6000)\left(\frac{2}{3}\right) = 2666.67$$
The second location will provide an expected profit of $2666.67.
The management should choose location 1 since it has a higher expected profit.

29. a. If each of the two stocks have equal weight then $w_1 = w_2 = 0.5$. The expected return on the investor's portfolio is
$$E = 0.5 \cdot 0.10 + 0.5 \cdot 0.15 = 0.125.$$
The investor expects a return of 12.5%.

b. To obtain a return of 14%, we need $E = 0.14$. Let x represent w_1 and let $1 - x$ represent w_2.
$$E = x \cdot 0.10 + (1 - x) \cdot 0.15 = 0.14$$
$$0.1x + 0.15 - 0.15x = 0.14$$
$$-0.05x = -0.01$$
$$x = 0.2$$
To realize a return of 14%, the investor's portfolio should be split with 20% Wal-Mart stock and 80% Viacom stock.

Chapter 7   Review Exercises

1. $\varnothing \subset \{0\},\ \varnothing \subseteq \{0\}$

3. none of these

5. none of these

7. $\{5\} \subset \{0,\ 5\},\ \{5\} \subseteq \{0,\ 5\}$

9. $\varnothing \subset \big[\{1,\ 2\} \cap \{3,\ 4,\ 5\}\big],$
$\varnothing = \big[\{1,\ 2\} \cap \{3,\ 4,\ 5\}\big]$

11. $\{1,\ 2\} \subset \big[\{1\} \cup \{2\}\big],\ \{1,\ 2\} \subseteq \big[\{1\} \cup \{2\}\big]$

13. $\big[\{1,\ 5\} \cap \{5,\ 6\}\big] \subset \{4,\ 5\},$
$\big[\{1,\ 5\} \cap \{5,\ 6\}\big] \subseteq \{4,\ 5\}$

15. $\big[\{6,\ 7,\ 8\} \cap \{6\}\big] \subseteq \{6\},$
$\big[\{6,\ 7,\ 8\} \cap \{6\}\big] = \{6\}$

17. a. $(A \cap B) \cup C = \{3,\ 6\} \cup \{6,\ 8,\ 9\}$
$= \{3,\ 6,\ 8,\ 9\}$

b. $(A \cap B) \cap C = \{3,\ 6\} \cap \{6,\ 8,\ 9\} = \{6\}$

c. $(A \cup B) \cap B = \{1,2,3,5,6,7,8\} \cap \{2,3,6,7\}$
$= \{2,3,6,7\} = B$

d. $B \cup \varnothing = B$

e. $A \cap \varnothing = \varnothing$

f. $(A \cup B) \cup C = \{1,\ 2,\ 3,\ 5,\ 6,\ 7,\ 8\} \cup \{6,\ 8,\ 9\}$
$= \{1,\ 2,\ 3,\ 5,\ 6,\ 7,\ 8,\ 9\}$

19. a.

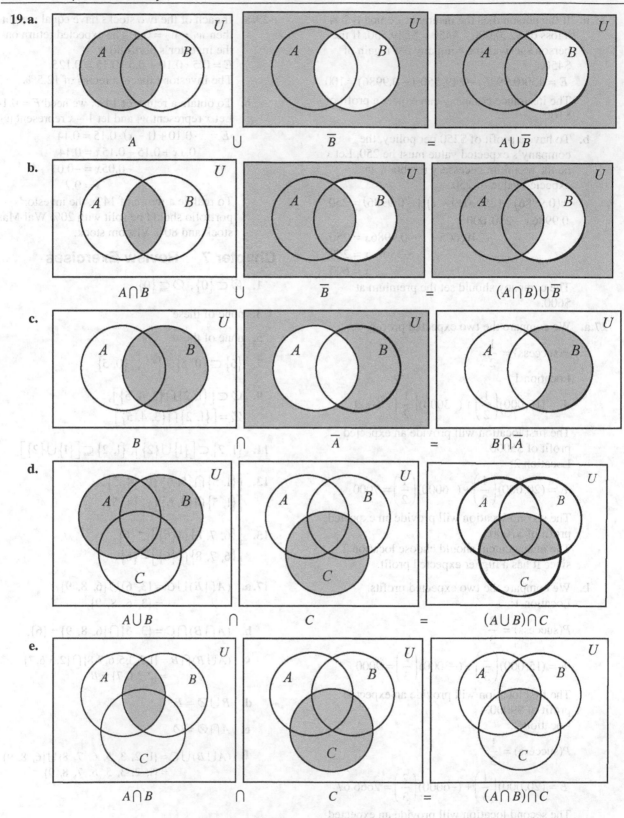

$$A \quad \cup \quad \overline{B} \quad = \quad A \cup \overline{B}$$

b.

$$A \cap B \quad \cup \quad \overline{B} \quad = \quad (A \cap B) \cup \overline{B}$$

c.

$$B \quad \cap \quad \overline{A} \quad = \quad B \cap \overline{A}$$

d.

$$A \cup B \quad \cap \quad C \quad = \quad (A \cup B) \cap C$$

e.

$$A \cap B \quad \cap \quad C \quad = \quad (A \cap B) \cap C$$

f.

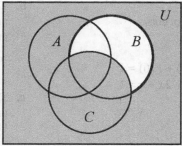

$\overline{B}$ $\overline{B} \cup C$ $\overline{(B \cup C)}$

21. $A \cap B = \varnothing$

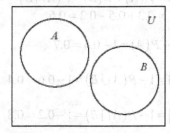

23. $A \cup V$ is the set of all states whose names begin with an A or end with a vowel.

25. $V \cap E$ is the set of all states whose names end with a vowel and which lie east of the Mississippi River.

27. $(A \cup V) \cap E$ is the set of all states whose names start with an A or end with a vowel and which lie east of the Mississippi.

29. $n(A \cup B) = n(A) + n(B) - n(A \cap B)$, so
$$n(A \cap B) = n(A) + n(B) - n(A \cup B)$$
$$= 24 + 12 - 33 = 3$$

31.a. $n(A \cup B) = n(A) + n(B) - n(A \cap B)$
$$= 3 + 17 - 0 = 20$$

b. Sets A and B are disjoint since their intersection has no elements.

33. $A = \{2, 4, 6\}$; $B = \{1, 2, 3\}$
$A \cap B = \{2\}$; $n(A \cap B) = 1$
$A \cup B = \{1, 2, 3, 4, 6\}$; $n(A \cup B) = 5$

35. We denote the sets of cars with heated seats, GPS, and a sun roof by S, G, and R respectively. From the data given we determine that
$$n(S) = 75 \qquad n(R) = 100$$
$$n(G) = 95 \quad n(G \cap \overline{R} \cap \overline{S}) = 40$$
$$n(S \cap R \cap G) = 20 \quad n(\overline{S} \cap P \cap G) = 10$$

$n(S \cap \overline{R} \cap \overline{G}) = 10$
$$n(R \cap G) = 50 \quad n(S \cap R) = 60$$
We next place the values on a Venn Diagram, so that we can answer the questions.

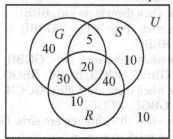

a. There were
$10 + 5 + 20 + 40 + 10 + 30 + 40 + 10 = 165$
cars sold in June.

b. 40 cars had only a GPS.

37. If $U = \{1, 2, 3, 4, 5\}$, $B = \{1, 4, 5\}$, and $A \cap B = \{1\}$, then A could equal any of the following sets: $\{1\}$, $\{1, 2\}$, $\{1, 3\}$, or $\{1, 2, 3\}$.

39. $S = \{0, 1, 2, 3, 4, 5\}$

41. $S = \{BB, BG, GB, GG\}$

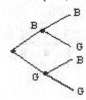

43. Let P represent penny, Q represent quarter, and D represent dime. $S = \{P, Q, D\}$
$$P(P) = \frac{4}{15} \qquad P(Q) = \frac{2}{5} \qquad P(D) = \frac{1}{3}$$

45. $S = \{1, 2, 3, 4, 5, 6\}$
Let x denote the probability a 1 occurs.
$P(1) = P(3) = P(4) = P(6) = x$
$P(2) = P(5) = 2x$
$P(1) + P(2) + P(3) + P(4) + P(5) + P(6) = 1$
$$x + 2x + x + x + 2x + x = 1$$
$$x = \frac{1}{8}$$

Thus, $P(1) = P(3) = P(4) = P(6) = \dfrac{1}{8}$ and

$$P(2) = P(5) = \frac{2}{8} = \frac{1}{4}.$$

47.a. The sample space is the number of girls in the family. $S = \{0, 1, 2, 3, 4\}$.
The simple event
0 occurs when there is no girl, BBBB.
1 occurs when there is 1 girl, GBBB, BGBB, BBGB, BBBG.
2 occurs when there are 2 girls, GGBB, GBGB, GBBG, BGGB, BGBG, BBGG.
3 occurs when there are 3 girls, GGGB, GGBG, GBGG, BGGG.
4 occurs when all 4 children are girls, GGGG.
We assign valid probabilities to each of these events, assuming it is equally likely for a child to be born G or B.

$$P(0) = \frac{1}{16}, \; P(1) = \frac{4}{16} = \frac{1}{4}, \; P(2) = \frac{6}{16} = \frac{3}{8},$$
$$P(3) = \frac{4}{16} = \frac{1}{4}, \; P(4) = \frac{1}{16}$$

b. i. $P(0) = \dfrac{1}{16}$

ii. $P(2) = \dfrac{3}{8}$

iii. $P(1) + P(2) + P(3) = \dfrac{1}{4} + \dfrac{3}{8} + \dfrac{1}{4} = \dfrac{7}{8}$

iv. $1 - P(4) = 1 - \dfrac{1}{16} = \dfrac{15}{16}$

49. Let W denote a white marble is chosen, Y denote a yellow marble is chosen, R denote a red marble is chosen, and B denote a blue marble is chosen.

a. $P(BB) = P(B \text{ on 1st}) \cdot P(B \text{ on 2nd})$
$$= \left(\frac{5}{14}\right)\left(\frac{4}{13}\right) = \frac{10}{91}$$

b. The probability exactly one is blue is the union of two mutually exclusive events. The probability is

$$P(B\overline{B} \cup \overline{B}B) = \left(\frac{5}{14}\right)\left(\frac{9}{13}\right) + \left(\frac{9}{14}\right)\left(\frac{5}{13}\right) = \frac{45}{91}.$$

c. The probability at least one is blue is the union of both are blue and exactly one is blue. Using the results from parts a and b,
$$P(\text{at least one is blue}) = \frac{10}{91} + \frac{45}{91} = \frac{55}{91}.$$

51.a. $P(A \cup B) = P(A) + P(B) - P(A \cap B)$
$$= 0.3 + 0.5 - 0.2 = 0.6$$

b. $P(\overline{A}) = 1 - P(A) = 1 - 0.3 = 0.7$

c. $P(\overline{A \cup B}) = 1 - P(A \cup B) = 1 - 0.6 = 0.4$

d. $P(\overline{A \cup B}) = 1 - P(A \cap B) = 1 - 0.2 = 0.8$

53.a. $P(\text{head}) = \dfrac{243}{300} = \dfrac{81}{100} = 0.81$

b. $P(\text{tail}) = \dfrac{57}{300} = \dfrac{19}{100} = 0.19$

55.a. $P(\overline{F}) = 1 - P(F) = 1 - 0.5 = 0.5$

b. $P(E \cup F) = P(E) + P(F) - P(E \cap F)$
$$= 0.63 + 0.50 - 0.35 = 0.78$$

c. E and F are not mutually exclusive. If they were, $P(E \cap F)$ would equal 0, but in this problem we are told that $P(E \cap F) = 0.35$.

57.a. $P(E \cup F) = P(E) + P(F) - P(E \cap F)$
$$= 0.2 + 0.6 - 0.1 = 0.7$$

b. $P(\overline{E}) = 1 - P(E) = 1 - 0.2 = 0.8$

c. $P(\overline{E \cap F}) = 1 - P(E \cap F) = 1 - 0.1 = 0.9$

59.a. $P(\overline{E}) = 1 - P(E) = 1 - 0.25 = 0.75$

b. $P(\overline{F}) = 1 - P(F) = 1 - 0.3 = 0.7$

c. $P(E \cap F) = P(E) + P(F) - P(E \cup F)$
$$= 0.25 + 0.3 - 0.55 = 0$$

d. $P(\overline{E \cap F}) = 1 - P(E \cap F) = 1 - 0 = 1$

e. $P(\overline{E} \cap \overline{F}) = P(\overline{E \cup F}) = 1 - P(E \cup F)$
$$= 1 - 0.55 = 0.45$$

f. $P(\overline{E} \cup \overline{F}) = P(\overline{E \cap F}) = 1 - P(E \cap F)$
$$= 1 - 0 = 1$$

61. a. No, the events are not equally likely.

b. The outcome 0, 0, 0 has the highest probability.

c. $P(F) = P(3, 3) \cdot \left(\dfrac{4}{8}\right) \cdot \left(\dfrac{3}{8}\right) \cdot \left(\dfrac{1}{8}\right)$
$$= (3 \cdot 2 \cdot 1) \cdot \left(\dfrac{4}{8}\right) \cdot \left(\dfrac{3}{8}\right) \cdot \left(\dfrac{1}{8}\right) = \dfrac{9}{64}$$

63. To count the number of different house styles that are available we use the Multiplication Principle. There are $3 \cdot 4 \cdot 6 = 72$ different styles of houses possible.

65. Each question on a true-false test has 2 possible answers, and answers can be used more than once. So, on a 10 question test, there are $2^{10} = 1024$ different ways the test can be answered.

67. Define E as the event, "Each person was born on a different day of the month."
$n(S) = 31^{15}$ and $n(E) = P(31, 15)$, so
$$P(E) = \dfrac{P(31, 15)}{31^{15}} = 0.017$$
The probability that each person in a room of 15 people have birthdays on different days of the month is 0.017.

69. Define the event E: A family has 4 boys.
$$P(E) = \left(\dfrac{1}{2}\right)^4 = \dfrac{1}{16}$$
The odds in favor of E are defined as the ratio $\dfrac{P(E)}{P(\overline{E})}$. The odds in favor of a family with 4 children to have 4 boys is $\dfrac{1}{15}$ or 1 to 15.

71. The odds for the Giants winning are 5:3, so the probability that the Giants will win is $\dfrac{5}{5+3} = \dfrac{5}{8}$ and the probability that the Giants will lose is $\dfrac{3}{5+3} = \dfrac{3}{8}$.

73. Let the random variable X denote the number of red balls chosen. $X = 0, 1,$ or 2.
$$P(X = 0) = \dfrac{C(2, 0) \cdot C(4, 2)}{C(6, 2)} = \dfrac{6}{15}$$
$$P(X = 1) = \dfrac{C(2, 1) \cdot C(4, 1)}{C(6, 2)} = \dfrac{8}{15}$$
$$P(X = 2) = \dfrac{C(2, 2) \cdot C(4, 0)}{C(6, 2)} = \dfrac{1}{15}$$
The expected value of the game is
$$E = 0\big(P(X = 0)\big) + 1\big(P(X = 1)\big) + 2\big(P(X = 2)\big)$$
$$= 0\left(\dfrac{6}{15}\right) + 1\left(\dfrac{8}{15}\right) + 2\left(\dfrac{1}{15}\right) = \dfrac{10}{15} \approx \$0.6667$$
The expected value of the game is 66.67 cents. Frank paid 70 cents, so he paid $3\dfrac{1}{3}$ cents too much.

75. The game pictured in the problem indicates that a person can win \$0.80 with probability $\dfrac{1}{6}$, \$0.30 with probability $\dfrac{1}{3}$, and \$0.10 with probability $\dfrac{1}{2}$. The expected value of the game is
$$E = 0.80\left(\dfrac{1}{6}\right) + 0.30\left(\dfrac{1}{3}\right) + 0.10\left(\dfrac{1}{2}\right) \approx 0.2833.$$
The game is not fair. Fair games have an expected value equal to 0.

77. In reality the player has paid the \$1.00 so if the color matches the bet, he wins only \$1.00. If the color does not match the bet he loses the \$1.00 that he bet.
$$E = (1)\left(\dfrac{18}{37}\right) + (-1)\left(\dfrac{19}{37}\right) = -\dfrac{1}{37} = -0.027$$
The game is not fair; a player can expect to lose 2.7 cents.

Chapter 7 Project

1. Under the present system there are $7 \cdot 10^5 = 700,000$ different producer numbers available. The first digit can be chosen from the set $\{0, 2, 3, 4, 5, 6, 7\}$, but each of the other digits have 10 possible choices.

3. If all possible producer numbers are allowed there are 10^{11} possible correct UPC codes. The last digit is the check digit.

5. In Mauritius there are only 4 digits to be split between the publisher block and the title block. So the following 3 possibilities exist. There can be:

- One digit for the publisher number and 3 digits for the title. This can be done $10 \cdot 10^3$ ways, which allows for $10^3 = 1000$ titles from each of 10 publishers or 10,000 titles.
- Two digits for the publisher number and 2 digits for the title. This can be done $10^2 \cdot 10^2$ ways, which allows for $10^2 = 100$ titles from each of 100 publishers or 10,000 titles.
- Three digits for the publisher number and 1 digit for the title. This can be done $10^3 \cdot 10$ ways, which allows for 10 titles from each of 1000 publishers or 10,000 titles.

Since only one ISBN number is assigned per title, we can use the counting formula.
$10,000 + 10,000 + 10,000 = 30,000$ titles can be published in Mauritius.

7. The restrictions placed on the publisher number limits the number of titles from English speaking areas. We consider the following 4 possibilities.

Number of Country Codes	Number of Publisher Codes	Number of Titles	Total Number of Titles
2	20	10^6	$2 \cdot 20 \cdot 10^6$ = 40,000,000
2	500	10^5	$2 \cdot 500 \cdot 10^5$ = 100,000,000
2	1500	10^4	$2 \cdot 1500 \cdot 10^4$ = 30,000,000
2	500	10^3	$2 \cdot 500 \cdot 10^3$ = 1,000,000

Using the counting formula, we find that there are
$40,000,000 + 100,000,000$
$+ 30,000,000 + 1,000,000$
$= 171,000,000$
possible titles from English speaking areas.

Mathematical Questions from Professional Exams

1. A
$$E = 0(0.1) + 1(0.15) + 2(0.2) + 3(0.4)$$
$$+ 4(0.1) + 5(0.05)$$
$$= 2.4 = \$2.40$$

3. B
$$E = 6000(0.2) + 8000(0.2) + 10,000(0.2)$$
$$+ 12,000(0.2) + 14,000(0.1)$$
$$+ 16,000(0.1)$$
$$= 10,200 \text{ units}$$

5. B or C

Chapter 8 Additional Probability Topics

8.1 Conditional Probability

1. There are 13 diamonds in a regular deck of cards. The probability of drawing a diamond on the first draw is $\dfrac{13}{52} = \dfrac{1}{4}$.

3. False. The conditional probability of the event E given the event F is denoted by $P(E|F)$.

5. $P(E) = 0.2 + 0.3 = 0.5$

7. $P(E|F) = \dfrac{P(E \cap F)}{P(F)} = \dfrac{0.3}{0.7} = \dfrac{3}{7} \approx 0.429$

9. $P(E \cap F) = 0.3$

11. $P(\overline{E}) = 1 - P(E) = 1 - 0.5 = 0.5$

13. $P(E|F) = \dfrac{P(E \cap F)}{P(F)} = \dfrac{0.1}{0.4} = \dfrac{1}{4} = 0.25$

$P(F|E) = \dfrac{P(E \cap F)}{P(E)} = \dfrac{0.1}{0.2} = \dfrac{1}{2} = 0.5$

15. $P(E|F) = \dfrac{P(E \cap F)}{P(F)}$

$P(F) = \dfrac{P(E \cap F)}{P(E|F)} = \dfrac{0.2}{0.4} = \dfrac{1}{2} = 0.5$

17. $P(E|F) = \dfrac{P(E \cap F)}{P(F)}$

$P(E \cap F) = P(F) \cdot P(E|F) = \dfrac{5}{13} \cdot \dfrac{4}{5} = \dfrac{4}{13}$

19. a. $P(F|E) = \dfrac{P(E \cap F)}{P(E)}$

$P(E) = \dfrac{P(E \cap F)}{P(F|E)} = \dfrac{\frac{1}{3}}{\frac{2}{3}} = \dfrac{1}{2}$

b. $P(E|F) = \dfrac{P(E \cap F)}{P(F)}$

$P(F) = \dfrac{P(E \cap F)}{P(E|F)} = \dfrac{\frac{1}{3}}{\frac{1}{2}} = \dfrac{2}{3}$

21. $P(C) = 0.7(0.9) + 0.3(0.2) = 0.69$

23. $P(C|A) = 0.9$

25. $P(C|B) = 0.2$

27. $P(E \cap F) = P(E) + P(F) - P(E \cup F)$
$= 0.5 + 0.4 - 0.8 = 0.1$

29. $P(F|E) = \dfrac{P(E \cap F)}{P(E)} = \dfrac{0.1}{0.5} = \dfrac{1}{5} = 0.2$

31. $P(E|\overline{F}) = \dfrac{P(E \cap \overline{F})}{P(\overline{F})} = \dfrac{0.4}{0.6} = \dfrac{2}{3} \approx 0.667$

A tree diagram helps to see this problem.

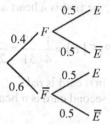

33. $S = \{\text{BBB, BBG, BGB, GBB, GGB, GBG,}$
$\text{BGG, GGG}\}; n(S) = 8$

a. Let E be the event, "The family has 2 girls."
$E = \{\text{GGB, GBG, BGG}\}$
Let F be the event, "The first child is a girl."
$F = \{\text{GBB, GGB, GBG, GGG}\}$

$P(F) = \dfrac{n(F)}{n(S)} = \dfrac{4}{8} = \dfrac{1}{2}$

$E \cap F = \{\text{GGB, GBG}\}$

$P(E \cap F) = \dfrac{2}{8} = \dfrac{1}{4}$

So the probability a family with 3 children has exactly 2 girls, given the first child is a girl is

$P(E|F) = \dfrac{P(E \cap F)}{P(F)} = \dfrac{\frac{1}{4}}{\frac{1}{2}} = \dfrac{1}{2}$.

b. Let E be the event, "The family has 1 girl."
$E = \{\text{GBB, BGB, BBG}\}$
Let F be the event, "The first child is a boy."
$F = \{\text{BBB, BBG, BGB, BGG}\}$

$P(F) = \dfrac{n(F)}{n(S)} = \dfrac{4}{8} = \dfrac{1}{2}$

$E \cap F = \{\text{BGB, BBG}\}$

$P(E \cap F) = \dfrac{2}{8} = \dfrac{1}{4}$

(continued on next page)

(continued)

So the probability a family with 3 children has exactly 1 girl, given the first child is a boy, is

$$P(E \mid F) = \frac{P(E \cap F)}{P(F)} = \frac{\frac{1}{4}}{\frac{1}{2}} = \frac{1}{2}.$$

35. The experiment consists of drawing two cards without replacement from a deck of 52 cards.

a. Define event E: The first card is a heart.
Define event F: The second card is red.

$$P(E) = \frac{13}{52} = \frac{1}{4}; \quad P(F \mid E) = \frac{25}{51}$$

The probability that when two cards are drawn without replacement, the first is a heart and the second is red is

$$P(E \cap F) = P(E) \cdot P(F \mid E) = \frac{1}{4} \cdot \frac{25}{51} = \frac{25}{204}$$

b. Define event G: The first card is red.
Define event H: The second card is a heart.

$$P(G) = \frac{26}{52} = \frac{1}{2}$$

$$P(H \mid G) = \frac{13}{51} \cdot \frac{1}{2} + \frac{12}{51} \cdot \frac{1}{2} = \frac{25}{102}$$

The probability that when two cards are drawn without replacement, the first is a red and the second is a heart is

$$P(G \cap H) = P(G) \cdot P(H \mid G) = \frac{1}{2} \cdot \frac{25}{102} = \frac{25}{204}$$

37. The experiment consists of drawing two balls without replacement from a box containing 6 balls. Define event E: A white ball and a yellow are drawn. The probability of choosing a white and a yellow ball without replacement can be considered as the union of two mutually exclusive events.

$$P(E) = P(\text{W first}) \cdot P(\text{Y} \mid \text{W first}) \\ \qquad + P(\text{Y first}) \cdot P(\text{W} \mid \text{Y first})$$

$$= \frac{3}{6} \cdot \frac{1}{5} + \frac{1}{6} \cdot \frac{3}{5} = \frac{1}{10} + \frac{1}{10} = \frac{1}{5}$$

39. The experiment consists of drawing a card from a deck of 52 cards; $n(S) = 52$.

a. Define event E to be "A red ace is drawn."
$n(E) = 2$

$$P(E) = \frac{n(E)}{n(S)} = \frac{2}{52} = \frac{1}{26}$$

b. Define event F to be "An ace is drawn."
$n(F) = 4$

$$P(E \mid F) = \frac{P(E \cap F)}{P(F)} = \frac{\frac{2}{52}}{\frac{4}{52}} = \frac{1}{2}$$

c. Define event G to be "A red card is picked."
$n(G) = 26$

$$P(E \mid G) = \frac{P(E \cap G)}{P(G)} = \frac{\frac{2}{52}}{\frac{26}{52}} = \frac{2}{26} = \frac{1}{13}$$

41. $P(E) = 0.40$ **43.** $P(H) = 0.24$

45. $P(E \cap H) = 0.10$ **47.** $P(G \cap H) = 0.08$

49. $P(E \mid H) = \dfrac{P(E \cap H)}{P(H)} = \dfrac{0.10}{0.24} = \dfrac{5}{12} \approx 0.417$

51. $P(G \mid H) = \dfrac{P(G \cap H)}{P(H)} = \dfrac{0.08}{0.24} = \dfrac{1}{3} \approx 0.333$

53. $P(E \mid G) = \dfrac{P(E \cap G)}{P(G)} = \dfrac{180}{180 + 60 + 20}$

$$= \frac{180}{260} = \frac{9}{13} \approx 0.692$$

The probability the customer likes the deodorant given he/she is from group I is $\dfrac{9}{13}$ or 0.692.

55. $P(H \mid E) = \dfrac{P(H \cap E)}{P(E)} = \dfrac{110}{180 + 110 + 55}$

$$= \frac{110}{345} = \frac{22}{69} \approx 0.219$$

The probability a customer is from group II given he/she likes the deodorant is $\dfrac{22}{69}$ or 0.319.

57. $P(F \mid G) = \dfrac{P(F \cap G)}{P(G)} = \dfrac{60}{180 + 60 + 20} = \dfrac{60}{260} = \dfrac{3}{13}$
$\qquad \approx 0.231$

The probability a customer does not like the deodorant given he/she is from group I is $\dfrac{3}{13}$ or 0.231.

59. $P(H \mid F) = \dfrac{P(H \cap F)}{P(F)} = \dfrac{85}{60 + 85 + 65} = \dfrac{85}{210} = \dfrac{17}{42}$
$\qquad \approx 0.405$

The probability a customer is from group II given he/she does not like the deodorant is $\dfrac{17}{42}$ or 0.405.

61. $P(F \mid I) = \dfrac{P(F \cap I)}{P(I)} = \dfrac{25}{30 + 25} = \dfrac{25}{55} = \dfrac{5}{11}$

≈ 0.4545

The probability the resident is female given he/she is an Independent is approximately 0.455.

63. $P(M \mid D) = \dfrac{P(M \cap D)}{P(D)} = \dfrac{50}{50 + 60} = \dfrac{50}{110} = \dfrac{5}{11}$

≈ 0.4545

The probability the resident is male given he/she is a Democrat is approximately 0.455.

65.

$P(M \mid R \cup I) = \dfrac{P(M \cap (R \cup I))}{P(R \cup I)}$

$= \dfrac{P[(M \cap R) \cup (M \cap I)]}{P(R \cup I)}$

$= \dfrac{P(M \cap R) + P(M \cap I)}{P(R) + P(I)}$

$= \dfrac{40 + 30}{40 + 30 + 30 + 25}$

$= \dfrac{70}{125} = \dfrac{14}{25} = 0.56$

The probability the resident is male given he/she is either a Republican or an Independent is 0.56.

67. a. $P(M) = \dfrac{1448}{2018} = \dfrac{724}{1009} \approx 0.7175$

b. $P(A) = \dfrac{666}{2018} = \dfrac{333}{1009} \approx 0.3300$

c. $P(F \cap B) = \dfrac{144}{2018} = \dfrac{72}{1009} \approx 0.0714$

d. $P(F \mid E) = \dfrac{P(F \cap E)}{P(E)} = \dfrac{102}{526} = \dfrac{51}{263} \approx 0.194$

e. $P(A \mid M) = \dfrac{P(A \cap M)}{P(M)} = \dfrac{342}{1448} = \dfrac{171}{724} = 0.2362$

f. $P(F \mid A \cup E) = \dfrac{P[F \cap (A \cup E)]}{P(A \cup E)}$

$= \dfrac{P(F \cap A) + P(F \cap E)}{P(A) + P(E)}$

$= \dfrac{324 + 102}{666 + 526} = \dfrac{426}{1192}$

$= \dfrac{213}{596} \approx 0.357$

g. $P(M \cap \overline{B}) = P[M \cap (A \cup E)]$

$= P(M \cap A) + P(M \cap E)$

$= \dfrac{342}{2018} + \dfrac{424}{2018} = \dfrac{766}{2018}$

$= \dfrac{383}{1009} \approx 0.380$

h. $P(F \mid \overline{E}) = P(F \mid A \cup B) = \dfrac{P[F \cap (A \cup B)]}{P(A \cup B)}$

$= \dfrac{P(F \cap A) + P(F \cap B)}{P(A) + P(B)}$

$= \dfrac{324 + 144}{666 + 826} = \dfrac{468}{1492} = \dfrac{117}{373}$

69. $P(B \mid Rh+) = \dfrac{P(B \cap Rh+)}{P(Rh+)}$

$= \dfrac{0.09}{0.39 + 0.31 + 0.09 + 0.03} = \dfrac{0.09}{0.82}$

≈ 0.110

71. $P(Rh+ \mid O) = \dfrac{P(Rh+ \cap O)}{P(O)} = \dfrac{0.39}{0.39 + 0.09}$

$= \dfrac{0.39}{0.48} = \dfrac{39}{48} = 0.8125$

73. Define event E as a household receives the mailing, and define event F as a household responds.

$P(E \cap F) = P(E) \cdot P(F \mid E)$

$= 0.20 \cdot 0.07 = 0.014$

75. a. $P(< 30) = 0.20 + 0.07 + 0.18 = 0.45$

45% of the purchases were made by people under 30 years old.

b. $P(> 50 \mid \geq \$500) = \dfrac{P(> 50 \cap \geq \$500)}{P(\geq \$500)}$

$= \dfrac{0.02}{0.18 + 0.10 + 0.02}$

$= \dfrac{0.02}{0.30} = \dfrac{1}{15} \approx 0.067$

The probability that a purchase of at least $500 was made by someone older than 50 is 0.067.

c. $P(30 - 50 \mid \$100 - \$499.99)$

$= \dfrac{P(30 - 50 \cap \$100 - \$499.99)}{P(\$100 - \$499.99)}$

$= \dfrac{0.21}{0.07 + 0.21 + 0.04} = \dfrac{0.21}{0.32} = \dfrac{21}{32} \approx 0.656$

There is 0.656 probability that a purchase between $100 and $499.99 was made by a person between 30 and 50 years old.

77. $P(\text{declined} \mid \text{high yield}) = \dfrac{533}{732 + 533 + 107}$

$= \dfrac{533}{1372} \approx 0.388$

79. $P(\text{52-week high} \mid \text{investment grade})$

$= \dfrac{200}{2062 + 1563 + 91} = \dfrac{200}{3716} \approx 0.054$

81. Define event H as an individual has health insurance.

a. $P(\overline{H} \mid {<}18) = \dfrac{P(\overline{H} \cap {<}18)}{P({<}18)} = \dfrac{8149}{74,403 + 8149}$

$= \dfrac{8149}{82,552} \approx 0.099$

The probability that a person who is under 18 years old has no health insurance is about 0.099.

b. $P({<}18 \mid \overline{H}) = \dfrac{P(\overline{H} \cap {<}18)}{P(\overline{H})}$

$= \dfrac{8149}{8149 + 26,037 + 10,784 + 686}$

$= \dfrac{8149}{45,656} \approx 0.178$

The probability that a person who has no health insurance is under 18 years old is about 0.178.

c. $P({<}18 \mid H)$

$= \dfrac{P(H \cap {<}18)}{P(H)}$

$= \dfrac{74,403}{74,403 + 110,676 + 77,237 + 36,790}$

$= \dfrac{74,403}{299,106} \approx 0.249$

The probability that a person who has health insurance is less than 18 years old is about 0.249.

83. The sample space, S, is the set of possible outcomes when 2 dice are rolled. $n(S) = 36$. We define 3 events: W: The player wins, $n(W) = 5$ (number of ways to roll 8), L: The player loses, $n(L) = 6$ (number of ways to roll 7), R: The player rolls again, $n(R) = 25$ (number of ways to roll a number other than 7 or 8). The player has already rolled an 8, so we can think of the next roll as the first roll. We want $P(W)$. This can happen on the first roll or it can happen on 2nd roll or it can happen on the 3rd roll, and so on. The probability of each individual outcome is determined using the Multiplication Principle. See the table below.

Player wins on roll	Outcome	Probability
1	W	$\dfrac{5}{36}$
2	RW	$\dfrac{25}{36} \cdot \dfrac{5}{36}$
3	RRW	$\dfrac{25}{36} \cdot \dfrac{25}{36} \cdot \dfrac{5}{36} = \left(\dfrac{25}{36}\right)^2 \cdot \dfrac{5}{36}$
4	RRRW	$\left(\dfrac{25}{36}\right)^3 \cdot \dfrac{5}{36}$
⋮	⋮	⋮

Each of these outcomes is mutually exclusive so we can use the addition rule.

$P(W) = P(W) + P(R) \cdot P(W)$
$\qquad + P(R) \cdot P(R) \cdot P(W)$
$\qquad + P(R) \cdot P(R) \cdot P(R) \cdot P(W)$
$\qquad + \cdots$

This is a geometric series with $a = \dfrac{5}{36}$ and

$r = \dfrac{25}{36}$. In Appendix A we learned that the sum

of a geometric series is $S = a\left(\dfrac{1 - r^n}{1 - r}\right)$. If n gets

larger $r^n = \left(\dfrac{25}{36}\right)^n$ becomes closer to 0. (Try it!)

In an infinite series, as we have here, one that

can continue forever, and $S = a\left(\dfrac{1}{1 - r}\right)$. The

probability of the player winning this game of craps is

$S = \left(\dfrac{5}{36}\right)\left(\dfrac{1}{1 - \frac{25}{36}}\right) = \left(\dfrac{5}{36}\right)\left(\dfrac{36}{11}\right) = \dfrac{5}{11} \approx 0.455.$

85. We use a tree diagram to determine the probability a person successfully completes training if no pre-test was administered. Let events E, F, and S be defined as follows: E: A person passes the test, F: A person fails the test, and S: A person successfully completes the training.

(*continued on next page*)

(continued)

$$P(S) = P(S \mid E) \cdot P(E) + P(S \mid F) \cdot P(F)$$
$$= (0.85)(0.70) + (0.40)(0.30)$$
$$= 0.715$$

87. Define the events, M, F, and S as follows. M: A person is male, F: A person is female, S: A person smokes. First we must determine the probability that a randomly selected person is male (or female).

$$P(M) = \frac{19.498}{19.498 + 22.435} \approx 0.465$$

$$P(F) = 1 - P(M) = 1 - 0.465 = 0.535$$

The probability a person 16 years of age or older randomly selected in Great Britain smokes is

$$P(S) = P(S \mid M) \cdot P(M) + P(S \mid F) \cdot P(F)$$
$$= (0.22)(0.465) + (0.21)(0.535) \approx 0.215$$

89. Given that $P(E) > 0$ and $P(F) > 0$,

$$P(F) \cdot P(E \mid F) = P(F) \cdot \frac{P(E \cap F)}{P(F)} = P(E \cap F)$$

$$= \frac{P(E)}{P(E)} \cdot P(E \cap F)$$

$$= P(E) \cdot \frac{P(E \cap F)}{P(E)}$$

$$= P(E) \cdot P(F \mid E)$$

91. $P(E \mid F) + P(\overline{E} \mid F) = \dfrac{P(E \cap F)}{P(F)} + \dfrac{P(\overline{E} \cap F)}{P(F)}$

$$= \frac{P(E \cap F) + P(\overline{E} \cap F)}{P(F)}.$$

Since $E \cap F$ and $\overline{E} \cap F$ are mutually exclusive events,

$$P(E \cap F) + P(\overline{E} \cap F) = P\big((E \cap F) \cup (\overline{E} \cap F)\big).$$

Using the distributive property, we have

$$(E \cap F) \cup (\overline{E} \cap F) = (E \cup \overline{E}) \cap F = S \cap F = F.$$

So, $P(E \cap F) + P(\overline{E} \cap F) = P(F)$ and

$$P(E \mid F) + P(\overline{E} \mid F) = \frac{P(F)}{P(F)} = 1.$$

93. We are told $P(E) > 0$ and $P(E \mid F) = P(E)$.

$$P(E \mid F) = \frac{P(F \cap E)}{P(F)} = P(E), \text{ so}$$

$$P(F \cap E) = P(E) \cdot P(F). \text{ Therefore,}$$

$$P(F) = \frac{P(F \cap E)}{P(E)} = \frac{P(E) \cdot P(F)}{P(E)} = P(F).$$

8.2 Independent Events

1. False. Mutually exclusive events are independent only if at least one event is the impossible event. If two events A and B are mutually exclusive, then $P(A \cap B) = 0$. If A and B are independent, then
$$P(A \cap B) = P(A) \cdot P(B).$$

3. $P(F \cap E) = P(E) \cdot P(F) = 0.4 \cdot 0.6 = 0.24$

5. $P(E \cup F) = P(E) + P(F) - P(E \cap F)$
$$= P(E) + P(F) - P(E)P(F)$$
$$= P(E) + P(F)\big(1 - P(E)\big)$$

$$P(F) = \frac{P(E \cup F) - P(E)}{1 - P(E)}$$

$$= \frac{0.3 - 0.2}{1 - 0.2} = \frac{0.1}{0.8} = \frac{1}{8} = 0.125$$

7. E and F are independent if
$$P(E \cap F) = P(E) \cdot P(F).$$
$$\frac{2}{9} \neq \left(\frac{4}{21}\right)\left(\frac{7}{12}\right) = \frac{1}{9}, \text{ so the events are not}$$
independent.

9.a. $P(E \mid F) = P(E) = 0.2$

b. $P(F \mid E) = P(F) = 0.4$

c. $P(E \cap F) = P(E) \cdot P(F) = 0.2 \cdot 0.4 = 0.08$

d. $P(E \cup F) = P(E) + P(F) - P(E \cap F)$
$$= 0.2 + 0.4 - 0.08 = 0.52$$

11. $P(E \cap F \cap G) = P(E) \cdot P(F) \cdot P(G)$
$$= \frac{2}{3} \cdot \frac{3}{7} \cdot \frac{2}{21} = \frac{4}{147}$$

13. $P(E \cap F) = P(E) + P(F) - P(E \cup F)$
$$= 0.3 + 0.2 - 0.4 = 0.1$$

$$P(E \mid F) = \frac{P(F \cap E)}{P(F)} = \frac{0.1}{0.2} = \frac{1}{2} = 0.5$$

$$P(E) \cdot P(F) = 0.3 \cdot 0.2 = 0.06 \neq P(E \cap F) = 0.1$$

Therefore, E and F are not independent.

15. The sample space is the set of all possible outcomes. Let R stand for the mouse goes right and L stand for the mouse goes left. $S = \{$RRR, RRL, RLR, LRR, RLL, LRL, LLR, LLL$\}$. On the first two runs $P(L) = P(R) = \frac{1}{2}$, and we are told that on the third try the mouse is twice as likely to choose L. From this we get $P(L) = \frac{2}{3}$ and $P(R) = \frac{1}{3}$. A tree diagram and the Multiplication Principle will help to assign the probabilities.

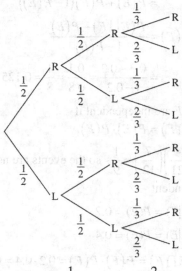

$$P(RRR) = \frac{1}{12} \quad P(RRL) = \frac{2}{12} \quad P(RLR) = \frac{1}{12}$$
$$P(RLL) = \frac{2}{12} \quad P(LRR) = \frac{1}{12} \quad P(LRL) = \frac{2}{12}$$
$$P(LLR) = \frac{1}{12} \quad P(LLL) = \frac{2}{12}$$

a. $P(E) = P(RRL \cup LRR) = P(RRL) + P(LRR)$
$$= \frac{2}{12} + \frac{1}{12} = \frac{3}{12} = \frac{1}{4}$$

b. $P(F) = P(LLL) = \frac{2}{12} = \frac{1}{6}$

c. $P(G) = P(LRR \cup LRL \cup LLR \cup LLL)$
$$= P(LRR) + P(LRL) + P(LLR) + P(LLL)$$
$$= \frac{1}{12} + \frac{2}{12} + \frac{1}{12} + \frac{2}{12} = \frac{6}{12} = \frac{1}{2}$$

d. $P(H) = P(RRR \cup RRL \cup LRR \cup LRL)$
$$= P(RRR) + P(RRL) + P(LRR) + P(LRL)$$
$$= \frac{1}{12} + \frac{2}{12} + \frac{1}{12} + \frac{2}{12} = \frac{6}{12} = \frac{1}{2}$$

17. Two marbles are chosen with replacement. Define the events R: The marble chosen is red. $P(R) = 0.60$.
W: The marble chosen is white. $P(W) = 0.40$. The events are independent.

a. Probability both marbles are red is
$$P(R \cap R) = P(R) \cdot P(R) = \frac{6}{10} \cdot \frac{6}{10} = \frac{36}{100} = \frac{9}{25}.$$

b. Probability exactly 1 of the marbles is red is
$$P(R \cap W) + P(W \cap R)$$
$$= P(R) \cdot P(W) + P(W) \cdot P(R)$$
$$= \frac{6}{10} \cdot \frac{4}{10} + \frac{4}{10} \cdot \frac{6}{10} = \frac{48}{100} = \frac{12}{25}.$$

19. Let H denote a child with heart disease, and $\overline{H}$ denote a child with no heart disease. The sample space is the set of all possible outcomes. The couple has two children.
$$S = \left\{ HH, H\overline{H}, \overline{H}H, \overline{H}\,\overline{H} \right\}$$
$$P(H) = \frac{3}{4}, \ P(\overline{H}) = \frac{1}{4}$$

a. $P(HH) = P(H \cap H) = P(H) \cdot P(H)$
$$= \frac{3}{4} \cdot \frac{3}{4} = \frac{9}{16}$$

b. $P(\overline{H}\,\overline{H}) = P(\overline{H} \cap \overline{H}) = P(\overline{H}) \cdot P(\overline{H})$
$$= \frac{1}{4} \cdot \frac{1}{4} = \frac{1}{16}$$

c. $P(H\overline{H} \cup \overline{H}H) = P(H \cap \overline{H}) + P(\overline{H} \cap H)$
$$= P(H) \cdot P(\overline{H}) + P(\overline{H}) \cdot P(H)$$
$$= \frac{3}{4} \cdot \frac{1}{4} + \frac{1}{4} \cdot \frac{3}{4} = \frac{6}{16} = \frac{3}{8}$$

21. Define the events G: The seed germinates and V: The seed chosen is a violet seed. The events are independent, so the probability a planted seed will grow into a violet is
$$P(G \cap V) = P(G) \cdot P(V) = 0.60 \cdot \frac{1}{3} = 0.20.$$

23.a. Because the inspectors are independent, the probability both inspectors miss a defective piece of furniture is $0.20^2 = 0.04$.

b. To find the number of inspectors needed to assure that the probability of missing a defect is less than 0.01, solve the equation $0.20^n = 0.01$. Write the exponential equation in logarithmic form and then use a change of base formula.

$$n = \log_{0.20} 0.01 = \frac{\log 0.01}{\log 0.20} = 2.861$$

Efraim Furniture Company should hire 3 inspectors to insure the probability of a missed defect is less than 0.01.

25. Since the probability that the two stocks increase in value is independent, the probability both stocks improve is $0.60^2 = 0.36$. The probability that at least one stock will not increase in value is

1 – probability both stocks increase in value
$$= 1 - 0.36 = 0.64.$$

27. a. The total of all people listed in the table is 344,762.

$$P(<18 \text{ years old}) = \frac{74,403 + 8149}{344,762}$$
$$= \frac{82,552}{344,762} \approx 0.239$$

$P(<18 \text{ years old} \mid \text{no health ins})$
$$= \frac{8149}{8149 + 26,037 + 10,784 + 686} = \frac{8149}{45,656}$$
$$\approx 0.178$$

b. The events are not independent. If they were independent then $P(<18$ years old $\mid$ no health insurance) and $P(<18$ years old) would be equal.

29. a. The failures are not mutually exclusive because P(both pumps fail) $\neq 0$.

b. The probability at least one pump fails is
$$1 - P(\text{neither pump fails}) = 1 - 0.95^2$$
$$= 0.0975.$$

c. The failures in the pumps are independent because
$$P(\text{one pump fails}) \cdot P(\text{other pump fails})$$
$$= P(\text{both pumps fail})$$
$$0.05 \cdot 0.05 = 0.0025$$

31. a. $P(\text{burglarized} \mid \text{suburban}) = \frac{30}{146} = \frac{15}{73} \approx 0.205$

b. $P(\text{rural} \mid \text{vehicle theft}) = \frac{4}{28} = \frac{1}{7} \approx 0.143$

c. If rural and vehicle theft are independent then $P(\text{rural} \mid \text{vehicle theft}) = P(\text{rural})$.
$$P(\text{rural}) = \frac{134}{500} = 0.268$$
$$\neq P(\text{rural} \mid \text{vehicle theft}) \approx 0.143$$
The events are not independent.

d. The events "Urban" and "Burglary" are independent if
$$P(\text{urban}) \cdot P(\text{burglary}) = P(\text{urban} \cap \text{burglary})$$
$$P(\text{urban}) \cdot P(\text{burglary}) = \frac{220}{500} \cdot \frac{100}{500}$$
$$= \frac{11}{125} = 0.088$$
$$P(\text{urban} \cap \text{burglary}) = \frac{44}{500} = \frac{11}{125} = 0.088$$
The events are independent.

33. Let V: The first voter votes for the candidate, and W: The second voter votes for the candidate.
$$P(V) = P(W) = \frac{2}{3}.$$

a. $P(V \cap W) = P(V) \cdot P(W) = \frac{2}{3} \cdot \frac{2}{3} = \frac{4}{9}$

b. $P(\overline{V} \cap \overline{W}) = P(\overline{V}) \cdot P(\overline{W}) = \frac{1}{3} \cdot \frac{1}{3} = \frac{1}{9}$

c. $P(V\overline{W} \cup \overline{V}W) = P(V\overline{W}) + P(\overline{V}W)$
$$= \frac{2}{3} \cdot \frac{1}{3} + \frac{1}{3} \cdot \frac{2}{3} = \frac{4}{9}$$

35. If components are tested individually n components need n tests. If they tested together, then we need 1 test for n components if they are all good, and $n + 1$ tests if the grouped test fails. The probability of needing only one test is p^n and the probability of needing $n + 1$ tests is $1 - p^n$. From this we can get the expected number of tests, E.
$$E = 1(p^n) + (n+1)(1 - p^n)$$
$$= p^n + n - np^n + 1 - p^n$$
$$= (n+1) - np^n$$
The expected number of tests saved is
$$n - \left[(n+1) - np^n \right] = np^n - 1.$$

If we are testing n components the number of tests saved per component becomes
$$\frac{np^n - 1}{n} = p^n - \frac{1}{n}.$$
The optimal group size is that which saves the most tests per component.

a. We are asked to find the optimal group size if the probability that a component is good is 0.8.

Group Size	Expected Tests Saved per Component $p = 0.8$	Percent Saving
2	$0.8^2 - \dfrac{1}{2} = 0.14$	14%
3	$0.8^3 - \dfrac{1}{3} = 0.1787$	17.9%
4	$0.8^4 - \dfrac{1}{4} = 0.1596$	16.0%
5	$0.8^5 - \dfrac{1}{5} = 0.12768$	12.8%
6	$0.8^6 - \dfrac{1}{6} = 0.09547$	9.5%
7	$0.8^7 - \dfrac{1}{7} = 0.06686$	6.7%
8	$0.8^8 - \dfrac{1}{8} = 0.042772$	4.3%

The optimal group size is 3. Its percent saving is 17.9%.

b. We are asked to find the optimal group size if the probability that a component is good is 0.95.

Group Size	Expected Tests Saved per Component $p = 0.95$	Percent Saving
2	$0.95^2 - \dfrac{1}{2} = 0.4025$	40.3%
3	$0.95^3 - \dfrac{1}{3} = 0.5240$	52.4%
4	$0.95^4 - \dfrac{1}{4} = 0.5645$	56.5%
5	$0.95^5 - \dfrac{1}{5} = 0.5738$	57.4%
6	$0.95^6 - \dfrac{1}{6} = 0.5684$	56.8%
7	$0.95^7 - \dfrac{1}{7} = 0.55548$	55.5%
8	$0.95^8 - \dfrac{1}{8} = 0.53842$	53.8%

The optimal group size is 5. Its percent saving is 57.4%.

c. If the probability a component is good is 0.99, then the savings incurred by grouping the components before testing is significant. The table below shows the savings.

Group Size	Expected Tests Saved per Component $p = 0.99$	Percent Saving
2	$0.99^2 - \dfrac{1}{2} = 0.4801$	48.01%
3	$0.99^3 - \dfrac{1}{3} = 0.6370$	63.70%
4	$0.99^4 - \dfrac{1}{4} = 0.7106$	71.06%
5	$0.99^5 - \dfrac{1}{5} = 0.7510$	75.10%
6	$0.99^6 - \dfrac{1}{6} = 0.7748$	77.48%
7	$0.99^7 - \dfrac{1}{7} = 0.7892$	78.92%
8	$0.99^8 - \dfrac{1}{8} = 0.7977$	79.77%
9	$0.99^9 - \dfrac{1}{9} = 0.8024$	80.24%
10	$0.99^{10} - \dfrac{1}{10} = 0.80438$	80.438%
11	$0.99^{11} - \dfrac{1}{11} = 0.80443$	80.443%
12	$0.99^{12} - \dfrac{1}{12} = 0.80305$	80.305%
13	$0.99^{13} - \dfrac{1}{13} = 0.80060$	80.060%

The optimal group size is 11. Its percent saving is 80.44%.

37. If the probability a test is positive is p, the probability the test is negative is $(1 - p)$.

a. The test for a pooled sample will be positive if at least one person tests positive. So,
P(pooled test is positive) $= 1 - P$(all tests are negative) $= 1 - (1 - p)^{20}$.

b. If the pooled test is negative then 1 test suffices for 20 allergens, but if the pooled test is positive then 21 tests must be done. The expected number of tests necessary using the pooled method is
$$E = 1(1 - p)^{20} + 21\left[1 - (1 - p)^{20}\right]$$
$$= (1 - p)^{20} + 21 - 21(1 - p)^{20}$$
$$= 21 - 20(1 - p)^{20}$$

c. A pooled sampling saves approximately $20 - (21 - 20(1 - p)^{20}) = 20(1 - p)^{20} - 1$ tests per individual.

39. If events E and F are mutually exclusive, then $P(E \cap F) = 0$. If events E and F are independent, then $P(E \cap F) = P(E) \cdot P(F)$. If the events are both mutually exclusive and independent then by substitution, $P(E) \cdot P(F) = 0$. The zero property of multiplication indicates that either $P(E) = 0$ or $P(F) = 0$ or both.

41. If E and F are independent events, $P(E \cap F) = P(E) \cdot P(F)$. We want to show that $P(\overline{E} \cap \overline{F}) = P(\overline{E}) \cdot P(\overline{F})$. From De Morgan's properties, $\overline{E} \cap \overline{F} = \overline{E \cup F}$, and $P(\overline{E \cup F}) = 1 - P(E \cup F)$. so,

$$P(\overline{E} \cap \overline{F}) = P(\overline{E \cup F}) = 1 - P(E \cup F)$$
$$= 1 - [P(E) + P(F) - P(E \cap F)].$$

Since E and F are independent, we can substitute $P(E) \cdot P(F)$ for $P(E \cap F)$.

$$P(\overline{E} \cap \overline{F}) = 1 - [P(E) + P(F) - P(E) \cdot P(F)]$$
$$= 1 - P(E) - P(F) + P(E) \cdot P(F)$$

Factoring by grouping, we get

$$P(\overline{E} \cap \overline{F}) = 1 - [P(E) + P(F) - P(E) \cdot P(F)]$$
$$= [1 - P(E)] - [1 - P(E)] \cdot P(F)$$
$$= [1 - P(E)][1 - P(F)]$$
$$= P(\overline{E}) \cdot P(\overline{F})$$

So events $\overline{E}$ and $\overline{F}$ are also independent.

43. From the definition of conditional probability, we know that $P(E \mid F) = \dfrac{P(E \cap F)}{P(F)}$ or $P(E \cap F) = P(F) \cdot P(E \mid F)$. We also know that $P(F \mid E) = \dfrac{P(E \cap F)}{P(E)}$ or $P(E \cap F) = P(E) \cdot P(F \mid E)$. Then we have $P(E \cap F) = P(F) \cdot P(E \mid F) = P(E) \cdot P(F \mid E)$. Since E is independent of F, $P(E \mid F) = P(E)$,

so $\begin{aligned} P(E) \cdot P(F \mid E) &= P(F) \cdot P(E) \Rightarrow \\ P(F \mid E) &= P(F). \end{aligned}$

Therefore, F is independent of E.

45. Answers will vary. Sample answers are given.

a. If M is the event "takes a math class" and S is the event "takes a science class". Then M and S are independent but not mutually exclusive since a student could take both classes.

b. See example 4 in the text. The events E and G are not independent, but they are mutually exclusive.

c. Roll two dice, let A be the event that both dice show a 1, 2, or 3. Let B be the event that the sum of the two dice is greater than 7. A and B are mutually exclusive because it is impossible for both events to happen at the same time. They are not independent because if A happens you know that B cannot happen. Knowing something about one of the events tells you about the other.

8.3 Bayes' Theorem

1. If A and B are two disjoint sets, then $A \cap B = \varnothing$.

3. True   **5.** $P(E \mid A) = 0.4$

7. $P(E \mid B) = 0.2$ **9.** $P(E \mid C) = 0.7$

11. $\begin{aligned} P(E) &= P(E \cap A) + P(E \cap B) + P(E \cap C) \\ &= 0.4(0.3) + 0.2(0.6) + 0.7(0.1) = 0.31 \end{aligned}$

13. $P(A \mid E) = \dfrac{P(A \cap E)}{P(E)} = \dfrac{P(E \mid A) \cdot P(A)}{P(E)}$
$$= \dfrac{(0.4)(0.3)}{0.31} = \dfrac{12}{31} \approx 0.387$$

15. $P(C \mid E) = \dfrac{P(C \cap E)}{P(E)} = \dfrac{P(E \mid C) \cdot P(C)}{P(E)}$
$$= \dfrac{(0.7)(0.1)}{0.31} = \dfrac{7}{31} \approx 0.226$$

17. $P(B \mid E) = \dfrac{P(B \cap E)}{P(E)} = \dfrac{P(E \mid B) \cdot P(B)}{P(E)}$
$$= \dfrac{(0.2)(0.6)}{0.31} = \dfrac{12}{31} \approx 0.387$$

19. $\begin{aligned} P(E) &= P(E \cap A_1) + P(E \cap A_2) \\ &= P(A_1) \cdot P(E \mid A_1) + P(A_2) \cdot P(E \mid A_2) \\ &= 0.4 \cdot 0.03 + 0.6 \cdot 0.02 = 0.024 \end{aligned}$

21. $\begin{aligned} P(E) &= P(E \cap A_1) + P(E \cap A_2) + P(E \cap A_3) \\ &= P(A_1) \cdot P(E \mid A_1) + P(A_2) \cdot P(E \mid A_2) \\ &\quad + P(A_3) \cdot P(E \mid A_3) \\ &= 0.6 \cdot 0.01 + 0.2 \cdot 0.03 + 0.2 \cdot 0.02 = 0.016 \end{aligned}$

23. $P(A_1 \mid E) = \dfrac{P(A_1 \cap E)}{P(E)} = \dfrac{P(A_1) \cdot P(E \mid A_1)}{P(E)}$
$$= \dfrac{(0.4)(0.03)}{0.024} = \dfrac{12}{24} = 0.5$$

(continued on next page)

(continued)

$$P(A_2 \mid E) = \frac{P(A_2 \cap E)}{P(E)} = \frac{P(A_2) \cdot P(E \mid A_2)}{P(E)}$$

$$= \frac{(0.6)(0.02)}{0.024} = \frac{12}{24} = 0.5$$

25. $P(A_1 \mid E) = \dfrac{P(A_1 \cap E)}{P(E)} = \dfrac{P(A_1) \cdot P(E \mid A_1)}{P(E)}$

$$= \frac{(0.6)(0.01)}{0.016} = \frac{3}{8} = 0.375$$

$$P(A_2 \mid E) = \frac{P(A_2 \cap E)}{P(E)} = \frac{P(A_2) \cdot P(E \mid A_2)}{P(E)}$$

$$= \frac{(0.2)(0.03)}{0.016} = \frac{3}{8} = 0.375$$

$$P(A_3 \mid E) = \frac{P(A_3 \cap E)}{P(E)} = \frac{P(A_3) \cdot P(E \mid A_3)}{P(E)}$$

$$= \frac{(0.2)(0.02)}{0.016} = \frac{1}{4} = 0.250$$

27. Define the events: U_1: Jar 1 is selected, U_2: Jar 2 is selected, and U_3: Jar 3 is selected. Since the jar is selected at random,

$$P(U_1) = P(U_2) = P(U_3) = \frac{1}{3}.$$

a. The probability the ball is red, $P(R)$, is
$$P(R) = P(R \cap U_1) + P(R \cap U_2) + P(R \cap U_3)$$
$$= P(U_1) \cdot P(R \mid U_1) + P(U_2) \cdot P(R \mid U_2)$$
$$+ P(U_3) \cdot P(R \mid U_3)$$
$$= \left(\frac{1}{3}\right)\left(\frac{5}{16}\right) + \left(\frac{1}{3}\right)\left(\frac{3}{16}\right) + \left(\frac{1}{3}\right)\left(\frac{7}{16}\right)$$
$$= \frac{15}{48} = \frac{5}{16} = 0.3125$$

b. The probability the ball is white, $P(W)$, is
$$P(W) = P(W \cap U_1) + P(W \cap U_2) + P(W \cap U_3)$$
$$= P(U_1) \cdot P(W \mid U_1) + P(U_2) \cdot P(W \mid U_2)$$
$$+ P(U_3) \cdot P(W \mid U_3)$$
$$= \left(\frac{1}{3}\right)\left(\frac{6}{16}\right) + \left(\frac{1}{3}\right)\left(\frac{4}{16}\right) + \left(\frac{1}{3}\right)\left(\frac{5}{16}\right)$$
$$= \frac{15}{48} = \frac{5}{16} = 0.3125$$

c. The probability the ball is blue, $P(B)$, is
$$P(B) = P(B \cap U_1) + P(B \cap U_2) + P(B \cap U_3)$$
$$= P(U_1) \cdot P(B \mid U_1) + P(U_2) \cdot P(B \mid U_2)$$
$$+ P(U_3) \cdot P(B \mid U_3)$$

$$P(B) = \left(\frac{1}{3}\right)\left(\frac{5}{16}\right) + \left(\frac{1}{3}\right)\left(\frac{9}{16}\right) + \left(\frac{1}{3}\right)\left(\frac{4}{16}\right)$$

$$= \frac{18}{48} = \frac{3}{8} = 0.375$$

d. $P(U_1 \mid R) = \dfrac{P(U_1 \cap R)}{P(R)} = \dfrac{P(U_1) \cdot P(R \mid U_1)}{P(R)}$

$$= \frac{\left(\frac{1}{3}\right)\left(\frac{5}{16}\right)}{\left(\frac{5}{16}\right)} = \frac{1}{3} \approx 0.333$$

If a ball is red, then the probability that it came from Jar I is 0.333.

e. $P(U_2 \mid B) = \dfrac{P(U_2 \cap B)}{P(B)} = \dfrac{P(U_2) \cdot P(B \mid U_2)}{P(B)}$

$$= \frac{\left(\frac{1}{3}\right)\left(\frac{9}{16}\right)}{\left(\frac{3}{8}\right)} = \frac{1}{2} = 0.500$$

If a ball is blue, then the probability that it came from Jar II is 0.5.

f. $P(U_3 \mid W) = \dfrac{P(U_3 \cap W)}{P(W)} = \dfrac{P(U_3) \cdot P(W \mid U_3)}{P(W)}$

$$= \frac{\left(\frac{1}{3}\right)\left(\frac{5}{16}\right)}{\left(\frac{5}{16}\right)} = \frac{1}{3} \approx 0.333$$

If a ball is white, then the probability it came form Jar III is 0.333.

29. Define the events A_1: a person is male, A_2: a person is female, and E: a person is color-blind. We will use a tree diagram to help determine the probabilities.

$P(A_1) = 0.491$, $P(A_2) = 0.509$,

$$P(E \mid A_1) = \frac{1}{12} \approx 0.0833,$$

$$P(E \mid A_2) = \frac{1}{250} = 0.004$$

$$P(E) = P(E \cap A_1) + P(E \cap A_2)$$
$$= P(A_1) \cdot P(E \mid A_1) + P(A_2) \cdot P(E \mid A_2)$$
$$= 0.491(0.0833) + 0.509(0.004) \approx 0.043$$

(continued on next page)

(*continued*)

$$P(A_1 \mid E) = \frac{P(A_1 \cap E)}{P(E)} = \frac{P(A_1) \cdot P(E \mid A_1)}{P(E)}$$

$$= \frac{0.491 \cdot 0.0833}{0.043} \approx 0.952$$

The probability that a color-blind person is male is 0.952.

31. Define the events E: the recipient sailed on Castaway before, M: the recipient is male, F: the recipient is female.

$$P(M) = \frac{6500}{10,000} = 0.65, \ P(F) = \frac{3500}{10,000} = 0.35,$$

$$P(E \mid M) = 0.8, \ P(E \mid F) = 0.4$$

a. Since we are looking for the probability a recipient who books a cruise had never sailed before, we need $\overline{E}$. Before finding $P(\overline{E})$, we note $P(\overline{E} \mid M) = 1 - P(E \mid M) = 1 - 0.8 = 0.2$ and $P(\overline{E} \mid F) = 1 - P(E \mid F) = 1 - 0.4 = 0.6$.

$$\begin{aligned} P(\overline{E}) &= P(\overline{E} \cap M) + P(\overline{E} \cap F) \\ &= P(M) \cdot P(\overline{E} \mid M) + P(F) \cdot P(\overline{E} \mid F) \\ &= 0.65 \cdot 0.2 + 0.35 \cdot 0.6 = 0.34 \end{aligned}$$

The probability that the customer has never cruised with Castaway is 0.34.

b. $P(M \mid \overline{E}) = \dfrac{P(M \cap \overline{E})}{P(\overline{E})} = \dfrac{P(M) \cdot P(\overline{E} \mid M)}{P(\overline{E})}$

$$= \frac{0.65 \cdot 0.2}{0.34} = \frac{13}{34} \approx 0.38$$

If the customer has never cruised with Castaway, the probability he is male is 0.38.

33. Define the events E: the plane is late; S: she is on Southwest. $P(S) = 0.7$, $P(\overline{S}) = 0.3$,

$$P(E \mid S) = 1 - P(\overline{E} \mid S) = 1 - 0.845 = 0.155,$$

$$P(E \mid \overline{S}) = 1 - P(\overline{E} \mid \overline{S}) = 1 - 0.813 = 0.187$$

$$\begin{aligned} P(E) &= P(E \cap S) + P(E \cap \overline{S}) \\ &= P(S) \cdot P(E \mid S) + P(\overline{S}) \cdot P(E \mid \overline{S}) \\ &= 0.7 \cdot 0.155 + 0.3 \cdot 0.187 = 0.1646 \end{aligned}$$

So, the probability she is on Southwest given the plane is late is

$$P(S \mid E) = \frac{P(S \cap E)}{P(E)} = \frac{P(S) \cdot P(E \mid S)}{P(E)}$$

$$= \frac{0.70 \cdot 0.155}{0.1646} \approx 0.659.$$

Since the plane is late, the probability she is on Southwest is 0.659.

35. Define the events E: there is negative feedback, R: the survey is from Riverside, S: the survey is from Springfield, C: the survey is from Centerville.

$$P(R) = 0.4, \ P(S) = 0.25, \ P(C) = 0.35,$$

$$P(E \mid R) = 0.08, \ P(E \mid S) = 0.12,$$

$$P(E \mid C) = 0.09$$

a. $\begin{aligned} P(E) &= P(R) \cdot P(E \mid R) + P(S) \cdot P(E \mid S) \\ &\qquad + P(C) \cdot P(E \mid C) \\ &= 0.4 \cdot 0.08 + 0.25 \cdot 0.12 + 0.35 \cdot 0.09 \\ &= 0.0935 \end{aligned}$

The probability that the survey has negative feedback is 0.0935.

b. $P(R \mid E) = \dfrac{P(R \cap E)}{P(E)} = \dfrac{P(R) \cdot P(E \mid R)}{P(E)}$

$$= \frac{0.4 \cdot 0.08}{0.0935} \approx 0.342$$

If the survey contains negative feedback, the probability that it is from the Riverside restaurant is 0.342.

c. $P(S \mid E) = \dfrac{P(S \cap E)}{P(E)} = \dfrac{P(S) \cdot P(E \mid S)}{P(E)}$

$$= \frac{0.25 \cdot 0.12}{0.0935} \approx 0.321$$

If the survey contains negative feedback, the probability that it is from the Springfield restaurant is 0.321.

37. Define the events E: a tax return is audited, M: an individual's adjusted gross income $\geq$ million.

$$P(E \mid \overline{M}) = \frac{1}{100}, \ P(E \mid M) = \frac{1}{16}, \ P(M) = \frac{1}{250}$$

$$P(M \mid E) = \frac{P(M \cap E)}{P(E)}$$

$$= \frac{P(M) \cdot P(E \mid M)}{P(M) \cdot P(E \mid M) + P(\overline{M}) \cdot P(E \mid \overline{M})}$$

$$P(M \mid E) = \frac{\dfrac{1}{250} \cdot \dfrac{1}{16}}{\dfrac{1}{250} \cdot \dfrac{1}{16} + \dfrac{249}{250} \cdot \dfrac{1}{100}} = \frac{25}{1021} \approx 0.025$$

If a tax return is audited, the probability that the filer's adjusted gross income is $1,000,000 or higher is 0.025.

39. Define the events E: an employee is female, M: occupation is classified management and related, S: occupation is classified service, O: occupation is classified sales and office, N: occupation is classified as natural resources, etc., P: occupation is classified production, etc.
$P(M) = 0.363$, $P(S) = 0.168$, $P(O) = 0.245$, $P(N) = 0.102$, $P(P) = 0.122$, $P(E|M) = 0.508$, $P(E|S) = 0.572$, $P(E|O) = 0.632$, $P(E|N) = 0.042$, $P(E|P) = 0.224$

a. $P(E) = P(M) \cdot P(E|M) + P(S) \cdot P(E|S) + P(O) \cdot P(E|O) + P(N) \cdot P(E|N) + P(P) \cdot P(E|P)$
$= 0.363 \cdot 0.508 + 0.168 \cdot 0.572 + 0.245 \cdot 0.632 + 0.102 \cdot 0.042 + 0.122 \cdot 0.224$
≈ 0.467

46.7% of employed persons in the U.S. are women.

b. $P(M|E) = \dfrac{P(M \cap E)}{P(E)} = \dfrac{P(M) \cdot P(E|M)}{P(E)}$
$= \dfrac{0.363 \cdot 0.508}{0.467} \approx 0.395$

If an employed woman is selected, the probability that her occupation is classified as management, professional, and related is 0.395.

c. $P(S|E) = \dfrac{P(S \cap E)}{P(E)} = \dfrac{P(S) \cdot P(E|S)}{P(E)}$
$= \dfrac{0.168 \cdot 0.572}{0.467} \approx 0.206$

If an employed woman is selected, the probability that her occupation is classified as service is 0.206.

d. $P(O|E) = \dfrac{P(O \cap E)}{P(E)} = \dfrac{P(O) \cdot P(E|O)}{P(E)}$
$= \dfrac{0.245 \cdot 0.632}{0.467} \approx 0.332$

If an employed woman is selected, the probability that her occupation is classified as sales and office is 0.332.

41. Define events D: a person is a Democrat, R: a person is a Republican, I: a person is an Independent, E: a person voted.
$P(E) = P(E \cap D) + P(E \cap R) + P(E \cap I)$
$= P(D) \cdot P(E|D) + P(R) \cdot P(E|R) + P(I) \cdot P(E|I)$
$= 0.55 \cdot 0.35 + 0.30 \cdot 0.65 + 0.15 \cdot 0.75$
$= 0.5$
The probability that a person voted is 0.5.

a. $P(D|E) = \dfrac{P(D \cap E)}{P(E)} = \dfrac{P(D) \cdot P(E|D)}{P(E)}$
$= \dfrac{(0.55)(0.35)}{0.5} = 0.385$

The probability that a voter was a Democrat was 0.385.

b. $P(R|E) = \dfrac{P(R \cap E)}{P(E)} = \dfrac{P(R) \cdot P(E|R)}{P(E)}$
$= \dfrac{(0.30)(0.65)}{0.5} = 0.39$

The probability that a voter was a Republican was 0.390.

c. $P(I|E) = \dfrac{P(I \cap E)}{P(E)} = \dfrac{P(I) \cdot P(E|I)}{P(E)}$
$= \dfrac{(0.15)(0.75)}{0.5} = 0.225$

The probability that a voter was an Independent was 0.225.

43. Define the events A_1: the soil is rock, A_2: the soil is clay, A_3: the is soil is sand, and E: the geological test is positive.
$P(A_1) = 0.53$, $P(A_2) = 0.21$, $P(A_3) = 0.26$, $P(E|A_1) = 0.35$, $P(E|A_2) = 0.48$, $P(E|A_3) = 0.75$
$P(E) = P(A_1) \cdot P(E|A_1) + P(A_2) \cdot P(E|A_2) + P(A_3) \cdot P(E|A_3)$
$= 0.53 \cdot 0.35 + 0.21 \cdot 0.48 + 0.26 \cdot 0.75$
$= 0.4813$

a. $P(A_1|E) = \dfrac{P(A_1 \cap E)}{P(E)} = \dfrac{P(A_1) \cdot P(E|A_1)}{P(E)}$
$= \dfrac{(0.53)(0.35)}{0.4813} \approx 0.385$

The probability that the soil is rock given the test is positive is 0.385.

b. $P(A_2 \mid E) = \dfrac{P(A_2 \cap E)}{P(E)} = \dfrac{P(A_2)\cdot P(E\mid A_2)}{P(E)}$

$= \dfrac{(0.21)(0.48)}{0.4813} \approx 0.209$

There is a 0.209 probability that the soil is clay given that the test is positive.

c. $P(A_3 \mid E) = \dfrac{P(A_3 \cap E)}{P(E)} = \dfrac{P(A_3)\cdot P(E\mid A_3)}{P(E)}$

$= \dfrac{(0.26)(0.75)}{0.4813} \approx 0.405$

The probability is 0.405 that the soil is sand given that the test came out positive.

45. Define the additional event E: a person votes Republican.

$P(N)=0.40,\ P(S)=0.10,\ P(M)=0.25,$
$P(W)=0.25,\ P(E\mid N)=0.40,\ P(E\mid S)=0.56,$
$P(E\mid M)=0.48,\ P(E\mid W)=0.52$

a. $P(E) = P(E\cap N)+P(E\cap S)$
$\qquad +P(E\cap M)+P(E\cap W)$
$= P(N)\cdot P(E\mid N)+P(S)\cdot P(E\mid S)$
$\qquad +P(M)\cdot P(E\mid M)$
$\qquad +P(W)\cdot P(E\mid W)$
$= 0.40\cdot 0.40 + 0.10\cdot 0.56 + 0.25\cdot 0.48$
$\qquad +0.25\cdot 0.52$
$= 0.466$

The probability that a person chosen at random votes Republican is 0.466.

b. $P(N\mid E) = \dfrac{P(N\cap E)}{P(E)} = \dfrac{P(N)\cdot P(E\mid N)}{P(E)}$

$= \dfrac{(0.40)(0.40)}{0.466} \approx 0.343$

If a person has voted Republican, there is a 34.3% probability the person is from the Northeast.

47. Define the events A_1: the nurse forgets to give Mr. Brown his pill, A_2: the nurse remembers to give Mr. Brown his pill, and E: Mr. Brown dies.

$P(A_1)=\dfrac{2}{3},\ P(A_2)=\dfrac{1}{3},\ P(E\mid A_1)=\dfrac{3}{4},$

$P(E\mid A_2)=\dfrac{1}{3}$

$P(E) = P(A_1)\cdot P(E\mid A_1)+P(A_2)\cdot P(E\mid A_2)$
$= \dfrac{2}{3}\cdot\dfrac{3}{4}+\dfrac{1}{3}\cdot\dfrac{1}{3}=\dfrac{11}{18}$

$P(A_1 \mid E) = \dfrac{P(A_1 \cap E)}{P(E)} = \dfrac{P(A_1)\cdot P(E\mid A_1)}{P(E)}$

$= \dfrac{\frac{2}{3}\cdot\frac{3}{4}}{\frac{11}{18}} = \dfrac{9}{11} \approx 0.818$

The probability the nurse forgot to give Mr. Brown his pill given that he died is 0.818.

49. Define the events E: an accident occurs, A: a driver is young driver.

$P(A)=0.13,\ P(\overline{A})=1-P(A)=1-0.13=0.87,$
$P(E\mid\overline{A})=0.024$

a. Since young drivers have 2.6 times more accidents than older drivers, the probability a young driver is in an accident is

$P(E\mid A)=2.6\cdot P(E\mid\overline{A})$
$= 2.6\cdot 0.024 = 0.0624$

$P(A\mid E) = \dfrac{P(A\cap E)}{P(E)}$

$= \dfrac{P(A)\cdot P(E\mid A)}{P(A)\cdot P(E\mid A)+P(\overline{A})\cdot P(E\mid\overline{A})}$

$= \dfrac{0.13\cdot 0.0624}{0.13\cdot 0.0624+0.87\cdot 0.024} = 0.280$

b. If 20% of the insureds are young drivers, then 80% are older drivers.

$P(\overline{A}\mid E) = \dfrac{P(\overline{A})\cdot P(E\mid\overline{A})}{P(A)\cdot P(E\mid A)+P(\overline{A})\cdot P(E\mid\overline{A})}$

$= \dfrac{0.87\cdot 0.024}{0.13\cdot 0.0624+0.87\cdot 0.024} \approx 0.720$

The probability that the driver was 25 or older is 0.720.

51. Define the events A_1: a student has the HIV virus, A_2: a student does not have the HIV virus, and E: the Elias test is positive.

a. $P(A_1)=0.002,\ P(A_2)=0.998,$
$P(E\mid A_1)=0.998,\ P(E\mid A_2)=0.002.$
$P(E) = P(A_1)\cdot P(E\mid A_1)+P(A_2)\cdot P(E\mid A_2)$
$= 0.002\cdot 0.998+0.998\cdot 0.002 \approx 0.004$

$P(A_1\mid E) = \dfrac{P(A_1\cap E)}{P(E)} = \dfrac{P(A_1)\cdot P(E\mid A_1)}{P(E)}$

$= \dfrac{(0.002)(0.998)}{0.004} = 0.499$

A student with a positive Elias test has a probability of 0.5 of having the HIV virus.

b. $P(A_1) = 0.05$, $P(A_2) = 0.95$,
$P(E \mid A_1) = 0.998$, $P(E \mid A_2) = 0.002$.
$$P(E) = P(A_1) \cdot P(E \mid A_1) + P(A_2) \cdot P(E \mid A_2)$$
$$= 0.05 \cdot 0.998 + 0.95 \cdot 0.002 \approx 0.0518$$
$$P(A_1 \mid E) = \frac{P(A_1 \cap E)}{P(E)} = \frac{P(A_1) \cdot P(E \mid A_1)}{P(E)}$$
$$= \frac{(0.05)(0.998)}{0.0518} \approx 0.963$$

In a high risk area a person with a positive Elias test has a 0.963 probability of having the HIV virus.

c. Define event F: A person has 2 positive Elias tests. Assuming, from part (a) that $P(A_1) = 0.002$ and that the tests are independent, we have
$$P(F \mid A_1) = \left[P(E \mid A_1)\right]^2 = 0.998^2 \approx 0.996$$
and
$$P(F \mid A_2) = \left[P(E \mid A_2)\right]^2$$
$$= 0.002^2 = 0.000004.$$
$$P(F) = P(A_1) \cdot P(F \mid A_1) + P(A_2) \cdot P(F \mid A_2)$$
$$= 0.002 \cdot 0.996 + 0.998 \cdot 0.000004$$
$$\approx 0.002$$
$$P(A_1 \mid F) = \frac{P(A_1 \cap F)}{P(F)} = \frac{P(A_1) \cdot P(F \mid A_1)}{P(F)}$$
$$= \frac{(0.002)(0.996)}{0.002} = 0.996$$

A student with 2 positive Elias tests has a 99.6% chance of having the HIV virus.

53. When F is a subset of E, $E \cap F = F$, and $P(E \cap F) = P(F)$. By the definition of conditional probability, $P(E \mid F) = \frac{P(E \cap F)}{P(F)}$, provided $P(F) \neq 0$. So, $P(E \mid F) = \frac{P(F)}{P(F)} = 1$.

8.4   Permutations

1. $10 \cdot 9 \cdot 8 \cdot 7 = 5040$ different 4-digit numbers can be formed from the given set if no digit is to be repeated.

3. $0! = \underline{1}$; $3! = 3 \cdot 2 \cdot 1 = \underline{6}$

5. False. The number of ordered arrangements of r objects chosen from n objects, in which the n objects are distinct and repetition is allowed is n^r.

7. $\frac{5!}{2!} = \frac{5 \cdot 4 \cdot 3 \cdot 2!}{2!} = 60$

9. $\frac{10!}{8!} = \frac{10 \cdot 9 \cdot 8!}{8!} = 90$

11. $\frac{9!}{8!} = \frac{9 \cdot 8!}{8!} = 9$

13. $\frac{8!}{2! \, 6!} = \frac{8 \cdot 7 \cdot 6!}{2 \cdot 1 \cdot 6!} = 28$

15. $P(7, \, 2) = 7 \cdot 6 = 42$

17. $P(8, \, 7) = \frac{8!}{(8-7)!} = 8 \cdot 7 \cdot 6 \cdot 5 \cdot 4 \cdot 3 \cdot 2 = 40,320$

19. $P(6, \, 0) = \frac{6!}{(6-0)!} = \frac{6!}{6!} = 1$

21. $\frac{8!}{(8-3)! \, 3!} = \frac{8!}{5! \, 3!} = \frac{8 \cdot 7 \cdot 6 \cdot 5!}{5! \cdot 3 \cdot 2 \cdot 1} = 56$

23. $\frac{6!}{(6-6)! \, 6!} = \frac{6!}{0! \, 6!} = 1$

25. The ordered arrangements of length 3 formed from the letters a, b, c, d, and e are:
abc, abd, abe, acb, acd, ace, adb, adc, ade, aeb, aec, aed, bac, bad, bae, bca, bcd, bce, bda, bdc, bde, bea, bec, bed, cab, cad, cae, cba, cbd, cbe, cda, cdb, cde, cea, ceb, ced, dab, dac, dae, dba, dbc, dbe, dca, dcb, dce, dea, deb, dec, eab, eac, ead, eba, ebc, ebd, eca, ecb, ecd, eda, edb, edc
$P(5, 3) = 5 \cdot 4 \cdot 3 = 60$

27. The ordered arrangements of length 3 formed from the objects 1, 2, 3, and 4 are:
123, 124, 132, 134, 142, 143, 213, 214, 231, 234, 241, 243, 312, 314, 321, 324, 341, 342, 412, 413, 421, 423, 431, 432
$P(4, 3) = 4 \cdot 3 \cdot 2 = 24$

29. Each letter of the code has 4 possible symbols, so there are $4^2 = 16$ possible two-letter codes using the letters A, B, C, and D.

31. Each digit of the number has two possible values, so there are $2^3 = 8$ possible three-digit numbers formed using the digits 0 and 1.

33. Four people can be lined up in $4! = 4 \cdot 3 \cdot 2 \cdot 1 = 24$ ways.

35. There are $5 \cdot 4 \cdot 3 = 60$ different three-letter codes that can be formed from the 5 letters A, B, C, D, and E if no letter can be used more than once. This is the same as finding $P(5, 3) = 60$.

37. To find how many ways there are to seat 5 people in 8 chairs, find the number of ways to arrange 8 objects taken 5 at a time. This is $P(8, 5) = 8 \cdot 7 \cdot 6 \cdot 5 \cdot 4 = 6720$. So there are 6720 ways to seat 5 people in 8 chairs.

39. Since the sets of symbols one-letter long, two-letters long, and three-letters long are disjoint, we can add the number of elements in each set.

$$\begin{pmatrix} \text{maximum number} \\ \text{of companies} \\ \text{in NYSE} \end{pmatrix}$$

$$= n\begin{pmatrix} \text{companies with} \\ \text{1-letter symbol} \end{pmatrix} + n\begin{pmatrix} \text{companies with} \\ \text{2- letter symbol} \end{pmatrix}$$

$$+ n\begin{pmatrix} \text{companies with} \\ \text{3 -letter symbol} \end{pmatrix}$$

$$= 26 + 26^2 + 26^3 = 18,278$$

So, there can be 18,278 different companies listed on the New York Stock Exchange.

41. This is an ordered arrangement where repetition is allowed, so the number of ways the firm can purchase the printers is $7^3 = 343$.

43. This is the number of permutations of 14 shows taken 4 at a time.

$$P(14, 4) = \frac{14!}{(14-4)!} = \frac{14!}{10!}$$
$$= 14 \cdot 13 \cdot 12 \cdot 11 = 24,024$$

The network can create 24,024 different comedy line-ups.

45. This is the number of permutations of 11 businesses taken 6 at a time.

$$P(11, 6) = \frac{11!}{(11-6)!} = \frac{11!}{5!} = 332,640$$

There are 332,640 different promotion schedules possible.

47. She needs the number of permutations of 21 facilities taken 8 at a time.

$$P(21, 8) = \frac{21!}{(21-8)!} = \frac{21!}{13!} = 8,204,716,800$$

There are 8,204,716,800 ways that she can schedule the first month's itinerary.

49. This is a permutation in which 2 different days are selected from a possible 365 days without a repetition. It is given by

$$P(365, 2) = \frac{365!}{(365-2)!} = \frac{365 \cdot 364 \cdot 363!}{363!}$$
$$= 365 \cdot 364 = 132,860$$

So, 2 people can each have a different birthday 132,860 different ways.

51. a. SUNDAY has 6 letters, and none of them are repeated. They can be arranged $P(6, 6) = 6! = 720$ different ways.

b. If the letter S must come first, then we are really only arranging 5 letters. This can be done $P(5, 5) = 5! = 120$ different ways.

c. If the letter S must come first and the letter Y must come last, then only four letters are being arranged. It is the permutation of 4 objects which can be done $P(4, 4) = 4! = 24$ ways.

53. Here we have to select 8 of the 12 children to whom we will distribute the books and then distribute the books among them. This is the number of permutations

$$P(12, 8) = \frac{12!}{(12-8)!}$$
$$= \frac{12 \cdot 11 \cdot 10 \cdot 9 \cdot 8 \cdot 7 \cdot 6 \cdot 5 \cdot 4!}{4!}$$
$$= 19,958,400$$

The 8 books can be distributed to the 12 children 19,958,400 different ways.

55. We need to choose the three winning lottery tickets from the 1500 tickets sold. Because the prizes differ, the order the tickets is chosen is important. The winning tickets can be chosen in

$$P(1500, 3) = \frac{1500!}{(1500-3)!}$$
$$= \frac{1500 \cdot 1499 \cdot 1498 \cdot 1497!}{1497!}$$
$$= 3,368,253,000 \text{ different ways.}$$

57. The number of ways to choose the president, vice-president, secretary, and treasurer from a club with 15 members (assuming no one individual holds more than one position) is given by $P(15, 4)$.

$$P(15, 4) = \frac{15!}{(15-4)!} = \frac{15 \cdot 14 \cdot 13 \cdot 12 \cdot 11!}{11!} = 32,760$$

So, the 4 officers can be chosen 32,760 different ways.

8.5 Combinations

1. A <u>combination</u> is an arrangement, without regard to order, of r objects chosen from n distinct objects without repetition, where $r \leq n$.

3. False. If $r \leq n$, then $P(n, r) = r!C(n, r)$.

5. $C(6, 4) = \dfrac{6!}{(6-4)!\, 4!} = \dfrac{6 \cdot 5 \cdot 4!}{2!\, 4!} = \dfrac{6 \cdot 5}{2} = 15$

7. $C(7, 2) = \dfrac{7!}{(7-2)!\ 2!} = \dfrac{7 \cdot 6 \cdot 5!}{5!\ 2!} = \dfrac{7 \cdot 6}{2 \cdot 1} = 21$

9. $C(5, 1) = \dfrac{5!}{(5-1)!\ 1!} = \dfrac{5 \cdot 4!}{4!\ 1!} = 5$

11. $C(8, 6) = \dfrac{8!}{(8-6)!\ 6!} = \dfrac{8 \cdot 7 \cdot 6!}{2!\ 6!} = \dfrac{8 \cdot 7}{2 \cdot 1} = 28$

13. The combinations of 5 objects a, b, c, d, and e taken three at a time are: abc, abd, abe, acd, ace, ade, bcd, bce, bde, cde

$$C(5, 3) = \dfrac{5!}{(5-3)!\ 3!} = \dfrac{5 \cdot 4 \cdot 3!}{2!\ 3!} = \dfrac{5 \cdot 4}{2 \cdot 1} = 10$$

15. The combinations of 4 objects 1, 2, 3, and 4 taken 3 at a time are: 123, 124, 134, 234

$$C(4, 3) = \dfrac{4!}{(4-3)!\ 3!} = \dfrac{4 \cdot 3!}{1!\ 3!} = 4$$

17. A committee of 4 students chosen from a pool of 7 students can be formed $C(7, 4) = 35$ ways.

$$C(7, 4) = \dfrac{7!}{(7-4)!\ 4!} = \dfrac{7 \cdot 6 \cdot 5 \cdot 4!}{3!\ 4!} = \dfrac{7 \cdot 6 \cdot 5}{3 \cdot 2 \cdot 1} = 35$$

19. The math department needs to select 4 professors without regard to order from a department of 17 eligible teachers. This can be done in

$$C(17, 4) = \dfrac{17!}{(17-4)!\ 4!} = \dfrac{17 \cdot 16 \cdot 15 \cdot 14 \cdot 13!}{13!\ 4!}$$
$$= \dfrac{17 \cdot 16 \cdot 15 \cdot 14}{4 \cdot 3 \cdot 2 \cdot 1} = 2380 \text{ ways.}$$

21. A committee is an unordered selection of people. We are interested in a committee of 3 members chosen from the 20 members of the Math Club. The committee can be selected in

$$C(20, 3) = \dfrac{20!}{(20-3)!\ 3!} = \dfrac{20 \cdot 19 \cdot 18 \cdot 17!}{17!\ 3!}$$
$$= \dfrac{20 \cdot 19 \cdot 18}{3 \cdot 2 \cdot 1} = 1140 \text{ ways.}$$

23. A bit is either a 0 or a 1. An 8-bit string is a list of 8 digits, all of which are either 0s or 1s. To determine how many 8-bit strings have exactly three 1s we need to select 3 positions to place the 1s. This can be done in

$$C(8, 3) = \dfrac{8!}{(8-3)!\ 3!} = \dfrac{8 \cdot 7 \cdot 6 \cdot 5!}{5!\ 3!} = \dfrac{8 \cdot 7 \cdot 6}{3 \cdot 2 \cdot 1} = 56$$

ways.

25. In the word ECONOMICS there are 9 letters, but they are not all distinct. There are 2 C's, 2 O's, and one each of the remaining 5 letters. We want the number of permutations of 9 objects, not all of which are distinct. The number of 9-letter words that can be formed is given by

$$\dfrac{9!}{2!\ 2!\ 1!\ 1!\ 1!\ 1!\ 1!} = 90,720.$$

27. There are 12 colored lights to be arranged in a string, but the colors are not all distinct. There are 3 reds, 4 yellows, and 5 blues. We want the number of permutations of 12 objects, not all of which are distinct. The number of different colored light arrangements possible is

$$\dfrac{12!}{3!\ 4!\ 5!} = \dfrac{12 \cdot 11 \cdot 10 \cdot 9 \cdot 8 \cdot 7 \cdot 6 \cdot 5!}{3 \cdot 2 \cdot 1 \cdot 4 \cdot 3 \cdot 2 \cdot 1 \cdot 5!} = 27,720.$$

29. This problem consists of two tasks: selecting the boys for the committee which can be done in $C(4, 2)$ ways and selecting the girls for the committee which can be done $C(8, 3)$ ways. Then by the Multiplication Principle, we find that the committee can be formed in

$$C(4, 2) \cdot C(8, 3) = \dfrac{4!}{(4-2)!\ 2!} \cdot \dfrac{8!}{(8-3)!\ 3!}$$
$$= \dfrac{4 \cdot 3 \cdot 2!}{2 \cdot 1 \cdot 2!} \cdot \dfrac{8 \cdot 7 \cdot 6 \cdot 5!}{5! \cdot 3 \cdot 2 \cdot 1}$$
$$= 6 \cdot 56 = 336 \text{ ways.}$$

31. Once on a team the children are no longer distinct, so placing people on teams is much like forming words in which all the letters are not distinct. The 12 children can be placed on 3 teams, a first having 3 players, a second having 5 players and a third having 4 players in

$$\dfrac{12!}{3! \cdot 5! \cdot 4!} = \dfrac{12 \cdot 11 \cdot 10 \cdot 9 \cdot 8 \cdot 7 \cdot 6 \cdot 5!}{3 \cdot 2 \cdot 1 \cdot 5! \cdot 4 \cdot 3 \cdot 2 \cdot 1}$$
$$= 1 \cdot 11 \cdot 10 \cdot 9 \cdot 4 \cdot 7 \cdot 1 = 27,720 \text{ ways.}$$

33. There are two tasks here. The first is to choose the 5 basic numbers. The number of ways of choosing these numbers is $C(55, 5)$. Then the Powerball number is chosen. By the Multiplication Principle, the number of distinct Powerball tickets is

$$C(55, 5) \cdot 42 = \dfrac{55!}{50! \cdot 5!} \cdot 42 = 3,478,761 \cdot 42$$
$$= 146,107,962.$$

35. We first select one member from each division, then we pool the remaining faculty to choose the 2 members at large. Finally we use the Multiplication Principle to find the number of negotiating teams possible.

$$C(25, 1) \cdot C(23, 1) \cdot C(15, 1) \cdot C(8, 1)$$
$$\cdot C(19, 1) \cdot C(85, 2)$$
$$= 25 \cdot 23 \cdot 15 \cdot 8 \cdot 19 \cdot \frac{85!}{83! \cdot 2!} = 4,680,270,000$$

There are 4,680,270,000 different negotiating teams possible.

37. We want to select 8 accounts from 58 accounts without regard to the order of selection. This can be done $C(58, 8) = 1,916,797,311$ different ways.

39. The 100 senators can be placed on the committees in

$$\frac{100!}{22! \cdot 13! \cdot 10! \cdot 5! \cdot 16! \cdot 17! \cdot 17!}$$
$$= 1.157 \times 10^{76} \text{ ways.}$$

41. Each experiment consists of choosing 2 stocks from 10 stocks which is done $C(10, 2) = 45$ ways.

a. Define the event E as both stock prices are greater than \$50.

$$P(E) = \frac{C(3, 2)}{C(10, 2)} = \frac{3}{45} = \frac{1}{15}$$

b. Define the event F as neither stock price is above \$50. This is equivalent to choosing 2 stocks from the 7 with prices less than \$50.

$$P(F) = \frac{C(7, 2)}{C(10, 2)} = \frac{\frac{7!}{2! \cdot 5!}}{45} = \frac{21}{45} = \frac{7}{15}$$

c. Define the event G as at least one stock is priced above \$50.

$$P(G) = 1 - P(F) = 1 - \frac{7}{15} = \frac{8}{15}$$

43. a. The Puppies of the Dow are AT&T, Home Depot, Kraft, Pfizer, and Verizon.

b. Define event E as both have yields above 4%. Three of the five Puppies of the Dow have yields above 4%.

$$P(E) = \frac{C(3, 2)}{C(5, 3)} = \frac{3}{10}$$

c. Define event F as exactly one will have a yield above 4%.

$$P(F) = \frac{C(3, 1) \cdot C(2, 1)}{C(5, 3)} = \frac{6}{10} = \frac{3}{5}$$

d. Define event G as neither has a yield above 4%.

$$P(G) = \frac{C(2, 2)}{C(5, 3)} = \frac{1}{10}$$

e. Define event H as at least one will have a yield above 4%.

$$P(H) = P(\overline{G}) = 1 - P(G) = 1 - \frac{1}{10} = \frac{9}{10}$$

45. The experiment consists of choosing 5 accounts from a list of 35 accounts. The accounts can be partitioned into two groups: 32 that are correct and 3 that contain errors. The number of elements in the sample space is $C(35, 5) = 324,632$.

a. The probability of choosing 5 accounts without errors is

$$\frac{C(32, 5) \cdot C(3, 0)}{C(35, 5)} = \frac{\frac{32!}{5! \cdot 27!} \cdot 1}{\frac{35!}{5! \cdot 30!}} = \frac{116}{187} \approx 0.620.$$

b. The probability of choosing exactly 1 account with errors is

$$\frac{C(32, 4) \cdot C(3, 1)}{C(35, 5)} = \frac{\frac{32!}{4! \cdot 28!} \cdot 3}{324,632} = \frac{435}{1309} \approx 0.332.$$

c. The probability of choosing at least 2 accounts containing errors is easiest done by using the complement. The probability of choosing at least 2 accounts containing errors is 1 minus the probability of choosing fewer than 2 accounts with errors.

$$1 - \left[P\left(\begin{array}{c}5 \text{ accounts} \\ \text{with 0 errors}\end{array}\right) + P\left(\begin{array}{c}4 \text{ accounts} \\ \text{with 0 errors}\end{array}\right)\right]$$
$$= 1 - 0.620 - 0.332 = 0.048$$

47. The codes are formed by a letter, 5 numbers, and 2 final letters. Repetition is allowed and order is important.

a. There are
$$26 \cdot 10 \cdot 10 \cdot 10 \cdot 10 \cdot 10 \cdot 26 \cdot 26 = 26^3 \cdot 10^5$$
$$= 1,757,600,000 \text{ different SIM card codes.}$$

b. The number of codes that end in AA is
$26 \cdot 10^5 = 2,600,000$. So the probability a code
ends in AA is

$$\frac{n(\text{codes ending in AA})}{n(\text{codes})} = \frac{2,600,000}{1,757,600,000}$$
$$= \frac{26}{17576} \approx 0.00148.$$

c. The number of codes that begin with A and
end with A is the same as the number that end
in AA. So the probability a code begins and
ends in A is 0.00148.

d. The number of codes with no repeated letter
and no repeated number is given by
$P(26, 3) \cdot P(10, 5) = 15,600 \cdot 30,240$
$= 471,744,000$.
The probability a code has no repeated letter
and no repeated number is

$$\frac{P(26, 3) \cdot P(10,5)}{26^3 \cdot 10^5} = \frac{471,744,000}{1,757,600,000}$$
$$\approx 0.268.$$

e. The number of codes that have no repeated
letter and all five of the same numbers is
$P(26, 3) \cdot 10 = 15,600 \cdot 10 = 156,000$.
The probability a code has no repeated letter
and all five of the same number is

$$\frac{156,000}{1,757,600,000} \approx 0.0000888.$$

49. The number of elements in the sample space S is
equal to the number of combinations of 50
refrigerators taken 5 at a time is

$$C(50, 5) = \frac{50!}{5! \cdot 45!} = 2,118,760.$$

Define E as the event, "5 refrigerators are
defective."

$$P(E) = \frac{C(6, 5)}{C(50, 5)} = \frac{6}{2,118,760} \approx 2.83 \times 10^{-6}$$

Define F as the event, "at least 2 refrigerators
are defective."

$$P(F) = 1 - P(0 \text{ defective} \cup 1 \text{ defective})$$
$$= 1 - \left[\begin{array}{c} \dfrac{C(6, 0) \cdot C(44, 5)}{2,118,760} \\[2mm] + \dfrac{C(6,1) \cdot C(44,4)}{2,118,760} \end{array} \right]$$
$$= 1 - \left[\dfrac{1,086,008}{2,118,760} + \dfrac{6 \cdot (135,751)}{2,118,760} \right]$$
$$\approx 0.103$$

51. The sample space consists of all four digit
permutations (with repetition).
$n(S) = 10^4 = 10,000.$

a. Define event E as the last two digits are 0s.
$n(E) = 10^2 = 100$. The probability of E is

$$P(E) = \frac{n(E)}{n(S)} = \frac{100}{10,000} = \frac{1}{100} = 0.01.$$

b. Define event F as the last two digits are 0s,
but the first digit is not 0. $n(F) = 9 \cdot 10 = 90$.
The probability of F is

$$P(F) = \frac{n(F)}{n(S)} = \frac{90}{10,000} = \frac{9}{1000} = 0.009.$$

53. The experiment consists of choosing a
committee of 7 senators from the 100 in the
Senate. The number of elements in the sample
space is $n(S) = C(100, 7)$.

a. Define the event E as the committee has
only Democratic members. (There are 59
Democrats.)

$$P(E) = \frac{n(E)}{n(S)} = \frac{C(59,7)}{C(100,7)} \approx 0.0213$$

b. Define the event F as the committee has
only Republican members. (There are 41
Republicans.)

$$P(F) = \frac{n(F)}{n(S)} = \frac{C(41, 7)}{C(100, 7)} \approx 0.0014$$

c. Define the event G as the committee has 4
Democrats and 3 Republicans.

$$P(G) = \frac{n(G)}{n(S)} = \frac{C(59, 4) \cdot C(41, 3)}{C(100,7)} \approx 0.3031$$

55. The experiment consists of choosing 2 printers
from 100. The sample space is all combinations
of the 2 printers that can be chosen.
$n(S) = C(100, 2)$. Define event E as the two
printers chosen work. $n(E) = C(95, 2)$. The
probability the shipment is rejected is $P(\overline{E})$.

$$P(\overline{E}) = 1 - P(E) = 1 - \frac{C(95,2)}{C(100,2)}$$
$$= 1 - \frac{893}{990} = \frac{97}{990} \approx 0.0980$$

57. The sample space, S, consists of all the possible
combinations of 5 cards; $n(S) = C(52, 5)$.

a. Define event E: All are hearts.
$n(E) = C(13, 5)$.

$$P(E) = \frac{n(E)}{n(S)} = \frac{C(13, 5)}{C(52, 5)} \approx 0.0005$$

b. Define event F: Exactly 4 are spades. The number of elements in F is determined as follows: Exactly 4 spades means 4 spades and 1 non-spade. The 4 spades can be chosen in $C(13, 4)$ ways. The non-spade can be chosen in $C(39, 1)$ ways. Using the Multiplication Principle, $C(F) = C(13, 4) \cdot C(39, 1)$.

$$P(F) = \frac{n(F)}{n(S)} = \frac{C(13, 4) \cdot C(39, 1)}{C(52, 5)} = 0.0107$$

c. Define event G: Exactly 2 are clubs. The number of elements in E is determined as follows: Exactly 2 clubs means 2 clubs and 3 non-clubs. The 2 clubs can be chosen $C(13, 2)$ ways and the 3 non-clubs can be chosen $C(39, 3)$ ways. Using the Multiplication Principle, $C(G) = C(13, 2) \cdot C(39, 3)$.

$$P(G) = \frac{n(G)}{n(S)} = \frac{C(13, 2) \cdot C(39, 3)}{C(52, 5)} = 0.2743.$$

59. The sample space, S, consists of all the possible combinations of 5 cards; $n(S) = C(52, 5)$.

a. Define event E: The hand is a royal flush; $n(E) = 4$ (One from each of 4 suits.).

$$P(E) = \frac{n(E)}{n(S)} = \frac{4}{C(52, 5)} \approx 1.539 \times 10^{-6}$$

b. Define event F: A hand consists of a straight flush. The number of elements in F is determined as follows: Each suit has 9 straight flushes, (cards A, 2, 3, 4, 5; 2, 3, 4, 5, 6; 3, 4, 5, 6, 7; ... ; 9, 10, J, Q, K) and there are 4 suits. Using the Multiplication Principle, $n(F) = 4 \cdot 9 = 36$.

$$P(F) = \frac{n(F)}{n(S)} = \frac{36}{C(52, 5)} \approx 1.385 \times 10^{-5}$$

c. Define the event G: A hand contains 4 of a kind. G is equivalent to 4 cards of one face value and 1 of another.
$n(G) = C(13, 1) \cdot C(4, 4) \cdot C(48, 1) = 624$

$$P(G) = \frac{n(G)}{n(S)} = \frac{624}{C(52, 5)} \approx 2.401 \times 10^{-4}$$

d. Define event H: A hand contains one pair and one triple of the same face values. The number of elements in H can be determined as follows. First, choose two suits for the pair, $C(4, 2) = 6$ ways, and then choose a face value for the pair, $C(13, 1) = 13$. Next, choose three suits for the triple, $C(4, 3) = 4$, and then choose a face value for the triple, $C(12, 1) = 12$. (One face value had been used to make the pair.)

Finally, use the Multiplication Principle to find

$$n(H) = C(4, 2) \cdot C(13, 1) \cdot C(4, 3) \cdot C(12, 1)$$
$$= 6 \cdot 13 \cdot 4 \cdot 12 = 3744.$$

$$P(H) = \frac{n(H)}{n(S)} = \frac{3744}{C(52, 5)} = 0.0014$$

e. Define event K: The hand contains 5 nonconsecutive cards from a single suit. The number of elements in K is determined as follows. First, choose the suit to be used, $C(4, 1) = 4$, and then choose 5 cards from the suit, $C(13, 5) = 1287$. Next, eliminate the cards that are in order, that is, the royal flushes from part (a) and the straight flushes from part (b).
$n(K) = 4 \cdot C(13, 5) - (4 + 36) = 5108$

$$P(K) = \frac{n(K)}{n(S)} = \frac{5108}{C(52, 5)} = 0.0020$$

f. Define event L: The hand consists of 5 cards in order. The number of elements in L is determined as follows. First, there are 10 straights. (A, 2, 3, 4, 5; 2, 3, 4, 5, 6; 3, 4, 5, 6, 7; ... ; 10, J, Q, K, A). Each card in the straight can come from any one of 4 suits, so can be chosen 4^5 ways. Next, eliminate the cards that are from the same suits, that is, the royal flushes from part (a) and the straight flushes from part (b).
$n(K) = 10 \cdot 4^5 - (4 + 36) = 10{,}200$

$$P(L) = \frac{n(L)}{n(S)} = \frac{10{,}200}{C(52, 5)} = 0.0039$$

8.6   The Binomial Probability Model

1. False. In a Bernoulli trial the probability of each outcome remains the same for each trial.

3. In a Bernoulli trial, the probability of exactly k successes in n trial is

$$b = (n, k; p) = \binom{n}{k} p^k q^{n-k} = \frac{n!}{k!(n-k)!} p^k q^{n-k},$$

where p is the probability of success and $q = 1 - p$ is the probability of failure.

5. True

7. $b(7, 4; 0.20) = \binom{7}{4}(0.2)^4 (0.8)^3$
$$= (35)(0.0016)(0.512) = 0.0287$$

9. $b(15, 8; 0.80) = \binom{15}{8}(0.80)^8 (0.2)^7$

$= (6435)(0.16777)(1.28 \times 10^{-5})$

≈ 0.01382

11. $b\left(15, 10; \frac{1}{2}\right) = \binom{15}{10}\left(\frac{1}{2}\right)^{10}\left(\frac{1}{2}\right)^5 \approx 0.09164$

13. $b(15, 3; 0.3) + b(15, 2; 0.3) + b(15, 1; 0.3)$
$\qquad + b(15, 0; 0.3)$

$= \binom{15}{3}(0.3)^3 (0.7)^{12} + \binom{15}{2}(0.3)^2 (0.7)^{13}$

$+ \binom{15}{1}(0.3)^1 (0.7)^{14} + \binom{15}{0}(0.3)^0 (0.7)^{15}$

$\approx 0.17004 + 0.09156 + 0.03052 + 0.00475$

$= 0.29687$

15. $n = 3, k = 2, p = \frac{1}{3}$

$b\left(3, 2; \frac{1}{3}\right) = \binom{3}{2}\left(\frac{1}{3}\right)^2\left(\frac{2}{3}\right)^1 = \frac{2}{9} \approx 0.2222$

17. $n = 3, k = 0, p = \frac{1}{6}$

$b\left(3, 0; \frac{1}{6}\right) = \binom{3}{0}\left(\frac{1}{6}\right)^0\left(\frac{5}{6}\right)^3 = \frac{125}{216} \approx 0.5787$

19. $n = 5, k = 3, p = \frac{2}{3}$

$b\left(5, 3; \frac{2}{3}\right) = \binom{5}{3}\left(\frac{2}{3}\right)^3\left(\frac{1}{3}\right)^2 = \frac{80}{243} \approx 0.3292$

21. $n = 10, k = 6, p = 0.3$

$b(10, 6; 0.3) = \binom{10}{6}(0.3)^6 (0.7)^4 \approx 0.0368$

23. $n = 12, k = 9, p = 0.8$

$b(12, 9; 0.8) = \binom{12}{9}(0.8)^9 (0.2)^3 \approx 0.2362$

25. $n = 8, p = 0.30$

The probability P of at least 5 successes is the probability of 5 or 6 or 7 or 8 successes. Since the events are mutually exclusive we can add the probabilities.

$P = b(8, 5; 0.30) + b(8, 6; 0.30)$
$\qquad + b(8, 7; 0.30) + b(8, 8; 0.30)$

$= \binom{8}{5}(0.30)^5 (0.70)^3 + \binom{8}{6}(0.30)^6 (0.70)^2$

$+ \binom{8}{7}(0.30)^7 (0.70)^1 + \binom{8}{8}(0.30)^8 (0.70)^0$

$\approx 0.0467 + 0.0100 + 0.0012 + 0.0001$

$= 0.0580$

27. $n = 8, k = 1; p = 0.5$
$P(\text{exactly 1 head}) = b(8, 1; 0.5)$

$= \binom{8}{1}(0.5)^1 (0.5)^7 \approx 0.03125$

29. $n = 8; p = 0.5$

$P\binom{\text{at least}}{\text{5 tails}} = P\binom{\text{exactly}}{\text{5 tails}} + P\binom{\text{exactly}}{\text{6 tails}}$

$\qquad + P\binom{\text{exactly}}{\text{7 tails}} + P\binom{\text{exactly}}{\text{8 tails}}$

$= b(8, 5; 0.5) + b(8, 6; 0.5)$
$\qquad + b(8, 7; 0.5) + b(8, 8; 0.5)$

$= \binom{8}{5}(0.5)^5 (0.5)^3 + \binom{8}{6}(0.5)^6 (0.5)^2$

$\qquad + \binom{8}{7}(0.5)^7 (0.5)^1$

$\qquad + \binom{8}{8}(0.5)^8 (0.5)^0$

≈ 0.3633

31. $n = 8; p = 0.5$

$P\binom{\text{at least}}{\text{1 head}} = P\binom{\text{exactly}}{\text{1 head}} + P\binom{\text{exactly}}{\text{2 heads}} + \cdots$

$\qquad + P\binom{\text{exactly}}{\text{8 heads}}$

$= 1 - P(0 \text{ heads})$

The intersection of "at least 1 head" and "exactly 2 heads" is "exactly 2 heads."

$P\left(\left.\begin{matrix}\text{exactly} \\ \text{2 heads}\end{matrix}\right| \begin{matrix}\text{at least} \\ \text{1 head}\end{matrix}\right) = \frac{P\binom{\text{exactly}}{\text{2 heads}}}{1 - P(0 \text{ heads})}$

$= \frac{b(8, 2; 0.5)}{1 - b(8, 0; 0.5)} \approx 0.1098$

33. The probability of rolling a sum of 7 with two dice is $\frac{1}{6}$. So, we have $n = 5, k = 2, p = \frac{1}{6}$.

$P\binom{\text{exactly 2}}{\text{sums of 7}} = b\left(5, 2; \frac{1}{6}\right) \approx 0.1608$

35. a.

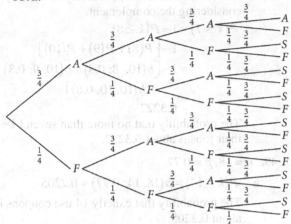

b. $P\begin{pmatrix} \text{exactly 2 successes,} \\ \text{2 failures} \end{pmatrix}$

$= P(SSFF) + P(SFSF) + P(SFFS)$

$\quad + P(FSSF) + P(FSFS) + P(FFSS)$

$= \left(\dfrac{3}{4}\right)\left(\dfrac{3}{4}\right)\left(\dfrac{1}{4}\right)\left(\dfrac{1}{4}\right) + \left(\dfrac{3}{4}\right)\left(\dfrac{1}{4}\right)\left(\dfrac{3}{4}\right)\left(\dfrac{1}{4}\right)$

$\quad + \left(\dfrac{3}{4}\right)\left(\dfrac{1}{4}\right)\left(\dfrac{1}{4}\right)\left(\dfrac{3}{4}\right) + \left(\dfrac{1}{4}\right)\left(\dfrac{3}{4}\right)\left(\dfrac{3}{4}\right)\left(\dfrac{1}{4}\right)$

$\quad + \left(\dfrac{1}{4}\right)\left(\dfrac{3}{4}\right)\left(\dfrac{1}{4}\right)\left(\dfrac{3}{4}\right) + \left(\dfrac{1}{4}\right)\left(\dfrac{1}{4}\right)\left(\dfrac{3}{4}\right)\left(\dfrac{3}{4}\right)$

$= 6\left(\dfrac{9}{256}\right) = \dfrac{27}{128}$

c. $P\begin{pmatrix} \text{exactly 2 successes,} \\ \text{2 failures} \end{pmatrix} = b\left(4, 2, \dfrac{3}{4}\right)$

$= \dbinom{4}{2}\left(\dfrac{3}{4}\right)^2\left(\dfrac{1}{4}\right)^2$

$= \dfrac{27}{128}$

37. $n = 8, p = 0.05$

a. $P(\text{exactly 1 defective}) = b(8, 1; 0.05) \approx 0.2793$

b. $P(\text{exactly 2 defective}) = b(8, 2; 0.05) \approx 0.0515$

c. $P(\text{at least 1 defective}) = 1 - P(0 \text{ defective})$

$= 1 - b(8, 0; 0.05)$

$\approx 1 - 0.6634 = 0.3366$

d. $P\begin{pmatrix} \text{fewer than} \\ \text{3 defective} \end{pmatrix}$

$= P(0 \text{ defective}) + P(1 \text{ defective})$

$\quad + P(2 \text{ defective})$

$= b(8, 0; 0.05) + b(8, 1; 0.05)$

$\quad + b(8, 2; 0.05)$

≈ 0.9942

39. a. When exactly 3 are boys, $n = 6, k = 3, p = 0.5$, and $b(6, 3; 0.5) = 0.3125$. The probability that a family with 6 children has exactly 3 boys and 3 girls is 0.3125.

b. The probability at least 5 are boys means that either or 5 or 6 of the children are boys.

$P\begin{pmatrix} \text{at least} \\ \text{5 boys} \end{pmatrix} = b(6, 5; 0.5) + b(6, 6; 0.5)$

≈ 0.1094

The probability a family with 6 children has at least 5 boys is 0.1094.

c. We now consider having a girl a success. The probability of a girl is 0.5, the probability at least 2 but fewer than 4 are girls is

$P(2 \le \text{girls} < 4) = P\begin{pmatrix} \text{exactly} \\ \text{2 girls} \end{pmatrix} + P\begin{pmatrix} \text{exactly} \\ \text{3 girls} \end{pmatrix}$

$= b(6, 2; 0.5) + b(6, 3; 0.5)$

≈ 0.5469

41. $n = 20, p = 0.5$

a. $P\begin{pmatrix} \text{student guesses} \\ \text{all correct} \end{pmatrix} = b(20, 20; 0.5)$

$\approx 9.537 \times 10^{-7}$

b. The student needs 12 correct to pass, but passes if he/she gets at least 12 correct. So the probability a student passes is the probability of getting 12, 13, 14, …, 20 correct.

$P(\text{passing}) = b(20, 12; 0.5) + b(20, 13; 0.5)$

$\quad + b(20, 14; 0.5) + b(20, 15; 0.5)$

$\quad + b(20, 16; 0.5) + b(20, 17; 0.5)$

$\quad + b(20, 18; 0.5) + b(20, 19; 0.5)$

$\quad + b(20, 20; 0.5)$

≈ 0.2517

c. The odds in favor of passing are

$\dfrac{P(\text{passing})}{P(\text{failing})} \approx \dfrac{0.252}{0.748} = \dfrac{252}{748} = \dfrac{63}{187}$

or about 1 to 3.

43. $n = 15, p = 0.12$

   a. It is easier to find the probability of $k \geq 5$ by considering the complement.

$$P(\text{at least } 5) = 1 - P(k \leq 4)$$
$$= 1 - \left[P(0) + P(1) + P(2) + P(3) + P(4) \right]$$
$$= 1 - \left[b(15, 0; 0.12) + b(15, 1; 0.12) \right.$$
$$+ b(15, 2; 0.12) + b(15, 3; 0.12)$$
$$\left. + b(15, 4; 0.12) \right]$$
$$\approx 1 - 0.9735 = 0.0265$$

The probability that at least 5 of those surveyed think it is acceptable to cheat is about 0.0265.

   b. $P(k < 3) = P(0) + P(1) + P(2)$
$$= b(15, 0; 0.12) + b(15, 1; 0.12)$$
$$+ b(15, 2; 0.12)$$
$$\approx 0.7346$$

The probability that fewer than 3 people surveyed think it is acceptable to cheat is about 0.7346.

   c. $P(0) = b(15, 0; 0.12) \approx 0.1470$

The probability that no one surveyed think it is acceptable to cheat is about 0.1470.

45. $n = 8, p = 0.10$

   a. $P(k = 2) = b(8, 2; 0.10) \approx 0.1488$

The probability that two of the returns were filed on April 15 is about 0.1488.

   b. $P(k = 5) = b(8, 5; 0.10) \approx 0.0004$

The probability that five of the returns were filed on April 15 is about 0.0004.

   c. $P(k = 0) = b(8, 0; 0.10) \approx 0.4305$

The probability that none of the returns were filed on April 15 is about 0.4305.

47. $n = 10, p = 0.8$

   a. $P(k = 10) = b(10, 10; 0.8) \approx 0.1074$

The probability that all like their jobs is about 0.1074.

   b. $P(k = 9) = b(10, 9; 0.8) \approx 0.2684$

The probability that exactly nine like their jobs is about 0.2684.

   c. It is easier to find the probability of $k \leq 7$ by considering the complement.

$$P(k \leq 7) = 1 - P(k \geq 8)$$
$$= 1 - \left[P(8) + P(9) + P(10) \right]$$
$$= 1 - \left[b(10, 8; 0.8) + b(10, 9; 0.8) \right.$$
$$\left. + b(10, 10; 0.8) \right]$$
$$\approx 0.3222$$

The probability that no more than seven like their jobs is about 0.3222.

49. $n = 18, p = 0.77$

   a. $P(k = 14) = b(18, 14; 0.77) \approx 0.2205$

The probability that exactly 14 use coupons is about 0.2205.

   b. $P(k \geq 14)$
$$= P(14) + P(15) + P(16) + P(17) + P(18)$$
$$= b(18, 14; 0.77) + b(18, 15; 0.77)$$
$$+ b(18, 16; 0.77) + b(18, 17; 0.77)$$
$$+ b(18, 18; 0.77)$$
$$\approx 0.5988$$

The probability that at least 16 use coupons is about 0.5988.

   c. $P(k \geq 16) = P(16) + P(17) + P(18)$
$$= b(18, 14; 0.77) + b(18, 15; 0.77)$$
$$+ b(18, 16; 0.77)$$
$$\approx 0.1813$$

The probability that at least 14 use coupons is about 0.1813.

51. $n = 12, p = 0.15$

   a. $P(k = 0) = b(12, 0; 0.15) \approx 0.1422$

The probability that none of the 12 claims involves fraud is about 0.1422.

   b. $P(k = 6) = b(12, 6; 0.15) \approx 0.0040$

The probability that exactly 6 of the 12 claims involves fraud is about 0.0040.

   c. $P(k \geq 1) = 1 - P(0) \approx 1 - 0.1422 = 0.8578$

The probability that at least one of the 12 claims involves fraud is about 0.8578.

53. $n = 44, p = 0.12$

   a. $P(k = 2) = b(44, 2; 0.12) \approx 0.0635$

The probability that two passengers do not show is about 0.0635.

   b. $P(k = 0) = b(44, 0; 0.12) \approx 0.0036$

The probability that there are 0 no-shows is about 0.0036.

c. $P\begin{pmatrix} \text{at least 2} \\ \text{no-shows} \end{pmatrix} = 1 - P(k \geq 2)$

$= 1 - \left[P(0) + P(1) \right]$

$= 1 - \left[b(44, 0; 0.12) \right.$

$\left. + b(44, 1; 0.12) \right]$

$\approx 1 - 0.0253 = 0.9747$

The probability that at least two passengers do not show is about 0.9747.

55. $n = 4, p = 0.250$

a. $P\begin{pmatrix} \text{at least} \\ \text{2 hits} \end{pmatrix}$

$= P\begin{pmatrix} \text{exactly} \\ \text{2 hits} \end{pmatrix} + P\begin{pmatrix} \text{exactly} \\ \text{3 hits} \end{pmatrix} + P\begin{pmatrix} \text{exactly} \\ \text{4 hits} \end{pmatrix}$

$= b(4, 2; 0.250) + b(4, 3; 0.250)$

$+ b(4, 4; 0.250)$

≈ 0.2617

A player with a 0.250 batting average has a 0.2617 probability of getting at least 2 hits in 4 at bats.

b. $P\begin{pmatrix} \text{at least} \\ \text{1 hit} \end{pmatrix} = 1 - P(\text{no hits})$

$= 1 - b(4, 0; 0.250) = 1 - 0.3164$

≈ 0.6836

The probability that the batter gets at least 1 hit is about 0.6836.

57. $n = 8, k = 8, p = 0.40$
$P(\text{all 8 voters prefer Ms. Moran}) = b(8, 8; 0.40)$
$\approx 0.0007.$
The probability that all eight voters prefer Ms. Moran is about 0.07%.

59. $n = 10, k = 4, p = 0.23$
$P(\text{exactly 4 deaths are due to heart attack})$
$= b(10, 4; 0.23) \approx 0.1225.$
There is a 0.1225 probability that 4 of the next 10 unexpected deaths are due to a heart attack.

61.a. $n = 6, p = 0.5$

$P\begin{pmatrix} \text{identifies} \\ \text{at least 5} \end{pmatrix} = P\begin{pmatrix} \text{identifies} \\ \text{exactly 5} \end{pmatrix} + P\begin{pmatrix} \text{identifies} \\ \text{exactly 6} \end{pmatrix}$

$= b(6, 5; 0.5) + b(6, 6; 0.5)$

≈ 0.1094

The probability of correctly identifying at least 5 out of 6 cups of coffee if merely guessing is 0.1094.

b. $n = 6, p = 0.8$

$P\begin{pmatrix} \text{identifies} \\ \text{fewer than 5} \end{pmatrix}$

$= 1 - P\begin{pmatrix} \text{identifies} \\ \text{at least 5} \end{pmatrix}$

$= 1 - \left[P\begin{pmatrix} \text{identifies} \\ \text{exactly 5} \end{pmatrix} + \begin{pmatrix} \text{identifies} \\ \text{exactly 6} \end{pmatrix} \right]$

$= 1 - \left[b(6, 5; 0.8) + b(6, 6; 0.8) \right]$

$\approx 1 - 0.6554 = 0.3446$

The probability her claim is rejected when she really has the ability to identify the coffee is about 0.3446.

63. $n = 10, p = 0.123$

a. $P(\text{exactly 4 are 65 or older}) = b(10, 4; 0.123)$
≈ 0.0219
The probability that exactly four of the ten people are 65 or older is about 0.0219.

b. $P(\text{no one is 65 or older}) = b(10, 0; 0.123)$
≈ 0.2692
The probability that none of the ten people are 65 or older is about 0.2692.

c. $P\begin{pmatrix} \text{at most 5} \\ \geq 65 \text{ years old} \end{pmatrix}$

$= P(0) + P(1) + P(2) + P(3) + P(4) + P(5)$

$= b(10, 0; 0.123) + b(10, 1; 0.123)$

$+ b(10, 2; 0.123) + b(10, 3; 0.123)$

$+ b(10, 4; 0.123) + b(10, 5; 0.123)$

≈ 0.9995

The probability that at most five of the ten people are 65 or older is about 0.9995.

65. This is an example of 2000 Bernoulli trials, where $p = \dfrac{1}{6}$ is the probability of success.

$E = np = 2000\left(\dfrac{1}{6}\right) \approx 333.333$

We expect 333.333 fives in 2000 rolls of a fair die.

67. This can be considered an example of 500 Bernoulli trials, where $p = 0.02$ is the probability a light bulb is defective.
$E = np = 500(0.02) = 10$
The shop owner expects 10 defective light bulbs in a shipment of 500.

69. This is a Bernoulli process, with $n = 20$ and $p = 0.14$.
$E = np = (20)(0.14) = 2.8$
We expect about 3 (2.8) of the adults surveyed to say that health care costs are paramount.

71. This is an example of 500 Bernoulli trials, where $p = 0.002$ is the probability of having an unfavorable reaction to the drug.
$E = np = (500)(0.002) = 1$
The doctor can expect 1 patient to have an unfavorable reaction to the drug.

73. The code is of length 15, and corrects 1 error. The probability a digit is transmitted correctly is 0.98. So we have $n = 15$, $p = 0.98$.
The probability a message is received correctly will be the probability it was received without error plus the probability if was received with one corrected error.
$b(15, 15; 0.98) + b(15, 14; 0.98) \approx 0.9647$
The message will be received correctly about 96.47% of the time using this Hamming code.

75. The code is of length 31, and corrects up to 2 errors. The probability a digit is transmitted correctly is 0.97. So we have $n = 31$, $p = 0.97$
The probability a message is received correctly will be the probability it was received without error or with one or two corrected errors.
$b(31, 31; 0.97) + b(31, 30; 0.97)$
$\quad + b(31, 29; 0.97) \approx 0.9349$
The probability the code is correctly received is about 0.9349.

77. Estimates will vary, but all should be close to the actual theoretical probabilities. The actual values of $P(k)$ are obtained from $b(4, k; 0.2)$ where k denotes the number of heads obtained in 4 tosses of the coin. Here since the probability of getting a tail is 0.80, the probability of getting a head is 0.20.

k	Estimate of $P(k)$	Actual $P(k)$
0		0.4096
1		0.4096
2		0.1536
3		0.0256
4		0.0016

79. Estimates will vary, but they should be close to the actual theoretical probability.
The actual value of P(exactly 3 heads)
$= b(8, 3; 0.2) \approx 0.1468$.

Number of Heads, k	Actual $P(k)$
0	0.1678
1	0.3355
2	0.2936
3	0.1468
4	0.0459

5	0.0092
6	0.0011
7	0.0001
8	0.0000

Chapter 8   Review Exercises

1. $P(E \mid A) = 0.82$

3. $P(E \mid B) = 0.10$

5. $P(A \cap E) = 0.90 \cdot 0.82 = 0.738$

7. $P(B \cap E) = 0.10 \cdot 0.10 = 0.01$

9. $P(E) = P(E \cap A) + P(E \cap B)$
$\quad = P(A) \cdot P(E \mid A) + P(B)(E \mid B)$
$\quad = 0.90 \cdot 0.82 + 0.10 \cdot 0.10 = 0.748$
$P(A \mid E) = \dfrac{P(E \cap A)}{P(E)} = \dfrac{P(A) \cdot P(E \mid A)}{P(E)}$
$\quad = \dfrac{(0.90)(0.82)}{0.748} \approx 0.9866$

11. $P(E) = P(E \cap A) + P(E \cap B)$
$\quad = P(A) \cdot P(E \mid A) + P(B)(E \mid B)$
$\quad = 0.90 \cdot 0.82 + 0.10 \cdot 0.10 = 0.748$
$P(B \mid E) = \dfrac{P(E \cap B)}{P(E)} = \dfrac{P(B) \cdot P(E \mid B)}{P(E)}$
$\quad = \dfrac{(0.10)(0.10)}{0.748} \approx 0.0134$

13. $P(E \mid A) = 0.5$ **15.** $P(E \mid B) = 0.4$

17. $P(E \mid C) = 0.3$

19. $P(A \cap E) = 0.4 \cdot 0.5 = 0.2$

21. $P(B \cap E) = 0.5 \cdot 0.4 = 0.2$

23. $P(C \cap E) = 0.1 \cdot 0.3 = 0.03$

25. $P(E) = P(E \cap A) + P(E \cap B) + P(E \cap C)$
$\quad = P(A) \cdot P(E \mid A) + P(B)(E \mid B)$
$\qquad\qquad\qquad + P(C) \cdot P(E \mid C)$
$\quad = 0.4 \cdot 0.5 + 0.5 \cdot 0.4 + 0.1 \cdot 0.3 = 0.43$

27. $P(A \mid E) = \dfrac{P(E \cap A)}{P(E)} = \dfrac{P(A) \cdot P(E \mid A)}{P(E)}$
$\quad = \dfrac{(0.4)(0.5)}{0.43} \approx 0.4651$

29. $P(B \mid E) = \dfrac{P(E \cap B)}{P(E)} = \dfrac{P(B) \cdot P(E \mid B)}{P(E)}$

$= \dfrac{(0.5)(0.4)}{0.43} \approx 0.4651$

31. $P(C \mid E) = \dfrac{P(E \cap C)}{P(E)} = \dfrac{P(C) \cdot P(E \mid C)}{P(E)}$

$= \dfrac{(0.1)(0.3)}{0.43} \approx 0.0698$

33. $0! = 1$

35. $\dfrac{7!}{4!} = \dfrac{7 \cdot 6 \cdot 5 \cdot 4!}{4!} = 210$

37. $\dfrac{12!}{11!} = \dfrac{12 \cdot 11!}{11!} = 12$

39. When choosing committees order is not important so a committee of 3 people can be chosen from 5 people in

$C(5, 3) = \binom{5}{3} = \dfrac{5!}{(5-3)! \, 3!} = 10$ ways.

41. When books are arranged on a shelf different orderings mean different arrangements. So 3 books can be arranged in $P(3, 3) = 3! = 6$ ways.

43. a. The number of 3-digit words that can be formed without repeating a digit from the symbols 1, 2, 3, 4, 5, 6 is given by $P(6, 3) = 6 \cdot 5 \cdot 4 = 120$. So, there are 120 words in this language.

b. If the same letters in different a order indicate the identical word, then there are only $C(6, 3) = 20$ possible words in this language.

45. a. A committee of 3 boys and 4 girls can be formed from 7 boys and 6 girls in $C(7, 3) \cdot C(6, 4) = 35 \cdot 15 = 525$ different ways.

b. If the committee must have at least one person of each sex, the committee could consist of 1 boy and 6 girls or 2 boys and 5 girls or 3 boys and 4 girls or 4 boys and 3 girls or 5 boys and 2 girls or 6 boys and 1 girl. Since each configuration is mutually exclusive, we will determine the number of ways the committee can be formed by using the Multiplication Principle 6 times and adding the results.

$C(7, 1) \cdot C(6, 6) + C(7, 2) \cdot C(6, 5)$
$+ C(7, 3) \cdot C(6, 4) + C(7, 4) \cdot C(6, 3)$
$+ C(7, 5) \cdot C(6, 2) + C(7, 6) \cdot C(6, 1)$
$= 7 \cdot 1 + 21 \cdot 6 + 35 \cdot 15 + 35 \cdot 20$
$\qquad + 21 \cdot 15 + 7 \cdot 6$
$= 7 + 126 + 525 + 700 + 315 + 42$
$= 1715$

There are 1715 different committees.

47. Since the books of the same subject must be kept together, there are 4! ways to arrange the history books, 5! ways to arrange the English books, and 6! ways to arrange the mathematics books. However, we must also decide the order the three subjects should be arranged on the shelf. This can be done 3! ways. Then by the Multiplication Principle there are $3! \cdot 4! \cdot 5! \cdot 6! = 12{,}441{,}600$ ways to arrange the books on the shelf.

49. When choosing a committee, order is not important. So a committee of 8 boys and 5 girls can be formed form a group of 10 boys and 11 girls in $C(10, 8) \cdot C(11, 5) = 45 \cdot 462 = 20{,}790$ ways.

51. We want to choose, without regard to order, 4 plums from a box of 25. Five of the plums are rotten, so we can partition the box into 20 good plums and 5 rotten plums.

a. We can choose only good plums in $C(20, 4) = 4845$ ways.

b. To determine how many ways there are to choose 3 good plums and 1 rotten plum, we use the Multiplication Principle. Three good plums and 1 rotten plum can be chosen in $C(20, 3) \cdot C(5, 1) = 1140 \cdot 5 = 5700$ ways.

c. The easiest way to find how many ways there are to get at least 1 rotten plum when choosing 4 plums is to subtract the number of ways we can get all good plums (part a) from the total number of ways we can choose 4 plums from the box of 25. This can be done in $C(25, 4) - C(20, 4) = 12{,}650 - 4845 = 7805$ ways.

53. We need to find the number of permutations of 6 items where 2 of them are not distinct. The number of 6 letter words that can be made from the word FINITE is $\dfrac{6!}{1! \, 2! \, 1! \, 1! \, 1!} = \dfrac{720}{2} = 360$.

55. We need to find the number of permutations of 10 items 5 of which are not distinct. The number of ways of arranging the 10 books on the shelf when there are 3 copies of one book and 2 copies of another is

$$\frac{10!}{3!\ 2!\ 1!\ 1!\ 1!\ 1!\ 1!} = \frac{3,628,800}{12} = 302,400.$$

57. Since a 6 is 3 times more likely to appear than any other number, the probabilities assigned to each of the outcomes in the sample space are

$$P(1) = P(2) = P(3) = P(4) = P(5) = \frac{1}{8};\ P(6) = \frac{3}{8}$$

$E = \{(3,1), (3,2), (3,3), (3,4), (3,5), (3,6)\}$
$F = \{(1,6), (2,6), (3,6), (4,6), (5,6), (6,6)\}$

$$E \cap F = \{(3, 6)\};\ P(E \cap F) = \frac{3}{64}$$

$$P(E) \cdot P(F) = \frac{1}{8} \cdot \frac{3}{8} = \frac{3}{64} = P(E \cap F),\ \text{so the}$$

events are independent.

59. We define events E: a person has blue eyes, and F: a person has brown eyes, and G: a person is left handed. $P(E) = 0.25$, $P(F) = 0.75$, $P(G\,|\,E) = 0.10$, and $P(G\,|\,F) = 0.05$.

a. $P(E \cap G) = P(G\,|\,E) \cdot P(E)$
$= 0.10 \cdot 0.25 = 0.025$
The probability that a person is blue-eyed and left handed is 0.025.

b. Notice that $P(E) + P(F) = 0.25 + 0.75 = 1.00$. This indicates that no other eye-color is possible.
$P(G) = P(G \cap E) + P(G \cap F)$
$= 0.025 + P(G\,|\,F) \cdot P(F)$
$= 0.025 + 0.05 \cdot 0.75 = 0.0625$
The probability that a person is left handed is 0.0625.

c. $P(E\,|\,G) = \frac{P(E \cap G)}{P(G)} = \frac{0.025}{0.0625} = \frac{2}{5}$
The probability that a person is blue-eyed given the person is left handed is 0.40.

61. Define events E: a student took Form A, F: a student took Form B, G: a student scored over 80%, H: a student scored under 80%.

a. $P(E\,|\,G) = \frac{P(E \cap G)}{P(G)} = \frac{8}{20} = \frac{2}{5} = 0.40$
The probability that a student who scored over 80% took form A is 0.40.

b. $P(G\,|\,E) = \frac{P(E \cap G)}{P(E)} = \frac{8}{40} = \frac{1}{5} = 0.20$
The probability that a student who took form A scored over 80% is 0.20.

c. To show that G and E are independent, we need to show that $P(E \cap G) = P(E) \cdot P(G)$.

$P(E \cap G) = \frac{8}{100} = 0.08$

$P(E) = \frac{40}{100} = 0.4;\ P(G) = \frac{20}{100} = 0.2$
$P(E) \cdot P(G) = 0.4 \cdot 0.2 = 0.08 = P(E \cap G)$
Thus, G and E are independent events.

d. To show that G and F are independent, we need to show that $P(F \cap G) = P(F) \cdot P(G)$.

$P(F \cap G) = \frac{12}{100} = 0.12$

$P(F) = \frac{60}{100} = 0.6;\ P(G) = \frac{20}{100} = 0.2$
$P(F) \cdot P(G) = 0.6 \cdot 0.2 = 0.12 = P(E \cap G)$
Thus, G and F are independent events.

63. E and F are independent events, so
$$P(E\,|\,F) = \frac{P(E \cap F)}{P(F)} = \frac{0.2}{0.4} = \frac{1}{2} = 0.5$$

65. Define E: the basketball player makes a free throw. $P(E) = 0.7$. Then, the probability that the player misses a free throw is
$$P(\overline{E}) = 1 - P(E) = 1 - 0.7 = 0.3.$$

a. The probability that the player misses a free throw and then makes 3 in a row is
$P(\overline{E} \cap E \cap E \cap E)$
$= P(\overline{E}) \cdot P(E) \cdot P(E) \cdot P(E)$
$= 0.3 \cdot 0.7^3 = 0.1029.$

b. The probability that the player makes ten free throws in a row is
$\left[P(E) \right]^{10} = 0.7^{10} \approx 0.0282.$

67. The probability that at least one person gets the correct letter is the complement of the probability everyone gets an incorrect letter. Define event E: everyone get an incorrect letter.

$P(E) = \frac{n(E)}{n(S)} = \frac{2}{3!} = \frac{1}{3}$. So, the probability at

least one person gets the correct letter is

$$1 - P(E) = 1 - \frac{1}{3} = \frac{2}{3}.$$

69. The number of elements in the sample space S is equal to the number of ways 4 jars can be chosen from 72 jars of jam, or $C(72, 4) = 1,028,790$.

  a. Define E as the event, "All 4 jars are underweight." E can occur $C(10, 4) = 210$ ways.
  $$P(E) = \frac{C(10, 4)}{C(72, 4)} = \frac{210}{1,028,790} \approx 0.0002$$

  b. Define F as the event, "2 jars are underweight." F can occur in $C(10, 2) \cdot C(62, 2) = 85,095$ ways.
  $$P(F) = \frac{C(10, 2) \cdot C(62, 2)}{C(72, 4)}$$
  $$= \frac{85,095}{1,028,790} \approx 0.083$$

  c. Define G as the event, "At most 1 jar is underweight." The event G is equivalent to selecting either 0 or 1 underweight jars. Since the events are mutually exclusive, the sum of their probabilities will give the probability of G.
  $$P(G) = P\binom{0 \text{ underweight}}{\text{jars}} + P\binom{1 \text{ underweight}}{\text{jar}}$$
  $$= \frac{C(10, 0) \cdot C(62, 4)}{C(72, 4)} + \frac{C(10, 1) \cdot C(62, 3)}{C(72, 4)}$$
  $$\approx 0.5422 + 0.3676 = 0.9098 \approx 0.910$$

71. Define the events
  A_1: The item comes from factory 1.
  A_2: The item comes from factory 2.
  A_3: The item comes from factory 3.
  E: The item is defective.
  We make a tree diagram to illustrate the problem.

  $$P(E) = P(E \cap A_1) + P(E \cap A_2) + P(E \cap A_3)$$
  $$= 0.55 \cdot 0.01 + 0.30 \cdot 0.02 + 0.15 \cdot 0.03$$
  $$= 0.016$$

  a. $P(A_1 \mid E) = \dfrac{P(E \cap A_1)}{P(E)} = \dfrac{P(A_1) \cdot P(E \mid A_1)}{P(E)}$
  $$= \frac{(0.55)(0.01)}{0.016} \approx 0.3438$$

  b. $P(A_2 \mid E) = \dfrac{P(E \cap A_2)}{P(E)} = \dfrac{P(A_2) \cdot P(E \mid A_2)}{P(E)}$
  $$= \frac{(0.30)(0.02)}{0.016} = 0.375$$

  c. $P(A_3 \mid E) = \dfrac{P(E \cap A_3)}{P(E)} = \dfrac{P(A_3) \cdot P(E \mid A_3)}{P(E)}$
  $$= \frac{(0.15)(0.03)}{0.016} \approx 0.2813$$

73. This is a Bernoulli experiment. We have $n = 5$ and $p = 0.2$.

  a. The probability that none of the people chosen will purchase the product is
  $$P\binom{\text{exactly } 0}{\text{purchases}} = b(5, 0; 0.2)$$
  $$= C(5, 0)(0.2)^0 (0.8)^5$$
  $$= 1 \cdot 1 \cdot 0.32768 \approx 0.3277.$$

  b. The probability that exactly 3 will purchase the product is
  $$P\binom{\text{exactly } 3}{\text{purchases}} = b(5, 3; 0.2)$$
  $$= C(5, 3)(0.2)^3 (0.8)^2$$
  $$= 10 \cdot 0.008 \cdot 0.64 = 0.0512.$$

75. This is a Bernoulli experiment with $n = 7$, the number of times the coin is tossed, and $p = 0.5$, the probability of tossing a head. The expected value of a Bernoulli experiment is given by $E = np$. The expected number of heads when a coin is tossed 7 times is
  $E = np = (7)(0.5) = 3.5$
  We expect 3.5 heads when we toss a coin 7 times.

Chapter 8 Project

1. $P(E \mid H)$ is the probability a person who tossed a head is truly overweight. This probability will give the proportion of overweight students in the school.

3. $P(E \mid T) = 1$ Since respondent was instructed to answer "Yes" if a coin toss resulted in a tail. (No one who flipped a tail answered, "No.")
  $$P(E) = P(H) \cdot P(E \mid H) + P(T) \cdot P(E \mid T)$$
  $$P(E \mid H) = \frac{P(E) - P(T) \cdot P(E \mid T)}{P(H)}$$

5. $P(A \mid H) = \dfrac{P(A \cap H)}{P(H)}$

7. $P(C\,|\,H) = \dfrac{P(C \cap H)}{P(H)}$

Mathematical Questions from Professional Exams

1. B.

$P \cap Q = \varnothing$ means that $P(P \cap Q) = 0$.

However, we are told that $P(P) > 0$ and $P(Q) > 0$ so $P(P) \cdot P(Q) > 0$ which indicates P and Q are not independent.

3. B.

$\left(\dfrac{26}{52}\right)\left(\dfrac{25}{51}\right)\left(\dfrac{24}{50}\right) = \dfrac{2}{17}$

5. B.

The probability that all are not the same is the complement of the probability that the same face shows each time.

$P\left(\begin{array}{c}\text{at least}\\ \text{1 different}\end{array}\right) = 1 - P\left(\begin{array}{c}\text{all faces}\\ \text{the same}\end{array}\right) > 0.999$

$P\left(\begin{array}{c}\text{all faces}\\ \text{the same}\end{array}\right) < 0.001$

$\left(\dfrac{1}{6}\right)^n < 0.001$

$n \log\left(\dfrac{1}{6}\right) < \log(0.001)$

$n > \dfrac{\log(0.001)}{\log\left(\dfrac{1}{6}\right)} > 3.9$

7. B.

Define E: the first roll is an even number, and F: the sum of the rolls is 8.

$P(E\,|\,F) = \dfrac{P(E \cap F)}{P(E)} = \dfrac{\frac{3}{36}}{\frac{1}{2}} = \dfrac{1}{6}$

9. C.

$P(S \cap T) = P(S \cap \overline{T}) = P(\overline{S} \cap T) = p$

$P(S \cup T) = P(S) + P(T) - P(S \cap T)$

$\qquad = \left[P(S \cap \overline{T}) + P(S \cap T)\right]$

$\qquad + \left[P(\overline{S} \cap T) + P(S \cap T)\right]$

$\qquad - P(S \cap T)$

$\qquad = p + p + p + p - p = 3p$

11. D.

$P\left(\begin{array}{c}\text{no fewer than 1 and}\\ \text{no more than 9}\end{array}\right)$

$= 1 - \left[P(0\ \text{heads}) + P(10\ \text{heads})\right]$

$= 1 - \left[\dfrac{1}{2^{10}} + \dfrac{1}{2^{10}}\right] = 1 - 2\left(\dfrac{1}{2^{10}}\right) = 1 - \dfrac{1}{2^{9}}$

Chapter 9 Statistics

9.1 Introduction to Statistics: Data and Sampling

1. A measurable characteristic is called a <u>variable</u>.

3. False. A discrete variable can assume only a finite set of values or as many values as there are whole numbers.

5. The variable is the number of heads thrown, and it is discrete.

7. The variable is the miles per gallon (or average gas mileage), and it is continuous.

9. The variable is the time of waiting in line, and it is continuous.

11. The variable is the number of flights, and it is discrete.

13. The variable is the number of people crossing the intersection, and it is discrete.

15. The variable is length of time, and it is continuous.

17. Answers will vary. All answers should include a method to choose a group of viewers for which each viewer of the program has an equal chance of being chosen.

19. Answers will vary. All answers should include a method to choose a sample in which each member of the population has an equal chance of being selected.

21. Answers will vary. All answers should include a method to choose a sample in which each member of the population has an equal chance of being selected.

23.–25. Answers will vary. All answers should give examples of possible bias.

9.2 Representing Qualitative Data Graphically: Bar Graphs; Pie Charts

1. Two popular ways to display data that can be separated in categories are in <u>bar graphs</u> and <u>pie charts</u>.

3. False. In a pie chart a circle is divided into sectors, one sector for each category represented by the data. The size of each sector is proportional to the size of the category.

5. a.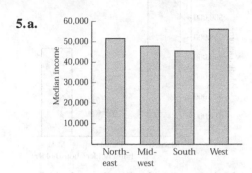

b. The west has the highest median income.

c. The south has the lowest median income.

d. Answers will vary. All discussions should conclude that pie charts are inappropriate.

7. a.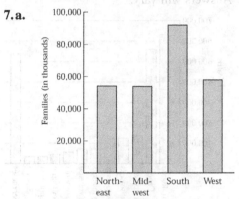

b.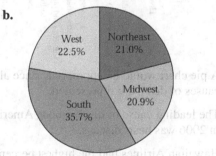

c. Answers will vary.

d. The south has the most families.

e. The midwest has the fewest families.

9.a.

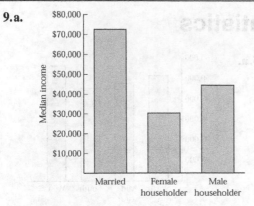

b. Married–couple families have the highest median income.

c. Female householder–no spouse families have the lowest median income.

d. Answers will vary.

11.a.

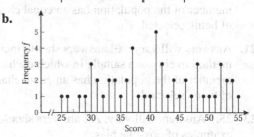

b. A pie chart would be appropriate since all causes of death are represented.

c. The leading cause of death among Americans in 2006 was heart disease.

13.a. Hawaiian Airlines had the highest percentage of on-time flights.

b. Comair had the lowest percentage of on-time flights.

c. About 85% of United Airlines' flights are on time.

15.a. Housing, fuel, and utilities make up the largest component of the CPI. This sector comprises 42% of the CPI.

b. The smallest component of the CPI is other goods and services. This sector comprises 3% of the CPI.

c. Answers will vary.

9.3 Organizing and Displaying Quantitative Data

1.a. The point $(-3, 4)$ is located in quadrant II.

b. The point $(1, 5)$ is located in quadrant I.

3. When grouping data, the range is divided into <u>class intervals</u> of equal size.

5.a.

Score	Frequency	Score	Frequency
25	1	41	5
26	1	42	3
28	1	43	1
29	1	44	2
30	3	45	1
31	2	46	2
32	1	47	1
33	2	48	3
34	2	49	1
35	1	50	1
36	2	51	1
37	4	52	3
38	1	53	2
39	1	54	2
40	1	55	1

b.

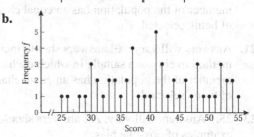

c., f.

Class Interval	Frequency	Cumulative Frequency
24–25.9	1	1
26–27.9	1	2
28–29.9	2	4
30–31.9	5	9
32–33.9	3	12
34–35.9	3	15
36–37.9	6	21
38–39.9	2	23

(*continued on next page*)

(*continued*)

Class Interval	Frequency	Cumulative Frequency
40–41.9	6	29
42–43.9	4	33
44–45.9	3	36
46–47.9	3	39
48–49.9	4	43
50–51.9	2	45
52–53.9	5	50
54–55.9	3	53

g.

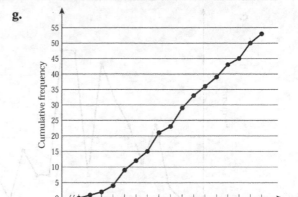

d.

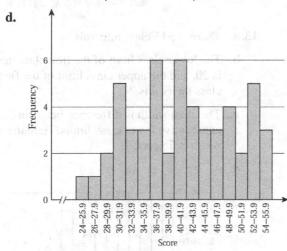

e.

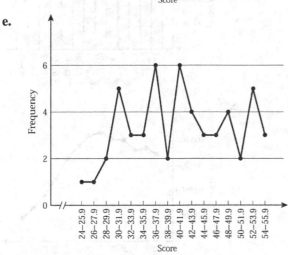

7.a., d.

Class Interval	Frequency	Cumulative Frequency
50–54.9	1	1
55–59.9	6	7
60–64.9	3	10
65–69.9	6	16
70–74.9	8	24
75–79.9	11	35
80–84.9	2	37
85–89.9	12	49
90–94.9	12	61
95–99.9	2	63
100–104.9	2	65
105–109.9	4	69
110–114.9	0	69
115–119.9	2	71

b.

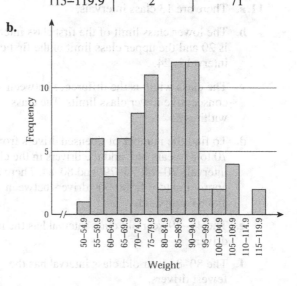

c.

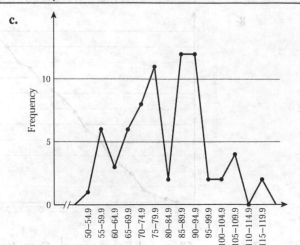

e.

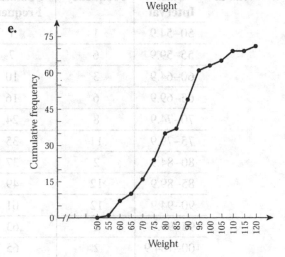

9. Since the data is clustered on the left giving the graph a tail on the right, this distribution is skewed right.

11.a. There are 13 class intervals.

b. The lower class limit of the first class interval is 20 and the upper class limit of the first class interval is 24.

c. The class width is the difference between consecutive lower class limits. The class width is 5.

d. To find the number of licensed drivers from 70 to 84 years old, add the drivers in the class intervals 70–74, 75–79, and 80–84. There are approximately 1,500,000 drivers between 70 and 84 years old.

e. The 45–49 year old class interval has the most drivers.

f. The 80–84 year old class interval has the fewest drivers.

g. The distribution is clustered on the left and has a longer tail on the right. The distribution is skewed right.

h.

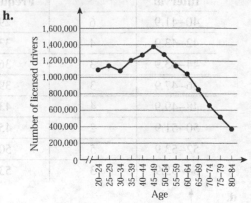

13.a. There are 13 class intervals

b. The lower class limit of the first class interval is 20, and the upper class limit of the first class interval is 24.

c. The class width is difference between consecutive lower class limits. Here the class width is 5 years.

d.

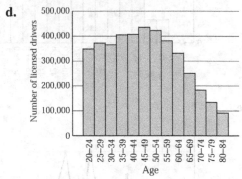

e.

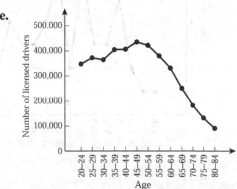

f. The age group 45–49 has the most licensed drivers.

g. The fewest licensed drivers are in the 80–84 year-old age group.

15. a. There are 19 class intervals.

  b. The lower class limit of the first class interval is $0, and the upper class limit of the first class interval is $1999.

  c. The class width is difference between consecutive lower class limits. Here the class width is $2000.

  d.

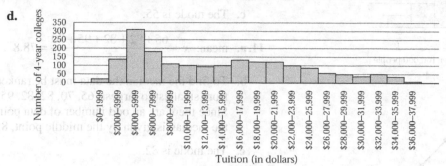

Tuition (in dollars)

  e.

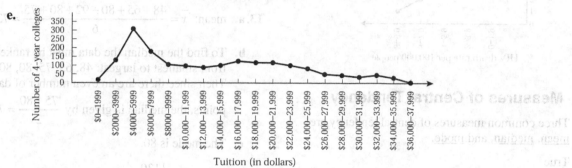

Tuition (in dollars)

17. a.

	Class Interval	Frequency
1	11.9–12.4	3
2	12.5–13.0	1
3	13.1–13.6	3
4	13.7–14.2	5
5	14.3–14.8	1
6	14.9–15.4	2
7	15.5–16.0	2
8	16.1–16.6	3

b.

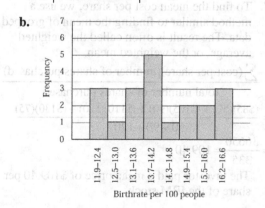

Birthrate per 100 people

c.

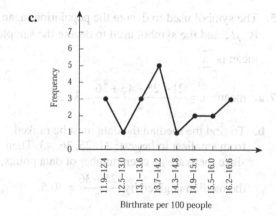

Birthrate per 100 people

19. a.

Class Interval	Frequency
0.0–1.9	7
2.0–3.9	6
4.0–5.9	4
6.0–7.9	2
8.0–9.9	1

b.

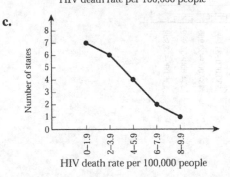

c.

HIV death rate per 100,000 people

9.4 Measures of Central Tendency

1. Three common measures of central tendency are <u>mean</u>, <u>median</u>, and <u>mode</u>.

3. True

5. The symbol used to denote the population mean is $\underline{\mu}$, and the symbol used to denote the sample mean is $\overline{x}$.

7.a. mean: $\overline{x} = \dfrac{21 + 25 + 43 + 36}{4} = 31.25$

   b. To find the median the data must be ranked from smallest to largest: 21, 25, 36, 43. Then since there are an even number of data points, the median is given by $\dfrac{25 + 36}{2} = 30.5$.

   c. Since no value is repeated more than once, there is no mode.

9.a. mean: $\overline{x} = \dfrac{55 + 55 + 80 + 92 + 70}{5} = 70.4$

   b. To find the median the data must be ranked from smallest to largest: 55, 55, 70, 80, 92. Then since there are an odd number of data points, the median is given by the middle point, 70.

   c. The mode is 55.

11.a. mean: $\overline{x} = \dfrac{65 + 82 + 82 + 95 + 70}{5} = 78.8$

   b. To find the median the data must be ranked from smallest to largest: 65, 70, 82, 82, 95. Since there are an odd number of data points, the median is given by the middle point, 82.

   c. The mode is 82.

13.a. mean: $\overline{x} = \dfrac{48 + 65 + 80 + 92 + 80 + 75}{6} = 73.33$

   b. To find the median the data must be ranked from smallest to largest: 48, 65, 75, 80, 80, 92. Then since there are an even number of data points, the median is given by $\dfrac{75 + 80}{2} = 77.5$

   c. The mode is 80.

15.a. mean: $\overline{x} = \dfrac{1130}{40} = 28.25$

   b. To find the median the data must be ranked from smallest to largest. Since there is an even number of data points, the median is given by $\dfrac{27 + 27}{2} = 27$

   c. The modes are 24 and 25.

17. To find the mean cost per share, we use a method similar to finding the mean of grouped data. The result is often called the weighted average, or the weighted mean.

$$\overline{x} = \dfrac{\sum (\text{cost per share})(\text{number of shares purchased})}{\text{total number of shares purchased}}$$

$$\overline{x} = \dfrac{(\$85)(50) + (\$105)(90) + (\$110)(120) + (\$130)(75)}{50 + 90 + 120 + 75}$$

$$= \dfrac{36650}{335} = 109.40$$

The investor paid a mean price of $109.40 per share of the IBM stock.

19. a.

Class Interval	Midpoint, m_i	Frequency, f	$f_i m_i$
15–19	17.5	445	7787.5
20–24	22.5	1082	24,345
25–29	27.5	1208	33,220
30–34	32.5	962	31,265
35–39	37.5	499	18,712.5
40–44	42.5	105	4462.5
Total		4301	119,792.5

The mean age of a new mother in the United States in 2007 was $\bar{x} = \dfrac{119,792.5}{4301} = 27.85$ years.

b. To find the median of grouped data we use the following steps.

Step 1: Find the interval containing the median. The median is the middle value when the 4301 items are in ascending order. So the median in this example is the 2151st entry. The 2151st entry is in the interval 25–29.

Step 2: In the interval containing the median, count the number p of items remaining to reach the median. The first two intervals account for 1527 data points; there are $2151 - 1527 = 624$ points left, so $p = 624$.

Step 3: Calculate the interpolation factor. q is the frequency for the interval containing the median; $q = 1208$ and i is the size of the interval, $i = 5$, so the interpolation factor is $\dfrac{p}{q} \cdot i = \dfrac{624}{1208} \cdot 5 = 2.58$.

Step 4: The median M is $M = \begin{bmatrix} \text{lower limit of interval} \\ \text{containing the median} \end{bmatrix} + [\text{interpolation factor}] = 25 + 2.58 = 27.58$.

The median age of a new mother in 2007 was about 27.58 years.

21. a.

Class Interval	Midpoint, m_i	Frequency, f	$f_i m_i$
20–24	22.5	348,587	7,843,207.5
25–29	27.5	372,032	10,230,880
30–34	32.5	365,111	11,866,107.5
35–39	37.5	404,931	15,184,912.5
40–44	42.5	406,831	17,290,317.5
45–49	47.5	435,706	20,696,035
50–54	52.5	422,420	22,177,050
55–59	57.5	380,898	21,901,635
60–64	62.5	331,010	20,688,125
65–69	67.5	250,679	16,920,832.5
70–74	72.5	182,974	13,265,615
75–79	77.5	133,888	10,376,320
80–84	82.5	91,318	7,533,735
Total		4,126,385	195,974,772.5

The mean age of a driver in Tennessee is $\bar{x} = \dfrac{\Sigma f_i m_i}{n} = \dfrac{195,974.772.5}{4,126,385} = 47.49$ years.

b. To find the median of grouped data we use the following steps.

Step 1: Find the interval containing the median. The median is the middle value when the 4,126,385 items are in ascending order. So the median in this example is the 2,063,193rd entry. This entry is in the interval 45–49.

Step 2: In the interval containing the median, count the number p of items remaining to reach the median. The first five intervals account for 1,897,492 data points; there are 2,063,193 − 1,897,492 = 165,701 points left, so p = 165,701.

Step 3: Calculate the interpolation factor. q is the frequency for the interval containing the median; q = 435,706 and i is the size of the interval, i = 5, so the interpolation factor is $\dfrac{p}{q} \cdot i = \dfrac{165,701}{435,706} \cdot 5 = 1.90$.

Step 4: The median age of a driver in Tennessee is

$$M = \begin{bmatrix} \text{lower limit of interval} \\ \text{containing the median} \end{bmatrix} + \begin{bmatrix} \text{interpolation factor} \end{bmatrix} = 45 + 1.9 = 46.9 \text{ years}$$

23.

Class Interval	Midpoint, m_i	Frequency, f	$f_i m_i$
$0–1999	1000	19	19,000
$2000–3999	3000	132	396,000
$4000–5999	5000	308	1,540,000
$6000–7999	7000	180	1,260,000
$8000–9999	9000	106	954,000
$10,000–11,999	11,000	99	1,089,000
$12,000–13,999	13,000	89	1,157,000
$14,000–15,999	15,000	98	1,470,000
$16,000–17,999	17,000	125	2,125,000
$18,000–19,999	19,000	116	2,204,000
$20,000–21,999	21,000	117	2,457,000
$22,000–23,999	23,000	100	2,300,000
$24,000–25,999	25,000	83	2,075,000
$26,000–27,999	27,000	53	1,431,000
$28,000–29,999	29,000	45	1,305,000
$30,000–31,999	31,000	36	1,116,000
$32,000–33,999	33,000	46	1,518,000
$34,000–35,999	35,000	32	1,120,000
$36,000–37,999	37,000	6	222,000
Total		1790	25,758,000

The mean tuition at 4 year colleges in 2006–2007 is $\bar{x} = \dfrac{\Sigma f_i m_i}{n} = \dfrac{25,758,000}{1790} = \$14,389.94$.

25. a. mean: $\bar{x} = \dfrac{34,000 + 35,000 + 36,000 + 36,500 + 65,000}{5} = \dfrac{206,500}{5} = 41,300$

median: The median is the middle entry once the items are placed in ascending order. $M = \$36,000$

b. The median describes the situation more realistically because the data is skewed to the right. This means that 4 salaries are clustered, while the salary of $65,000 is much higher.

c. If you are among the four lower-paid members, use the mean to show that you are underpaid. If you are the one earning $65,000, use the median.

9.5 Measures of Dispersion

1. The range, variance, and standard deviation measure the <u>dispersion or spread</u> of the distribution.

3. The Empirical Rule states that if the distribution of the data is bell-shaped, about <u>68%</u> of the outcomes will be within one standard deviation of the mean.

5. False. The standard deviation is preferred to the variance because the units are not squared.

7. There are 7 scores, so $n = 7$. The mean of the scores is

$$\frac{4+5+9+9+10+14+25}{7} = \frac{76}{7} \approx 10.8571.$$

x_i	$x - \bar{x}$	$(x - \bar{x})^2$
4	−6.8571	47.0204
5	−5.8571	34.3061
9	−1.8571	3.4490
9	−1.8571	3.4490
10	−0.8571	0.7347
14	3.1429	9.8776
25	14.1429	200.0204
Total	0	298.8571

$$s = \sqrt{\frac{\sum (x_i - \bar{x})^2}{n-1}} = \sqrt{\frac{298.8571}{7-1}} \approx 7.058$$

9. There are 4 scores, so $n = 4$. The mean of the scores is $\frac{62+58+70+70}{4} = \frac{260}{4} = 65.$

x_i	$x - \bar{x}$	$(x - \bar{x})^2$
62	−3	9
58	−7	49
70	5	25
70	5	25
Total	0	108

$$s = \sqrt{\frac{\sum (x_i - \bar{x})^2}{n-1}} = \sqrt{\frac{108}{3}} = 6$$

11. There are 5 scores, so $n = 5$. The mean of the scores is $\frac{85+75+62+78+100}{5} = \frac{400}{5} = 80.$

x_i	$x - \bar{x}$	$(x - \bar{x})^2$
85	5	25
75	−5	25
62	−18	324
78	−2	4
100	20	400
Total	0	778

$$s = \sqrt{\frac{\sum (x_i - \bar{x})^2}{n-1}} = \sqrt{\frac{778}{4}} = 13.9463$$

13.

Class Interval	Midpoint, m_i	Frequency, f_i	$f_i m_i$	$m_i - \bar{x}$	$(m_i - \bar{x})^2$	$(m_i - \bar{x})^2 \cdot f_i$
10–16	13.5	1	13.5	−18.88	356.41	356.4086
17–23	20.5	3	61.5	−11.88	141.11	423.3168
24–30	27.5	10	275	−4.88	23.80	238.0257
31–37	34.5	12	414	2.12	4.50	53.9945
38–44	41.5	5	207.5	9.12	83.20	415.9826
45–51	48.5	2	97	16.12	259.89	519.7870
Total		33	1068.5			2007.5152

The mean is $\bar{x} = \frac{\sum f_i m_i}{n} = \frac{1068.5}{33} \approx 32.379.$ The standard deviation is

$$s = \sqrt{\frac{\sum (x_i - \bar{x})^2}{n-1}} = \sqrt{\frac{2007.5152}{32}} \approx 7.921.$$

15. $n = 6$. The mean of the lifetimes of the light bulbs is $\bar{x} = \dfrac{968 + 893 + 769 + 845 + 922 + 915}{6} = \dfrac{5312}{6} = 885.333$.

x_i	$x - \bar{x}$	$(x - \bar{x})^2$
968	82.667	6833.778
893	7.667	58.778
769	−116.333	13,533.444
845	−40.333	1626.778
922	36.667	1344.444
915	29.667	880.111
Total	0	24,277.333

$$s = \sqrt{\frac{\sum(x_i - \bar{x})^2}{n-1}} = \sqrt{\frac{24,277.333}{5}} = 69.681$$

17.a. The range is the difference between the largest value and the smallest value.
The range for the Yankees' ages is $40 - 23 = 17$ years.

b.

Age, x_i	Frequency, f_i	$f_i x_i$	$x_i - \bar{x}$	$(x_i - \bar{x})^2$	$(x_i - \bar{x})^2 \cdot f_i$
23	4	92	−5.25	27.5625	110.25
24	6	144	−4.25	18.0625	108.375
25	6	150	−3.25	10.5625	63.375
26	3	78	−2.25	5.0625	15.1875
27	5	135	−1.25	1.5625	7.8125
28	1	28	−0.25	0.0625	0.0625
29	4	116	0.75	0.5625	2.25
30	1	30	1.75	3.0625	3.0625
33	3	99	4.75	22.5625	67.6875
34	1	34	5.75	33.0625	33.0625
35	1	35	6.75	45.5625	45.5625
36	1	36	7.75	60.0625	60.0625
37	1	37	8.75	76.5625	76.5625
38	2	76	9.75	95.0625	190.125
40	1	40	11.75	138.0625	138.0625
Totals	40	1130			921.5

The standard deviation assuming sample data is $s = \sqrt{\dfrac{\sum(x_i - \bar{x})^2 \cdot f_i}{n-1}} = \sqrt{\dfrac{921.5}{39}} \approx 4.86$.

c. The standard deviation assuming population data is $\sigma = \sqrt{\dfrac{\sum(x_i - \mu)^2 \cdot f_i}{n}} = \sqrt{\dfrac{921.5}{40}} \approx 4.80$.

d. The data are population because it is the data for the entire population (the whole team).

19. a. These are population data since all of the mothers under age 45 are included.

b.

Class Interval	Midpoint, m_i	Frequency, f	$f_i m_i$	$x_i - \bar{x}$	$(x_i - \bar{x})^2$	$(x_i - \bar{x})^2 \cdot f_i$
10–14	12.5	6	75	−15.3309	235.0352	1410.2110
15–19	17.5	445	7787.5	−10.3309	106.7266	47,493.3375
20–24	22.5	1082	24,345	−5.3309	28.4180	30,748.3124
25–29	27.5	1208	33,220	−0.3309	0.1095	132.2352
30–34	32.5	962	31,265	4.6691	21.8009	20,972.4646
35–39	37.5	499	18,712.5	9.6691	93.4923	46,652.6733
40–44	42.5	105	4462.5	14.6691	215.1838	22,594.2952
Total		4307	119,867.5			170,003.5291

The mean age of a new mother in the United States in 2007 was $\mu = \dfrac{119,867.5}{4307} \approx 27.83$ years.

The standard deviation assuming population data is $\sigma = \sqrt{\dfrac{\sum (x_i - \bar{x})^2 \cdot f_i}{n}} = \sqrt{\dfrac{170,0003.5291}{4307}} \approx 6.28.$

21. a. These are population data since all earthquakes are included.

b.

Class Interval	Midpoint, m_i	Frequency, f	$f_i m_i$	$m_i - \mu$	$(m_i - \mu)^2$	$(m_i - \mu)^2 \cdot f_i$
0–0.9	0.5	21	10.5	−3.53	12.46	261.5145
1.0–1.9	1.5	26	39	−2.53	6.40	166.2775
2.0–2.9	2.5	3009	7522.5	−1.53	2.34	7033.5606
3.0–3.9	3.5	2899	10146.5	−0.53	0.28	810.9247
4.0–4.9	4.5	6908	31086	0.47	0.22	1533.1871
5.0–5.9	5.5	1776	9768	1.47	2.16	3843.5512
6.0–6.9	6.5	142	923	2.47	6.11	867.1059
7.0–7.9	7.5	16	120	3.47	12.05	192.7776
Total		14,797	59,615.5			14,708.8991

The mean magnitude worldwide in 2009 was $\mu = \dfrac{59,615.5}{14,797} \approx 4.03$ years.

c. The standard deviation assuming population data is $\sigma = \sqrt{\dfrac{\sum (x_i - \mu)^2 \cdot f_i}{n}} = \sqrt{\dfrac{14,708.8991}{14,797}} \approx 0.9970.$

23. a.

Class Interval	Midpoint m_i	Frequency f	$f_i m_i$	$x_i - \bar{x}$	$(x_i - \bar{x})^2$	$(x_i - \bar{x})^2 \cdot f_i$
20–24	22.5	348,587	7,843,207.5	−24.9931	624.6545	217,746,447.7432
25–29	27.5	372,032	10,230,880	−19.9931	399.7236	148,709,982.0637
30–34	32.5	365,111	11,866,107.5	−14.9931	224.7927	82,074,300.4665
35–39	37.5	404,931	15,184,912.5	−9.9931	99.8618	40,437,154.5762
40–44	42.5	406,831	17,290,317.5	−4.9931	24.9309	10,142,680.7496
45–49	47.5	435,706	20,696,035	0.0069	0.0000	20.8066
50–54	52.5	422,420	22,177,050	5.0069	25.0692	10,589,711.1136
55–59	57.5	380,898	21,901,635	10.0069	100.1383	38,142,461.3938
60–64	62.5	331,010	20,688,125	15.0069	225.2074	74,545,888.2221
65–69	67.5	250,679	16,920,832.5	20.0069	400.2765	100,340,903.7280
70–74	72.5	182,974	13,265,615	25.0069	625.3456	114,421,979.9787
75–79	77.5	133,888	10,376,320	30.0069	900.4147	120,554,719.6284
80–84	82.5	91,318	7,533,735	35.0069	1225.4838	111,908,727.4794
Total		4,126,385	195,974,772.5			1,069,614,977.9497

The mean age of a driver in Tennessee is $\bar{x} = \dfrac{\sum f_i m_i}{n} = \dfrac{195,974.772.5}{4,126,385} = 47.49$ years.

The standard deviation assuming sample data is $s = \sqrt{\dfrac{\sum (x_i - \bar{x})^2 \cdot f_i}{n}} = \sqrt{\dfrac{1,069,614,977.9497}{4,126,384}} \approx 16.1001$.

b. The standard deviation assuming population data is

$$\sigma = \sqrt{\dfrac{\sum (x_i - \mu)^2 \cdot f_i}{n}} = \sqrt{\dfrac{1,069,614,977.9497}{4,126,385}} \approx 16.1001.$$

c. Answers will vary.

25. a. These are population data; all 4-year colleges are represented.

b. From section 9.4 exercise 23, we have the mean tuition at 4 year colleges in 2006–2007 is

$$\mu = \frac{\sum f_i m_i}{n} = \frac{25,758,000}{1790} = \$14,389.94.$$

Class Interval	m_i	Frequency, f	$f_i m_i$	$x_i - \mu$	$(x_i - \mu)^2$	$(x_i - \mu)^2 \cdot f_i$
$0–1999	1,000	19	19,000	−13,389.9441	179,290,603.91	3,406,521,474.36
$2000–3999	3,000	132	396,000	−11,389.9441	129,730,827.38	17,124,469,213.82
$4000–5999	5,000	308	1,540,000	−9,389.9441	88,171,050.84	27,156,683,659.06
$6000–7999	7,000	180	1,260,000	−7,389.9441	54,611,274.30	9,830,029,374.86
$8000–9999	9,000	106	954,000	−5,389.9441	29,051,497.77	3,079,458,763.46
$10,000–11,999	11,000	99	1,089,000	−3,389.9441	11,491,721.23	1,137,680,401.98
$12,000–13,999	13,000	89	1,157,000	−1,389.9441	1,931,944.70	171,943,077.93
$14,000–15,999	15,000	98	1,470,000	610.0559	372,168.16	36,472,479.64
$16,000–17,999	17,000	125	2,125,000	2,610.0559	6,812,391.62	851,548,952.90
$18,000–19,999	19,000	116	2,204,000	4,610.0559	21,252,615.09	2,465,303,350.08
$20,000–21,999	21,000	117	2,457,000	6,610.0559	43,692,838.55	5,112,062,110.42
$22,000–23,999	23,000	100	2,300,000	8,610.0559	74,133,062.01	7,413,306,201.43

(continued on next page)

(continued)

Class Interval	m_i	Frequency, f	$f_i m_i$	$x_i - \mu$	$(x_i - \mu)^2$	$(x_i - \mu)^2 \cdot f_i$
$24,000–25,999	25,000	83	2,075,000	10,610.0559	112,573,285.48	9,343,582,694.67
$26,000–27,999	27,000	53	1,431,000	12,610.0559	159,013,508.94	8,427,715,973.91
$28,000–29,999	29,000	45	1,305,000	14,610.0559	213,453,732.41	9,605,417,958.24
$30,000–31,999	31,000	36	1,116,000	16,610.0559	275,893,955.87	9,932,182,411.29
$32,000–33,999	33,000	46	1,518,000	18,610.0559	346,334,179.33	15,931,372,249.31
$34,000–35,999	35,000	32	1,120,000	20,610.0559	424,774,402.80	13,592,780,889.49
$36,000–37,999	37,000	6	222,000	22,610.0559	511,214,626.26	3,067,287,757.56
Totals:		1790	25,758,000			147,685,818,994.41

$$\sigma = \sqrt{\frac{\sum (x_i - \mu)^2 \cdot f_i}{n}} = \sqrt{\frac{147,685,818,994.41}{1790}} \approx 9083.28$$

The standard deviation of the tuition is $9083.28.

27. We know that the mean of the IQ test is 100, the standard deviation is 15, and the distribution is symmetric. So the Empirical Rule applies.

a. Since 70 is two standard deviations below the mean and 130 is two standard deviations above the mean, the Empirical Rule states that approximately 95% of the IQ scores are between 70 and 130. So we conclude that approximately 95% of persons have an IQ score between 70 and 130.

b. Since 95% of the scores are between 70 and 130, $1 - 0.95 = 0.05 = 5\%$ of the scores are either below 70 or above 130.

c. Since the Empirical Rule requires the distribution to be roughly symmetric, we assume that the percent of scores over 130 is equal to the percent below 70. So, approximately 2.5% of IQ scores are above 130.

29. We are told the distribution of kidney weight is bell shaped with a mean of 325 grams and a standard deviation of 30 grams, so the Empirical Rule applies.

a. About 95% of kidneys will be between two standard deviations of the mean. That means 95% of the kidneys will weigh between $325 - 2(30) = 325 - 60 = 265$ grams and $325 + 2(30) = 325 + 60 = 385$ grams.

b. $\frac{325 - 235}{30} = \frac{90}{30} = 3$ and $\frac{415 - 325}{30} = \frac{90}{30} = 3$.

According to the Empirical Rule approximately 99.7% of the data lie within 3 standard deviations of the mean. So, approximately 99.7% of adult male kidneys weigh between 235 grams and 415 grams.

c. Kidneys weighing less than 235 grams or more than 415 grams are more than 3 standard deviations from the mean. So, $1 - 0.997 = 0.003$ or 0.3% of adult male kidneys weigh less than 235 grams or more than 415 grams.

d. $\frac{325 - 295}{30} = \frac{30}{30} = 1$ and $\frac{385 - 325}{30} = \frac{60}{30} = 2$.

Since 295 is one standard deviation below the mean, and 385 are two standard deviations above the mean from the Empirical Rule, we see that $0.34 + 0.34 + 0.135 = 0.815$ or 81.5% of the adult male kidneys weight between 295 and 385 grams.

31. We are told that $\mu = 25$ and $\sigma = 3$.

a. We want the outcome to be between 19 and 31, so $k = \mu - 19 = 25 - 19 = 6$, and according to Chebychev's theorem, the probability is at least

$$1 - \frac{\sigma^2}{k^2} = 1 - \frac{3^2}{6^2} = 1 - \frac{9}{36} = \frac{3}{4} = 0.75.$$

At least 75% of the outcomes are between 19 and 31.

b. We want the outcome to be between 20 and 30, so $k = \mu - 20 = 25 - 20 = 5$, and according to Chebychev's theorem, the probability is at least

$$1 - \frac{\sigma^2}{k^2} = 1 - \frac{3^2}{5^2} = 1 - \frac{9}{25} = \frac{16}{25} = 0.64.$$

At least 64% of the outcomes are between 20 and 30.

c. We want the outcome to be between 16 and 34, so $k = \mu - 16 = 25 - 16 = 9$, and according to Chebychev's theorem, the probability is at least $1 - \dfrac{\sigma^2}{k^2} = 1 - \dfrac{3^2}{9^2} = 1 - \dfrac{1}{9} = \dfrac{8}{9} \approx 0.889$.

At least 88.9% of the outcomes are between 16 and 34.

d. We want the outcome to be less than 19 or more than 31. This is the opposite event from part (a), so the probability is at most
$1 - 0.75 = 0.25$. At most 25% of the outcomes are less than 19 or greater than 31.

e. We want the outcome to be less than 16 or more than 34. This is the opposite event from part (c) so the probability is at most
$1 - 0.889 = 0.111$. At most 11.1% of the outcomes are less than 16 or more than 34.

33. We are told that $\mu = 6$ and $\sigma = 2$. We want to estimate the number of boxes that have between 0 and 12 defective watches. $k = \mu - 0 = 6 - 0 = 6$. According to Chebychev's theorem, the probability is at least

$1 - \dfrac{\sigma^2}{k^2} = 1 - \dfrac{2^2}{6^2} \approx 0.8889$ that there are between 0 and 12 defective watches in a box. So we expect at least

88.9% of the 1000 boxes or at least 889 boxes to have between 0 and 12 defective watches.

35.a. These are population data because they record all live births in the United States for the years listed.

Year	Births, f_i	$f_i - \mu$	$(f_i - \mu)^2$
2006	4,265,555	156,627.5	24,532,173,756.25
2005	4,138,349	29,421.5	865,624,662.25
2004	4,112,052	3,124.5	9,762,500.25
2003	4,089,950	−18,977.5	360,145,506.25
2002	4,021,726	−87,201.5	7,604,101,602.25
2001	4,025,933	−82,994.5	6,888,087,030.25
Totals: $n = 6$	24,653,565		40,259,895,057.5

b. $\mu = \dfrac{24,653,565}{6} \approx 4,108,927.5 \approx 4,108,928$

c. $\sigma = \sqrt{\dfrac{\sum(f_i - \mu)^2}{n}} = \sqrt{\dfrac{40,259,895,057.5}{6}} \approx 81,914.4829$

The standard deviation is 81,914.5.

d. The mean and standard deviations are exact because the data are not grouped.

e. Answers will vary.

9.6 The Normal Distribution

1. True. In a binomial probability model, each trial has only two possible outcomes, success or failure.

3. True

5. To standardize a normal random variable, we find a <u>Z-score</u>.

7. 0.4.

9. The mean is always at the center of the normal curve. By inspection, $\mu = 8$. Since 34.135% of the area under the curve lies between μ and σ, $\sigma = 10 - 8 = 2$.

11. The mean is always at the center of the normal curve. By inspection, $\mu = 18$. Since 68.27% of the area under the curve lies between $\mu - \sigma$ and $\mu + \sigma$, $18 + \sigma = 19 \Rightarrow \sigma = 1$.

13. $\mu = 13.1$, $\sigma = 9.3$

a. $x = 7$
$$Z = \frac{x - \mu}{\sigma} = \frac{7 - 13.1}{9.3} = -0.66$$

b. $x = 9$
$$Z = \frac{x - \mu}{\sigma} = \frac{9 - 13.1}{9.3} = -0.44$$

c. $x = 13$
$$Z = \frac{x - \mu}{\sigma} = \frac{13 - 13.1}{9.3} = -0.01$$

d. $x = 29$
$$Z = \frac{x - \mu}{\sigma} = \frac{29 - 13.1}{9.3} = 1.71$$

e. $x = 37$
$$Z = \frac{x - \mu}{\sigma} = \frac{37 - 13.1}{9.3} = 2.57$$

f. $x = 41$
$$Z = \frac{x - \mu}{\sigma} = \frac{41 - 13.1}{9.3} = 3$$

15. Using the Standard Normal Curve Table on the inside back cover of the text, we can find the area under the standard normal curve between the standard score, Z, and the mean.

a. $Z = 0.89$
Read down the table under Z until you reach the row beginning 0.8. Then read across the row until you reach the entry under the column marked 0.09. The area under the standard normal curve between 0 and $Z = 0.89$ is 0.3133.

b. $Z = 1.10$
Read down the table under Z until you reach the row beginning 1.1. The next entry in the row, 0.3642, represents the area under the standard normal curve between the mean and $Z = 1.10$.

c. $Z = 3.06$
Read down the table under Z until you reach the row beginning 3.0. Then read across the row until you reach the entry under the column marked 0.06. The area under the standard normal curve between 0 and $Z = 3.06$ is 0.4989.

d. $Z = -1.22$
There are no negative Z-scores on this table, but we use the symmetry of the normal curve to find the area under the curve between $Z = -1.22$ and the mean.

Read down the table under Z until you reach the row beginning 1.2. Then read across the row until you reach the entry under the column marked 0.02. Because of symmetry, 0.3888 is area under the standard normal curve between $Z = -1.22$ and 0, as well as between 0 and $Z = 1.22$.

e. $Z = 2.30$
Read down the table under Z until you reach the row beginning 2.3. The next entry in the row, 0.4893, represents the area under the standard normal curve between the mean and $Z = 2.30$.

f. $Z = -0.75$
There are no negative Z-scores on this table, but we use the symmetry of the normal curve to find the area under the curve between $Z = -0.75$ and the mean. Read down the table under Z until you reach the row beginning 0.7. Then read across the row until you reach the entry under the column marked 0.05. Because of symmetry, 0.2734 is area under the standard normal curve between $Z = -0.75$ and 0, as well as between 0 and $Z = 0.75$.

17. $Z = -0.5$, $A = 0.1915$
Since we want the area to the left of Z, subtract A from 0.5000.
Area $= 0.5000 - 0.1915 = 0.3085$

19. $Z_1 = -1.2$, $A_1 = 0.3849$
$Z_2 = 1.5$, $A_2 = 0.4332$
Since both Z-scores are on opposite sides of the mean, add the areas.
Area $= A_1 + A_2 = 0.3849 + 0.4332 = 0.8181$

21. To approximate the probability of obtaining between 285 and 315 successes in the 750 trials, we find the area under a normal curve from $x = 284.5$ to $x = 315.5$. We convert to Z-scores.
$$x = 284.5: Z_1 = \frac{x - \mu}{\sigma} = \frac{284.5 - 300}{13.4} = -1.16 \Rightarrow$$
$A_1 = 0.3770$
$$x = 315.5: Z_2 = \frac{x - \mu}{\sigma} = \frac{315.5 - 300}{13.4} = 1.16 \Rightarrow$$
$A_2 = 0.3770$
Since the values are on opposite sides of the mean, we add the areas.
Area $= A_1 + A_2 = 0.3770 + 0.3770 = 0.7540$
The approximate probability that there are between 285 and 315 successes is 0.7540. Using a TI-84, the approximate probability is 0.7521.

(continued on next page)

(continued)

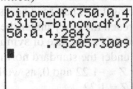

23. To approximate the probability of obtaining 300 or more successes in the 750 trials, we find the area under a normal curve to the right of $x = 299.5$.

$$Z = \frac{x - \mu}{\sigma} = \frac{299.5 - 300}{13.4} = -0.0373 \approx -0.04 \Rightarrow$$
$$A = 0.0160$$

So, the approximate probability of obtaining 300 or more successes is approximately
$0.5 + 0.0160 = 0.5160$

25. To approximate the probability of obtaining 325 or more successes in the 750 trials, we find the area under a normal curve to the right of $x = 324.5$. We first convert 324.5 to a Z-score.

$$Z = \frac{x - \mu}{\sigma} = \frac{324.5 - 300}{13.4} = 1.8284 \Rightarrow$$
$$A = 0.4664$$

We need the area to the right of $Z = 1.83$, so subtract A from 0.5000.
Area = 0.5000 − 0.4664 = 0.0336
The approximate probability of 325 or more successes is 0.0336. Using a TI-84, the approximate probability is 0.0343.

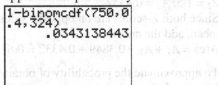

27. We use the Standard Normal Curve Table and the interpretation of a Z-score as the number of standard deviations the original score is from its mean to solve this problem.

A: A score exceeds $\mu + 1.6\sigma$. Here $Z = 1.6$; the area under the standard normal curve between 0 and 1.6 is 0.4452. We need the area to the right of $Z = 1.6$. So we subtract 0.4452 from 0.5, the area under the curve to the right of 0.
0.5000 − 0.4452 = 0.0548
So, 5.48% of the class will get a grade of A.

B: A score is between $\mu + 0.6\sigma$ and $\mu + 1.6\sigma$. We know that the area under the curve from 0 to $Z = 1.6$ is 0.4452. We find the area under the curve from 0 to $Z = 0.6$ is 0.2257. The area under the curve between the two Z-scores is the difference between the two areas, 0.4452 − 0.2257, which is 0.2195. So, 21.95% of the class will get a grade of B.

C: A score is between $\mu - 0.3\sigma$ and $\mu + 0.6\sigma$. We use symmetry and find the area under the curve between 0 and $Z = -0.3$ is 0.1179. We found in part (b) that the area under the curve from 0 to $Z = 0.6$ is 0.2257. Here since the Z-scores have opposite signs we add the two areas, 0.1179 + 0.2257, and get 0.3436. So, 34.36% of the class will get a grade of C.

D: A score is between $\mu - 1.4\sigma$ and $\mu - 0.3\sigma$. We use symmetry and find the areas under the curve between 0 and $Z = -1.4$, which is 0.4192, and 0 and $Z = -0.3$, which is 0.1179. Since both Z-scores are negative, the area between them is the difference between 0.4192 and 0.1179, which is 0.3013. So, 30.13% of the class will get a grade of D.

F: A score is below $\mu - 1.4\sigma$. Use the symmetry of the curve to determine that the area between 0 and $Z = -1.4$ is 0.4192. But we need the area to the left of Z, so we subtract 0.4192 from 0.5, the total area under the curve to the left of 0.
0.5000 − 0.4192 = 0.0808
So, 8.08% of the class will get a grade of F.
(Note: When you add all the percents you should get 100%, the entire class.)

29. We are told that $\bar{x} = 64$ and $S = 2$. We will convert the given heights to Z-scores to determine the percent of women in the required intervals. Then we will calculate how many of the 2000 women are in the interval.

a. Between 62 and 66 inches tall
These women are within 1 standard deviation of the mean, $64 - 2 = 62$ and $64 + 2 = 66$. This gives $Z_1 = -1.0$ which corresponds to $A_1 = 0.3413$, and $Z_2 = 1$ which corresponds to $A_2 = 0.3413$. Since the two Z-scores are on opposite sides of the mean, add the corresponding areas.
Area = $A_1 + A_2 = 0.3413 + 0.3413 = 0.6826$
So, approximately 68.26% of the women or 1365 of the 2000 women sampled will be between 62 and 66 inches tall.

b. Between 60 and 68 inches tall
These women are within 2 standard deviations of the mean, $64 - 2(2) = 60$ and $64 + 2(2) = 68$. This gives $Z_1 = -2.0$ which corresponds to $A_1 = 0.4772$, and $Z_2 = 2.0$ which corresponds to $A_2 = 0.4772$. Since the two Z-scores are on opposite sides of the mean, add the corresponding areas.
Area $= A_1 + A_2 = 0.4772 + 0.4772 = 0.9544$
So, approximately 95.44% of the women or 1909 of the 2000 women sampled will be between 60 and 68 inches tall.

c. Between 58 and 70 inches tall
These women are within 3 standard deviations of the mean, $64 - 3(2) = 58$ and $64 + 3(2) = 70$. This gives $Z_1 = -3.0$ which corresponds to $A_1 = 0.4987$, and $Z_2 = 3.0$ which corresponds to $A_2 = 0.4987$. Since the two Z-scores are on opposite sides of the mean, add the corresponding areas.
Area $= A_1 + A_2 = 0.4987 + 0.4987 = 0.9974$
So, approximately 99.74% of the women or 1995 of the 2000 women sampled will be between 58 and 70 inches tall.

d. More than 70 inches tall
These women are more than 3 standard deviations from the mean. This gives $Z = 3.0$ which corresponds to $A = 0.4987$. We must subtract 0.4987 from 0.5000 to find the percent of the population more than 3 standard deviations from the mean.
Area $= 0.5000 - 0.4987 = 0.0013$
So, approximately 0.13% of the women or about 3 of the 2000 women sampled will be more than 70 inches tall.

e. Shorter than 58 inches.
These women are more than 3 standard deviations below the mean. This gives $Z = -3.0$ which corresponds to $A = 0.4987$. We must subtract 0.4987 from 0.5000 to find the percent of the population more than 3 standard deviations from the mean.
Area $= 0.5000 - 0.4987 = 0.0013$
So, approximately 0.13% of the women or about 3 of the 2000 women sampled will be shorter than 58 inches tall.

31. We are told that $\mu = 130$ and $\sigma = 5.2$ pounds.
  a. Convert 142 pounds to a Z-score.
  $$Z = \frac{x - \mu}{\sigma} = \frac{142 - 130}{5.2} = 2.308 \Rightarrow A = 0.4896$$
  We are interested in the area under the normal curve that to the right of Z, so we subtract A from 0.5000.

$0.5000 - 0.4896 = 0.0104$
Approximately 1.04% of the students or 1 student weighs at least 142 pounds.

  b. To find the range of weights that includes the middle 70% of the students, we need to find the Z-score corresponding to $A = 0.35$. (Because of symmetry 35% of the students will weigh more than the mean, and 35% will weigh less than the mean.) Looking in the body of the Standard Normal Curve Table, we find that 0.3508 is closest to $A = 0.3500$, and 0.3508 corresponds to $Z = 1.04$ and $Z = -1.04$. Using $Z = \pm 1.04$, $\mu = 130$, and $\sigma = 5.2$, we solve the following.

$$Z = \frac{x - \mu}{\sigma} \qquad Z = \frac{x - \mu}{\sigma}$$
$$1.04 = \frac{x - 130}{5.2} \qquad -1.04 = \frac{x - 130}{5.2}$$
$$5.408 = x - 130 \qquad -5.408 = x - 130$$
$$x = 135.408 \qquad x = 124.592$$

So, we would expect 70% of the students to weight between 124.59 and 135.41 pounds.

33. We are given that $\mu = 40$ months and $\sigma = 7$ months. Standardize 28 months and 42 months by finding the Z-scores corresponding to 28 and 40.

$$Z_1 = \frac{x_1 - \mu}{\sigma} = \frac{28 - 40}{7} \approx -1.71 \Rightarrow A_1 = 0.4564$$
$$Z_2 = \frac{x_2 - \mu}{\sigma} = \frac{42 - 40}{7} \approx 0.29 \Rightarrow A_2 = 0.1141$$

Now find the area under the standard normal curve between the two Z-scores. Since the Z-scores are on opposite sides of the mean, add the areas.
Area $= A_2 + A_1 = 0.4564 + 0.1141 = 0.5705$
So, 57.05% of the clothing can be expected to last between 28 and 42 months.

35. We are given that $\mu = 10,000$ and $\sigma = 1000$ people.
  a. The lowest 70% of the attendance figures includes the 50% that are less than the mean and the 20% that are between the mean and some positive value of Z. We look in the body of the Standard Normal Curve Table for the number closest to 0.2000. We find 0.2000 is almost half way between 0.1985 and 0.2019, which correspond to $Z = 0.52$ and $Z = 0.53$ respectively.

(continued on next page)

(continued)

We will use an approximate $Z = 0.525$. Using $Z = 0.525$, $\mu = 10{,}000$ and $\sigma = 1000$, solve

$$Z = \frac{x - \mu}{\sigma}$$

$$0.525 = \frac{x - 10000}{1000}$$

$$525 = x - 10000 \Rightarrow x = 10{,}525$$

Attendance lower than 10,525 will be in the lowest 70% of the figures.

b. To find the percent of attendance figures that falls between 8500 and 11,000 persons, find the Z-score for each and determine the area under the standard normal curve between the two Z-scores.

$$Z_1 = \frac{x_1 - \mu}{\sigma} = \frac{8500 - 10000}{1000} = -1.5 \Rightarrow$$
$$A_1 = 0.4332$$

$$Z_2 = \frac{x_2 - \mu}{\sigma} = \frac{11000 - 10000}{1000} = 1.0 \Rightarrow$$
$$A_2 = 0.3413$$

Since the Z-scores are on opposite sides of the mean, add the areas.
Area $= 0.4332 + 0.3413 = 0.7745$
Approximately 77.45% of the attendance figures are between 8500 and 11,000 persons.

c. Here we are looking for the percent of attendance figures that are more than 11,500 or less than 8500. First find the Z-scores and areas corresponding to $x = 11{,}500$ and $x = 8500$.

$$Z_1 = \frac{x_1 - \mu}{\sigma} = \frac{11500 - 10000}{1000} = 1.5 \Rightarrow$$
$$A_1 = 0.4332$$

From part (b), $Z_2 = -1.5$ and $A_2 = 0.4332$. Now find the area under the standard normal curve outside the two Z-scores. Since the Z-scores are on opposite sides of the mean, add these areas.
Area $= (0.5000 - 0.4332) + (0.5000 - 0.4332)$
$\qquad = 0.1336$
Approximately 13.36% of the attendance figures differ from the mean by 1500 persons or more.

37. Transform each score to a standard score and then compare.
Colleen's score was 76; Colleen's standard

score is $Z = \dfrac{x - \mu}{\sigma} = \dfrac{76 - 82}{7} = -0.857$

Mary's score was 89; Mary's standard score is
$$Z = \frac{x - \mu}{\sigma} = \frac{89 - 93}{2} = -2.0$$
Kathleen's score was 21; Kathleen's standard
score is $Z = \dfrac{x - \mu}{\sigma} = \dfrac{21 - 24}{9} = -0.33$
Kathleen has the highest relative standing.

38. We are told that the mean grade is 75.0 and the standard deviation is 10.0. To determine how many students are in the class, convert 68.0 and 82.0 to Z-scores and find the percent of area between them.

$x_1 = 68$: $Z_1 = \dfrac{x_1 - \mu}{\sigma} = \dfrac{68.0 - 75.0}{10.0} = -0.70 \Rightarrow$
$A_1 = 0.2580$

$x_2 = 82$: $Z_2 = \dfrac{x_2 - \mu}{\sigma} = \dfrac{82.0 - 75.0}{10.0} = 0.70 \Rightarrow$
$A_2 = 0.2580$

Since the Z-scores are on opposite sides of the mean, the area is the sum of A_1 and A_2.
Area $= 0.2580 + 0.2580 = 0.5160$
Approximately 51.6% of the class scored between 68 and 82 on the final exam. This corresponds to the 15 students who got Cs. If we let x represent the number of students in Mathematics 135, then we solve the equation $0.516x = 15$. We find $x = 29.07$. There are 29 students in Mathematics 135.

39. a. To find the line chart and the frequency curve, we need to first find the frequency distribution for the experiment. It is binomial with $n = 15$ and $p = 0.3$. The distribution follows.

Number of Heads, k	Probability $b(15, k, 0.3)$
0	.005
1	.031
2	.092
3	.170
4	.219
5	.206
6	.147
7	.081
8	.035
9	.012
10	.003
11	.0006
12	.0001
13	< 0.0001
14	< 0.0001
15	< 0.0001

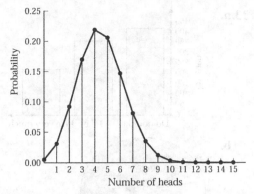

b. The distribution is skewed right.

c. Mean: $\mu = np = 15(0.3) = 4.5$

Standard deviation:

$\sigma = \sqrt{np(1-p)} = \sqrt{15(0.3)(0.7)} = 1.775$

41. This is a binomial distribution which we will approximate with a normal curve. We are told that the player's lifetime batting average is 0.250; this is his probability of success, and that he will bat 300 times; this is n. We first find the mean and the standard deviation of the distribution.

Mean: $\mu = np = 300(0.250) = 75$ hits

Standard deviation:

$\sigma = \sqrt{np(1-p)} = \sqrt{300(0.25)(0.75)} = 7.5$

a. To approximate the probability that he gets at least 80 but no more than 90 hits, we find the area under a normal curve from $x = 79.5$ to $x = 90.5$. Convert 79.5 and 90.5 to Z-scores.

$x_1 = 79.5: Z_1 = \dfrac{x-\mu}{\sigma} = \dfrac{79.5-75}{7.5} = 0.6 \Rightarrow$

$A_1 = 0.2257$

$x_2 = 90.5: Z_2 = \dfrac{x-\mu}{\sigma} = \dfrac{90.5-75}{7.5} = 2.07 \Rightarrow$

$A_2 = 0.4808$

Since both 79.5 and 90.5 are greater than the mean, the area between the points is the difference between A_1 and A_2.

Area $= 0.4808 - 0.2257 = 0.2551$

The approximate probability of having at least 80, but no more than 90 hits is 0.2551. Using a TI-84, the approximate probability is 0.2510.

```
binomcdf(300,0.2
5,90)-binomcdf(3
00,0.25,79)
      .2509559258
```

b. To approximate the probability that 85 or more hits occur, we find the area under a normal curve to the right of $x = 84.5$. We first convert 84.5 to a Z-score.

$Z = \dfrac{x-\mu}{\sigma} = \dfrac{84.5-75}{7.5} = 1.27 \Rightarrow A = 0.3980$

Since the area to the right of the mean is 0.5, we subtract A from 0.5 to obtain the area to the right of Z.

Area $= 0.5000 - 0.3980 = 0.1020$

The approximate probability of 85 or more hits occurring is 0.1020. Using a TI-84, the approximate probability is 0.1037.

```
1-binomcdf(300,0
.25,84)
      .1037265101
```

43. This is a binomial distribution which we will approximate with a normal curve. We are told that 3% of the packages do not seal properly; so we will let $p = 0.03$. 500 packages will be selected, so $n = 500$. We first find the mean and the standard deviation of the distribution.

Mean: $\mu = np = 500(0.03) = 15$ packages

Standard deviation:

$\sigma = \sqrt{np(1-p)} = \sqrt{500(0.03)(0.97)} \approx 3.8$

To approximate the probability that at least 10 packages are not properly sealed, we find the area under a normal curve to the right of $x = 9.5$. Converting 9.5 to a Z-score gives

$Z = \dfrac{x-\mu}{\sigma} = \dfrac{9.5-15}{3.8} = -1.45 \Rightarrow A = 0.4265.$

Since we want the area to the right of $x = 9.5$, we add 0.5 to A.

Area $= 0.5000 + 0.4265 = 0.9265$

The approximate probability of selecting at least 10 unsealed packages is 0.9265.

45. $y = \dfrac{1}{\sqrt{2\pi}} e^{-(1/2)x^2}$

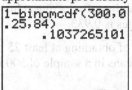

[−3, 3] by [−0.25, 0.5]

The function assumes its maximum at $x = 0$.

47. From Problem 43, we have a binomial distribution with $n = 500$ and $p = 0.03$. We want the probability that at least 10 bags of jelly beans are not sealed properly. The probability at least 10 bags are improperly sealed will be given by 1 minus the opposite event, the probability that at most 9 bags are improperly sealed.

```
1-binomcdf(500,0
.03,9)
      .9330714394
```

The exact probability of selecting at least 10 unsealed packages is 0.933.

49. The probability of obtaining at least 25 bags of unsealed jelly beans in a sample of 500 is 0.010.

Chapter 9 Review Exercises

1. circumference of head; continuous

3. number of people; discrete

5. number of defective products; discrete

7. Answers will vary. All answers should include a method to choose a sample of 100 students from the population for which each student has an equal chance of being chosen.

9. Answers will vary. All answers should give examples of possible bias.

11. a.

b.
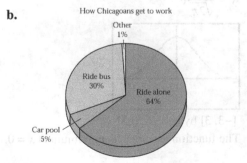

c. Answers will vary.

13. a.
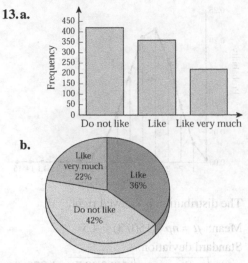

b.
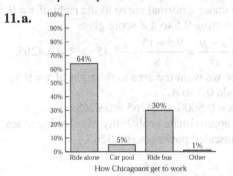

c. Answers will vary.

15. a. American Indians made up the smallest percentage of college enrollment with 1%.

b. Asian-Americans were overrepresented in four-year colleges. They were 4% of the general population, but 6% of the college student population.

c. Approximately, $(0.10)(9,160,973) = 916,097$ Hispanics were enrolled in four-year colleges in 2007.

17. a. The highest level of education that most Americans have obtained is high school graduate.

b. Approximately $38,000,000 + 20,000,000 = 58,000,000$ Americans have a bachelor's degree.

c. Approximately 28,000,000 Americans do not have a high school diploma.

d. Approximately $30,000,000 + 20,000,000 = 50,000,000$ Americans have gone to college but do not have a bachelor's degree.

19. a.

Score	Frequency	Score	Frequency
21	2	73	2
33	1	74	1
41	2	75	1
42	1	77	1
44	1	78	2
48	1	80	4
52	2	82	1
55	1	83	1
60	1	85	2
62	1	87	2
63	2	89	1
66	2	90	3
68	1	91	1
69	1	92	1
70	2	95	1
71	1	100	2
72	2		

The range of scores is $100 - 21 = 79$.

b.

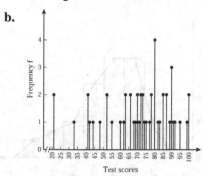

c.,f.

Class Interval	Frequency	Cumulative Frequency
20–29.9	2	2
30–39.9	1	3
40–49.9	5	8
50–59.9	3	11

c.,f.

Class Interval	Frequency	Cumulative Frequency
60–69.9	8	19
70–79.9	12	31
80–89.9	11	42
90–99.9	6	48
100–109.9	2	50

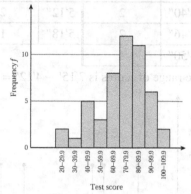

d.

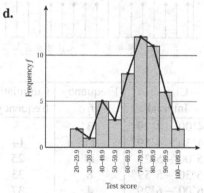

e. The distribution is skewed left.

g.

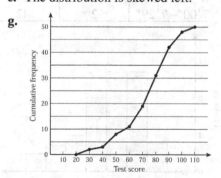

21. a.

Time	Frequency	Time	Frequency	Time	Frequency	Time	Frequency
4′12″	1	4′52″	1	5′20″	3	6′02″	1
4′15″	1	4′56″	1	5′31″	2	6′10″	1
4′22″	1	5′01″	1	5′37″	1	6′12″	1
4′30″	2	5′02″	1	5′40″	2	6′30″	1
4′36″	1	5′06″	2	5′43″	1	6′32″	1
4′39″	1	5′08″	1	5′48″	1	6′40″	1
4′40″	2	5′12″	2	5′55″	1	7′05″	1
4′46″	2	5′18″	1	6′01″	1	7′15″	1
4′50″	1						

The range of scores is 7′15″ − 4′12″ = 3′3″ or 3 minutes, 3 seconds.

b.

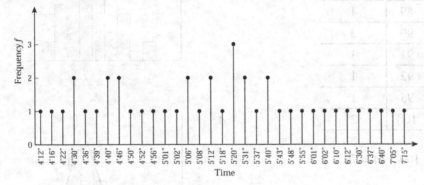

c., g.

Class Interval	Frequency f_i	Cumulative frequency
4′00″– 4′29″	3	3
4′30″– 4′59″	11	14
5′00″– 5′29″	11	25
5′30″– 5′59″	8	33
6′00″– 6′29″	4	37
6′30″– 6′59″	3	40
7′00″– 7′29″	2	42

e.

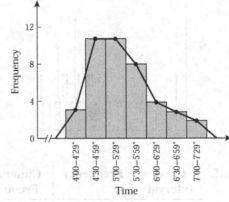

f. Since the distribution has a tail that extends to the right, it is skewed right.

d.

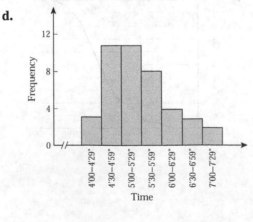

h.

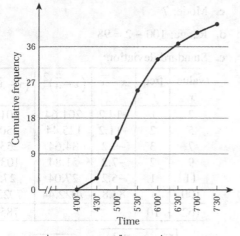

d.

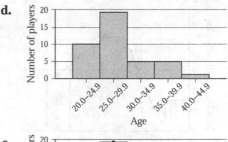

e.

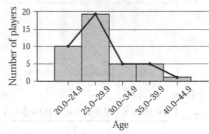

f. The distribution is right-skewed.

23. a.

Age	Frequency	Age	Frequency
23	4	32	0
24	6	33	3
25	6	34	1
26	3	35	1
27	5	36	1
28	1	37	1
29	4	38	2
30	1	39	0
31	0	40	1

h.

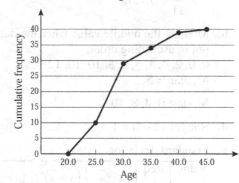

b.

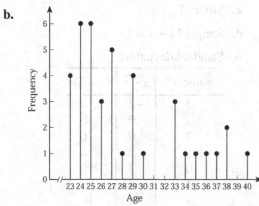

25. a.

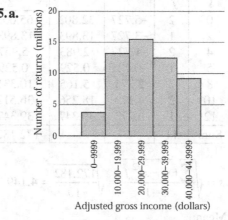

c, g.

Class Interval	Frequency	Cumulative Frequency
20–24.9	10	10
25–29.9	19	29
30–34.9	5	34
35–39.9	5	39
40–44.9	1	40

There are five class intervals.

b.

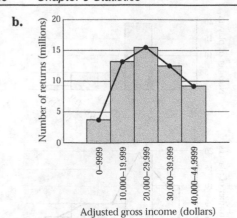

Adjusted gross income (dollars)

27. a. Mean:
$$\bar{x} = \frac{12+10+8+2+0+4+10+5+4+8+0}{11}$$
$$= \frac{63}{11} = 5.7273$$

b. Median: the middle value when the data are put in ascending order.
0, 0, 2, 4, 4, 5, 8, 8, 10, 10, 12
Median = 5

c. Modes: 0, 4, 8, 10

d. Range: $12 - 0 = 12$

e. Standard deviation:

value x	freq f	$x-\bar{x}$	$(x-\bar{x})^2$	$(x-\bar{x})^2 f$
0	2	−5.727	32.802	65.603
2	1	−3.727	13.893	13.893
4	2	−1.727	2.983	5.967
5	1	−0.727	0.529	0.529
8	2	2.273	5.165	10.331
10	2	4.273	18.256	36.512
12	1	6.273	39.347	39.347
Sum	11			172.182

$$s = \sqrt{\frac{\sum(x_i-\bar{x})^2 f}{n-1}} = \sqrt{\frac{172.182}{10}} = 4.149$$

29. a. Mean:
$$\bar{x} = \frac{2+5+5+7+7+7+9+9+11+100}{8}$$
$$= \frac{162}{10} = 16.2$$

b. Median: the middle value when the data are put in ascending order.
2, 5, 5, 7, 7, 7, 9, 9, 11, 100
$$\text{Median} = = \frac{7+7}{2} = 7$$

c. Mode: 7

d. Range: $100 - 2 = 98$

e. Standard deviation:

value x	freq f	$x-\bar{x}$	$(x-\bar{x})^2$	$(x-\bar{x})^2 f$
2	1	−14.2	201.64	201.64
5	2	−11.2	125.44	250.88
7	3	−9.2	84.64	253.92
9	2	−7.2	51.84	103.68
11	1	−5.2	27.04	27.04
100	1	83.8	7022.44	7022.44
162	10			7859.6

$$s = \sqrt{\frac{\sum(x_i-\bar{x})^2 f}{n-1}} = \sqrt{\frac{7859.6}{9}} = 29.5515$$

31. a. Mean:
$$\bar{x} = \frac{5+7+7+9+10+11+1+6+2+12}{10} = 7$$

b. Median: the middle value when the data are put in ascending order.
1, 2, 5, 6, 7, 7, 9, 10, 11, 12
$$\text{Median} = = \frac{7+7}{2} = 7$$

c. Mode: 7

d. Range: $12 - 1 = 11$

e. Standard deviation:

value x	$x-\bar{x}$	$(x-\bar{x})^2$
5	−2	4
7	0	0
7	0	0
9	2	4
10	3	9
11	4	16
1	−6	36
6	−1	1
2	−5	25
12	5	25
Sum = 70		120

$$S = \sqrt{\frac{\sum(x_i-\bar{x})^2}{n-1}} = \sqrt{\frac{120}{9}} = 3.6515$$

33. a. Answers may vary. For this problem, we will assume we have sample data; it is implied that Joe has played more than 7 rounds of golf.

b. Mean:
$$\bar{x} = \frac{74+72+76+81+77+76+73}{7} = 75.57$$

c.

value x	$x - \overline{x}$	$(x - \overline{x})^2$
74	-1.571	2.469
72	-3.571	12.755
76	0.429	0.184
81	5.429	29.469
77	1.429	2.041
76	0.429	0.184
73	-2.571	6.612
Sum = 529		53.714

$$S = \sqrt{\frac{\sum (x_i - \overline{x})^2}{n-1}} = \sqrt{\frac{53.714}{6}} = 2.99$$

35.a.

Class Interval	Midpoint m_i	Frequency f	$f_i m_i$	$x_i - \overline{x}$	$(x_i - \overline{x})^2$	$(x_i - \overline{x})^2 \cdot f_i$
0–4	2.5	10,412	26,030.0	-36.2327	1312.81	13,668,943.12
5–9	7.5	10,073	75,547.5	-31.2327	975.48	9,826,009.39
10–14	12.5	9751	121,887.5	-26.2327	688.15	6,710,181.79
15–19	17.5	10,486	183,505.0	-21.2327	450.83	4,727,366.17
20–24	22.5	10,446	235,035.0	-16.2327	263.50	2,752,517.97
25–29	27.5	10,562	290,455.0	-11.2327	126.17	1,332,638.90
30–34	32.5	9780	317,850.0	-6.2327	38.85	379,916.10
35–39	37.5	10,185	381,937.5	-1.2327	1.52	15,475.96
40–44	42.5	10,487	445,697.5	3.7673	14.19	148,839.31
45–49	47.5	11,535	547,912.5	8.7673	76.87	886,649.34
50–54	52.5	11,083	581,857.5	13.7673	189.54	2,100,663.63
55–59	57.5	9770	561,775.0	18.7673	352.21	3,441,116.31
60–64	62.5	8234	514,625.0	23.7673	564.89	4,651,269.49
65–69	67.5	6273	423,427.5	28.7673	827.56	5,191,277.83
70–74	72.5	4925	357,062.5	33.7673	1140.23	5,615,644.05
75–79	77.5	4176	323,640.0	38.7673	1502.91	6,276,133.59
80–84	82.5	3524	290,730.0	43.7673	1915.58	6,750,499.73
85–89	87.5	2395	209,562.5	48.7673	2378.25	5,695,913.71
90–94	92.5	1077	99,622.5	53.7673	2890.93	3,113,526.58
95–99	97.5	319	31,102.5	58.7673	3453.60	1,101,697.95
100–104	102.5	55	5,527.5	61.7673	3815.20	209,836.14
Total		155,548	6,024,790.0			84,596,117.06

The mean age of a female is $\mu = \dfrac{\sum f_i m_i}{n} = \dfrac{6,024,790}{155,548} \approx 38.73$ years.

b. To find the median of grouped data we use the following steps.
Step 1: Find the interval containing the median. The median is the middle value when the 155,548 items are in ascending order. So the median in this example is the mean of the 77,774[th] and 77,775[th] entries. These entries are in the interval 35–39.
Step 2: In the interval containing the median, count the number p of items remaining to reach the median. The first two intervals account for 71,510 data points; there are $77,774 - 71,510 = 6264$ points left, so $p = 6264$.
Step 3: Calculate the interpolation factor. q is the frequency for the interval containing the median; $q = 10,185$ and i is the size of the interval, $i = 5$, so the interpolation factor is $\dfrac{p}{q} \cdot i = \dfrac{6264}{10,185} \cdot 5 \approx 3.08$.

Step 4: The median salary $M = \begin{bmatrix}\text{lower limit of interval}\\ \text{containing the median}\end{bmatrix} + [\text{interpolation factor}] = 35 + 3.08 = 38.08$.

c. The standard deviation assuming population data is $\sigma = \sqrt{\dfrac{\sum (x_i - \mu)^2 \cdot f_i}{n}} = \sqrt{\dfrac{84,596,117.06}{155,548}} \approx 23.32$.

37. a. 99.7% of the data lie between 3 standard deviations of the mean.
$\mu - 3\sigma = 600 - 3 \cdot 53 = 441$
$\mu + 3\sigma = 600 + 3 \cdot 53 = 759$
So, 99.7% of the light bulbs have lifetimes between 441 and 759 hours.

b. We use the Empirical Rule and compute
$\dfrac{494 - 600}{53} = -2$ and $\dfrac{706 - 600}{53} = 2$.
For a bell-shaped distribution, approximately 95% of the data fall between 2 standard deviations of the mean. Approximately 95% of the light bulbs have a lifetime between 494 and 706 hours.

c. We use the Empirical Rule and compute
$\dfrac{547 - 600}{53} = -1$ and $\dfrac{706 - 600}{53} = 2$
Because of the symmetry of a bell-shaped distribution, $\dfrac{1}{2} \cdot 0.68 = 0.34$ of the data lie between the mean and one standard deviation and $\dfrac{1}{2} \cdot 0.95 = 0.475$ of the data lie between the mean and two standard deviations. So, $0.34 + 0.475 = 0.815$ or 81.5% of the light bulbs will last between 547 and 706 hours.

d. From part (a), 441 is three standard deviations. The manufacturer replaces all light bulbs that burn out before this time. Since 99.7% of the data lie between three standard deviations of the mean, and since the distribution is symmetric, the firm expects to replace
$\dfrac{1}{2}(1 - 0.997) = \dfrac{1}{2} \cdot 0.003 = 0.0015 = 0.15\%$ of the light bulbs.

38. a. 99.7% of the data lie between 3 standard deviations of the mean.
$\mu - 3\sigma = 4302 - 3 \cdot 340 = 3282$
$\mu + 3\sigma = 4302 + 3 \cdot 340 = 5322$
So, 99.7% of the cartridges print between 3282 and 5322 pages.

b. We use the Empirical Rule to compute
$\dfrac{3622 - 4302}{340} = -2$ and $\dfrac{4982 - 4302}{340} = 2$
For a bell-shaped distribution, approximately 95% of the data fall between 2 standard deviations of the mean. So 95% of the cartridges print between 3622 and 4982 pages.

c. From part (b), we know that 3622 is two standard deviations below the mean. Because the distribution is symmetric, the Empirical Rule, states that
$\dfrac{1}{2}(1 - 0.95) = \dfrac{1}{2} \cdot 0.05 = 0.025 = 2.5\%$ of the cartridges print fewer than 3622 pages. The company will replace 2.5% of the toner cartridges.

39. $\mu = 12$, $\sigma = 0.05$

We want to estimate the probability a jar contains between 11.9 and 12.1 ounces, so $k = 12.1 - \mu = 12.1 - 12 = 0.1$ or $k = \mu - 11.9 = 12 - 11.9 = 0.1$. By Chebychev's theorem, the probability that a jar has between 11.9 and 12.1 ounces is at least

$$1 - \frac{\sigma^2}{k^2} = 1 - \frac{0.05^2}{0.1^2} = 0.75.$$

We expect at least 75% of the jars, or $0.75(1000) = 750$ jars, to contain between 11.9 and 12.1 ounces.

41. $\mu = 10$, $\sigma = 0.25$

We want to estimate the probability a bag weighs less than 9.5 or more than 10.5 pounds. $k = 10.5 - \mu = 10.5 - 10 = 0.5$ or $k = \mu - 9.5 = 10 - 9.5 = 0.5$. By Chebychev's theorem, the probability a bag weighs between 9.5 and 10.5 pounds is at least

$$1 - \frac{\sigma^2}{k^2} = 1 - \frac{0.25^2}{0.5^2} = 0.75.$$

The probability a bag weighs less than 9.5 or more than 10.5 pounds is less than $1 - 0.75 = 0.25 = 25\%$.

43. $Z = \frac{x - \mu}{\sigma} = \frac{8 - 10}{3} = -\frac{2}{3} = -0.667$

45. $Z = \frac{x - \mu}{\sigma} = \frac{8 - 1}{5} = \frac{7}{5} = 1.40$

47. $Z = \frac{x - \mu}{\sigma} = \frac{60 - 55}{3} = \frac{5}{3} = 1.667$

49. $Z_1 = -1.35$, $A_1 = 0.4115$
$Z_2 = -2.75$, $A_2 = 0.4970$
Since both Z-scores are on the same side as the mean, subtract the areas.
Area $= A_2 - A_1 = 0.4970 - 0.4115 = 0.0855$

51. $Z_1 = -0.75$, $A_1 = 0.2734$
$Z_2 = 2.1$, $A_2 = 0.4821$
Since the Z-scores are on opposite sides of the mean, add the areas.
Area $= A_1 + A_2 = 0.2734 + 0.4821 = 0.7555$

53. $\mu = 25$, $\sigma = 5$

a. We need to find the area under a normal curve between $x = 20$ and $x = 30$.

$x = 20$: $Z_1 = \frac{x - \mu}{\sigma} = \frac{20 - 25}{5} = -1.00 \Rightarrow$

$A_1 = 0.3413$

$x = 30$: $Z_2 = \frac{x - \mu}{\sigma} = \frac{30 - 25}{5} = 1.00 \Rightarrow$

$A_2 = 0.3413$
Since the values are on opposite sides of the mean, we add the areas.
Area $= A_1 + A_2 = 0.3413 + 0.3413 = 0.6826$
So, 68.26% of the scores fall between 20 and 30.

b. We need to find the area under a normal curve to the right of $x = 35$.
$x = 35$:

$$Z = \frac{x - \mu}{\sigma} = \frac{35 - 25}{5} = 2 \Rightarrow A = 0.4772$$

Since the area to the right of the mean is 0.5, the area to the right of $x = 35$ is $0.5000 - 0.4772 = 0.0228$. So, 2.28% of the scores are above 35.

55. $\mu = 14$, $\sigma = 1.25$
We need the area under a normal curve to the left of 10 years, 4 months which is equivalent to $10\frac{1}{3} = \frac{31}{3}$ years. We convert $x = \frac{31}{3}$ to a Z-score.

$$Z = \frac{x - \mu}{\sigma} = \frac{\frac{31}{3} - 14}{1.25} = -2.93 \Rightarrow A = 0.4983$$

The area to the left of x is $0.5000 - 0.4983 = 0.0017$.
Approximately 0.17% of dogs will die before reaching the age of 10 years 4 months.

57. We need to convert both test grades to standard scores, and then compare the grades.
Mathematics: score: 89; $\mu = 79$; $\sigma = 5$

$$Z = \frac{x - \mu}{\sigma} = \frac{89 - 79}{5} = 2.0$$

Sociology: score: 79; $\mu = 72$, $\sigma = 35$

$$Z = \frac{x - \mu}{\sigma} = \frac{79 - 72}{3.5} = 2.0$$

Bob scored equally well on both exams.

59. We need to find the area under a normal curve between $x = 30$ and $x = 50$. We convert these value to Z-scores, using $\mu = 40$ and

$\sigma = \sqrt{25} = 5$.

$x = 30$:

$$Z_1 = \frac{x - \mu}{\sigma} = \frac{30 - 40}{5} = -2 \Rightarrow A_1 = 0.4772$$

$x = 50$:

$$Z_2 = \frac{x - \mu}{\sigma} = \frac{50 - 40}{5} = 2 \Rightarrow A_2 = 0.4772$$

Since the values are on opposite sides of the mean, we add the areas.
Area $= A_1 + A_2 = 0.4772 + 0.4772 = 0.9544$
So, the probability that this week's production will be between 30 and 50 is 0.9544.

61. This is a binomial distribution which we will approximate using a normal curve with a correction for continuity. Since we are interested in the probability of more than 160 successes, we let $x = 160.5$, and convert x to a Z-score. We have $p = 0.7$, the probability of a positive result, and $n = 200$, the number of blood samples.
$\mu = np = 200(0.7) = 140$ and

$$\sigma = \sqrt{np(1-p)} = \sqrt{200(0.7)(0.3)} = 6.48$$

$x = 160.5$: $Z = \dfrac{x - \mu}{\sigma} = \dfrac{160.5 - 140}{6.48} = 3.16 \Rightarrow$

$A = 0.4992$
The probability of obtaining more than 160 positive results is $0.5000 - 0.4992 = 0.0008$.

63. This is a binomial distribution which we will approximate using a normal curve with corrections for continuity. We obtain the area under a normal curve between $x = 19.5$ and $x = 30.5$ by converting both to Z-scores. We have $p = 0.05$, the probability the gate does not open, and $n = 500$, the number of times the gate is opened.
$\mu = np = 500(0.05) = 25$ and

$$\sigma = \sqrt{np(1-p)} = \sqrt{500(0.05)(0.95)} = 4.87$$

$x = 19.5$:

$$Z_1 = \frac{x - \mu}{\sigma} = \frac{19.5 - 25}{4.87} = -1.13 \Rightarrow A_1 = 0.3708$$

$x = 30.5$

$$Z_2 = \frac{x - \mu}{\sigma} = \frac{30.5 - 25}{4.87} = 1.13 \Rightarrow A_2 = 0.3708$$

Since the values are on opposite sides of the mean, we add the areas.
Area $= A_1 + A_2 = 0.3708 + 0.3708 = 0.7416$
The probability that the toll gate fails to work between 20 and 30 times is 0.7416.

Using a TI-84, the approximate probability is 0.2510.

Chapter 9 Project

1. Mean high temperature: $\dfrac{2237}{30} \approx 74.6°$ F

Mean low temperature: $\dfrac{1834}{30} \approx 61.1°$ F

Total rainfall: 10.06 inches

3. Mean monthly temperature:

$$\frac{\sum \frac{1}{2}(\text{daily high} + \text{daily low})}{30}$$

$$= \frac{\frac{1}{2}\sum(\text{daily high} + \text{daily low})}{30}$$

$$= \frac{\sum(\text{daily high} + \text{daily low})}{60}$$

$$= \frac{\sum(\text{daily high}) + \sum(\text{daily low})}{60}$$

Mean of mean high and mean low temperatures:

$$\frac{1}{2}\left[\left(\frac{\sum \text{daily high}}{30}\right) + \left(\frac{\sum \text{daily low}}{30}\right)\right]$$

$$= \frac{1}{2}\left[\frac{\sum(\text{daily high} + \text{daily low})}{30}\right]$$

$$= \frac{\sum(\text{daily high} + \text{daily low})}{60}$$

5. Modal high temperatures: 71° and 80°
modal low temperature: 60°.

7. High temperatures:

x_i	$x_i - \bar{x}$	$(x_i - \bar{x})^2$	x_i	$x_i - \bar{x}$	$(x_i - \bar{x})^2$
71	−3.6	12.72	70	−4.6	20.85
80	5.4	29.52	68	−6.6	43.12
72	−2.6	6.59	65	−9.6	91.52
68	−6.6	43.12	77	2.4	5.92
60	−14.6	212.19	71	−3.6	12.72
76	1.4	2.05	78	3.4	11.79
82	7.4	55.25	75	0.4	0.19
75	0.4	0.19	80	5.4	29.52
71	−3.6	12.72	76	1.4	2.05
67	−7.6	57.25	79	4.4	19.65
67	−7.6	57.25	83	8.4	71.12
80	5.4	29.52	81	6.4	41.39
73	−1.6	2.45	82	7.4	55.25
73	−1.6	2.45	81	6.4	41.39
72	−3.6	12.72	84	9.4	88.99
				Sum:	1065.37

$$\sigma = \sqrt{\frac{\sum(x-74.6)^2}{30}} = \sqrt{\frac{1065.37}{30}} \approx 5.96$$

Low temperatures:

x_i	$x_i - \bar{x}$	$(x_i - \bar{x})^2$	x_i	$x_i - \bar{x}$	$(x_i - \bar{x})^2$
53	−8.1	66.15	55	−6.1	37.62
60	−1.1	1.28	60	−1.1	1.28
54	−7.1	50.88	61	−0.1	0.02
56	−5.1	26.35	63	1.9	3.48
55	−6.1	37.62	63	1.9	3.48
60	−1.1	1.28	64	2.9	8.22
66	4.9	23.68	66	4.9	23.68
60	−1.1	1.28	65	3.9	14.95
59	−2.1	4.55	64	2.9	8.22
59	−2.1	4.55	67	5.9	34.42
59	−2.1	4.55	63	1.9	3.48
67	5.9	34.42	64	2.9	8.22
61	−0.1	0.02	65	3.9	14.95
57	−4.1	17.08	67	5.9	34.42
56	−5.1	26.35	65	3.9	14.95
				Sum:	511.47

$$\sigma = \sqrt{\frac{\sum(x-61.1)^2}{30}} = \sqrt{\frac{511.47}{30}} \approx 4.13$$

Since for the low temperatures, mode < median < mean, the distribution is not symmetric, but is skewed right. This means that there are a few high temperatures that are influencing the mean.

9.

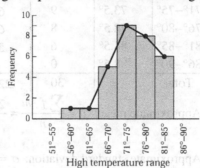

High temperature range

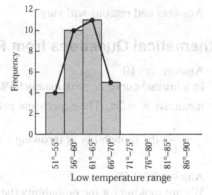

Low temperature range

11.

Class Interval	Midpoint m_i	Frequency f	$f_i m_i$	$x_i - \bar{x}$	$(x_i - \bar{x})^2$	$(x_i - \bar{x})^2 \cdot f_i$
51°–55°	53.5°	0	0	−21.9	478.15	0.00
56°–60°	58.5°	1	58.5	−16.9	284.48	284.48
61°–65°	63.5°	1	63.5	−11.9	140.82	140.82
66°–70°	68.5°	5	342.5	−6.9	47.15	235.76
71°–75°	73.5°	9	661.5	−1.9	3.48	31.36
76°–80°	78.5°	8	628	3.1	9.82	78.54
81°–85°	84.5°	6	507	9.1	83.42	500.51
86°–90°	88.5°	0	0	13.1	172.48	0.00
Total		30	2261			1271.47

Approximate mean: $\mu = \dfrac{\sum f \cdot m}{n} = \dfrac{2261}{30} \approx 75.4°$

Approximate standard deviation: $\sigma = \sqrt{\dfrac{\sum f \cdot (m - \mu)^2}{n}} = \sqrt{\dfrac{1271.41}{30}} \approx 6.51°$

Answers and reasons will vary.

Mathematical Questions from Professional Exams

1. Answer: (e) 10
    In a normal curve, approximately 95% of the area under the curve lies within 2 standard deviations of the mean, so $K = 2\sigma$. The experiment is binomial, meaning $\sigma = \sqrt{np(1-p)}$.

    $p = \dfrac{1}{6}$, the probability of throwing a sum of 7; $n = 180$ throws; $K = 2\sqrt{(180)\left(\dfrac{1}{6}\right)\left(\dfrac{5}{6}\right)} = 10$.

3. Answer (c) 0.68
    We are looking for the probability the balls are within 1 standard deviation of the mean weight of the 100 balls. The area under a normal curve between $\mu - \sigma$ and $\mu + \sigma$ is approximately 0.

Chapter 10 Markov Chains; Games

10.1 Markov Chains and Transition Matrices

1. $\dfrac{1}{4}$

3. The dimension of the matrix $\begin{bmatrix} 2 & 3 & 4 & 5 \end{bmatrix}$ is 1×4.

5. False. In transition matrices, all the entries are between 0 and 1 inclusive and the sum of the entries in every row is 1.

7. In a Markov chain with m states the intial probability distribution is a row vector of dimension $\underline{1 \times m}$.

9. a. The entry $\dfrac{1}{4}$ represents the probability that an object in state 2 will move to state 1.

   b. We use the tree diagram below.

   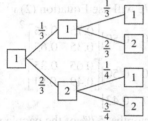

   The probability distribution after one observation is $v^{(1)} = \begin{bmatrix} \dfrac{1}{3} & \dfrac{2}{3} \end{bmatrix}$.

   The probability distribution after two observations is

   from state 1 to $1 = \dfrac{1}{3}\left(\dfrac{1}{3}\right) + \dfrac{2}{3}\left(\dfrac{1}{4}\right)$

   $= \dfrac{1}{9} + \dfrac{1}{6} = \dfrac{5}{18}$

   from state 1 to $2 = \dfrac{1}{3}\left(\dfrac{2}{3}\right) + \dfrac{2}{3}\left(\dfrac{3}{4}\right)$

   $= \dfrac{2}{9} + \dfrac{1}{2} = \dfrac{13}{18}$

   $v^{(2)} = \begin{bmatrix} \dfrac{5}{18} & \dfrac{13}{18} \end{bmatrix}$.

   c. We use the tree diagram below.

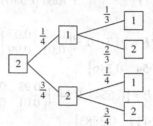

The probability distribution after one observation is $v^{(1)} = \begin{bmatrix} \dfrac{1}{4} & \dfrac{3}{4} \end{bmatrix}$.

The probability distribution after two observations is

from state 2 to $1 = \dfrac{1}{4}\left(\dfrac{1}{3}\right) + \dfrac{3}{4}\left(\dfrac{1}{4}\right)$

$= \dfrac{1}{12} + \dfrac{3}{16} = \dfrac{13}{48}$

from state 2 to $2 = \dfrac{1}{4}\left(\dfrac{2}{3}\right) + \dfrac{3}{4}\left(\dfrac{3}{4}\right)$

$= \dfrac{1}{6} + \dfrac{9}{16} = \dfrac{35}{48}$

$v^{(2)} = \begin{bmatrix} \dfrac{13}{48} & \dfrac{35}{48} \end{bmatrix}$.

11. Using Equation (2) on page 582 in your text, we have $v^{(2)} = v^{(0)}P^2 = \begin{bmatrix} \dfrac{1}{4} & \dfrac{3}{4} \end{bmatrix}\begin{bmatrix} \dfrac{1}{3} & \dfrac{2}{3} \\ \dfrac{1}{4} & \dfrac{3}{4} \end{bmatrix}^2$

$= \begin{bmatrix} \dfrac{1}{4} & \dfrac{3}{4} \end{bmatrix}\begin{bmatrix} \dfrac{5}{18} & \dfrac{13}{18} \\ \dfrac{13}{48} & \dfrac{35}{48} \end{bmatrix} = \begin{bmatrix} \dfrac{157}{576} & \dfrac{419}{576} \end{bmatrix}$

13. $v^{(1)} = v^{(0)}P$

$= \begin{bmatrix} 0.25 & 0.25 & 0.5 \end{bmatrix}\begin{bmatrix} 0.7 & 0.2 & 0.1 \\ 0.6 & 0.2 & 0.2 \\ 0.4 & 0.1 & 0.5 \end{bmatrix}$

$= \begin{bmatrix} 0.525 & 0.15 & 0.325 \end{bmatrix}$

15. To be a transition matrix, all entries must be non-negative and the sum of the entries in a row must equal 1.

$\begin{bmatrix} 0.2 & a & 0.4 \\ b & 0.6 & 0.3 \\ 0 & c & 0 \end{bmatrix}$ $\quad\begin{aligned} a &= 1 - 0.2 - 0.4 = 0.4 \\ b &= 1 - 0.6 - 0.3 = 0.1 \\ c &= 1 \end{aligned}$

17. If in Example 3, $v^{(0)} = [0.7 \quad 0.3]$ then we can get the probability distribution after 5 years by using Equation (1) $v^{(k)} = v^{(k-1)}P$ (text p. 580), five times.

$$v^{(1)} = v^{(0)}P = [0.7 \quad 0.3]\begin{bmatrix} 0.93 & 0.07 \\ 0.01 & 0.99 \end{bmatrix}$$
$$= [0.654 \quad 0.346]$$

$$v^{(2)} = v^{(1)}P = [0.654 \quad 0.346]\begin{bmatrix} 0.93 & 0.07 \\ 0.01 & 0.99 \end{bmatrix}$$
$$= [0.6117 \quad 0.3883]$$

$$v^{(3)} = v^{(2)}P = [0.6117 \quad 0.3883]\begin{bmatrix} 0.93 & 0.07 \\ 0.01 & 0.99 \end{bmatrix}$$
$$= [0.5727 \quad 0.4273]$$

$$v^{(4)} = v^{(3)}P = [0.5727 \quad 0.4273]\begin{bmatrix} 0.93 & 0.07 \\ 0.01 & 0.99 \end{bmatrix}$$
$$= [0.5369 \quad 0.4631]$$

$$v^{(5)} = v^{(4)}P = [0.5369 \quad 0.4631]\begin{bmatrix} 0.93 & 0.07 \\ 0.01 & 0.99 \end{bmatrix}$$
$$= [0.5040 \quad 0.4960]$$

After 5 years 50.4% of the residents live in the city and 49.6% live in the suburbs.

19. a. This is a Markov chain because it represents a sequence of experiments each of which results in one of two states, and the probability of being in a particular state depends only on the previous state.

b. $P = \begin{array}{c} R \\ C \end{array}\begin{array}{cc} R & C \\ \begin{bmatrix} 0.90 & 0.10 \\ 0.05 & 0.95 \end{bmatrix} \end{array}$

c. $P^2 = \begin{array}{c} R \\ C \end{array}\begin{array}{cc} R & C \\ \begin{bmatrix} 0.815 & 0.185 \\ 0.0925 & 0.9075 \end{bmatrix} \end{array}$

$P^3 = \begin{array}{c} R \\ C \end{array}\begin{array}{cc} R & C \\ \begin{bmatrix} 0.74275 & 0.25725 \\ 0.128625 & 0.871375 \end{bmatrix} \end{array}$

21. a. This is a Markov chain because it represents a sequence of experiments each of which results in one of two states, and the probability of being in a particular state depends only on the previous state.

b. $P = \begin{bmatrix} \frac{1}{2} & 0 & 0 & \frac{1}{2} & 0 & 0 & 0 & 0 & 0 \\ 0 & \frac{1}{2} & 0 & 0 & \frac{1}{2} & 0 & 0 & 0 & 0 \\ 0 & 0 & \frac{1}{2} & 0 & 0 & \frac{1}{2} & 0 & 0 & 0 \\ \frac{1}{3} & 0 & 0 & \frac{1}{3} & 0 & 0 & \frac{1}{3} & 0 & 0 \\ 0 & \frac{1}{3} & 0 & 0 & \frac{1}{3} & 0 & 0 & \frac{1}{3} & 0 \\ 0 & 0 & 0 & 0 & 0 & 1 & 0 & 0 & 0 \\ 0 & 0 & 0 & 0 & 0 & 0 & 1 & 0 & 0 \\ 0 & 0 & 0 & 0 & 0 & 0 & 0 & 1 & 0 \\ 0 & 0 & 0 & 0 & 0 & 0 & 0 & 0 & 1 \end{bmatrix}$

23. This is a Markov chain with a transition matrix

$$P = \begin{array}{c} \text{Pinot Grigio} \\ \text{Other White} \end{array}\begin{array}{cc} \text{Pinot Grigio} & \text{Other White} \\ \begin{bmatrix} 0.75 & 0.25 \\ 0.35 & 0.65 \end{bmatrix} \end{array}$$

If 50% of the people currently drink Pinot Grigio, the initial probability distribution is $v^{(0)} = [0.50 \quad 0.50]$. To find what percentage will drink Pinot Grigio after 2 months, we need to find $v^{(2)}$. We will use Equation (2) $v^{(2)} = v^{(0)}P^2$.

$$v^{(2)} = [0.50 \quad 0.50]\begin{bmatrix} 0.75 & 0.25 \\ 0.35 & 0.65 \end{bmatrix}^2$$
$$= [0.50 \quad 0.50]\begin{bmatrix} 0.65 & 0.35 \\ 0.49 & 0.51 \end{bmatrix}$$
$$= [0.57 \quad 0.43]$$

After two months, 57% of the people will drink Pinot Grigio.

25. $v^{(10)} = v^{(0)}P^{10}$

$$= [0.4 \quad 0.3 \quad 0.1 \quad 0.2]\begin{bmatrix} 0.5 & 0.2 & 0.1 & 0.2 \\ 0.3 & 0.3 & 0.2 & 0.2 \\ 0.1 & 0.5 & 0.1 & 0.3 \\ 0.25 & 0.25 & 0.25 & 0.25 \end{bmatrix}^{10}$$
$$= [0.3200 \quad 0.2892 \quad 0.1631 \quad 0.2277]$$

27. a. $P = \begin{bmatrix} \frac{1}{2} & 0 & 0 & \frac{1}{2} & 0 & 0 \\ 0 & \frac{1}{2} & 0 & 0 & \frac{1}{2} & 0 \\ 0 & 0 & \frac{1}{2} & 0 & 0 & \frac{1}{2} \\ \frac{1}{2} & 0 & 0 & \frac{1}{2} & 0 & 0 \\ 0 & 0 & 0 & 0 & 0 & 1 \\ 0 & 0 & 0 & 0 & 0 & 1 \end{bmatrix}$

b. If the mouse is initially in room 2, the initial probability distribution is

$$v^{(0)} = [0 \quad 1 \quad 0 \quad 0 \quad 0 \quad 0].$$

c. After 10 stages the probability distribution will be

$$v^{(10)} = v^{(0)}P^{10}$$

$$= \begin{bmatrix} 0 & 1 & 0 & 0 & 0 & 0 \end{bmatrix} \begin{bmatrix} \frac{1}{2} & 0 & 0 & \frac{1}{2} & 0 & 0 \\ 0 & \frac{1}{2} & 0 & 0 & \frac{1}{2} & 0 \\ 0 & 0 & \frac{1}{2} & 0 & 0 & \frac{1}{2} \\ \frac{1}{2} & 0 & 0 & \frac{1}{2} & 0 & 0 \\ 0 & 0 & 0 & 0 & 0 & 1 \\ 0 & 0 & 0 & 0 & 0 & 1 \end{bmatrix}^{10}$$

$$= \begin{bmatrix} 0 & 0.001 & 0 & 0 & 0.001 & 0.998 \end{bmatrix}$$

d. The mouse will most likely be in room 6.

29. Since A is a transition matrix all of its entries are non-negative, and the sum of the rows is one.

$$a_{11} + a_{12} = 1 \qquad a_{21} + a_{22} = 1$$

Since $u = \begin{bmatrix} u_1 & u_2 \end{bmatrix}$ is a probability row vector its entries are non-negative and the sum of the entries is one.

$$uA = \begin{bmatrix} u_1 & u_2 \end{bmatrix} \begin{bmatrix} a_{11} & a_{12} \\ a_{21} & a_{22} \end{bmatrix}$$

$$= \begin{bmatrix} u_1 a_{11} + u_2 a_{21} & u_1 a_{12} + u_2 a_{22} \end{bmatrix}$$

If uA is a probability vector then its entries must be non-negative, which they are because they are the sums and products of non-negative numbers, and the entries must sum to 1.

$$u_1 a_{11} + u_2 a_{21} + u_1 a_{12} + u_2 a_{22} = 1$$
$$u_1 a_{11} + u_1 a_{12} + u_2 a_{21} + u_2 a_{22}$$
$$= u_1 (a_{11} + a_{12}) + u_2 (a_{11} + a_{12})$$
$$= u_1 (1) + u_2 (1) \quad \text{(from above)}$$
$$= 1 \quad \text{(since } u \text{ is a probability row vector)}$$

Thus, uA is a probability vector.

31. The matrix given cannot be the transition matrix for a Markov chain because it has a negative entry. All entries in a transition matrix must be between 0 and 1, and the sum of the entries in every row must equal 1.

33. If A is a transition matrix then $A^{(2)}$, $A^{(3)}$, …, $A^{(n)}$ are also transition matrices.

10.2 Regular Markov Chains

1. $\begin{cases} x_1 + x_2 = 1 & (1) \\ 0.07x_1 - 0.01x_2 = 0 & (2) \end{cases}$

Solve equation (1) for x_1, then substitute the expression for x_1 into equation (2) and solve for x_2.

$$x_1 + x_2 = 1 \Rightarrow x_1 = 1 - x_2$$
$$0.07(1 - x_2) - 0.01x_2 = 0$$
$$0.07 - 0.07x_2 - 0.01x_2 = 0$$
$$-0.08x_2 = -0.07 \Rightarrow x_2 = \frac{7}{8}$$

Now substitute $x_2 = \frac{7}{8}$ into equation (1) and solve for x_1.

$$x_1 + \frac{7}{8} = 1 \Rightarrow x_1 = \frac{1}{8}.$$

3. False. A Markov chain is regular if some power of the transition matrix P has positive entries.

5. P is not regular; the product of row 2 and column 1 is always 0. So $p_{21} = 0$ for every power of P.

7. P is not regular; the product of row 1 and column 2 is always 0. So $p_{12} = 0$ for every power of P.

9. P is regular.

$$P^2 = \begin{bmatrix} \frac{7}{36} & \frac{13}{36} & \frac{4}{9} \\ \frac{1}{8} & \frac{3}{8} & \frac{1}{2} \\ \frac{5}{24} & \frac{1}{3} & \frac{11}{24} \end{bmatrix} \text{ has all positive entries.}$$

11. P is not regular. The product of row 1 and column 3 is always 0. (So are the products of row 2 and column 3; row 3 and column 1; and row 3 and column 2). So $p_{13} = 0$, $p_{23} = 0$, $p_{31} = 0$, and $p_{32} = 0$ for every power of P.

13. P is regular.

$$P^2 = \begin{bmatrix} \frac{1}{3} & \frac{3}{16} & \frac{13}{48} & \frac{5}{24} \\ \frac{1}{3} & \frac{3}{16} & \frac{13}{48} & \frac{5}{24} \\ \frac{5}{18} & \frac{1}{6} & \frac{5}{18} & \frac{5}{18} \\ \frac{1}{3} & \frac{3}{16} & \frac{13}{48} & \frac{5}{24} \end{bmatrix}$$

has all positive entries.

15. Define $\mathbf{t} = \begin{bmatrix} t_1 & t_2 \end{bmatrix}$. We want to find $\mathbf{t}$ such that $\mathbf{t}P = \mathbf{t}$.

$$\begin{bmatrix} t_1 & t_2 \end{bmatrix} \begin{bmatrix} \dfrac{1}{3} & \dfrac{2}{3} \\ \dfrac{1}{2} & \dfrac{1}{2} \end{bmatrix} = \begin{bmatrix} t_1 & t_2 \end{bmatrix}$$

This leads to the system of equations

$$\begin{cases} \dfrac{1}{3}t_1 + \dfrac{1}{2}t_2 = t_1 \\ \dfrac{2}{3}t_1 + \dfrac{1}{2}t_2 = t_2 \\ t_1 + t_2 = 1 \end{cases} \text{ or } \begin{cases} -\dfrac{2}{3}t_1 + \dfrac{1}{2}t_2 = 0 \\ \dfrac{2}{3}t_1 - \dfrac{1}{2}t_2 = 0 \\ t_1 + t_2 = 0 \end{cases}$$

The first two equations are equivalent so we actually have a system of 2 equations,

$$\begin{cases} \dfrac{2}{3}t_1 - \dfrac{1}{2}t_2 = 0 & \text{(1)} \\ t_1 + t_2 = 1 & \text{(2)} \end{cases}$$

We will solve by substitution, substituting $t_2 = 1 - t_1$. Equation (1) becomes

$$\dfrac{2}{3}t_1 - \dfrac{1}{2}(1 - t_1) = 0$$
$$4t_1 - 3 + 3t_1 = 0$$
$$7t_1 = 3 \Rightarrow t_1 = \dfrac{3}{7}$$

Back-substituting t_1 in equation (2), we have $t_2 = 1 - \dfrac{3}{7} = \dfrac{4}{7}$. Thus, the fixed probability vector is $\mathbf{t} = \begin{bmatrix} \dfrac{3}{7} & \dfrac{4}{7} \end{bmatrix}$.

17. Define $\mathbf{t} = \begin{bmatrix} t_1 & t_2 \end{bmatrix}$. We want to find $\mathbf{t}$ such that $\mathbf{t}P = \mathbf{t}$.

$$\begin{bmatrix} t_1 & t_2 \end{bmatrix} \begin{bmatrix} \dfrac{1}{4} & \dfrac{3}{4} \\ \dfrac{1}{2} & \dfrac{1}{2} \end{bmatrix} = \begin{bmatrix} t_1 & t_2 \end{bmatrix}$$

This leads to the system of equations

$$\begin{cases} \dfrac{1}{4}t_1 + \dfrac{1}{2}t_2 = t_1 \\ \dfrac{3}{4}t_1 + \dfrac{1}{2}t_2 = t_2 \\ t_1 + t_2 = 1 \end{cases} \text{ or } \begin{cases} -\dfrac{3}{4}t_1 + \dfrac{1}{2}t_2 = 0 \\ \dfrac{3}{4}t_1 - \dfrac{1}{2}t_2 = 0 \\ t_1 + t_2 = 1 \end{cases}$$

The first two equations are equivalent so we actually have a system of 2 equations,

$$\begin{cases} \dfrac{3}{4}t_1 - \dfrac{1}{2}t_2 = 0 & \text{(1)} \\ t_1 + t_2 = 1 & \text{(2)} \end{cases}$$

We will solve by substitution, substituting $t_2 = 1 - t_1$. Equation (1) becomes

$$\dfrac{3}{4}t_1 - \dfrac{1}{2}(1 - t_1) = 0$$
$$3t_1 - 2 + 2t_1 = 0$$
$$5t_1 = 2 \Rightarrow t_1 = \dfrac{2}{5}$$

Back-substituting t_1 in equation (2), we have $t_2 = 1 - \dfrac{2}{5} = \dfrac{3}{5}$. Thus, the fixed probability vector is $\mathbf{t} = \begin{bmatrix} \dfrac{2}{5} & \dfrac{3}{5} \end{bmatrix}$.

19. Define $\mathbf{t} = \begin{bmatrix} t_1 & t_2 & t_3 \end{bmatrix}$. We want to find $\mathbf{t}$ such that $\mathbf{t}P = \mathbf{t}$.

$$\begin{bmatrix} t_1 & t_2 & t_3 \end{bmatrix} \begin{bmatrix} \dfrac{1}{3} & \dfrac{1}{3} & \dfrac{1}{3} \\ \dfrac{1}{2} & \dfrac{1}{4} & \dfrac{1}{4} \\ \dfrac{1}{4} & \dfrac{1}{4} & \dfrac{1}{2} \end{bmatrix} = \begin{bmatrix} t_1 & t_2 & t_3 \end{bmatrix}$$

This leads to the system of equations

$$\begin{cases} \dfrac{1}{3}t_1 + \dfrac{1}{2}t_2 + \dfrac{1}{4}t_3 = t_1 \\ \dfrac{1}{3}t_1 + \dfrac{1}{4}t_2 + \dfrac{1}{4}t_3 = t_2 \\ \dfrac{1}{3}t_1 + \dfrac{1}{4}t_2 + \dfrac{1}{2}t_3 = t_3 \\ t_1 + t_2 + t_3 = 1 \end{cases} \text{ or }$$

$$\begin{cases} -\dfrac{2}{3}t_1 + \dfrac{1}{2}t_2 + \dfrac{1}{4}t_3 = 0 \\ \dfrac{1}{3}t_1 - \dfrac{3}{4}t_2 + \dfrac{1}{4}t_3 = 0 \\ \dfrac{1}{3}t_1 + \dfrac{1}{4}t_2 - \dfrac{1}{2}t_3 = 0 \\ t_1 + t_2 + t_3 = 1 \end{cases}$$

The first equation is equivalent to the sum of the second two equations, so we have a system of 3 equations which we will solve by elimination.

$$\begin{cases} \dfrac{1}{3}t_1 - \dfrac{3}{4}t_2 + \dfrac{1}{4}t_3 = 0 & \text{(1)} \\ \dfrac{1}{3}t_1 + \dfrac{1}{4}t_2 - \dfrac{1}{2}t_3 = 0 & \text{(2)} \\ t_1 + t_2 + t_3 = 1 & \text{(3)} \end{cases}$$

Multiplying (1) and (2) by 12 gives

$$\begin{cases} 4t_1 - 9t_2 + 3t_3 = 0 & \text{(1)} \\ 4t_1 + 3t_2 - 6t_3 = 0 & \text{(2)} \\ t_1 + t_2 + t_3 = 1 & \text{(3)} \end{cases}$$

(continued on next page)

(continued)

Multiply equation (3) by −3, then add the resulting equation to equation (1).

$$4t_1 - 9t_2 + 3t_3 = 0 \quad (1)$$
$$-3t_1 - 3t_2 - 3t_3 = -3 \quad -3 \times (3)$$
$$\overline{\qquad t_1 - 12t_2 \qquad = -3} \quad (4)$$

Multiply equation (3) by 6, then add the resulting equation to equation (2).

$$4t_1 + 3t_2 - 6t_3 = 0 \quad (1)$$
$$6t_1 + 6t_2 + 6t_3 = 6 \quad 6 \times (3)$$
$$\overline{10t_1 + 9t_2 \qquad = 6} \quad (5)$$

Now solve the system consisting of equations (4) and (5) using substitution.

$$\begin{cases} t_1 = 12t_2 - 3 & (4) \\ 10t_1 + 9t_2 = 6 & (5) \end{cases}$$

$$10(12t_2 - 3) + 9t_2 = 6$$

$$129t_2 = 36 \Rightarrow t_2 = \frac{36}{129} = \frac{12}{43}.$$

Back- substituting into (4), we get

$$t_1 = \frac{144}{43} - \frac{129}{43} = \frac{15}{43}.$$

Substituting the values for t_1 and t_2 into equation

(3) gives $t_3 = 1 - \frac{15}{43} - \frac{12}{43} = \frac{16}{43}$. The fixed

probability vector is $\mathbf{t} = \begin{bmatrix} \dfrac{15}{43} & \dfrac{12}{43} & \dfrac{16}{43} \end{bmatrix}$.

21. a. The transition matrix P is

$$\begin{array}{cccc} & A & B & C \\ P = \begin{array}{c} A \\ B \\ C \end{array} & \left[\begin{array}{ccc} 0.7 & 0.15 & 0.15 \\ 0.1 & 0.8 & 0.1 \\ 0.2 & 0.2 & 0.6 \end{array} \right] \end{array}$$

b. We need to find the fixed probability vector $\mathbf{t}$ to learn her long term distribution of brands.

$$\begin{bmatrix} t_1 & t_2 & t_3 \end{bmatrix} \begin{bmatrix} 0.7 & 0.15 & 0.15 \\ 0.1 & 0.8 & 0.1 \\ 0.2 & 0.2 & 0.6 \end{bmatrix} = \begin{bmatrix} t_1 & t_2 & t_3 \end{bmatrix}$$

This leads to the system of equations

$$\begin{cases} 0.7t_1 + 0.1t_2 + 0.2t_3 = t_1 \\ 0.15t_1 + 0.8t_2 + 0.2t_3 = t_2 \\ 0.15t_1 + 0.1t_2 + 0.6t_3 = t_3 \\ t_1 + t_2 + t_3 = 1 \end{cases} \text{ or }$$

$$\begin{cases} -0.3t_1 + 0.1t_2 + 0.2t_3 = 0 \\ 0.15t_1 - 0.2t_2 + 0.2t_3 = 0 \\ 0.15t_1 + 0.1t_2 - 0.4t_3 = 0 \\ t_1 + t_2 + t_3 = 1 \end{cases}$$

The first equation is equivalent to the sum of the next two, so we have a system of three equations which we solve by elimination.

$$\begin{cases} 0.15t_1 - 0.2t_2 + 0.2t_3 = 0 & (1) \\ 0.15t_1 + 0.1t_2 - 0.4t_3 = 0 & (2) \\ t_1 + t_2 + t_3 = 1 & (3) \end{cases}$$

Multiply equation (3) by (−0.15), then add the resulting equation to equation (1).

$$0.15t_1 - 0.20t_2 + 0.20t_3 = 0 \quad (1)$$
$$-0.15t_1 - 0.15t_2 - 0.15t_3 = -0.15 \quad -0.15 \times (3)$$
$$\overline{\qquad -0.35t_2 + 0.05t_3 = -0.15} \quad (4)$$

Multiply equation (3) by (−0.15), then add the resulting equation to equation (2).

$$0.15t_1 + 0.1t_2 - 0.4t_3 = 0 \quad (2)$$
$$-0.15t_1 - 0.15t_2 - 0.15t_3 = -0.15 \quad -0.15 \times (3)$$
$$\overline{\qquad -0.05t_2 - 0.55t_3 = -0.15} \quad (5)$$

Now solve the system consisting of equations (4) and (5).

$$\begin{cases} -0.35t_2 + 0.05t_3 = -0.15 & (4) \\ -0.05t_2 - 0.55t_3 = -0.15 & (5) \end{cases}$$

This system can be simplified to

$$\begin{cases} -7t_2 + t_3 = -3 & (4) \\ t_2 + 11t_3 = 3 & (5) \end{cases}$$

Solving (5) for t_2 and then substituting that value into (4), we have

$$\begin{cases} -7t_2 + t_3 = -3 & (4) \\ t_2 = 3 - 11t_3 & (5) \end{cases}$$

$$-7(3 - 11t_3) + t_3 = -3$$

$$78t_3 = 18 \Rightarrow t_3 = \frac{18}{78} = \frac{3}{13}.$$

Back-substituting in (5) gives

$t_2 = -11\left(\dfrac{3}{13}\right) + \dfrac{39}{13} = \dfrac{6}{13}$. Substitute the

values for t_2 and t_3 in equation (1) to solve for t_1.

$$t_1 = 1 - \frac{6}{13} - \frac{3}{13} = \frac{4}{13}$$

Thus, the fixed probability vector is

$\mathbf{t} = \begin{bmatrix} \dfrac{4}{13} & \dfrac{6}{13} & \dfrac{3}{13} \end{bmatrix}$. In the long run the

grocer's stock will consist of $\dfrac{4}{13}$ or 30.8%

brand A, $\dfrac{6}{13}$ or 46.2% brand B, and $\dfrac{3}{13}$ or

23.1% brand C detergent.

23. a. First we form the transition matrix P, letting C, L, and S stand for Conservative, Labour, and Socialist respectively.

$$P = \begin{matrix} & \begin{matrix} C & L & S \end{matrix} \\ \begin{matrix} C \\ L \\ S \end{matrix} & \begin{bmatrix} 0.7 & 0.3 & 0 \\ 0.4 & 0.5 & 0.1 \\ 0.2 & 0.4 & 0.4 \end{bmatrix} \end{matrix}$$

Since we are interested in how the grandson of a Labourite votes, we need $v^{(2)}$ when the initial probability distribution is $v^{(0)} = [0 \ 1 \ 0]$.

$$v^{(2)} = v^{(0)}P^2 = \begin{bmatrix} 0 & 1 & 0 \end{bmatrix}\begin{bmatrix} 0.7 & 0.3 & 0 \\ 0.4 & 0.5 & 0.1 \\ 0.2 & 0.4 & 0.4 \end{bmatrix}^2$$

$$= \begin{bmatrix} 0.5 & 0.41 & 0.09 \end{bmatrix}$$

The probability the grandson of a Labourite votes Socialist is 0.09.

b. The long term membership distribution is the fixed probability vector $\mathbf{t}$. We find $\mathbf{t}$ that satisfies the equation $\mathbf{t}P = \mathbf{t}$.

$$\begin{bmatrix} t_1 & t_2 & t_3 \end{bmatrix}\begin{bmatrix} 0.7 & 0.3 & 0 \\ 0.4 & 0.5 & 0.1 \\ 0.2 & 0.4 & 0.4 \end{bmatrix} = \begin{bmatrix} t_1 & t_2 & t_3 \end{bmatrix}$$

This leads to the system of equations

$$\begin{cases} 0.7t_1 + 0.4t_2 + 0.2t_3 = t_1 \\ 0.3t_1 + 0.5t_2 + 0.4t_3 = t_2 \\ \quad\quad 0.1t_2 + 0.4t_3 = t_3 \\ t_1 + t_2 + t_3 = 1 \end{cases} \text{ or }$$

$$\begin{cases} -0.3t_1 + 0.4t_2 + 0.2t_3 = 0 \\ 0.3t_1 - 0.5t_2 + 0.4t_3 = 0 \\ \quad\quad 0.1t_2 - 0.6t_3 = 0 \\ t_1 + t_2 + t_3 = 1 \end{cases}$$

The first equation is equivalent to the sum of the next two, so we have a system of three equations which we solve by elimination.

$$\begin{cases} 0.3t_1 - 0.5t_2 + 0.4t_3 = 0 \quad (1) \\ \quad\quad 0.1t_2 - 0.6t_3 = 0 \quad (2) \\ t_1 + t_2 + t_3 = 1 \quad (3) \end{cases}$$

Multiply equation (1) by 10 and add the resulting equation to -3 times equation (3).

$$\begin{array}{rl} 3t_1 - 5t_2 + 4t_3 = 0 & 10\times(1) \\ -3t_1 - 3t_2 - 3t_3 = -3 & -3\times(3) \\ \hline -8t_2 + t_3 = -3 \text{ or } t_3 = 8t_2 - 3 & (4) \end{array}$$

Substitute this expression for t_3 into (2) and solve for t_2.

$$0.1t_2 - 0.6(8t_2 - 3) = 0$$
$$-4.7t_2 = -1.8 \Rightarrow t_2 = \frac{18}{47}$$

Back-substituting into (4), we have

$$t_3 = 8\left(\frac{18}{47}\right) - 3 = \frac{3}{47}.$$

Substitute the values for t_2 and t_3 in (3) and solve for t_1. $t_1 = 1 - \frac{18}{47} - \frac{3}{47} = \frac{26}{47}$. Thus, the fixed probability vector is

$$\mathbf{t} = \begin{bmatrix} \frac{26}{47} & \frac{18}{47} & \frac{3}{47} \end{bmatrix}. \text{ In the long run,}$$

$$\frac{26}{47} = 55.3\% \text{ will vote Conservative.}$$

25. a. This situation forms a Markov chain because her service calls can be thought of as a sequence of experiments each of which results in one of a finite number of states, and the movement to the next state depends only on the current state. The transition matrix P is given by

$$P = \begin{matrix} & \begin{matrix} U_1 & U_2 & U_3 \end{matrix} \\ \begin{matrix} U_1 \\ U_2 \\ U_3 \end{matrix} & \begin{bmatrix} 0 & 1 & 0 \\ \frac{3}{4} & 0 & \frac{1}{4} \\ \frac{3}{4} & \frac{1}{4} & 0 \end{bmatrix} \end{matrix}$$

b. The next month's probability vector is given by $v^{(1)} = v^{(0)}P$.

$$v^{(1)} = \begin{bmatrix} \frac{1}{3} & \frac{1}{3} & \frac{1}{3} \end{bmatrix}\begin{bmatrix} 0 & 1 & 0 \\ \frac{3}{4} & 0 & \frac{1}{4} \\ \frac{3}{4} & \frac{1}{4} & 0 \end{bmatrix}$$

$$= \begin{bmatrix} \frac{1}{2} & \frac{5}{12} & \frac{1}{12} \end{bmatrix}$$

c. In the long run the distribution will be given by the fixed probability vector $\mathbf{t}$. $\mathbf{t}$ must satisfy the equation $\mathbf{t}P = \mathbf{t}$.

$$\begin{bmatrix} t_1 & t_2 & t_3 \end{bmatrix}\begin{bmatrix} 0 & 1 & 0 \\ \frac{3}{4} & 0 & \frac{1}{4} \\ \frac{3}{4} & \frac{1}{4} & 0 \end{bmatrix} = \begin{bmatrix} t_1 & t_2 & t_3 \end{bmatrix}$$

Using a graphing calculator, we can find $\mathbf{t}$ by computing P^n for a large value of n. P^{40} is large enough to find $\mathbf{t}$.

$$\mathbf{t} = \begin{bmatrix} 0.4286 & 0.4571 & 0.1143 \end{bmatrix}$$

In the long run she sells at university 1, 42.9% of the time, at university 2, 45.7% of the time, and at university 3, 11.4% of the time.

27. We enter the transition matrix P, into the graphing utility and calculate powers of the matrix until the rows are identical to 4 decimal places. The resulting row, rounded to three decimal places, is the fixed probability vector **t**.

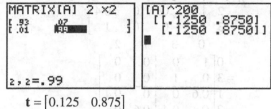

$$\mathbf{t} = \begin{bmatrix} 0.125 & 0.875 \end{bmatrix}$$

29. We enter the transition matrix P, into the graphing utility and calculate powers of the matrix until the rows are identical to 5 decimal places. The resulting row, rounded to three decimal places is the fixed probability vector **t**.

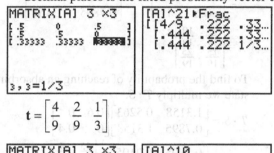

$$\mathbf{t} = \begin{bmatrix} \dfrac{4}{9} & \dfrac{2}{9} & \dfrac{1}{3} \end{bmatrix}$$

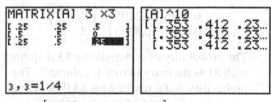

$$\mathbf{t} = \begin{bmatrix} 0.353 & 0.412 & 0.235 \end{bmatrix}$$

31. We enter the transition matrix P, into the graphing utility and calculate powers of the matrix until the rows are identical to 5 decimal places. The resulting row, rounded to three decimal places is the fixed probability vector **t**.

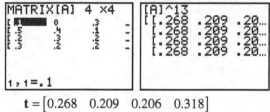

$$\mathbf{t} = \begin{bmatrix} 0.268 & 0.209 & 0.206 & 0.318 \end{bmatrix}$$

33. A fixed probability vector of $\begin{bmatrix} \dfrac{1}{3} & 0 & \dfrac{1}{3} & \dfrac{1}{3} \end{bmatrix}$ cannot be from a transition matrix of a regular Markov chain. The definition of a regular Markov chain is that there is a power of P for which all entries are positive. After that power all higher powers will also be positive.

10.3 Absorbing Markov Chains

1. To find the inverse of $\begin{bmatrix} 2 & -3 \\ -1 & 5 \end{bmatrix}$ first augment the matrix with I_2, then use row operations to obtain the reduced row echelon form of the matrix.

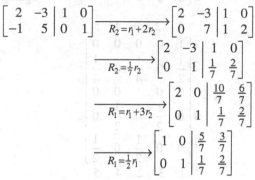

The inverse is $\begin{bmatrix} \dfrac{5}{7} & \dfrac{3}{7} \\ \dfrac{1}{7} & \dfrac{2}{7} \end{bmatrix}$.

3. True

5. State 1 is an absorbing state, and it's possible to move from state 2 to state 1. So this is an absorbing Markov chain.

7. There are no absorbing states. So this transition matrix is not from an absorbing Markov chain.

9. States 1 and 3 are absorbing states, and it possible to move from state 2 to both absorbing states. So this is an absorbing Markov chain.

11. State 1 is an absorbing state, but it is impossible to move from either state 2 or state 3 to state 1. So this is not an absorbing Markov chain.

13. There are no absorbing states. So this transition matrix is not from an absorbing Markov chain.

15. States 2 and 3 are absorbing states, and it possible to move from both states 1 and 4 to the absorbing states. So this is an absorbing Markov chain.

17. $P = \begin{array}{c} \\ 1 \\ 2 \\ 3 \end{array} \begin{array}{c} \begin{array}{ccc} 1 & 2 & 3 \end{array} \\ \begin{bmatrix} 1 & 0 & 0 \\ \dfrac{1}{3} & \dfrac{1}{3} & \dfrac{1}{3} \\ 0 & 0 & 1 \end{bmatrix} \end{array} = \begin{array}{c} \\ 1 \\ 3 \\ 2 \end{array} \begin{array}{c} \begin{array}{ccc} 1 & 3 & 2 \end{array} \\ \left[\begin{array}{cc|c} 1 & 0 & 0 \\ 0 & 1 & 0 \\ \hline \dfrac{1}{3} & \dfrac{1}{3} & \dfrac{1}{3} \end{array} \right] \end{array}$

19.
$$P = \begin{array}{c}1\\2\\3\\4\end{array}\begin{array}{cccc}1&2&3&4\\ \frac{1}{4}&\frac{1}{4}&\frac{1}{4}&\frac{1}{4}\\ 0&1&0&0\\ \frac{1}{2}&0&\frac{1}{2}&0\\ 0&0&0&1\end{array} = \begin{array}{c}2\\4\\1\\3\end{array}\left[\begin{array}{cc|cc}2&4&1&3\\1&0&0&0\\0&1&0&0\\ \hline \frac{1}{4}&\frac{1}{4}&\frac{1}{4}&\frac{1}{4}\\ 0&0&\frac{1}{2}&\frac{1}{2}\end{array}\right]$$

21.
$$P = \begin{array}{c}1\\2\\3\\4\\5\end{array}\left[\begin{array}{ccccc}1&2&3&4&5\\1&0&0&0&0\\ \frac{1}{2}&0&\frac{1}{4}&0&\frac{1}{4}\\ 0&0&1&0&0\\ 0&\frac{1}{2}&0&\frac{1}{2}&0\\ 0&0&\frac{1}{2}&0&\frac{1}{2}\end{array}\right]$$

$$= \begin{array}{c}1\\3\\2\\4\\5\end{array}\left[\begin{array}{cc|ccc}1&3&2&4&5\\1&0&0&0&0\\0&1&0&0&0\\ \hline \frac{1}{2}&\frac{1}{4}&0&0&\frac{1}{4}\\ 0&0&\frac{1}{2}&\frac{1}{2}&0\\ 0&\frac{1}{2}&0&0&\frac{1}{2}\end{array}\right]$$

23.a. The expected number of times the person will have $3 given that she started with $1 is the entry in row 1, column 3. If she started with $1 we would expect her to have $3 about 0.5 time. The expected number of times the person will have $3 given that she started with $2 is the entry in row 2, column 3. If she started with $2 we would expect her to have $3 about 1 time.

b. The expected number of games a person will play before absorption is the sum of the entries in the row corresponding to her starting value. A person who starts with $3 can expect to play $0.5 + 1.0 + 1.5 = 3.0$ games before absorption.

25. First we construct the transition matrix P and subdivide it so the absorbing states appear first.

$$P = \begin{array}{c}0\\1\\2\\3\end{array}\left[\begin{array}{cccc}0&1&2&3\\1&0&0&0\\0.6&0&0.4&0\\0&0.6&0&0.4\\0&0&0&1\end{array}\right]$$

$$= \begin{array}{c}0\\3\\1\\2\end{array}\left[\begin{array}{cc|cc}0&3&1&2\\1&0&0&0\\0&1&0&0\\ \hline 0.6&0&0&0.4\\0&0.4&0.6&0\end{array}\right]$$

Next we find the fundamental matrix T.

$$T = [I_2 - Q]^{-1} = \left[\begin{bmatrix}1&0\\0&1\end{bmatrix} - \begin{bmatrix}0&0.4\\0.6&0\end{bmatrix}\right]^{-1}$$

$$= \begin{bmatrix}1&-0.4\\-0.6&1\end{bmatrix}^{-1} = \begin{bmatrix}1.3158&0.5263\\0.7895&1.3158\end{bmatrix}$$

$$= \begin{bmatrix}\frac{25}{19}&\frac{10}{19}\\ \frac{15}{19}&\frac{25}{19}\end{bmatrix}$$

To find the probability of reaching an absorbing state we multiply $T \cdot S$.

$$T \cdot S = \begin{bmatrix}1.3158&0.5263\\0.7895&1.3158\end{bmatrix}\begin{bmatrix}0.6&0\\0&0.4\end{bmatrix}$$

$$= \begin{bmatrix}0.7895&0.2105\\0.4737&0.5263\end{bmatrix} = \begin{bmatrix}\frac{15}{19}&\frac{4}{19}\\ \frac{9}{19}&\frac{10}{19}\end{bmatrix}$$

The probability of accumulating $3 if starting with $1 is the entry in row 1, column 2. The probability that a player wins $3 if he started with $1 is 0.2105 or $\frac{4}{19}$. The probability of accumulating $3 if starting with $2 is 0.5263 or $\frac{10}{19}$.

27. a. We construct the transition matrix P and subdivide it so the absorbing states appear first.

	$0	$1000	$2000	$3000	$4000
$0	1	0	0	0	0
$1000	0.6	0	0.4	0	0
$P=$ $2000	0.6	0	0	0	0.4
$3000	0	0	0.6	0	0.4
$4000	0	0	0	0	1

	$0	$4000	$1000	$2000	$3000
$0	1	0	0	0	0
$4000	0	1	0	0	0
$=$ $1000	0.6	0	0	0.4	0
$2000	0.6	0.4	0	0	0
$3000	0	0.4	0	0.6	0

The fundamental matrix T is given by $T = \left[I_3 - Q \right]^{-1} = \begin{bmatrix} 1 & -0.4 & 0 \\ 0 & 1 & 0 \\ 0 & -0.6 & 1 \end{bmatrix}^{-1} = \begin{bmatrix} 1 & 0.4 & 0 \\ 0 & 1 & 0 \\ 0 & 0.6 & 1 \end{bmatrix}$.

Since Colleen started with $1000 the expected number of wagers placed before the game ends is found by summing the entries in row 1 of T. She expects to place 1.4 wagers before the game ends.

b. We need to the product of T and S.

$$T \cdot S = \begin{bmatrix} 1 & 0.4 & 0 \\ 0 & 1 & 0 \\ 0 & 0.6 & 1 \end{bmatrix} \begin{bmatrix} 0.6 & 0 \\ 0.6 & 0.4 \\ 0 & 0.4 \end{bmatrix} = \begin{bmatrix} 0.84 & 0.16 \\ 0.6 & 0.4 \\ 0.36 & 0.64 \end{bmatrix}$$

The probability Colleen is wiped out is the entry in row 1, column 1 of matrix $T \cdot S$. She is wiped out with probability 0.84.

c. The probability that Colleen wins enough to purchase the car is $1 - 0.84 = 0.16$. (It can also be found in row 1, column 2 of the matrix $T \cdot S$.)

29. Define the states of the system as

N:	No tank survives.	A:	Tank A alone survives.
B:	Tank B alone survives.	C:	Tank C alone survives.
AB:	Tanks A and B survive.	AC:	Tanks A and C survive.
BC:	Tanks B and C survive.	ABC:	All three tanks survive.

Since the game ends when either one tank or no tanks are left, states N, A, B, and C are absorbing states. We will set up the transition matrix P with this in mind and put those states first. We then need to determine the probabilities of moving from state to state in the nonabsorbing states. The rules of the game state that each tank aims at its strongest opponent. If strength is defined as the probability of hitting the target, then A has strength $\frac{1}{3}$, B has strength $\frac{1}{2}$, and C has strength $\frac{1}{6}$. At the start of the game all tanks are active. A tank survives a round if either it is not fired upon or it is missed.

Row ABC: A fires on B, B fires on A, C fires on B; possible resulting states and their probabilities: (Note that C must survive since no tank fires on it.)

ABC: P(all three tanks miss their targets)

$= P(\text{ABC} \mid \text{ABC}) = \frac{2}{3} \cdot \frac{1}{2} \cdot \frac{5}{6} = \frac{5}{18}$

BC: P(B hits and both A and C miss)

$= \frac{1}{2} \cdot \frac{2}{3} \cdot \frac{5}{6} = \frac{5}{18}$

AC: P(either A or C hit and B misses)

$= \left(1 - \frac{2}{3} \cdot \frac{5}{6}\right)\frac{1}{2} = \frac{2}{9}$

C: P(either A or C hit and B hits)

$= \left(1 - \frac{2}{3} \cdot \frac{5}{6}\right)\frac{1}{2} = \frac{2}{9}$

Row AB: A fires on B, B fires on A. Possible resulting states and their probabilities.

AB: P(both tanks miss) $= \frac{2}{3} \cdot \frac{1}{2} = \frac{1}{3}$

B: P(A misses and B hits) $= \frac{2}{3} \cdot \frac{1}{2} = \frac{1}{3}$

A: P(A hits and B misses) $= \frac{1}{3} \cdot \frac{1}{2} = \frac{1}{6}$

N: P(both hit) $= \frac{1}{3} \cdot \frac{1}{2} = \frac{1}{6}$

(continued on next page)

(continued)

Row AC: A fires on C, C fires on A. Possible resulting states and their probabilities:

AC: $P(\text{both tanks miss}) = \dfrac{2}{3} \cdot \dfrac{5}{6} = \dfrac{5}{9}$ C: $P(\text{A misses and C hits}) = \dfrac{2}{3} \cdot \dfrac{1}{6} = \dfrac{1}{9}$

A: $P(\text{A hits and C misses}) = \dfrac{1}{3} \cdot \dfrac{5}{6} = \dfrac{5}{18}$ N: $P(\text{both hit}) = \dfrac{1}{3} \cdot \dfrac{1}{6} = \dfrac{1}{18}$

Row BC: B fires on C, C fires on B. Possible resulting states and their probabilities:

BC: $P(\text{both tanks miss}) = \dfrac{1}{2} \cdot \dfrac{5}{6} = \dfrac{5}{12}$ C: $P(\text{B misses and C hits}) = \dfrac{1}{2} \cdot \dfrac{1}{6} = \dfrac{1}{12}$

B: $P(\text{B hits and C misses}) = \dfrac{1}{2} \cdot \dfrac{5}{6} = \dfrac{5}{12}$ N: $P(\text{both tanks hit}) = \dfrac{1}{2} \cdot \dfrac{1}{6} = \dfrac{1}{12}$

$$
P = \begin{array}{c} \\ N \\ A \\ B \\ C \\ ABC \\ AB \\ AC \\ BC \end{array}
\begin{array}{c}
\begin{array}{cccccccc} N & A & B & C & ABC & AB & AC & BC \end{array} \\
\left[\begin{array}{cccc|cccc}
1 & 0 & 0 & 0 & 0 & 0 & 0 & 0 \\
0 & 1 & 0 & 0 & 0 & 0 & 0 & 0 \\
0 & 0 & 1 & 0 & 0 & 0 & 0 & 0 \\
0 & 0 & 0 & 1 & 0 & 0 & 0 & 0 \\ \hline
0 & 0 & 0 & \frac{2}{9} & \frac{5}{18} & 0 & \frac{2}{9} & \frac{5}{18} \\
\frac{1}{6} & \frac{1}{6} & \frac{1}{3} & 0 & 0 & \frac{1}{3} & 0 & 0 \\
\frac{1}{18} & \frac{5}{18} & 0 & \frac{1}{9} & 0 & 0 & \frac{5}{9} & 0 \\
\frac{1}{12} & 0 & \frac{5}{12} & \frac{1}{12} & 0 & 0 & 0 & \frac{5}{12}
\end{array} \right]
\end{array}
$$

a. There are 8 states in the chain.

b. Four of the 8 states are absorbing.

c. To find the expected number of rounds fired we need the fundamental matrix T.

$$
T = [I_4 - Q]^{-1} = \begin{bmatrix}
\frac{13}{18} & 0 & -\frac{2}{9} & -\frac{5}{18} \\
0 & \frac{2}{3} & 0 & 0 \\
0 & 0 & \frac{4}{9} & 0 \\
0 & 0 & 0 & \frac{7}{12}
\end{bmatrix}^{-1}
= \begin{bmatrix}
\frac{18}{13} & 0 & \frac{9}{13} & \frac{60}{91} \\
0 & \frac{3}{2} & 0 & 0 \\
0 & 0 & \frac{9}{4} & 0 \\
0 & 0 & 0 & \frac{12}{7}
\end{bmatrix}
$$

Since all three tanks took part in the battle, the expected number of rounds fired is the sum of the entries in row 1 of T. We would expect about 2.74 rounds of fire.

d. The product TS will give the probability A survives, or wins the game.

$$
TS = \begin{bmatrix}
\frac{17}{182} & \frac{5}{26} & \frac{25}{91} & \frac{40}{91} \\
\frac{1}{4} & \frac{1}{4} & \frac{1}{2} & 0 \\
\frac{1}{8} & \frac{5}{8} & 0 & \frac{1}{4} \\
\frac{1}{7} & 0 & \frac{5}{7} & \frac{1}{7}
\end{bmatrix}
$$

The probability that A survives is the entry in the first row (since all 3 tanks took part) and the second column (since A is the winner). We would expect A would win 19.2% of the time.

31. If I is an absorbing state, the transition matrix becomes

$$P = \begin{array}{c} \\ I \\ D \\ N \end{array}\begin{array}{ccc} I & D & N \\ \left[\begin{array}{c|cc} 1 & 0 & 0 \\ \hline 0.070 & 0.638 & 0.292 \\ 0.079 & 0.064 & 0.857 \end{array}\right] \end{array}$$

The fundamental matrix *T* will give the number of days until a decreasing stock begins to increase.

$$T = [I_2 - Q]^{-1} = \begin{bmatrix} 0.362 & -0.292 \\ -0.064 & 0.143 \end{bmatrix}^{-1}$$

$$= \begin{bmatrix} 4.3231 & 8.8276 \\ 1.9348 & 10.9438 \end{bmatrix}$$

A decreasing stock should begin to increase in $4.3231 + 8.8276 = 13.1507$ days.

33. a. The transition matrix is

$$P = \begin{array}{c} \\ R1 \\ R2 \\ R3 \\ R4 \\ R5 \end{array}\begin{array}{ccccc} R1 & R2 & R3 & R4 & R5 \\ \begin{bmatrix} 1 & 0 & 0 & 0 & 0 \\ 0.25 & 0.25 & 0.25 & 0 & 0.25 \\ 0.20 & 0.20 & 0.20 & 0.20 & 0.20 \\ 0.25 & 0 & 0.25 & 0.25 & 0.25 \\ 0 & 0 & 0 & 0 & 1 \end{bmatrix} \end{array}$$

b. Two rooms, 1 and 5, represent absorbing states in this Markov chain.

c. If the mouse starts in room 3, $v^{(0)} = [0\ 0\ 1\ 0\ 0]$. The probability the mouse is in room 5 after five steps is given by $v^{(5)}$.

$$v^{(5)} = v^{(0)} P^5$$
$$= [0\ 0\ 1\ 0\ 0]$$
$$\cdot \begin{bmatrix} 1 & 0 & 0 & 0 & 0 \\ 0.25 & 0.25 & 0.25 & 0 & 0.25 \\ 0.20 & 0.20 & 0.20 & 0.20 & 0.20 \\ 0.25 & 0 & 0.25 & 0.25 & 0.25 \\ 0 & 0 & 0 & 0 & 1 \end{bmatrix}^5$$
$$= [0.474\ 0.015\ 0.022\ 00.15\ 0.474]$$

The probability the mouse is in room 5 after five steps is 0.474.

d. The subdivided matrix is

$$= \begin{array}{c} \\ R1 \\ R5 \\ R2 \\ R3 \\ R4 \end{array}\begin{array}{ccccc} R1 & R5 & R2 & R3 & R4 \\ \left[\begin{array}{cc|ccc} 1 & 0 & 0 & 0 & 0 \\ 0 & 1 & 0 & 0 & 0 \\ \hline 0.25 & 0.25 & 0.25 & 0.25 & 0 \\ 0.20 & 0.20 & 0.20 & 0.20 & 0.20 \\ 0.25 & 0.25 & 0 & 0.25 & 0.25 \end{array}\right] \end{array}$$

e. $I_2 = \begin{bmatrix} 1 & 0 \\ 0 & 1 \end{bmatrix}$, $S = \begin{bmatrix} 0.25 & 0.25 \\ 0.20 & 0.20 \\ 0.25 & 0.25 \end{bmatrix}$,

$$Q = \begin{bmatrix} 0.25 & 0.25 & 0 \\ 0.20 & 0.20 & 0.20 \\ 0 & 0.25 & 0.25 \end{bmatrix}$$

f. $T = [I_3 - Q]^{-1} = \begin{bmatrix} 0.75 & -0.25 & 0 \\ -0.20 & 0.80 & -0.20 \\ 0 & -0.25 & 0.75 \end{bmatrix}^{-1}$

$$= \begin{bmatrix} 1.4667 & 0.5 & 0.1333 \\ 0.4 & 1.5 & 0.4 \\ 0.1333 & 0.5 & 1.4667 \end{bmatrix}$$

g. $TS = \begin{bmatrix} 1.4667 & 0.5 & 0.1333 \\ 0.4 & 1.5 & 0.4 \\ 0.1333 & 0.5 & 1.4667 \end{bmatrix}\begin{bmatrix} 0.25 & 0.25 \\ 0.20 & 0.20 \\ 0.25 & 0.25 \end{bmatrix}$

$$= \begin{bmatrix} 0.5 & 0.5 \\ 0.5 & 0.5 \\ 0.5 & 0.5 \end{bmatrix}$$

h. Since the mouse started in room 3, the sum of the entries in row 2 of the fundamental matrix is the expected time until absorption. The mouse can stay out of the trap or away from the cheese for about 2.3 steps, with a 0.5 probability of ending in the trap.

10.4 Two-Person Games

1. 2×4

3. Katy's game matrix is

	one finger	two fingers
1 finger	-1	1
2 fingers	1	-1

where the entries are dimes.

5. Katy's game matrix is

$$\begin{array}{c} \\ 1 \\ 4 \\ 7 \end{array}\begin{array}{ccc} 1 & 4 & 7 \\ \begin{bmatrix} -2 & 5 & -8 \\ 5 & -8 & 11 \\ -8 & 11 & -14 \end{bmatrix} \end{array}$$

where the entries stand for number of dimes.

For exercises 7–16, recall that a game is strictly determined when there is an entry that is the smallest element in its row and the largest element in its column. That entry, if it exists, is called the value of the game.

7. The game is strictly determined; −2 is the smallest element in row 1 and the largest element in column 1. The value of the game is −2.

9. The game is strictly determined; 3 is the smallest element in row 1 and the largest element in column 2. The value of the game is 3.

11. The game is not strictly determined. No entry satisfies the criteria.

13. The game is strictly determined; 2 is the smallest element in row 3 and the largest element both in columns 1 and 2. The value of the game is 2.

15. The game is not strictly determined. No entry satisfies the criteria.

17. For the matrix to be strictly determined a must be the smallest value in row 1 and the largest value in column 1. To be the smallest value in row 1, a must be less than or equal to 3. To be the largest value in column 1 a must be greater than or equal to 0. There is no value of a that will make the entry in row 2, column 2 or in row 3, column 3 a saddle point. So, for the matrix to be strictly determined, $0 \le a \le 3$.

19. The game is strictly determined if one of the following are true.
    1. If a is the saddle point then $a \le 0$ to be the smallest value in row 1, and $a \ge 0$ to be the largest value in column 1, so $a = 0$.
    2. If b is the saddle point then $b \le 0$ to be the smallest value in row 2, and $b \ge 0$ to be the largest value in column 2, so $b = 0$.
    From the Zero Product Property, we can say that the game is strictly determined if and only if the product $ab = 0$.

10.5 Mixed Strategies

1. $\begin{bmatrix} 2 & -1 \\ -2 & 3 \end{bmatrix}\begin{bmatrix} 2 & -3 & 4 & 0 \\ 4 & 5 & -9 & 3 \end{bmatrix} = \begin{bmatrix} 0 & -11 & 17 & -3 \\ 8 & 21 & -35 & 9 \end{bmatrix}$

For exercises 3–13, the expected payoff E of player I is given by PAQ where P is player I's strategy and Q is player II's strategy.

3. $P = \begin{bmatrix} 0.30 & 0.70 \end{bmatrix}$, $Q = \begin{bmatrix} 0.40 \\ 0.60 \end{bmatrix}$

   $E = PAQ = \begin{bmatrix} 0.30 & 0.70 \end{bmatrix}\begin{bmatrix} 6 & 0 \\ -2 & 3 \end{bmatrix}\begin{bmatrix} 0.40 \\ 0.60 \end{bmatrix} = 1.42$

   The game is biased in favor of player 1 and has an expected payoff of $1.42.

5. $E = PAQ = \begin{bmatrix} \frac{1}{2} & \frac{1}{2} \end{bmatrix}\begin{bmatrix} 4 & 0 \\ 2 & 3 \end{bmatrix}\begin{bmatrix} \frac{1}{2} \\ \frac{1}{2} \end{bmatrix} = \begin{bmatrix} 3 & \frac{3}{2} \end{bmatrix}\begin{bmatrix} \frac{1}{2} \\ \frac{1}{2} \end{bmatrix}$

   $= \frac{9}{4}$

   The game is biased in favor of player 1 and has an expected payoff of 2.25.

7. $E = PAQ = \begin{bmatrix} \frac{1}{4} & \frac{3}{4} \end{bmatrix}\begin{bmatrix} 4 & 0 \\ 2 & 3 \end{bmatrix}\begin{bmatrix} \frac{1}{2} \\ \frac{1}{2} \end{bmatrix} = \begin{bmatrix} \frac{5}{2} & \frac{9}{4} \end{bmatrix}\begin{bmatrix} \frac{1}{2} \\ \frac{1}{2} \end{bmatrix}$

   $= \frac{19}{8} = 2.375$

   The game is biased in favor of player 1 and has an expected payoff of 2.375.

9. $E = PAQ = \begin{bmatrix} \frac{2}{3} & \frac{1}{3} \end{bmatrix}\begin{bmatrix} 4 & 0 \\ -3 & 6 \end{bmatrix}\begin{bmatrix} \frac{1}{3} \\ \frac{2}{3} \end{bmatrix} = \begin{bmatrix} \frac{5}{3} & 2 \end{bmatrix}\begin{bmatrix} \frac{1}{3} \\ \frac{2}{3} \end{bmatrix}$

   $= \frac{17}{9}$

   The game is biased in favor of player I and has an expected payoff of $\frac{17}{9}$.

11. $E = PAQ = \begin{bmatrix} \frac{1}{3} & \frac{1}{3} & \frac{1}{3} \end{bmatrix}\begin{bmatrix} 1 & 0 & 0 \\ 0 & 1 & 0 \\ 0 & 0 & 1 \end{bmatrix}\begin{bmatrix} \frac{1}{3} \\ \frac{1}{3} \\ \frac{1}{3} \end{bmatrix}$

    $= \begin{bmatrix} \frac{1}{3} & \frac{1}{3} & \frac{1}{3} \end{bmatrix}\begin{bmatrix} \frac{1}{3} \\ \frac{1}{3} \\ \frac{1}{3} \end{bmatrix} = \frac{1}{3}$

    The game is biased in favor of player I and has an expected payoff of $\frac{1}{3}$.

13. Player I's game matrix for matching pennies is

$$A = \begin{array}{c} \\ H \\ T \end{array} \begin{array}{cc} H & T \\ \left[\begin{array}{cc} 1 & -1 \\ -1 & 1 \end{array} \right] \end{array}.$$

In this game we are told each player chooses heads half the time, so $P = [0.5 \quad 0.5]$ and

$$Q = \begin{bmatrix} 0.5 \\ 0.5 \end{bmatrix}.$$

$$E = [0.5 \quad 0.5] \begin{bmatrix} 1 & -1 \\ -1 & 1 \end{bmatrix} \begin{bmatrix} 0.5 \\ 0.5 \end{bmatrix} = [0 \quad 0] \begin{bmatrix} 0.5 \\ 0.5 \end{bmatrix} = 0$$

The game has an expected payoff of 0. The game is fair.

15. Assume the matrix is not strictly determined. Then there is no element that is the smallest in its row and the largest in its column. That is, there is no saddle point. There are 3 possible relations between a_{11} and a_{12} in row 1. They are $a_{11} < a_{12}$, $a_{11} > a_{12}$ or $a_{11} = a_{12}$.

1. If $a_{11} < a_{12}$ then $a_{11} < a_{21}$ otherwise it would be the smallest in the row, largest in the column, a saddle point. This forces $a_{21} > a_{22}$ so that a_{21} is not the saddle point, and $a_{12} > a_{22}$ preventing a_{22} from being the saddle point. So if the matrix is not strictly determined and $a_{11} < a_{12}$ then $a_{11} < a_{21}$, $a_{21} > a_{22}$, and $a_{12} > a_{22}$ are also true (statement (b)).

2. If $a_{11} > a_{12}$ then $a_{12} < a_{22}$ or a_{12} would be a saddle point. Since a_{22} is the largest in its column, it cannot be the smallest in row 2, so $a_{21} < a_{22}$. Finally, to prevent a_{21} from being the saddle point, $a_{11} > a_{21}$. So if the matrix is not strictly determined and $a_{11} > a_{12}$ then $a_{12} < a_{22}$, $a_{21} < a_{22}$, and $a_{11} > a_{21}$ are also true (statement (a)).

3. If $a_{11} = a_{12}$ then either entry could be considered largest or smallest in row 1, and the matrix would be strictly determined.

10.6 Optimal Strategy in Two-Person Zero-Sum Games with 2 × 2 Matrices

1. To find the point of intersection, equate the two expressions for y and solve for x.
$$3x - 2 = -4x + 3$$
$$7x = 5 \Rightarrow x = \frac{5}{7}$$

Now substitute the value for x into either equation and solve for y.

$$y = 3\left(\frac{5}{7} \right) - 2 = \frac{1}{7}$$

The intersection of the two lines is $\left(\dfrac{5}{7}, \dfrac{1}{7} \right)$.

3. $A = \begin{bmatrix} 1 & 2 \\ 4 & 1 \end{bmatrix}$

Player I's strategy is $[p \quad 1-p]$.

If player II chooses column 1, player I expects to win $E_I = p + 4(1-p) = -3p + 4$ (1)

If player II chooses column 2, player I expects to win $E_I = 2p + 1(1-p) = p + 1$ (2)

The intersection of the lines represents player I's maximum payoff and optimal strategy. We solve (1) and (2) simultaneously.

$$-3p + 4 = p + 1 \Rightarrow 3 = 4p \Rightarrow \frac{3}{4} = p \text{ and}$$

$1 - p = \dfrac{1}{4}$. Thus, player I's optimal strategy is

$$P = \begin{bmatrix} \dfrac{3}{4} & \dfrac{1}{4} \end{bmatrix}.$$

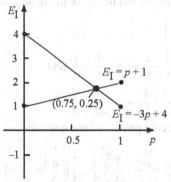

Using formula (5) to check the result, we have $P = [p_1 \quad p_2]$.

$$p_1 = \frac{1-4}{1+1-2-4} = \frac{-3}{-4} = \frac{3}{4} \text{ and}$$

$$p_2 = \frac{1-2}{1+1-2-4} = \frac{-1}{-4} = \frac{1}{4}$$

Player II's strategy is $\begin{bmatrix} q \\ 1-q \end{bmatrix}$.

If player I chooses row 1, player II expects to win $E_{II} = -q - 2(1-q) = q - 2$ (1)

If player I chooses row 2, player II expects to win $E_{II} = -4q - 1(1-q) = -3q - 1$ (2)

(continued on next page)

(continued)

The intersection of the lines represents player II's maximum payoff and optimal strategy. We solve (1) and (2) simultaneously.

$$q-2=-3q-1 \Rightarrow 4q=1 \Rightarrow q=\frac{1}{4} \text{ and}$$

$1-q=\frac{3}{4}$. Thus, player II's optimal strategy is

$$Q=\begin{bmatrix}\frac{1}{4}\\\frac{3}{4}\end{bmatrix}.$$

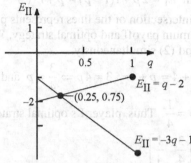

$E_{II}=q-2$

(0.25, 0.75)

$E_{II}=-3q-1$

Using formula (6) to check the result, we have

$$Q=\begin{bmatrix}q_1\\q_2\end{bmatrix}.$$

$$q_1=\frac{1-2}{1+1-2-4}=\frac{-1}{-4}=\frac{1}{4} \text{ and}$$

$$q_2=\frac{1-4}{1+1-2-4}=\frac{-3}{-4}=\frac{3}{4}$$

The expected payoff of the game is

$$E=PAQ=\begin{bmatrix}\frac{3}{4}&\frac{1}{4}\end{bmatrix}\begin{bmatrix}1&2\\4&1\end{bmatrix}\begin{bmatrix}\frac{1}{4}\\\frac{3}{4}\end{bmatrix}=\frac{7}{4}=1.75.$$

5. $A=\begin{bmatrix}-3&2\\1&0\end{bmatrix}$

Player I's strategy is $\begin{bmatrix}p&1-p\end{bmatrix}$.

If player II chooses column 1, player I expects to win $E_I=-3p+1(1-p)=-4p+1$ (1)

If player II chooses column 2, player I expects to win $E_I=2p+0(1-p)=2p$ (2)

The intersection of the lines represents player I's maximum payoff and optimal strategy. We solve (1) and (2) simultaneously.

$$-4p+1=2p \Rightarrow 1=6p \Rightarrow \frac{1}{6}=p \text{ and}$$

$$1-p=\frac{5}{6}.$$

(continued on next page)

Thus, player I's optimal strategy is

$$P=\begin{bmatrix}\frac{1}{6}&\frac{5}{6}\end{bmatrix}.$$

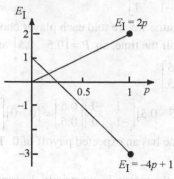

$E_I=2p$

$E_I=-4p+1$

Using formula (5) to check the result, we have
$P=\begin{bmatrix}p_1&p_2\end{bmatrix}$.

$$p_1=\frac{0-1}{-3+0-2-1}=\frac{-1}{-6}=\frac{1}{6} \text{ and}$$

$$p_2=\frac{-3-2}{-3+0-2-1}=\frac{-5}{-6}=\frac{5}{6}$$

Player II's strategy is $\begin{bmatrix}q\\1-q\end{bmatrix}$.

If player I chooses row 1, player II expects to win $E_{II}=3q-2(1-q)=5q-2$ (1)

If player I chooses row 2, player II expects to win $E_{II}=-1q+0(1-q)=-q$ (2)

The intersection of the lines represents player II's maximum payoff and optimal strategy. We solve (1) and (2) simultaneously.

$$5q-2=-q \Rightarrow -2=-6q \Rightarrow q=\frac{1}{3} \text{ and}$$

$$1-q=\frac{2}{3}.$$

Thus, player II's optimal strategy is $Q=\begin{bmatrix}\frac{1}{3}\\\frac{2}{3}\end{bmatrix}$.

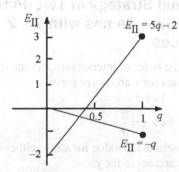

$E_{II}=5q-2$

$E_{II}=-q$

(continued on next page)

(*continued*)

Using formula (6) to check the result, we have

$$Q = \begin{bmatrix} q_1 \\ q_2 \end{bmatrix}.$$

$$q_1 = \frac{0-2}{-3+0-2-1} = \frac{-2}{-6} = \frac{1}{3} \text{ and}$$

$$q_2 = \frac{-3-1}{-3+0-2-1} = \frac{-4}{-6} = \frac{2}{3}$$

The expected payoff of the game is

$$E = PAQ = \begin{bmatrix} \frac{1}{6} & \frac{5}{6} \end{bmatrix} \begin{bmatrix} -3 & 2 \\ 1 & 0 \end{bmatrix} \begin{bmatrix} \frac{1}{3} \\ \frac{2}{3} \end{bmatrix} = \frac{1}{3}.$$

7. $A = \begin{bmatrix} 2 & -1 \\ -1 & 4 \end{bmatrix}$

Player I's strategy is $\begin{bmatrix} p & 1-p \end{bmatrix}$.

If player II chooses column 1, player I expects to win $E_I = 2p - 1(1-p) = 3p - 1$ (1)

If player II chooses column 2, player I expects to win $E_I = -1p + 4(1-p) = -5p + 4$ (2)

The intersection of the lines represents player I's maximum payoff and optimal strategy. We solve (1) and (2) simultaneously.

$$3p - 1 = -5p + 4 \Rightarrow 8p = 5 \Rightarrow p = \frac{5}{8} \text{ and}$$

$1 - p = \frac{3}{8}$. Thus, player I's optimal strategy is

$$P = \begin{bmatrix} \frac{5}{8} & \frac{3}{8} \end{bmatrix}.$$

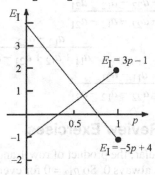

Using formula (5) to check the result, we have

$$P = \begin{bmatrix} p_1 & p_2 \end{bmatrix}.$$

$$p_1 = \frac{4+1}{2+4+1+1} = \frac{5}{8} \text{ and } p_2 = \frac{2+1}{2+4+1+1} = \frac{3}{8}$$

Player II's strategy is $\begin{bmatrix} q \\ 1-q \end{bmatrix}$.

If player I chooses row 1, player II expects to win $E_{II} = -2q + 1(1-q) = -3q + 1$ (1)

If player I chooses row 2, player II expects to win $E_{II} = q - 4(1-q) = 5q - 4$ (2)

The intersection of the lines represents player II's maximum payoff and optimal strategy. We solve (1) and (2) simultaneously.

$$-3q + 1 = 5q - 4 \Rightarrow 5 = 8q \Rightarrow q = \frac{5}{8} \text{ and}$$

$$1 - q = \frac{3}{8}.$$

Thus, player II's optimal strategy is $Q = \begin{bmatrix} \frac{5}{8} \\ \frac{3}{8} \end{bmatrix}.$

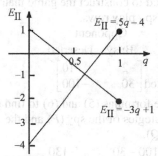

Using formula (6) to check the result, we have

$$Q = \begin{bmatrix} q_1 \\ q_2 \end{bmatrix}.$$

$$q_1 = \frac{4+1}{2+4+1+1} = \frac{5}{8} \text{ and } q_2 = \frac{2+1}{2+4+1+1} = \frac{3}{8}$$

The expected payoff of the game is

$$E = PAQ = \begin{bmatrix} \frac{5}{8} & \frac{3}{8} \end{bmatrix} \begin{bmatrix} 2 & -1 \\ -1 & 4 \end{bmatrix} \begin{bmatrix} \frac{5}{8} \\ \frac{3}{8} \end{bmatrix} = \frac{7}{8}.$$

9.a. $A = \begin{bmatrix} 4 & -1 \\ 0 & 3 \end{bmatrix}$

We will use formulas (5) and (6) to determine the Democrat's (who plays rows) and the Republican's (who plays columns) best strategies.

$$p_1 = \frac{3-0}{4+3+1-0} = \frac{3}{8}; \quad p_2 = 1 - p_1 = \frac{5}{8}$$

The Democrat's best strategy is to spend

$\frac{3}{8} = 37.5\%$ of the time on domestic issues and

$\frac{5}{8} = 62.5\%$ of the time on foreign issues.

(*continued on next page*)

(*continued*)

$$q_1 = \frac{3+1}{4+3+1-0} = \frac{4}{8} = \frac{1}{2}; \quad q_2 = 1 - q_1 = \frac{1}{2}$$

The Republican's best strategy is to spend 50% of the time on domestic issues and 50% of the time on foreign issues.

b. The expected payoff of the game is

$$PAQ = \begin{bmatrix} \frac{3}{8} & \frac{5}{8} \end{bmatrix} \begin{bmatrix} 4 & -1 \\ 0 & 3 \end{bmatrix} \begin{bmatrix} \frac{1}{2} \\ \frac{1}{2} \end{bmatrix} = \frac{3}{2}$$

The game favors the Democrat who expects to gain 1.5 units when optimal strategies are used.

11. First we need to construct the game matrix. We will let the spy play rows.

$$\begin{array}{cc} & \text{Opponent} \\ & \text{Busy} \quad \text{Deserted} \end{array}$$
$$\begin{array}{c} \text{Spy Busy} \\ \text{Deserted} \end{array} \begin{bmatrix} -2 & 10 \\ 30 & -100 \end{bmatrix}$$

We will use formulas (5) and (6) to find the optimal strategies of the spy (P) and the opponent (Q).

$$p_1 = \frac{-100 - 30}{-2 - 100 - 10 - 30} = \frac{-130}{-142} = 0.915$$

$$p_2 = 1 - p_1 = 0.085$$

The spy should select the busy exit 91.5% of the time and the deserted exit 8.5% of the time.

$$q_1 = \frac{-100 - 10}{-2 - 100 - 10 - 30} = \frac{-110}{-142} = 0.775$$

$$q_2 = 1 - q_1 = 0.225$$

The opponent should select the busy exit 77.5% of the time and the deserted exit 22.5% of the time.

The expected payoff of the game is given by

$$PAQ = \frac{(-2)(-100) - (10)(30)}{-2 - 100 - 10 - 30} = \frac{-100}{-142} = 0.704.$$

The game favors the spy whose optimal strategy is [0.915   0.085].

13. We define a game matrix for which Player I plays rows.

$$\begin{array}{c} & \text{Player II} \end{array}$$
$$\text{Player I} \begin{bmatrix} a_{11} & a_{12} \\ a_{21} & a_{22} \end{bmatrix}$$

First we prove formula (5). If Player II plays column 1, then Player I expects to win

$$E_{\text{I}} = a_{11}p + a_{21}(1 - p) = a_{11}p + a_{21} - a_{21}p \quad (1)$$

If Player II plays column 2, then Player I expects to win

$$E_{\text{I}} = a_{12}p + a_{22}(1 - p) = a_{12}p + a_{22} - a_{22}p \quad (2)$$

Solving the system of equations for p we have

$$a_{11}p + a_{21} - a_{21}p = a_{12}p + a_{22} - a_{22}p$$
$$a_{11}p - a_{21}p - a_{12}p + a_{22}p = a_{22} - a_{21}$$
$$(a_{11} - a_{21} - a_{12} + a_{22})p = a_{22} - a_{21}$$
$$p = \frac{a_{22} - a_{21}}{a_{11} - a_{21} - a_{12} + a_{22}}$$
$$= p_1$$

$$p_2 = 1 - p_1 = 1 - \frac{a_{22} - a_{21}}{a_{11} + a_{22} - a_{12} - a_{21}}$$

$$= \frac{a_{11} + a_{22} - a_{12} - a_{21}}{a_{11} + a_{22} - a_{12} - a_{21}} - \frac{a_{22} - a_{21}}{a_{11} + a_{22} - a_{12} - a_{21}}$$

$$p_2 = \frac{a_{11} - a_{12}}{a_{11} + a_{22} - a_{12} - a_{21}}$$

Next we prove formula (6). If Player 1 plays row 1, then Player II expects to win

$$E_{\text{II}} = -a_{11}q - a_{12}(1 - q) = -a_{11}q - a_{12} + a_{12}q$$

If Player I plays row 2, then Player I expects to

$$E_{\text{II}} = -a_{21}q - a_{22}(1 - q) = -a_{21}q - a_{22} + a_{22}q$$

Solving the system of equations for p we have

$$-a_{11}q - a_{12} + a_{12}q = -a_{21}q - a_{22} + a_{22}q$$
$$-a_{11}q + a_{12}q + a_{21}q - a_{22}q = a_{12} - a_{22}$$
$$(-a_{11} + a_{12} + a_{21} - a_{22})q = a_{12} - a_{22}$$
$$\frac{a_{12} - a_{22}}{-a_{11} + a_{12} + a_{21} - a_{22}} = q = q_1$$

$$q_2 = 1 - \frac{a_{12} - a_{22}}{-a_{11} + a_{12} + a_{21} - a_{22}}$$

$$= \frac{a_{11} + a_{22} - a_{12} - a_{21}}{a_{11} + a_{22} - a_{12} - a_{21}}$$

$$- \frac{a_{12} - a_{22}}{-a_{11} + a_{12} + a_{21} - a_{22}}$$

$$q_2 = \frac{a_{11} - a_{21}}{a_{11} + a_{22} - a_{12} - a_{21}}$$

Chapter 10 Review Exercises

1. P is not regular; the product of row 1 and column 2 is always 0. So $p_{12} = 0$ for every power of P.

3. P is not regular; the product of row 2 and column 1 is always 0. So $p_{21} = 0$ for every power of P.

5. P is regular.

$$P^3 = \begin{bmatrix} \dfrac{5}{12} & \dfrac{77}{288} & \dfrac{91}{288} \\[2mm] \dfrac{1}{3} & \dfrac{5}{12} & \dfrac{1}{4} \\[2mm] \dfrac{15}{64} & \dfrac{455}{1536} & \dfrac{721}{1536} \end{bmatrix}$$ has all positive

entries.

7. Define $\mathbf{t} = \begin{bmatrix} t_1 & t_2 \end{bmatrix}$. We want to find $\mathbf{t}$ such that $\mathbf{t}P = \mathbf{t}$.

$$\begin{bmatrix} t_1 & t_2 \end{bmatrix} \begin{bmatrix} \dfrac{1}{2} & \dfrac{1}{2} \\[2mm] \dfrac{2}{3} & \dfrac{1}{3} \end{bmatrix} = \begin{bmatrix} t_1 & t_2 \end{bmatrix}$$

This leads to the system of equations

$$\begin{cases} \dfrac{1}{2}t_1 + \dfrac{2}{3}t_2 = t_1 \\[2mm] \dfrac{1}{2}t_1 + \dfrac{1}{3}t_2 = t_2 \end{cases} \text{ or } \begin{cases} -\dfrac{1}{2}t_1 + \dfrac{2}{3}t_2 = 0 \\[2mm] \dfrac{1}{2}t_1 - \dfrac{2}{3}t_2 = 0 \end{cases}$$

These 2 equations are equivalent, so we have a system of two equations.

$$\begin{cases} -\dfrac{1}{2}t_1 + \dfrac{2}{3}t_2 = 0 & (1) \\[2mm] t_1 + \quad t_2 = 1 & (2) \end{cases}$$

Multiply (1) by 2, then add the resulting equation to (2) and solve for t_2.

$$-t_1 + \dfrac{4}{3}t_2 = 0 \quad 2 \times (1)$$
$$t_1 + \quad t_2 = 1 \quad (2)$$
$$\dfrac{7}{3}t_2 = 1 \Rightarrow t_2 = \dfrac{3}{7}$$

Substitute the value for t_2 into (2) and solve for t_1.

$$t_1 + \dfrac{3}{7} = 1 \Rightarrow t_1 = \dfrac{4}{7}$$

Therefore, $\mathbf{t} = \begin{bmatrix} \dfrac{4}{7} & \dfrac{3}{7} \end{bmatrix}$.

9. Define $\mathbf{t} = \begin{bmatrix} t_1 & t_2 & t_3 \end{bmatrix}$. We want to find $\mathbf{t}$ such that $\mathbf{t}P = \mathbf{t}$.

$$\begin{bmatrix} t_1 & t_2 & t_3 \end{bmatrix} \begin{bmatrix} \dfrac{1}{2} & 0 & \dfrac{1}{2} \\[2mm] \dfrac{1}{4} & \dfrac{1}{2} & \dfrac{1}{4} \\[2mm] \dfrac{1}{3} & \dfrac{1}{3} & \dfrac{1}{3} \end{bmatrix} = \begin{bmatrix} t_1 & t_2 & t_3 \end{bmatrix}$$

This leads to the system of equations

$$\begin{cases} \dfrac{1}{2}t_1 + \dfrac{1}{4}t_2 + \dfrac{1}{3}t_3 = t_1 \\[2mm] \dfrac{1}{2}t_2 + \dfrac{1}{3}t_3 = t_2 \\[2mm] \dfrac{1}{2}t_1 + \dfrac{1}{4}t_2 + \dfrac{1}{3}t_3 = t_3 \end{cases} \text{ or}$$

$$\begin{cases} -\dfrac{1}{2}t_1 + \dfrac{1}{4}t_2 + \dfrac{1}{3}t_3 = 0 \\[2mm] -\dfrac{1}{2}t_2 + \dfrac{1}{3}t_3 = 0 \\[2mm] \dfrac{1}{2}t_1 + \dfrac{1}{4}t_2 - \dfrac{2}{3}t_3 = 0 \end{cases}$$

The first equation is equivalent to the sum of the second two equations, so we have a system of 3 equations which we will solve by elimination.

$$\begin{cases} t_1 + \quad t_2 + \quad t_3 = 1 & (1) \\[2mm] -\dfrac{1}{2}t_2 + \dfrac{1}{3}t_3 = 0 & (2) \\[2mm] \dfrac{1}{2}t_1 + \dfrac{1}{4}t_2 - \dfrac{2}{3}t_3 = 0 & (3) \end{cases}$$

Multiply (3) by -2, then add the resulting equation to (1).

$$t_1 + \quad t_2 + \quad t_3 = 1 \quad (1)$$
$$-t_1 - \dfrac{1}{2}t_2 + \dfrac{4}{3}t_3 = 0 \quad -2 \times (3)$$
$$\rule{4cm}{0.4pt}$$
$$\dfrac{1}{2}t_2 + \dfrac{7}{3}t_3 = 1 \quad (4)$$

Now add equations (2) and (4).

$$-\dfrac{1}{2}t_2 + \dfrac{1}{3}t_3 = 0 \quad (2)$$
$$\dfrac{1}{2}t_2 + \dfrac{7}{3}t_3 = 1 \quad (4)$$
$$\rule{4cm}{0.4pt}$$
$$\dfrac{8}{3}t_3 = 1 \Rightarrow t_3 = \dfrac{3}{8}$$

Back-substituting into equation (2) gives

$$-\dfrac{1}{2}t_2 + \dfrac{1}{3}\left(\dfrac{3}{8}\right) = 0 \Rightarrow t_2 = \dfrac{1}{4}.$$

Back-substituting t_2 and t_3 into (1) gives

$$t_1 = 1 - \dfrac{1}{4} - \dfrac{3}{8} = \dfrac{3}{8}.$$ Therefore, the fixed

probability vector $\mathbf{t} = \begin{bmatrix} \dfrac{3}{8} & \dfrac{1}{4} & \dfrac{3}{8} \end{bmatrix}$.

11. Define $\mathbf{t} = \begin{bmatrix} t_1 & t_2 & t_3 \end{bmatrix}$. We want to find $\mathbf{t}$ such that $\mathbf{t}P = \mathbf{t}$.

$$\begin{bmatrix} t_1 & t_2 & t_3 \end{bmatrix} \begin{bmatrix} 0.7 & 0.1 & 0.2 \\ 0.6 & 0.1 & 0.3 \\ 0.4 & 0.2 & 0.4 \end{bmatrix} = \begin{bmatrix} t_1 & t_2 & t_3 \end{bmatrix}$$

This leads to the system of equations

$$\begin{cases} 0.7t_1 + 0.6t_2 + 0.4t_3 = t_1 \\ 0.1t_1 + 0.1t_2 + 0.2t_3 = t_2 \quad \text{or} \\ 0.2t_1 + 0.3t_2 + 0.4t_3 = t_3 \end{cases}$$

$$\begin{cases} -0.3t_1 + 0.6t_2 + 0.4t_3 = 0 \\ 0.1t_1 - 0.9t_2 + 0.2t_3 = 0 \\ 0.2t_1 + 0.3t_2 - 0.6t_3 = 0 \end{cases}$$

The first equation is equivalent to the sum of the second two equations, so we have a system of 3 equations which we will solve by elimination. First, we will multiply equations (2) and (3) by 10 to remove the decimals.

$$\begin{cases} t_1 + t_2 + t_3 = 1 \quad (1) \\ 0.1t_1 - 0.9t_2 + 0.2t_3 = 0 \quad (2) \Rightarrow \\ 0.2t_1 + 0.3t_2 - 0.6t_3 = 0 \quad (3) \end{cases}$$

$$\begin{cases} t_1 + t_2 + t_3 = 1 \quad (1) \\ t_1 - 9t_2 + 2t_3 = 0 \quad (2) \\ 2t_1 + 3t_2 - 6t_3 = 0 \quad (3) \end{cases}$$

Multiply (2) by -1, then add the resulting equation to (1).

$$\begin{array}{l} t_1 + t_2 + t_3 = 1 \quad (1) \\ \underline{-t_1 + 9t_2 - 2t_3 = 0} \quad -1 \times (2) \\ \quad\quad 10t_2 - t_3 = 1 \quad (4) \end{array}$$

Now multiply (1) by -2, then add the resulting equation to (3).

$$\begin{array}{l} -2t_1 - 2t_2 - 2t_3 = -2 \quad -2 \times (1) \\ \underline{2t_1 + 3t_2 - 6t_3 = 0} \quad (3) \\ \quad\quad t_2 - 8t_3 = -2 \quad (5) \end{array}$$

Now solve the system consisting of equations (4) and (5) by substitution. From (4) we have $t_3 = 10t_2 - 1$.

$$t_2 - 8(10t_2 - 1) = -2 \Rightarrow -79t_2 = -10 \Rightarrow t_2 = \frac{10}{79}$$

and $t_3 = 10\left(\frac{10}{79}\right) - 1 = \frac{21}{79}$. From equation (1),

we have $t_1 = 1 - \frac{10}{79} - \frac{21}{79} = \frac{48}{79}$.

Thus, the fixed probability vector is

$$\mathbf{t} = \begin{bmatrix} \dfrac{48}{79} & \dfrac{10}{79} & \dfrac{21}{79} \end{bmatrix} \approx \begin{bmatrix} 0.608 & 0.127 & 0.266 \end{bmatrix}.$$

13.a. This situation forms a Markov chain because the shifts in market shares can be thought of as a sequence of experiments each of which results in one of a finite number of states, and the movement to the next state depends only on the current state. The transition matrix P is given by

$$\begin{array}{cc} & \begin{array}{ccc} A & B & C \end{array} \\ P = \begin{array}{c} A \\ B \\ C \end{array} & \begin{bmatrix} 0.50 & 0.20 & 0.30 \\ 0.40 & 0.40 & 0.20 \\ 0.50 & 0.25 & 0.25 \end{bmatrix} \end{array}$$

b. We are told that currently each has $\dfrac{1}{3}$ of the market, so $v^{(0)} = \begin{bmatrix} \dfrac{1}{3} & \dfrac{1}{3} & \dfrac{1}{3} \end{bmatrix}$. We are looking for $v^{(1)}$, the probability distribution after one year.

$$v^{(1)} = v^{(0)}P = \begin{bmatrix} \dfrac{1}{3} & \dfrac{1}{3} & \dfrac{1}{3} \end{bmatrix} \begin{bmatrix} 0.50 & 0.20 & 0.30 \\ 0.40 & 0.40 & 0.20 \\ 0.50 & 0.25 & 0.25 \end{bmatrix}$$

$$= \begin{bmatrix} \dfrac{7}{15} & \dfrac{17}{60} & \dfrac{1}{4} \end{bmatrix} \approx \begin{bmatrix} 0.4667 & 0.2833 & 0.25 \end{bmatrix}$$

If the trend continues, after one year distributor A has 46.7% of the market, distributor B has 28.3% of the market, and distributor C has 25% of the market.

c. The distribution after 2 years is given by $v^{(2)} = v^{(1)}P$.

$$v^{(2)} = \begin{bmatrix} \dfrac{7}{15} & \dfrac{17}{60} & \dfrac{1}{4} \end{bmatrix} \begin{bmatrix} 0.50 & 0.20 & 0.30 \\ 0.40 & 0.40 & 0.20 \\ 0.50 & 0.25 & 0.25 \end{bmatrix}$$

$$\approx \begin{bmatrix} 0.47167 & 0.26917 & 0.25917 \end{bmatrix}$$

After 2 years distributor A will have 47.2% of the market, distributor B will have 26.9% of the market and distributor C will have 25.9% of the market.

d. The long run distribution of the market share is given by the fixed probability vector **t**.

$$\begin{bmatrix} t_1 & t_2 & t_3 \end{bmatrix} \begin{bmatrix} 0.50 & 0.20 & 0.30 \\ 0.40 & 0.40 & 0.20 \\ 0.50 & 0.25 & 0.25 \end{bmatrix} = \begin{bmatrix} t_1 & t_2 & t_3 \end{bmatrix}$$

This leads to the system of equations

$$\begin{cases} 0.5t_1 + 0.4t_2 + 0.5t_3 = t_1 \\ 0.2t_1 + 0.4t_2 + 0.25t_3 = t_2 \text{ or} \\ 0.3t_1 + 0.2t_2 + 0.25t_3 = t_3 \end{cases}$$

$$\begin{cases} -0.5t_1 + 0.4t_2 + 0.5t_3 = 0 \\ 0.2t_1 - 0.6t_2 + 0.25t_3 = 0 \\ 0.3t_1 + 0.2t_2 - 0.75t_3 = 0 \end{cases}$$

The first equation is equivalent to the sum of the second two equations, so we have a system of 3 equations which we will solve by elimination. First, we will multiply equations (2) and (3) by 100 to remove the decimals.

$$\begin{cases} t_1 + t_2 + t_3 = 1 & (1) \\ 0.2t_1 - 0.6t_2 + 0.25t_3 = 0 & (2) \Rightarrow \\ 0.3t_1 + 0.2t_2 - 0.75t_3 = 0 & (3) \end{cases}$$

$$\begin{cases} t_1 + t_2 + t_3 = 1 & (1) \\ 20t_1 - 60t_2 + 25t_3 = 0 & (2) \\ 30t_1 + 20t_2 - 75t_3 = 0 & (3) \end{cases}$$

We will solve the system using matrices.

$$\begin{bmatrix} 1 & 1 & 1 & | & 1 \\ 20 & -60 & 25 & | & 0 \\ 30 & 20 & -75 & | & 0 \end{bmatrix}$$

$$\xrightarrow[R_3 = -30r_1 + r_3]{R_2 = -20r_1 + r_2} \begin{bmatrix} 1 & 1 & 1 & | & 1 \\ 0 & -80 & 5 & | & -20 \\ 0 & -10 & -105 & | & -30 \end{bmatrix}$$

$$\xrightarrow{R_3 = -r_2 + 8r_3} \begin{bmatrix} 1 & 1 & 1 & | & 1 \\ 0 & -80 & 5 & | & -20 \\ 0 & 0 & -845 & | & -220 \end{bmatrix}$$

$$\xrightarrow{R_3 = -\frac{1}{845}r_3} \begin{bmatrix} 1 & 1 & 1 & | & 1 \\ 0 & -80 & 5 & | & -20 \\ 0 & 0 & 1 & | & \frac{44}{169} \end{bmatrix}$$

$$\xrightarrow[R_2 = -5r_3 + r_2]{R_1 = -r_3 + r_1} \begin{bmatrix} 1 & 1 & 0 & | & \frac{125}{169} \\ 0 & -80 & 0 & | & -\frac{3600}{169} \\ 0 & 0 & 1 & | & \frac{44}{169} \end{bmatrix}$$

$$\xrightarrow{R_2 = -\frac{1}{80}r_2} \begin{bmatrix} 1 & 1 & 0 & | & \frac{125}{169} \\ 0 & 1 & 0 & | & \frac{45}{169} \\ 0 & 0 & 1 & | & \frac{44}{169} \end{bmatrix}$$

$$\xrightarrow{R_1 = -r_2 + r_1} \begin{bmatrix} 1 & 0 & 0 & | & \frac{80}{169} \\ 0 & 1 & 0 & | & \frac{45}{169} \\ 0 & 0 & 1 & | & \frac{44}{169} \end{bmatrix}$$

Thus, the fixed probability vector is

$$\mathbf{t} = \begin{bmatrix} \dfrac{80}{169} & \dfrac{45}{169} & \dfrac{44}{169} \end{bmatrix}$$

$$\approx \begin{bmatrix} 0.4734 & 0.2663 & 0.2604 \end{bmatrix}.$$

In the long run, distributor A will have 47.3% of the market, distributor B will have 26.6% of the market and distributor C will have 26.0% of the market.

15. The Markov chain in not absorbing.

17. The Markov chain is absorbing. State (2) is the absorbing state. The subdivided transition matrix is

$$\begin{array}{c} \\ 2 \\ 1 \\ 3 \end{array} \begin{array}{ccc} 2 & 1 & 3 \\ \begin{bmatrix} 1 & 0 & 0 \\ \frac{1}{3} & 0 & \frac{2}{3} \\ \frac{1}{4} & \frac{3}{8} & \frac{3}{8} \end{bmatrix} \end{array}$$

19. The Markov chain is not absorbing.

21. The Markov chain is absorbing. States (1) and (4) are absorbing states. The subdivided transition matrix is

$$\begin{array}{c} \\ 1 \\ 4 \\ 2 \\ 3 \end{array} \begin{array}{cccc} 1 & 4 & 2 & 3 \\ \begin{bmatrix} 1 & 0 & 0 & 0 \\ 0 & 1 & 0 & 0 \\ 0.3 & 0.1 & 0.5 & 0.1 \\ 0.25 & 0.2 & 0.35 & 0.2 \end{bmatrix} \end{array}$$

23. a.

$$P = \begin{array}{c} \\ 0 \\ 1 \\ 2 \\ 3 \\ 4 \\ 5 \end{array} \begin{array}{cccccc} 0 & 1 & 2 & 3 & 4 & 5 \\ \left[\begin{array}{cccccc} 1 & 0 & 0 & 0 & 0 & 0 \\ 0.55 & 0 & 0.45 & 0 & 0 & 0 \\ 0 & 0.55 & 0 & 0.45 & 0 & 0 \\ 0 & 0 & 0.55 & 0 & 0.45 & 0 \\ 0 & 0 & 0 & 0.55 & 0 & 0.45 \\ 0 & 0 & 0 & 0 & 0 & 1 \end{array}\right] \end{array}$$

$$\rightarrow \begin{array}{c} \\ 0 \\ 5 \\ 1 \\ 2 \\ 3 \\ 4 \end{array} \begin{array}{cccccc} 0 & 5 & 1 & 2 & 3 & 4 \\ \left[\begin{array}{cccccc} 1 & 0 & 0 & 0 & 0 & 0 \\ 0 & 1 & 0 & 0 & 0 & 0 \\ 0.55 & 0 & 0 & 0.45 & 0 & 0 \\ 0 & 0 & 0.55 & 0 & 0.45 & 0 \\ 0 & 0 & 0 & 0.55 & 0 & 0.45 \\ 0 & 0.45 & 0 & 0 & 0.55 & 0 \end{array}\right] \end{array}$$

b. Since the gambler began with $2, we use the second row of the fundamental matrix T which was given in the text. He expects to have $1 on the average of 1.298 times, $2 on the average of 2.361 times, $3 on the average of 1.412 times and $4 on the average of 0.635 time.

c. Since he started with $2 the expected length of the game is the sum of the entries in the 2nd row.

$$E = 1.298406 + 2.360738 + 1.411737$$
$$+ 0.635282$$
$$= 5.706163$$

The gambler should expect to play 5.71 games before absorption.

d. To find the probability the gambler loses all his money or wins $5, we look at the product of matrices T and S.

$$TS = \begin{bmatrix} 1.584 & 1.062 & 0.635 & 0.286 \\ 1.298 & 2.361 & 1.412 & 0.635 \\ 0.949 & 1.725 & 2.361 & 1.062 \\ 0.522 & 0.949 & 1.298 & 1.584 \end{bmatrix}$$
$$\cdot \begin{bmatrix} 0.55 & 0 \\ 0 & 0 \\ 0 & 0 \\ 0 & 0.45 \end{bmatrix}$$
$$= \begin{bmatrix} 0.871 & 0.129 \\ 0.714 & 0.286 \\ 0.522 & 0.478 \\ 0.287 & 0.713 \end{bmatrix}$$

The probability the gambler loses all his money is 0.714. This is the entry in row 2 (he started out with $2), column 1 (he ended up with $0) of the product *TS*. The probability the gambler wins $5 is 0.286. We can get this either from finding the difference of 1 and 0.714 or from row 2, column 2 of the product *TS*.

For exercises 25–29, recall that a game is strictly determined when there is an entry that is the smallest element in its row and the largest element in its column. That entry, if it exists, is called the value of the game.

25. The game is not strictly determined; no entry satisfies the criteria.

27. The game is strictly determined; 9 is smallest element in row 2 and largest element in column 1. The value of the game is 9.

29. The game is strictly determined; 12 is the smallest element in row 3 and the largest element in column 3. The value of the game is 12.

For exercises 31–35, recall that expected payoff of player 1 is given by *PAQ* where *P* is player I's strategy and *Q* is player II's strategy.

31. $E = PAQ = \begin{bmatrix} \frac{1}{3} & \frac{2}{3} \end{bmatrix} \begin{bmatrix} -1 & 1 \\ 1 & -1 \end{bmatrix} \begin{bmatrix} 1 \\ 0 \end{bmatrix}$

$$= \begin{bmatrix} \frac{1}{3} & -\frac{1}{3} \end{bmatrix} \begin{bmatrix} 1 \\ 0 \end{bmatrix} = \frac{1}{3}$$

Player I should expect to win $\frac{1}{3}$.

33. $E = PAQ = \begin{bmatrix} 0.5 & 0.5 \end{bmatrix} \begin{bmatrix} 4 & 2 \\ -2 & 3 \end{bmatrix} \begin{bmatrix} 0.5 \\ 0.5 \end{bmatrix}$

$$= \begin{bmatrix} 1 & 2.5 \end{bmatrix} \begin{bmatrix} 0.5 \\ 0.5 \end{bmatrix} = 1.75$$

Player I should expect to win 1.75.

35. $E = PAQ = \begin{bmatrix} \frac{1}{3} & \frac{1}{3} & \frac{1}{3} \end{bmatrix} \begin{bmatrix} 1 & 4 & -3 \\ 2 & -5 & -1 \\ 1 & -2 & 3 \end{bmatrix} \begin{bmatrix} 1 \\ 0 \\ 0 \end{bmatrix}$

$$= \begin{bmatrix} \frac{4}{3} & -1 & -\frac{1}{3} \end{bmatrix} \begin{bmatrix} 1 \\ 0 \\ 0 \end{bmatrix} = \frac{4}{3}$$

Player I should expect to win $\frac{4}{3}$.

37. We will use formulas (5) and (6) from section 10.6 to find the optimal strategy. The optimal strategy for player I is $P = [p_1 \quad p_2]$ where

$$p_1 = \frac{a_{22} - a_{21}}{a_{11} + a_{22} - a_{12} - a_{21}} = \frac{-1-1}{-1-1-1-1} = \frac{1}{2}$$

and

$$p_2 = \frac{a_{11} - a_{12}}{a_{11} + a_{22} - a_{12} - a_{21}} = \frac{-1-1}{-1-1-1-1} = \frac{1}{2}.$$

So player I's optimal strategy is $P = \begin{bmatrix} \dfrac{1}{2} & \dfrac{1}{2} \end{bmatrix}$.

The optimal strategy for player II is $Q = \begin{bmatrix} q_1 \\ q_2 \end{bmatrix}$

where

$$q_1 = \frac{a_{22} - a_{12}}{a_{11} + a_{22} - a_{12} - a_{21}} = \frac{-1-1}{-1-1-1-1} = \frac{1}{2}$$

and

$$q_2 = \frac{a_{11} - a_{21}}{a_{11} + a_{22} - a_{12} - a_{21}} = \frac{-1-1}{-1-1-1-1} = \frac{1}{2}.$$

So player II's optimal strategy is $Q = \begin{bmatrix} \dfrac{1}{2} \\ \dfrac{1}{2} \end{bmatrix}$.

39.a. We will assume the economy is an active opponent working against the investor. We will use formula (5) to get the investor's optimal strategy $P = [p_1 \quad p_2]$.

$$p_1 = \frac{a_{22} - a_{21}}{a_{11} + a_{22} - a_{12} - a_{21}}$$
$$= \frac{0 - 18}{-5 + 0 - 20 - 18} = \frac{18}{43}$$

and $p_2 = \dfrac{a_{11} - a_{12}}{a_{11} + a_{22} - a_{12} - a_{21}}$
$$= \frac{-5 - 20}{-5 + 0 - 20 - 18} = \frac{25}{43}.$$

Player I's optimal strategy is

$$P = \begin{bmatrix} \dfrac{18}{43} & \dfrac{25}{43} \end{bmatrix} \approx [0.4186 \quad 0.5814].$$

The investor should invest in A 41.9% of the time and in investment B 58.1% of the time.

b. The expected payoff corresponding to the optimal strategy is given by the product PAQ. However, since we did not need to find Q in this problem, we will use formula (7) to calculate the expected payoff.

$$E = PAQ = \frac{a_{11}a_{22} - a_{12}a_{21}}{a_{11} + a_{22} - a_{12} - a_{21}}$$
$$= \frac{(-5)(0) - (20)(18)}{-5 + 0 - 20 - 18} = \frac{-360}{-43} = 8.372$$

The investor should expect to gain 8.37% on the investment.

41. Assuming the county is an active opponent working against the investor, we will use equation (5) to get the investor's optimal strategy $P = [p_1 \quad p_2]$.

$$p_1 = \frac{a_{22} - a_{21}}{a_{11} + a_{22} - a_{12} - a_{21}}$$
$$= \frac{0.80 - 0.20}{0.70 + 0.80 - 0.30 - 0.20} = 0.60$$

$$p_2 = \frac{a_{11} - a_{12}}{a_{11} + a_{22} - a_{12} - a_{21}}$$
$$= \frac{0.70 - 0.3}{0.70 + 0.80 - 0.30 - 0.30} = 0.40.$$

The real estate developer's optimal strategy is $P = [0.6 \quad 0.4]$. The developer should use 60% of the land for a shopping center and 40% of the land for houses.

Chapter 10 Project

1.

	College	High School	Elementary
College	0.50	0.48	0.02
High School	0.12	0.76	0.12
Elementary	0.06	0.68	0.26

3. The sum across each row is 1.

5. Using P^2 from exercise 4, the probability that the grandchild of a high school graduate finishes college is 0.1584.

7. Using a graphing calculator, we can find **t** by computing P^n for a large value of n. P^{25} is large enough to find **t**.

$$\mathbf{t} = [0.18209 \quad 0.69954 \quad 0.11836]$$

The long-run distribution of highest educational attainment is 18.2% college level, 70.0% high school, and 11.8% elementary level.

Chapter 11 Logic

11.1 Propositions

1. A declarative sentence that can be meaningfully classified as either true or false is called a <u>proposition</u>.

3. False. The inclusive disjunction of two statements p and q is true provided at least one of the propositions p, q is true.

In exercises 5–11, recall that a proposition is a declarative sentence that can be meaningfully classified as either true or false.

5. "The cost of shell egg futures was up on June 18, 1990." is a declarative sentence. It is either true or false. The statement is a proposition.

7. "What a portfolio!" is not a declarative sentence. It is not a proposition.

9. "The earnings of XYZ Company doubled last year." is a declarative sentence. It is either true or false. The statement is a proposition.

11. "Jones is guilty of murder in the first degree." is a declarative sentence. It is either true or false. The statement is a proposition.

13. The negation of "A fox is an animal" is "A fox is not an animal."

15. The negation of, "I am buying stocks" is "I am not buying stocks."

17. The negation of, "No one wants to buy my house" is "Someone wants to buy my house."

19. The negation of "Some people have no car" is "Everybody has a car."

21. John is an economics major, or John is a sociology minor (or both).

23. John is an economics major and a sociology minor.

25. John is not an economics major, or John is not a sociology minor (or both).

27. John is not an economics major or John is a sociology minor (or both).

1. The associative property of addition of real numbers states that $a + (b + c) = (a + b) + c$.

3. True

5. For a proposition p, the idempotent properties state that $p \wedge p \equiv p$ and $p \vee p \equiv p$.

7.

p	q	$\sim q$	$p \vee \sim q$
T	T	F	T
T	F	T	T
F	T	F	F
F	F	T	T

9.

p	q	$\sim p$	$\sim q$	$\sim p \wedge \sim q$
T	T	F	F	F
T	F	F	T	F
F	T	T	F	F
F	F	T	T	T

11.

p	q	$\sim p$	$\sim p \wedge q$	$\sim(\sim p \wedge q)$
T	T	F	F	T
T	F	F	F	T
F	T	T	T	F
F	F	T	F	T

13.

p	q	$\sim p$	$\sim q$	$\sim p \vee \sim q$	$\sim(\sim p \vee \sim q)$
T	T	F	F	F	T
T	F	F	T	T	F
F	T	T	F	T	F
F	F	T	T	T	F

15.

p	q	$\sim q$	$p \vee \sim q$	$(p \vee \sim q) \wedge p$
T	T	F	T	T
T	F	T	T	T
F	T	F	F	F
F	F	T	T	F